8th Conference on Emerging Energy & Process Technology (CONCEPT 2019)

IOP Conference Series: Materials Science and Engineering
Volume 808

Kuala Lumpur, Malaysia
27 - 28 November 2019

Editor:

Nurfatehah Wahyuny Che Jusoh

ISBN: 978-1-7138-3159-4
ISSN: 1757-8981

Printed from e-media with permission by:

Curran Associates, Inc.
57 Morehouse Lane
Red Hook, NY 12571

CURRAN ASSOCIATES INC. proceedings .com

Some format issues inherent in the e-media version may also appear in this print version.

Institute of Physics
Dirac House, Temple Back
Bristol BS1 6BE UK

Phone: 44 1 17 929 7481
Fax: 44 1 17 920 0979

techtracking@iop.org

Additional copies of this publication are available from:

Curran Associates, Inc.
57 Morehouse Lane
Red Hook, NY 12571 USA
Phone: 845-758-0400
Fax: 845-758-2633
Email: curran@proceedings.com
Web: www.proceedings.com

TABLE OF CONTENTS

Preface

The 8th Conference of Emerging Energy and Process technology (CONCEPT 2019) is an annual event organized by Institute of Future Energy of Universiti Teknologi Malaysia. The conference was successfully held in Kuala Lumpur, Malaysia in November 27-28, 2019. CONCEPT 2019 is an Energy and Chemical Process-based conference focus on three themes; Sustainable Energy, Process System, and Safety. The overarching objectives of this conference to stimulate a broader discussion of current technological innovation and invention and identify potential collaborative research among industry experts, academicians, or students. The conference also provides great opportunities for participants to meet experts, exchange ideas, share experiences and establish the ground for future research and business relations with local and global partners. CONCEPT 2019 Proceedings contains 42 selected papers that have been accepted for presentation in the conference. All articles had been reviewed by the researchers and experts in the field appointed by the conference technical committee.

The guest editor would like to sincerely thank the publication team for their efforts in preparing this proceeding. Finally, the publication committee wishes to thank all the authors for their contributions to CONCEPT 2019 proceedings and we look forward to welcoming more of your contributions to the CONCEPT 2020.

Editorial

Guest Editor: Nurfatehah Wahyuny Che Jusoh

Editorial members:

Arshad Ahmad

Adnan Ripin

Roshafima Rasit Ali

Saidatul Sophia Md Sha'rani

Guest Editor Email: nurfatehah@utm.my

Phone No. +603-2203 1324

PHOTOS OF CONCEPT 2019

IOP Conf. Series: Materials Science and Engineering **808** (2020) 011002 doi:10.1088/1757-899X/808/1/011002

Peer review statement

All papers published in this volume of *IOP Conference Series: Materials Science and Engineering* have been peer reviewed through processes administered by the proceedings Editors. Reviews were conducted by expert referees to the professional and scientific standards expected of a proceedings journal published by IOP Publishing.

Thermodynamic sensitivity analysis of CO_2 reforming of methane based on equilibrium predictions

A A Abdulrasheed[1,3], **A A Jalil**[1,2], **T J Siang**[1] and **H U Hambali**[1]

[1]School of Chemical and Energy Engineering, Faculty of Engineering, Universiti Teknologi Malaysia, 81310 UTM Johor Bahru, Johor, Malaysia
[2]Centre of Hydrogen Energy, Institute of Future Energy, Universiti Teknologi Malaysia, 81310 UTM Johor Bahru, Johor, Malaysia
[3]Department of Chemical Engineering, Abubakar Tafawa Balewa University, PMB 0248, Bauchi, Bauchi State, Nigeria

*aishahaj@utm.my

Abstract. Thermodynamic sensitivity analysis was conducted to evaluate the selectivity dry reforming of methane (DRM) with respect to other competing side reactions in terms of Gibbs energy minimization and equilibrium reactor compositions. Spontaneity of each reaction was evaluated by their corresponding changes in Gibbs energy as a function of reactor temperature from 100 – 1000 °C. At temperatures above 700 °C, all reactions considered with the exception of the Boudouard reaction are spontaneous due to their negative deltaG values. Equilibrium compositions were computed for 40 kmol of feed (CO_2 + CH_4) to determine its sensitivity to occurrence of reverse water-gas-shift (RWGS) reaction and changes in feed CO_2/CH_4 ratio. The presence of RWGS reaction results to higher reactant conversion accompanied by more severe carbon formation. The H_2/CO ratio was however not affected by its occurrence. At CO_2/CH_4 feed ratio of 2, the product ratio was barely affected and at minimal carbon formation while CO_2/CH_4 ratio value of 0.5 produced a lot of carbon especially at high temperatures due to the prevalence of CH_4 cracking. DRM is thus more preferable at temperatures above 700 °C, devoid of RWGS and high feed CH_4.

1. Introduction

Production of essential long-chain hydrocarbons from synthetic gas via Fischer-Tropsch process has triggered proliferation of world syngas demand [1, 2]. Methane gas is the most widely used for syngas production by reforming with carbon dioxide, steam or partial oxidation because of its widespread availability and renewability potentials [3, 4]. From environmental and economic perspective, syngas generation from catalytic dry reforming of methane (DRM) is the most promising over others. This is due to fact that it produces syngas with a stoichiometric H_2/CO ratio of 1 (Equation 1) which is a more suitable feedstock for the Fischer-Tropsch synthesis and conversion and converts two gases with green house tendencies to value-added fuels and chemicals [5]. The major challenge of the industrial application of DRM process is the short lifespan of existing catalysts as a result active metal or support sintering and carbon deposition [6]. Carbon deposition occur as a consequence of carbon precipitating side reactions such as methane cracking (Equation 2), reverse water-gas-shift (Equation 3) and Boudouard reaction (Equation 4).

$$CH_4 + CO_2 \rightleftharpoons 2CO + 2H_2 \quad \Delta H^0_{298K} = 260.5 \text{ kJ/mol} \qquad (1)$$

$$CH_4 \rightleftharpoons C + 2H_2 \qquad \Delta H^0_{298K} = 75.0 \text{ kJ/mol} \qquad (2)$$

$$CO_2 + H_2 \rightleftharpoons CO + H_2O \qquad \Delta H^0_{298K} = 41.0 \text{ kJ/mol} \qquad (3)$$

$$2CO \rightleftharpoons C + CO_2 \qquad \Delta H^0_{298K} = -173.0 \text{ kJ/mol} \qquad (4)$$

Catalysts from noble metals such as Pt, Pd, Rh and Ru are known for their high resistance to carbon deposition at high activity, but their associated high cost and scarcity hampered their industrial application [7]. Beside noble metals, the potential other metals predominantly Ni, Co and Fe based catalysts have been investigated out of which Ni catalysts produced higher activity emulous to noble metals and thus, commonly used [8 - 11]. Despite its associated performance and economic viability, vulnerability of Ni-based catalysts to sintering and carbon deposition at high reaction temperature is the bottleneck of its DRM applications [12]. From the thermodynamics of DRM, high temperatures are necessary to attain the required syngas yield and reactant conversion, a condition at which carbon deposition is inevitable [13]. Carbon formed during DRM are detrimental to catalyst performance through different media namely: encapsulation of active crystallites phase, pore blockage, collapse of pore structure, clogging of reactor bed and diffusion limitation [14, 15].

The objective of this paper is to conduct a theoretical thermodynamic studies of DRM reaction alongside competing side reactions such as methane decomposition, reverse water-gas-shift, Boudouard reaction and carbon gasification reactions aided by carbon dioxide or steam. The simulation was conducted using theoretical equilibrium computations and Gibbs free energy minimization on HSC Chemistry 6.0 software.

2. Experimental

The thermodynamic analysis was also conducted to investigate the effects of temperature on DRM in the presence of other competitive side reactions at equilibrium conditions. The thermodynamic computations were carried out with the aid of HSC Chemistry 6.0 software, utilizing the reaction equations and equilibrium compositions interface of the software. The effect of reaction temperature on Gibbs free energy (ΔG) minimization for all reactions was also computed to evaluate the spontaneity of DRM in comparison to other side reactions.

3. Results and discussions

A thermodynamic feasibility studies was conducted using the HSC Chemistry 6.0 software to evaluate the effect of change in Gibbs free energy (ΔG) on the spontaneity of DRM and major competitive side reactions such as methane cracking, RWGS, Boudouard and carbon gasification reactions using carbon dioxide or steam. The effect of temperature on ΔG for all chemical reactions from 100 – 1000 ^{o}C are depicted in Figure 1. The ΔG with negative values implies the spontaneity of the system towards favouring forward reaction while positive values are indicative of non-spontaneous chemical reactions [16]. A zero value of ΔG signifies a reaction in the state of equilibrium with no visible conversion. From the figure, it was observed that reaction temperature significantly affects the ΔG of each reaction pathway.

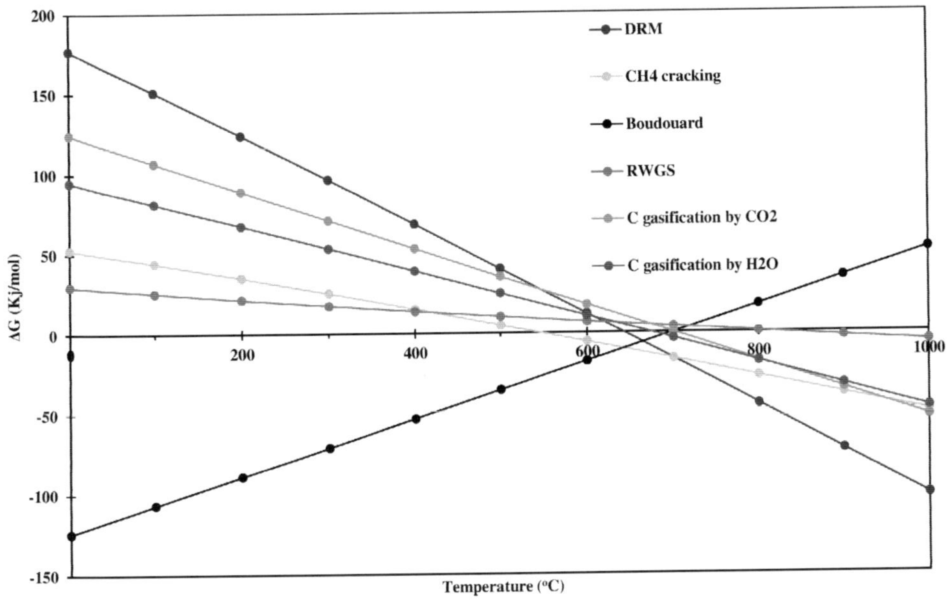

Figure 1. Thermodynamic analysis to determine spontaneity of DRM and side reactions.

All reactions considered with the exception of the Boudouard reaction are highly endothermic particularly at low temperatures which is attributed to the high stability of their respective reactants. DRM becomes thermodynamically favourable when the reaction temperature is increased to about 700 °C, below which only methane cracking is favourable (~550 °C). It should also be noted that each reaction attains equilibrium at different temperatures, with DRM and methane cracking reaching equilibrium at temperatures approximately 150 °C and 250 °C lower than the equilibrium temperature of RWGS reaction respectively. Boudouard reaction being highly exothermic was spontaneous at low temperatures [17], with equilibrium attained at 700 °C and temperatures above 700 °C are unfavourable for the reaction. The results obtained agrees with the thermodynamic computations conducted by [18]. The DRM above 700 °C is therefore considered the most suitable, with lower feasibility of Boudouard reaction, methane cracking and RWGS. The feasibility of carbon gasification by CO_2 or steam are also greatly enhanced in this temperature range.

Figure 2 shows the thermodynamics of the equilibrium compositions of all desired and side products at atmospheric pressure as a function of reaction temperature. The molar composition of reactants and products are determined from the Gibbs energy minimisation simulations at equimolar mixture of CO_2(g) and CH_4(g), with H_2(g), CO(g), C(s) and H_2O(g) taken as components in the equilibrium mixture [19]. Methane cracking and RWGS occur simultaneously alongside the main DRM reaction producing equilibrium stream compositions and H_2/CO ratios as a function of temperature. The impact of RWGS based on H_2O produced was investigated by making comparison of equilibrium plot with water (Figure 2A) with that without water formation as obtained in Figure 2B. The presence of RGWS reduces the amount of CO_2 in the system by about 50 % especially at temperatures below 800 °C due to high CO_2 consumption. High CO_2 composition in the RWGS free system produced less carbon deposits (Figure 2B), which is attributed to the superior carbon gasification by CO_2 over gasification using steam produced from RWGS. Methane composition was also higher in the absence of RWGS because of the absence of H_2 consumption which can adjust the equilibrium towards forward reaction for less CH_4 conversion.

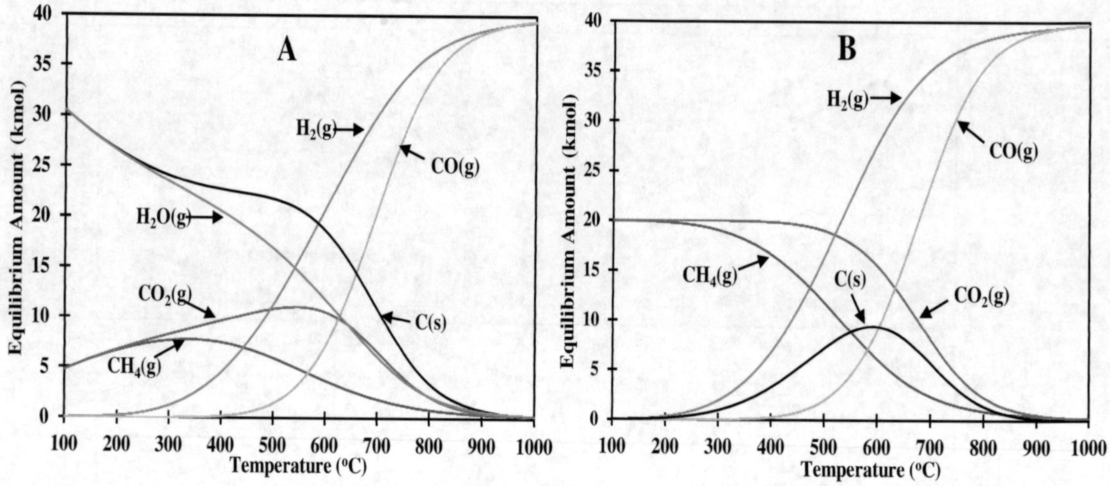

Figure 2. Thermodynamic plots for CH_4 + CO_2 reaction at 1 atm, temperature of 100 – 1000 °C and inlet feed ratio of CO_2:CH_4 = 1 with equilibrium reactor product compositions obtained (A) with the presence of RWGS and (B) without RWGS reaction.

For both systems, H_2/CO ratio became constant at a value of 1.0 for temperatures above 750 °C. Below the 750 °C, H_2/CO ratio increases without RWGS as a result of lower CO production. In summary, the occurrence of RWGS was found to significantly affect the equilibrium conversions and composition during DRM. Therefore, tuning this reaction selectivity by the appropriate catalyst development is crucial towards obtaining the desired results.

The effect of variations in the feed (CH_4/CO_2) composition on the product distribution was investigated by feeding a total of 60 kmol of reactants at a ratio of 1:2 and 2:1 as a function of temperature (Figure 3). From the figure, changing the stoichiometric amount of the reactants thermodynamically affects the product selectivity. Higher amount of CO_2/CH_4 in the feed produces syngas with higher CO content than H_2 (Figure 3A). Similar trend was obtained with higher feed CH_4 producing more H_2 (Figure 3B). Carbon deposition is however more prominent on increasing CH_4 in the feed, especially at temperatures above 700 °C. This is indicative of the carbon deposition predominantly from methane cracking as validated by the rapid increment in H_2/CO ratio. This further confirms that methane cracking contributes the highest amount of carbon deposits compared to the Boudouard reaction [20]. Similar approach was conducted to analyse and compare carbon deposition from different reforming reactions [21] and comparative study of types of carbon deposited during steam and dry reforming [17]. For a high CO_2 feed, an increase in the amount of CO_2 was noticed in the system producing larger amount of H_2O at temperatures above 700 °C attributed to more occurrence of RWGS [22]. It can be deduced that beside influencing the product selectivity, altering the stoichiometric feed ratio can also be detrimental to the stability of DRM catalyst. High CH_4 feed produces more carbon deposits while high CO_2 feed produces more water that can affect the thermal stability of catalysts [23].

IOP Conf. Series: Materials Science and Engineering **808** (2020) 012001 doi:10.1088/1757-899X/808/1/012001

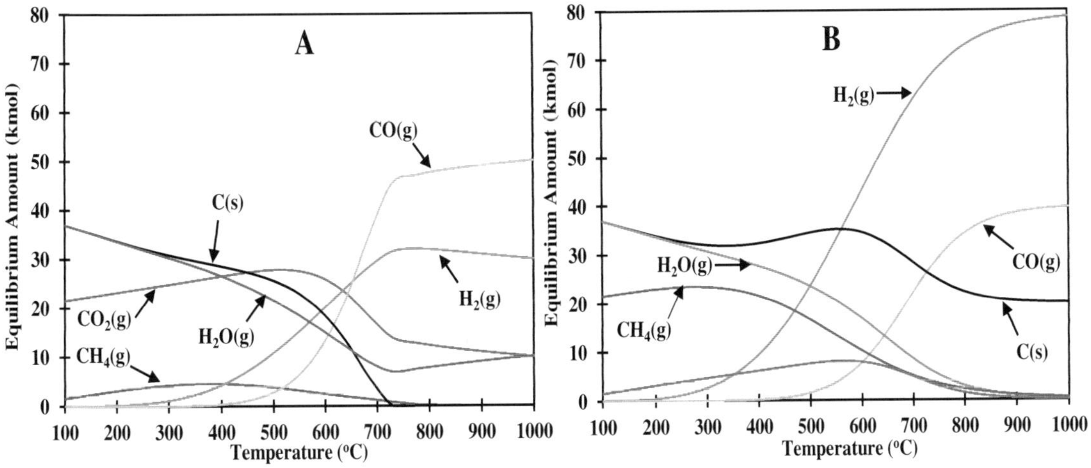

Figure 3. Thermodynamic plots for DRM at different inlet feed ratio with equilibrium reactor product compositions obtained (A) with the CO_2:CH_4 = 2 and (B) with the CO_2:CH_4 = 0.5.

4. Conclusion

Thermodynamic analysis was conducted for DRM reaction and other side reactions to evaluate and compare the spontaneity of each reaction as a function of reaction temperature. It was observed that dry reforming of methane is most affected by methane cracking and RWGS reaction than it is by Boudouard reaction. This is because the three reactions are spontaneous at high temperatures above 700 °C. Boudouard reaction was however found to be less detrimental due to its spontaneity found to exist only at temperatures lower than reaction temperature. From the thermodynamic sensitivity analysis conducted, it was observed that RWGS reaction does not significantly affect the syngas H_2/CO ratio, but was found to cause more carbon formation. This is attributed to the increased favourability of CH_4 cracking caused by consumption of CO_2 and H_2. The sensitivity of products composition to changes in CO_2:CH_4 ratio in the feed also revealed that syngas H_2/CO ratio is more affected by low CO_2:CH_4 and is accompanied by more carbon formation. It can thus be concluded that, for a dynamic upstream feed composition, higher CO_2 in the feed than the stoichiometric amount has less consequences on the product ratio and catalyst life span. Higher amount of CH_4 in the feed favours more CH_4 cracking for increased H_2 production and carbon deposition.

Acknowledgments

This work was supported by the Universiti Teknologi Malaysia Transdisciplinary Research Grant (Grant No. 06G53) and the National Economic Empowerment and Development Strategy (NEEDS) Assessment Intervention Fund of Abubakar Tafawa Balewa University Bauchi, Nigeria.

References

[1] Abdulrasheed A A, Jalil A A, Triwahyono S, Zaini M A A, Gambo Y and Ibrahim M 2018 Surface modification of activated carbon for adsorption of SO_2 and NO_X: A review of existing and emerging technologies *Renew. Sustain. Energy Rev.* **94** 1067–85.

[2] Abdel N, Aramouni K, Touma J G, Tarboush BA, Zeaiter J and Ahmad M N 2018 Catalyst design for dry reforming of methane : Analysis review *Renew. Sustain. Energy Rev.* **82** 2570–85.

[3] Du X, Zhang D, Shi L, Gao R and Zhang J 2012 Morphology Dependence of Catalytic Properties of Ni/CeO_2 Nanostructures for Carbon Dioxide Reforming of Methane *J. Phys. Chem. C*.**116**

10009–16.

[4] Gambo Y, Jalil A A, Triwahyono S, Abdulrasheed A A 2018 Recent advances and future prospect in catalysts for oxidative coupling of methane to ethylene: A review *J. Ind. Eng. Chem.* **59** 218–29.

[5] Theofanidis S A, Batchu R, Galvita V V, Poelman H and Marin GB 2016 Carbon gasification from Fe-Ni catalysts after methane dry reforming *Appl. Catal. B. Environ.* **185** 42–55.

[6] Estephane J, Aouad S, Hany S, El Khoury B, Gennequin C, El Zakhem H, et al. 2015 CO_2 reforming of methane over Ni-Co/ZSM5 catalysts. Aging and carbon deposition study *Int. J. Hydrogen Energy* **40** 9201–8.

[7] Ren J, Li M, Chinghuan A, Cheng K, and Chen Y 2019 IR spectroscopic measurement of hydrogen production kinetics in methane dry reforming *Int. J. Hydrogen Energy* **44** 9866-72

[8] Pakhare D and Spivey J 2014 A review of dry (CO_2) reforming of methane over noble metal catalysts *Chem. Soc. Rev.* **43** 7813–37.

[9] Li M and van Veen AC 2018 Tuning the catalytic performance of Ni-catalysed dry reforming of methane and carbon deposition via Ni-CeO_{2-x} interaction *Appl. Catal. B. Environ.* **237** 641–8.

[10] Hambali U H., Jalil A A, Siang T J, Abdulrasheed A A, Fatah N A A, Hussain I, et al. 2019 Effect of ZSM-5 Acidity in Enhancement of Methanol-to-Olefins Process J. *Energy Saf. Technol.* **2** 9–13.

[11] Sidik S M, Triwahyono S, Jalil A A, Aziz M A A, Fatah N A A and Teh L P 2016 Tailoring the properties of electrolyzed Ni/mesostructured silica nanoparticles (MSN) via different Ni-loading methods for CO_2 reforming of CH_4 *J. CO_2 Util.* **13** 71–80.

[12] Abdulrasheed A, Jalil A A, Gambo Y, Ibrahim M, Hambali H U and Shahul Hamid M Y 2019 A review on catalyst development for dry reforming of methane to syngas: Recent advances. *Renew. Sustain. Energy Rev.***108** 175–93.

[13] Abdulrasheed A A, Jalil A A, Hamid M Y S, Siang T J, Fatah N A A, Izan S M, et al. 2019 Dry reforming of methane to hydrogen-rich syngas over robust fibrous KCC-1 stabilized nickel catalyst with high activity and coke resistance. *Int. J. Hydrogen Energy* 1–13.

[14] Abdullah B, Abd Ghani N A, Vo D V N 2017 Recent advances in dry reforming of methane over Ni-based catalysts *J. Clean Prod.* **162** 170–85.

[15] Jalil A A, Gambo Y, Ibrahim M, Abdulrasheed A A, Hassan N S, Nawawi M G M, et al.2019 Platinum-promoted fibrous silica Y zeolite with enhanced mass transfer as a highly selective catalyst for n-dodecane hydroisomerization *Int. J. Energy Res.* 1–16.

[16] Kale G R, Kulkarni B D and Joshi A R 2010 Thermodynamic study of combining chemical looping combustion and combined reforming of propane *Fuel* **89** 3141–6.

[17] Giehr A, Maier L, Schunk S A and Deutschmann O 2018 Thermodynamic Considerations on the Oxidation State of Co/γ-Al_2O_3 and Ni/γ-Al_2O_3 Catalysts under Dry and Steam Reforming Conditions *Chem. Cat. Chem.* **10** 751–7.

[18] Myint M N Z, Yan B, Wan J, Zhao S and Chen J G 2016 Reforming and oxidative dehydrogenation of ethane with CO_2 as a soft oxidant over bimetallic catalysts *J. Catal.* **343** 168–77.

[19] Pakhare D, Shaw C, Haynes D, Shekhawat D and Spivey J 2013 Effect of reaction temperature on activity of Pt- and Ru-substituted lanthanum zirconate pyrochlores ($La_2Zr_2O_7$) for dry (CO_2) reforming of methane (DRM) *J. CO_2 Util.* **1**:37–42.

[20] Horlyck J, Lawrey C, Lovell E C, Amal R and Scott J 2018 Elucidating the impact of Ni and Co loading on the selectivity of bimetallic NiCo catalysts for dry reforming of methane *Chem. Eng. J.* **352** 572–80.

[21] Kumar N, Wang Z, Kanitkar S and Spivey J J 2016 Methane reforming over Ni-based pyrochlore catalyst: deactivation studies for different reactions *Appl. Petrochemical Res.* **6** 201–7.

[22] Tsai H L and Wang C S 2008 Thermodynamic equilibrium prediction for natural gas dry reforming in thermal plasma reformer *J. Chinese Inst. Eng.* **31** 891–6.

[23] Gholami R, Alyani M and Smith K 2015 Deactivation of Pd Catalysts by Water during Low Temperature Methane Oxidation Relevant to Natural Gas Vehicle Converters *Calaysts* 5 561–94.

IOP Conf. Series: Materials Science and Engineering **808** (2020) 012002 doi:10.1088/1757-899X/808/1/012002

Cellulose conversion into renewable chemicals and fuels over supported metal catalysts

N S Hassan[1], A A Jalil[1,2*]

[1]School of Chemical and Energy Engineering' Faculty of Engineering, Universiti Teknologi Malaysia, 81310 UTM Johor Bahru, Johor, Malaysia.
[2]Centre of Hydrogen Energy, Institute of Future Energy, Universiti Teknologi Malaysia, 81310 UTM Johor Bahru, Johor, Malaysia

*aishahaj@utm.my

Abstract. Nowadays, the development of biomass as an alternative energy source has been explored for the production of energy, chemicals and transportation fuels. Cellulose is an abundant, low-cost and renewable energy source that can be transformed into solid, liquid and gas products by pyrolysis. Sustainable, low cost and efficient catalysts are needed for catalytic cellulose conversion by pyrolysis. Among heterogeneous catalysts, supported metal catalysts have been extensively studied which demonstrated distinct reactivity and/or selectivity in a variety of chemical reactions. This paper reviews and provides a summary of the cellulose conversion by supported metal catalysts into value added chemicals and fuels. Emphases will be put on the reaction mechanism and key factors that exert effect on the catalytic performance.

1. Introduction

The diminishing of fossil fuel resources has initiated more researchers to use renewable resource for the production of energy, chemicals, and transportation fuels [1]. Lignocellulose composed of 30–55 wt% of cellulose, 25–30 wt% of lignin and 25–30 wt% of hemicellulose [2]. Being the most abundant and low cost renewable energy source, the usage of cellulose become more popular to produce chemicals and fuels. It consists of glucose units that are linked by β-1,4-glycosidic bonds with the intermolecular and intramolecular hydrogen-bond networks [3]. Even though large amount of cellulose are synthesized through photosynthesis of plants (800,000 billion metric tons), merely 0.2 billion metric tons is being used as feedstock in pharmaceuticals, paper, and food industry [4]. Thus, new advances has been created to limit the underused cellulose that is squandered each year.

Nowadays, the cellulose conversion into valuable chemicals and fuels has attracted worldwide interest [5]. Generally, the potential chemicals and fuels can be produced by different chemical processes. There are two routes for conversion of cellulose, which are chemo-catalytic through glucose by hydrolysis of cellulose and direct or one-pass catalytic process in the presence of a multifunctional catalyst. Recent research has paid much attention to the latter routes in order to enhance the conversion of cellulose. The typical processes can be categorized as gasification, pyrolysis, liquefaction and combustion. The latter method is the most direct and easiest process to convert organic matter to heat, carbon dioxide and water by adding an oxidant. Meanwhile, the gasification method involves the heating process and using an oxidant to produce syngas, which contains a mixture of carbon monoxide and hydrogen. Compared to combustion, gasification has many advantages, which include converting the low-value material into electricity and vehicle fuels. However, gasification process requires complex and more expensive technologies compared to conventional petrochemical processes.

Pyrolysis is a thermal heating in an inert atmosphere, which produces the char, gas, and a liquid composed of a mixture of hundreds of oxygenated organic compounds. These unsaturated compounds cause low stability under storage conditions, which prompts their need to be upgraded. Currently, catalytic treatment has been the focus of many researchers to produce a liquid fuel. Therefore, it is necessary to design selective catalysts and optimize upgrading process in order to maximize the yield of high value added chemicals and minimize the formation of undesired compounds. In addition, the application of cellulose conversion in the chemical industry was hindered as a result the low reactivity of cellulose, accompanied by poor solubility in solvents and stable chemical structure. Based on those facts, the cellulose conversion by heterogeneous catalysis plays important role for the production of a chemicals and fuels such as hydroxymethylfurfural (HMF), glucose and sorbitol for further bio-refinery procedures. It offer advantages in terms of easy separation, reusability and anticorrosion with the ability to provide practical conveniences in a continuous system. Another merit of heterogeneous catalysts is their stability to severe reaction environments; some can withstand temperatures as high as 1300 °C and pressure up to 350 bar.

2. Catalytic Conversion of Cellulose to Renewable Chemicals

Lately, many researchers have focused on heterogeneous catalysis for cellulose conversion into valuable chemicals, particularly glucose, polyols, 5-hydroxy-methylfurfural (5-HMF), lactic acid and levulinic acid (Scheme 1) [6]. Different catalysts and reaction temperatures varies the major product from glucose to polyols. Glucose as the primary product when the reaction temperature below 423 K, meanwhile under pressurized hydrogen gas (H_2) and hydrogenation catalyst (supported ruthenium (Ru), platinum (Pt), iridium (Ir) and etc) were led to sorbitol production. For catalyst function, the acid site that originate from acidic groups can enhance the production of glucose by cleavage the C-C bonds. Similarly, basic catalyst such as calcium oxide also can act as C-C bond cleavage agent, however, the isomerization of glucose to fructose resulting in glycerol rather than polyols as the major product. Based on those facts, this review can stimulate a more profound comprehension with respect to the current catalytic conversion of cellulose into platform chemicals in the biorefinery.

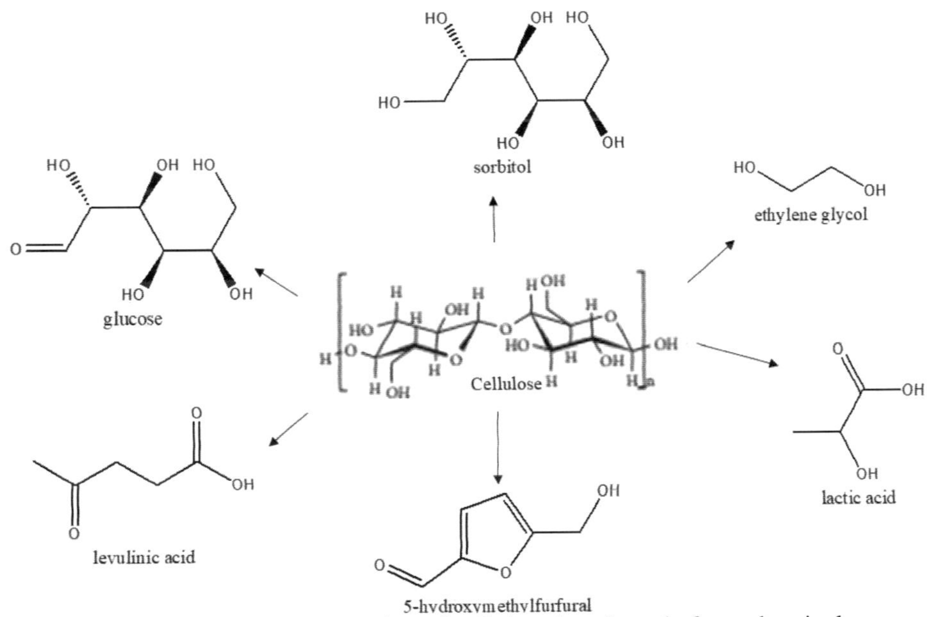

Scheme 1. Catalytic conversion of cellulose into key platform chemicals.

2.1. Glucose

The production of glucose has become more attractive in catalysis study and fuel industry due to its broad applications. Glucose is a versatile chemical to produce value added products such as bioethanol, 5-HMF, and biodegradable plastics. Heterogeneous catalysts have been developed under aqueous conditions. Hara and co-workers obtained 64% of water-soluble oligosaccharides (β-1,4-glucan) and 4% of glucose yield with 68% cellulose conversion at 373 K for 3 h using sulfonated carbonaceous materials [7]. The higher production of β-1,4-glucan may be due to the faster hydrolysis process as consequent of the higher adsorption capacity of β-1,4-glucan by phenolic hydroxyl groups. Simultaneously, the hydrogen bonding was formed as a consequence of the cleavage of glycosidic bonds by the sulfonic groups.

Supported metal catalysts have also been widely used for the hydrolysis of cellulose. The metal site promoted the hydrolysis of cellulose via heterolytic dissociation of hydrogen gas (H_2) as well as hydrogenation of glucose. Currently, mesoporous carbon materials (CMKs) as the catalyst support showed the excellent properties which enhanced the production of glucose. Fukuoka *et al.* successfully prepared the CMK supported Ru catalyst (Ru/CMK) for conversion of cellulose to glucose which gave the glucose and oligosaccharide yield were 34% and 5%, respectively [8]. Compared to bare CMK, only 21% yield of glucose and 22% yield of oligosaccharide were obtained in the hydrolysis process. This result showed that synergystic effect of CMK and Ru in the production of glucose, in which CMK played role in the conversion cellulose into oligosaccharides, meanwhile Ru was able to convert oligosaccharides into glucose (Scheme 2). By increasing Ru loading to 10 wt%, the yield of glucose was raised up to 31%. The Ru species with high valent can be proposed as a catalyst for hydrolysis due to its higher dispersion on CMK.

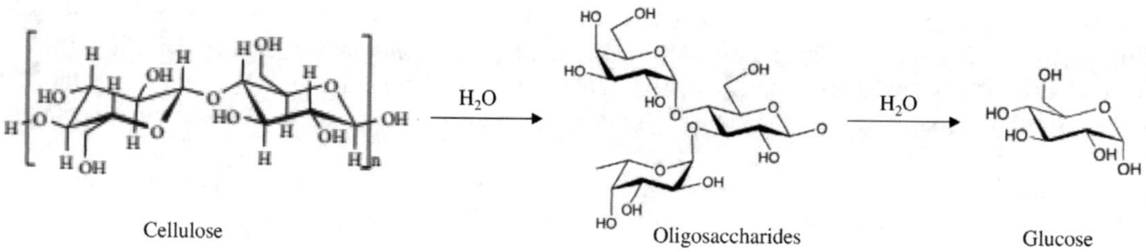

Scheme 2. Conversion of cellulose into glucose via soluble oligomers.

2.2. Polyols

The cellulose conversion into chemicals is important to the improvement of a sustainable society. In particular, the catalytic cellulose conversion to polyols is a significant part of biorefinery, which has pulled in more enthusiasm to fulfil vitality needs. In this section, the latest developments in the catalysis of cellulose conversion into ethylene glycol (EG) and sorbitol have been focused, which they are widely utilized in the food industry, beauty care products and pharmaceutical production [9]. Scheme 3 shows the cellulose transformation into sorbitol and EG by hydrogenation process of glucose and glycolaldehyde, respectively. Firstly, cellulose is hydrolysed to glucose, which can be directly hydrogenated to sorbitol or break its C-C bond to glycolaldehyde (GA) that is subsequently hydrogenated to EG.

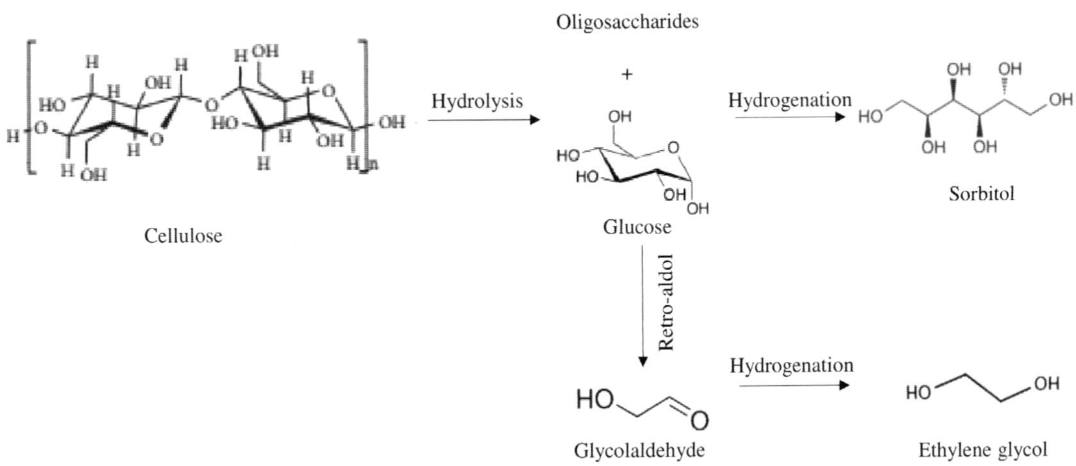

Scheme 3. Conversion of cellulose into sorbitol and ethylene glycol.

2.2.1. Ethylene glycol (EG). EG is an essential chemical, which is mainly used as a raw material for the production of polyester fibers and anti-freeze [9]. There are multi reactions involved in the production of EG such as of hydration, cracking and epoxidation in petroleum-based ethylene. The EG production by a viable process is essential to be accomplished due to its significance in chemical industry. Therefore, the cellulose was used as the starting material to produce EG. Numerous types of reactions are involved in cellulose transformation into EG such as hydrogenation, hydrolysis, condensation of retro-aldol, cracking, dehydration, decarbonylation, and hydration. About twenty compounds can be produced as the intermediate products as well as byproducts [10].

Table 1. EG yields from the conversion of cellulose by different catalyst.

Catalyst	Temperature (K), Pressure (MPa)	EG Yield (%)	Conversion of cellulose (%)	Ref
Ni-W/SBA-15	518, 6	75	100	[11]
Ru-W/CNT	478, 5	41	100	[13]
3Al-15W-3Ni	503, 4	76	100	[14]
Ni-W/CNF	518, 6	34	95	[15]
Ru/WO_3	513, 4	76	100	[16]

Table 1 shows the EG yields from the conversion of cellulose by different catalyst. The yield of EG about 75% was achieved with tungsten (W)-based bimetallic catalyst such as nickel (Ni)-W supported on SBA-15 (Ni-W/SBA-15), however the SBA-15 structure collapsed and catalyst could not be reused [11]. Thus, the higher resistance to acid and base attacking of carbon as a support is preferred under hydrothermal conditions [12]. Other than that, Ribeiro *et al.* reported that the production of EG from cellulose by Ru and W supported on multi-walled carbon nanotubes gave the higher yield about 41 % compared than the Ru and W monometallic catalysts which about 35 % at 5 MPa of H_2 and 478 K for 5h [13]. This was due to the synergistic effect of the W and Ru component, due to their close interaction of active sites for hydrogenation and retro aldol condensation reaction.

Hamdy *et al.* reported that the selective hydrogenolysis of cellulose into EG via hydrothermal and sol-gel process by 3Al-15W-3Ni catalyst gave a 100 % of cellulose conversion with 76% yield of EG under 4 MPa of H_2 pressure for 90 min at 503 K [14]. The presence of tri-metallic sites enhanced the production of EG, in which Al^{3+} acidic active species promote the hydrolysis of cellulose into glucose. The oxygen vacancies in the partially reduced WO_{3-x} adsorb the oxygen of glucose, which assist in the fast formation of glycolaldehyde [14]. Finally, glycolaldehyde was reduced to EG over the nanoparticles of metallic Ni.

Yang *et al.* synthesized a carbon nanofiber (CNF) supported bimetallic (Ni-W/CNF) catalyst gave 95% of cellulose conversion and 34% of EG yield under 6 MPa of H_2 pressure at 518 K for 2 h [15]. The higher catalytic activity most probably due to the excellent accessibility and uniform dispersion of metal species. Other than that, the W species acts as C-C cleavage agent before the intermediate glycolaldehyde hydrogenated onto EG by the catalysis of Ni. Next, Li *et al.* also found that effective hydrogenolysis for the cellulose conversion to EG by Ru supported on tungsten oxide (Ru/WO_3), leading to higher yield of EG about 76% with 100% cellulose conversion at 513 K for 2 h under 4 MPa of H_2 pressure [16]. This results increased due to the enhanced active site of W^{5+} as a Lewis base by addition of Ru species, thereby enhancing the retro-aldol condensation reaction of glucose by etching of H^+ for the hydrogenolysis of cellulose to EG (Scheme 4).

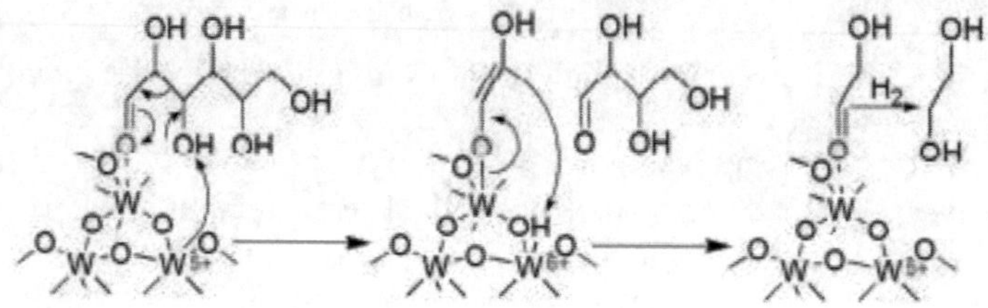

Scheme 4. W^{5+} species for retro-aldol condensation reaction in the production of EG. Adapted from [16].

2.2.2. Sorbitol. Sorbitol is the most valuable bio-based chemicals that can be used in food, textiles and pharmaceuticals and as starting material in synthesis of chemicals and fuels such as isosorbide, alkane and glycols [17]. There are two steps involves in the production of sorbitol which are hydrolysis of cellulose to glucose and then hydrogenation of glucose to sorbitol by acid and metal site, respectively [18]. Thus, the acid site, metal site and H^+ played the vital role in production of sorbitol. Recently, supported transition metals were used for the cellulose conversion to sorbitol. For example, the production of sorbitol was achieved by using Ni based catalyst, attributed to its low cost and high activity. However, they have shown low stability in the reaction medium [19]. Alternatively, various catalyst including Ru have shown as active catalyst for this reaction (Table 2).

Table 2. Sorbitol yields from the conversion of cellulose by different catalyst.

Catalyst	Temperature (K), Pressure (MPa)	Sorbitol Yield (%)	Conversion of cellulose (%)	Ref
Ru/$NbOPO_4$	443, 4	69.1	90	[20]
Ru/SiO_2-SO_3H	423, 4	61.2	90.5	[21]
Ru/CCD-SO_3H	453, 4	85.5	100	[22]
Ru/MCM-48	393, 2.5	89.6	100	[23]

Xi *et al.* prepared a novel mesoporous niobium phosphate-supported bifunctional Ru catalyst (Ru/$NbOPO_4$), which showed excellent performance with 69.1% of sorbitol yield after 24 h at 333 K and 4 MPa H_2 [20]. This can be attributed to the high acid amount, mesoporosity as well as synergy effect between $NbOPO_4$ for cellulose hydrolysis and Ru nanoparticle for hydrogenation. The yield of sorbitol about 61.2% was achieved with sulfonic acid-functionalized silica-supported Ru catalyst (Ru/SiO_2-SO_3H) for the hydrolytic and hydrogenation of cellulose to sorbitol after 10 h at 423 K and 4

MPa H_2 [21]. Similar study was done by Li *et al.*, using SO_3-H which was supported on carbonized cassava dregs (CCD) and gave the sorbitol yield of 85.5% and 100% of cellulose conversion at 453 K and 4 MPa H_2 for 10 h [22]. The SO_3H sites and Ru hydrogenation sites of the catalyst played an important role in hydrolytic hydrogenation of cellulose. Romero *et al.*, prepared a bifunctional catalyst using MCM-48 as the support and Ru as the metal phase. MCM-48 provides a high surface area acid structure with good diffusional properties for the hydrolytic step, while Ru acts as the active phase for the hydrogenation of D-Glucose. This catalyst showed the good yield of sorbitol about 89.56% with 100% cellulose conversion at 393 K and 2.5 MPa H_2 [23].

2.2.3. Hydroxymethylfurfural (5-HMF). 5-HMF is an important platform chemical to fuels and plastics since it has been a target in the biorefinery [24]. A 5-HMF yield of 53.3% was achieved over niobia/carbon catalyst (Nb/C) at 443 K for 8 h, accompanied with 24.5% glucose [24]. The appropriate Brønsted (BAS) or Lewis acid sites (LAS) was favoured the cellulose conversion to 5-HMF (Scheme 5). The LAS plays the important role in the isomerization of glucose to fructose. Other than that, similar study also showed that the Lewis acidic catalyst such as aluminium chloride and phosphoric acid (H_3PO_4) as Brønsted acidic catalyst were used for the production of 5-HMF from cellulose, leading to the highest yield of 5-HMF about 49.42% at 453 K for 2 h in a single-phase reaction system of water and 1,2-dimethoxyethane (Scheme 6) [25]. The H_3PO_4 acts as Bronsted acid to provide hydrogen ions for the hydrolysis and dehydration reactions. Other than that, an effective catalytic activity of novel bifunctional catalytic system, $CrCl_3$/[R_4N]ReO_4 gave the yield of 5-HMF about 49.3% was achieved at 423 K for 30 min. Compared to the single $CrCl_3$ catalyst, the bifunctional catalytic system of the catalyst exhibited superior catalytic performance. This may be due to the synergistic effect of both role of LAS and BAS [26].

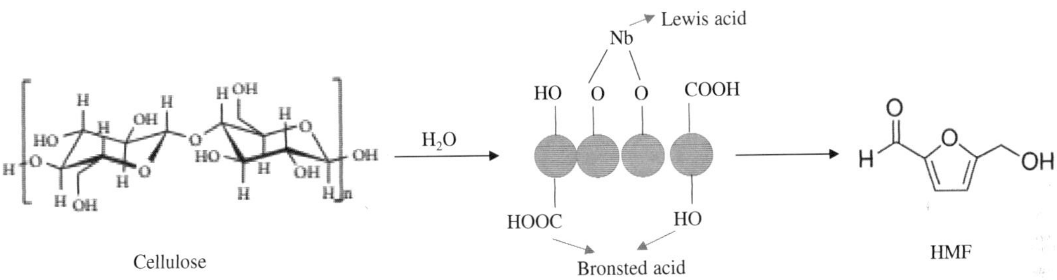

Scheme 5. Conversion of cellulose into HMF.

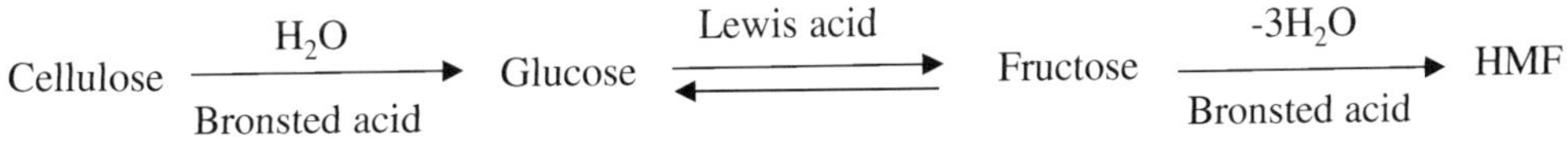

Scheme 6. Cellulose conversion to HMF catalysed by Lewis acid and Bronsted acid.

2.3 Levulinic acid
Levulinic acid (LA) is a versatile platform molecule, which can be transformed to various high-value chemicals. Thus, the conversion of cellulose into LA has been more focused by most researchers. Generally, the Bronsted or Lewis acids was used for production of LA. Xiang and co-workers reported that cellulose conversion into LA using hierarchically porous ETS-10-supported Ni catalyst (Ni-HMETS-10) in aqueous solution, led the highest yield of LA (91.0%) with complete conversion of

cellulose (100%). This reaction was achieved under 6 MPa of H_2 pressure at 473 for 6 h. The synergistic effect of LAS and hierarchical porous structures of ETS-10 in a H_2 atmosphere play the main role in
order to improve the production of LA [27]. Moderate Lewis acidic centers of the catalyst enabled the activation of the reaction substrates as well as intermediates, subsequently leading to the higher catalytic activity. Additionally, the incorporation of Ni has played a positive role in conversion growth, which proves to be a quite delightful and feasible strategy to take place of those widely used noble metals with promising activity.

2.4 Lactic acid

Lactic acid (LA) is an important platform chemical for the production of biodegradable plastics and starting materials for the pharmaceutical, chemical and industry [29]. The catalytic transformation of cellulose into lactic acid has become more interest in recent years. In order to achieve an effective of cellulose conversion into lactic acid, the activation of C–O and C–C bonds as well as multifunctional catalysts are mostly required, in which the latter requirement can combine various reactions comprising of hydrolysis, isomerization as well as retro-aldol condensation. Table 3 shows the catalyst used for conversion of cellulose into LA. The exceptional catalytic testing of the tin-modified niobium (SnO_2/Nb_2O_5) for the isomerization of triose sugars to LA, leading 98.7% of LA yield at 443 K after 4 h [29]. Basically, the promising yield of LA achieved from the conversion of dihydroxyacetone (DHA) in water in the presence of BAS as well as LAS (Scheme 7). As known, $Nb_2O_5.nH_2O$ has both strong BAS and LAS, which responsible for the dehydration of DHA to pyruvic aldehyde (PA), and the subsequent isomerization reaction of PA to LA occurred at the Nb and Sn sites. The strong interaction between these two sites makes the Sn/Nb_2O_5 catalysts suitable for the conversion of DHA to LA.

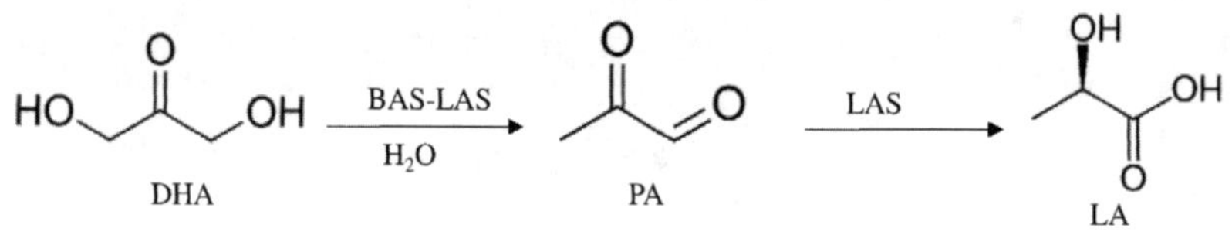

Scheme 7. Conversion of DHA in water into LA.

The isomerization of glucose over Pb–Sn-beta zeolite, that was prepared via ion exchange, led to 52% of LA yield [30]. Two steps are involved in this reaction, where firstly, the isomerization reaction of glucose took place in order to produce the fructose, and then followed by fructose condensation process to form the C_3 intermediates via retro-aldol reaction. Other than that, the catalyst also exhibited greater catalytic performance for both dehydration and isomerization of DHA and PA. In addition, this catalyst also inhibit the production of HMF. The effective usage of sugars as the starting material for production of lactic acid was achieved by sulfonated Sn-doped KIT-6 (Sn-KIT-6-Pr-SO_3H) catalyst in aqueous phase [31]. The catalyst showed high efficiency in lactic acid production (66.2 %) with cellulose conversion of 88.9%. Compare to Amberlyst15, HUSY catalyst, those catalysts exhibited superior performance in the production of LA.

Catalyst	Temperature (K), Pressure (MPa)	Lactic acid Yield (%)	Conversion of cellulose (%)	Ref
SnO_2/Nb_2O_5	433, 5	98.7	95	[29]
Pb–Sn-beta zeolite	463, 5	52	99	[30]
sulfonated Sn-doped KIT-6	483, 5	16.2	88.9	[31]

Table 3. Lactic acid yields from the conversion of cellulose by different catalyst.

3. Catalytic Conversion of Cellulose to Fuels

The depletion of fossil resources and increasing energy demand has initiated researchers to explore the usage of renewable biomass resources for production of transportation fuels. Some researches regarding conversion of cellulose into aliphatic and aromatic hydrocarbon were reported.

3.1 Aliphatic hydrocarbon

Direct cellulose conversion into fuels has been considered as most prospective ways for the production of transportation fuels. The production of aliphatic hydrocarbon (C_5 and C_6 alkanes) from conversion of cellulose is better owing to its low boiling point and hydrophobicity as well as low energy consumption for their separation. Scheme 8 shows the reaction pathway of C_3 and C_4 hydrocarbons formation of cellulose. The intermediate products such as 5-HMF, levulinic acid and pentanoic acid were formed during the cracking reaction of glucose into C_4 hydrocarbons. Table 4 shows the summary of catalyst used for conversion of cellulose into aliphatic hydrocarbon.

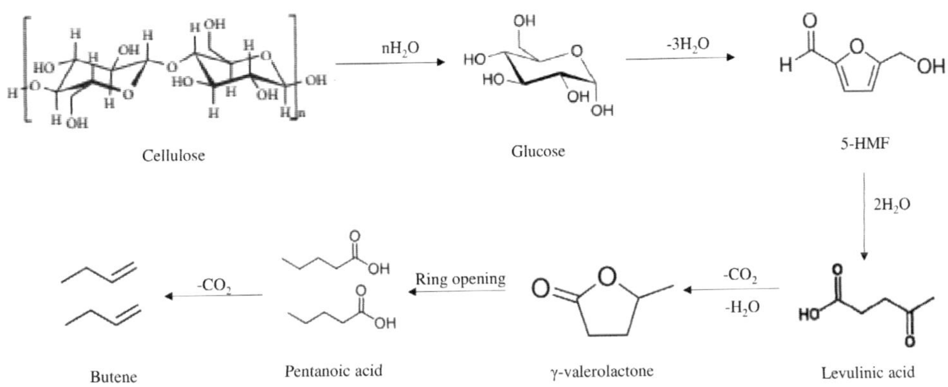

Scheme 8. Reaction pathway of C_3 and C_4 hydrocarbon formation from cellulose.

Reportedly, cellulose can be transformed into hydrocarbons using heterogeneous solid catalysts. The platinum supported on beta zeolite (Pt/H-beta) was used in one-pot cellulose conversion in aqueous solution with low selectivity of C_{5+} hydrocarbons about 11.2%, and higher C_3 and C_4 hydrocarbon production is about 46.8% at 443 K [32]. The Pt site and acid site in beta zeolite were the significant means for enhanced the production of hydrocarbon. A large pore system and effective acidity also contributed to the effective C_3 and C_4 hydrocarbons. Ogo *et al.* also reported one pot direct catalytic conversion of cellulose using platinum supported on USY zeolite (Pt/H-USY) catalyst at low temperature (443 K) that gave the 65.6 % of C_3 and C_4 hydrocarbons [33]. The results suggested that the acid site of H-USY zeolite solubilizes the cellulose; meanwhile Pt site cracked the glucose into hydrocarbons (Scheme 9).

IOP Conf. Series: Materials Science and Engineering **808** (2020) 012002 doi:10.1088/1757-899X/808/1/012002

Table 4. Aliphatic hydrocarbon yields from the conversion of cellulose by different catalyst.

Catalyst	Temperature (K), Pressure (MPa)	Product (% Yield)	Conversion of cellulose (%)	Ref
Pt/H-beta	443, 0	C_3-C_4 (46.8)	10	[32]
Pt/H-USY	443, 0	C_3-C_4 (65.6)	19.4	[33]
Pt/HZSM-5	673, 6.5	C_2-C_9 (89)	-	[34]
Ir–ReO_x/SiO_2+HZSM-5	463, 6	C_6 (83)	98	[35]
TSARu/C	493, 5	C_6 (82)	100	[36]
$LiTaMoO_6$/Ru/C	503, 6	C_6 (82)	-	[37]
BPO_4/Ru/C	403, 6	C_5-C_6(70.6)	100	[38]

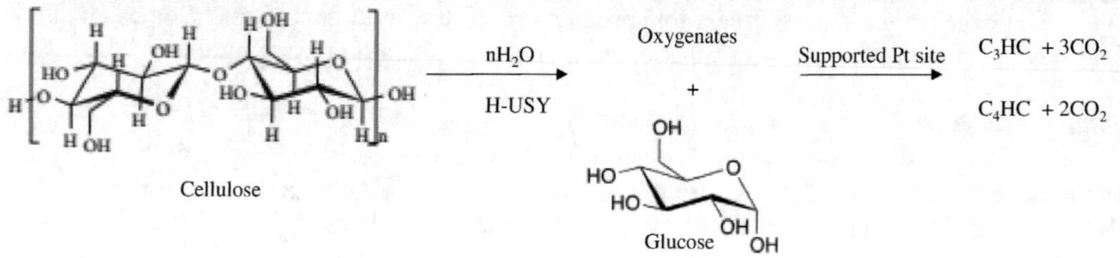

Scheme 9. Reaction pathway of hydrocarbon formation from cellulose over Pt/H-USY.

The Pt/HZSM-5 showed the yield of C_2-C_9 alkanes of about 89% when using the pretreated cellulose at 673 K with 6.5 MPa of H_2 atmosphere [34]. The pretreated cellulose in alcohol was lowering the molecular weight of cellulose with producing the intermediates products such as monosaccharide and disaccharide. Besides the pretreatment of cellulose, the reaction systems also influence the catalytic conversion of cellulose. The combination of metal-acid catalysts such as iridium-rhenium supported on silicon dioxide mixed with ZSM-5 zeolite (Ir–ReO_x/SiO_2+HZSM-5) or tungstosilicic acid (TSA) modified Ru/C catalysts was performed in biphasic reaction systems such as organic and water and exhibited higher yield of hexanes [35, 36]. The former catalyst performed in n-dodecane and water under 6 MPa of H_2. The HZSM-5 enhanced the hydrolysis of cellulose, meanwhile Ir–ReO_x/SiO_2 catalyst promotes the hydrogenation and hydrogenolysis of C-O bond, which gave the higher yield of n-hexane (C_6 paraffin) is about 83% at 463 K (Scheme 10). Other than that, the cellulose was directly transformed into n-hexane (82%) over Ru/C catalysts with H_2 pressure of 5.0 MPa at 493 K for 5 h [36].

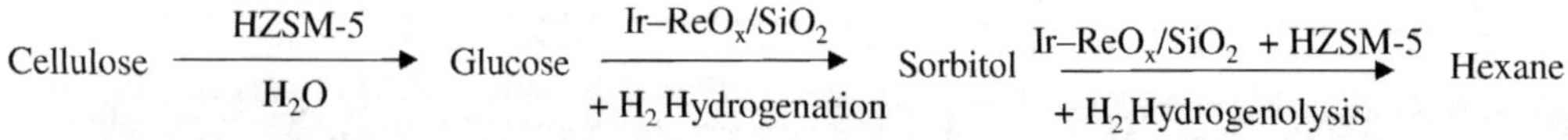

Scheme 10. Reaction pathway for conversion of cellulose over Ir–ReO_x/SiO_2+HZSM-5.

Compared to biphasic system or organic phase, the mono-phasic aqueous system is necessary to explore in order to make the catalytic conversion of cellulose into hexane more economical and environmentally friendly. Liu *et al.* has prepared a three-catalyst system, lithium tantalum molybdenum oxide ($LiTaMoO_6$) which is combined with Ru/C to transform cellulose into liquid alkanes in phosphoric acid (PA) solution giving a high yield (82%) [37]. The PA solution was used for the hydrolysis of cellulose, removal of oxygen as well as ring opening reaction. Other than that, the combining of boron phosphate with Ru/C catalyst in aqueous solution was used for C_5 and C_6 alkanes production (gasoline-range hydrocarbons) from cellulose and exhibited higher yield (70.6%) [38]. The presence of boric acid plays important effect on the formation of hexane from sorbitol, in which reduced the formation rate of isosorbite (intermediate products of sorbitol). Hence, the production of liquid hydrocarbon can be improved significantly.

3.2 Aromatic hydrocarbon

Monoaromatic compounds, especially benzene, toluene and xylene (BTX), are versatile chemicals in the petroleum industry, which are used as gasoline additives. They also used as solvents and raw materials for preparation of chemical products such as rubber, plastics, and fiber. The general pathway of aromatic production from oxygenated compounds is illustrated in Scheme 11. Table 5 shows the summarized aromatics hydrocarbon yields from the conversion of cellulose by different catalyst. Rezaie *et al.* reported that the combining of HZSM-5 with dealuminated HY showed higher yields of aromatic hydrocarbon (27%) compared to the individual HZSM-5 (20%) and HY (9%) at 798 K [39]. The formation of coke was lower when using the combination catalyst, HZSM-5 with dealuminated HY, may be due to the larger pore of HY, which is as consequence of dealumination process, hence allow the diffusion of larger molecules into zeolite. Other than that, high density of acid sites located in the small space in HZSM-5 pores also caused the lower coke formation. Thus, it can be concluded that the synergistic effect between pore structure and acidity play important roles in minimizing the formation of coke and maximize the production of desired products.

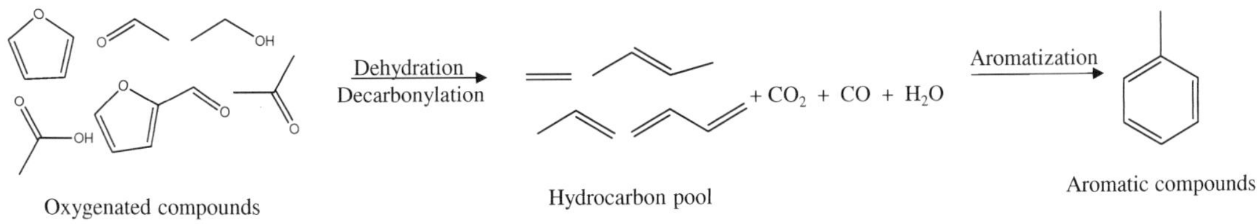

Scheme 11. General pathway towards aromatic hydrocarbon from oxygenated compounds.

Table 5. Aromatics hydrocarbon yields from the conversion of cellulose by different catalyst.

Catalyst	Temperature (K), Pressure (MPa)	BTX Product (% Yield)	Ref
HZSM-5 with dealuminated HY	773, 1	27	[39]
Ga/HZSM-5	773, -	53.7	[40]
Na_2CO_3- HZSM-5	573, -	38.2	[41]
Zn/HZSM-5	773,-	71.8	[42]
Ni/HF-HZSM-5	773, -	44.9	[43]

In viewpoint of the above requirement, metal modified and treated by alkali are the most approached methods by most researchers in order to improve the catalytic performance of zeolites, especially HZSM-5. The generation of more mesopore by alkaline treatment, can cause the large molecules more accessible into the zeolite. Differently, the micropore structure and acid sites are well preserved when using the alkali for desilication process. Dai *et al.* reported that the hierarchical HZSM-5 was prepared by desilication using 0.3 M sodium hydroxide (NaOH) solution. This catalyst showed the best catalytic performance, that gave 51.16 wt% of aromatic hydrocarbons as well as increases the BTX selectivity [40]. Qiao *et al.*, also treated the HZSM-5 zeolites with three different types of alkali, which are NaOH, sodium carbonate (Na_2CO_3) and tetrapropylammonium hydroxide (TPAOH) [41]. Among them, the highest yields of aromatic hydrocarbon (38.2%) was achieved by using the Na_2CO_3-treated HZSM-5. This treatment increases the acidity of catalyst as well as having the micro and mesopore structure, which can produced the lower coke.

Metal modification also influence in improving the yield of aromatic hydrocarbon. Zheng *et al.*, successfully synthesized the metal loaded HZSM-5 (M-HZSM-5), in which using the Zn, Ga, Ni, Co, Mg and Cu [42]. Among them, Zn-HZSM-5 exhibited the highest yield of toluene and xylenes (71.84%), meanwhile Ni-HZSM-5 produced the higher contents of benzene (29.16%). This is because modification with Zn led to the increase of Lewis acid and promoted the reaction process. The Zn has an ability in dehydrogenation and aromatization which can promote the intermediate product of the pyrolysis to form aromatic hydrocarbons. Other than that, Fe and Ga was also used for conversion of cellulose. This metal modification gave the different effect on aromatics production in which, Fe-HZSM-5 exhibited higher production of BTX (58.13 %), while Ga-HZSM-5 enhanced the production of both BTX (53.70%) and alkylbenzenes [40]. Besides, the highest light aromatic yield (44.9%) was obtained by nickel/hydrogen fluoride modified HZSM-5 (Ni/HF-HZSM-5) [43]. The Ni acted as acceptors of electrons and promote the production of olefin oligomerization, particularly ethylene oligomerization. In addition, Ni induced the secondary cracking reactions to dehydrogenate and deoxidize.

4. Conclusion

The conversion of cellulose into renewable chemicals and fuels is one of the hot topics in green and sustainable chemistry. Various types of supported metal catalysts can be designed for the cellulose conversion. In this review, supported Ru and Ni was acted as active sites that promoted the hydrolysis of cellulose via heterolytic dissociation of H_2 as well as hydrogenation of glucose to produce renewable chemicals such as ethylene glycol, sorbitol, 5-hydroxymethylfurfural, levulinic acid and lactic acid. The Pt and Ir catalysts promote the hydrogenation and hydrogenolysis of C-O bond to produce the aliphatic hydrocarbon. Meanwhile, the transition metal such as Zn and Ni have an ability in dehydrogenation and aromatization, which can promote the aromatic hydrocarbons.

Acknowledgments

This research study was sponsored by the Universiti Teknologi Malaysia through Professional Development Research University Grant (No. 04E73), UTM Transdiciplinary Research Grant (Grant No. 06G53) and Collaborative Research Grant (Grant No. 07G62).

References

[1] Li G, Liu W, Ye C, Li X, and Si CL 2018 *Int. J. Polym. Sci*
[2] Wang H, Guo Y, and Chang C 2016 *Appl. Catal., A* **523** 182
[3] Yabushita M, Kobayashi H, and Fukuoka A 2014 *Appl Catal., B* **145** 1
[4] Pengfei Y, Kobayashi H, and Fukuoka A 2011 *Chinese J Catal* **32** 716
[5] Serrano DP, Melero JA, Morales G, Iglesias J, and Pizarro P 2018 *Catal Rev* **60** 1
[6] Wang J, Xi J, and Wang Y, 2015*Green Chem* **17** 737
[7] Suganuma S, Nakajima K., Kitano M., Yamaguchi D., Kato H., Hayashi S. and Hara M 2008 *J. Am. Chem. Soc* **130** 12787
[8] Kobayashi H, Komanoya T, Hara K, and Fukuoka A, 2018 *Chem Sus Chem* **3** 440
[9] Yue H, Zhao Y, Ma X, and Gong J 2012 *Chem Soc Rev* **41** 4218
[10] Zheng M, Pang J, Sun R, Wang A, and Zhang T 2017 *ACS Catal* **7** 1939
[11] Zheng MY, Wang AQ, Ji N, Pang JF, Wang XD, and Zhang T 2010 *ChemSusChem* **3** 63
[12] Wang A, and Zhang T 2013 *Acc. Chem. Res* **46** 1377
[13] Ribeiro LS, Órfão JJM, and Pereira MFR 2018 *Bioresour Technol* **263** 402
[14] Hamdy MS, Eissa, MA, and Keshk SMAS 2017 *Green Chem* **19** 5144
[15] Yang Y, Zhang W, and Yang F 2016 *Green Chem* **18** 3949
[16] Li N, Zheng Y, Wei L, Teng H, and Zhou 2017 *J Green Chem* **19** 682

[17] Murillo L, López Granados M, Fierro JLP, and Mariscal R 2016 *Appl. Catal. B-Environ* **185** 141
[18] Ribeiro LS, Delgado JJ, Órfão, JJM, and Pereira MFR 2017 *Catal. Today* **279** 244
[19] Gallezot P, and Cerino PJ 2014 *J Catal* **146** 93
[20] Xi J, Zhang Y, and Xi Q 2013 *Appl Catal A* **459** 52
[21] Zhu W, Yang H, and Chen J, 2014 *Green Chem* **16** 1534
[22] Li Z, Liu Y, Liu C, Wu S, and Wei W 2019 *Bioresour Technol* **274** 190–197
[23] Romero A, Alonso E, and Sastre A 2016 *Micropor Mesopor Mat* **224**, 1
[24] Li X, Peng K, Xia Q, Liu X, and Wang Y 2018 *Chem. Eng. J* **332** 528
[25] Zhao Y, Wang S, Lin H, Chen J, and Xu H 2018 *RSC Adv* **8** 7235
[26] Bai X, Wang JY, Men N, and Zang SL 2018 *Catal Commun* **104** 37
[27] Xiang M, Liu J, Fu W, Tang T, and Wu D 2017 *ACS Sustain Chem Eng* **5** 5800
[28] Zhang X, Lin L, Zhang T, Liu H, and Zhang X 2016 *Chem. Eng. J* **284**, 934
[29] Wang X, Song Y, Huang L, and Wang H 2019 *Catal Sci Technol* **7**
[30] Xia M, Dong W, Gu M, Chang C, Shen Z, and Zhang Y 2018 *RSC Adv* **8** 8965
[31] Cai W, Chen Q, Xuan H, Li C, Yu H, Cui L, and Qu F 2019 *Korean J Chem Eng* **36**, 513
[32] Kato Y, and Sekine Y 2013 *Catal Lett* **143** 418
[33] Ogo S, Nishio T, Sekine H, Onda A, and Sekine Y 2016 *Fuel Process Technol* **141** 123
[34] Murata K, Liu Y, Inaba M, and Takahara I 201*0 Catal Lett* **140** 8
[35] Liu S, Tamura M, Nakagawa Y, and Tomishige K 2014 *ACS Sustain Chem Eng* **2** 1819
[36] Beeck BO, Dusselier M, Geboers J, Holsbeek J, Morre E, and Oswald S 2015 *Energ Environ Sci* **8** 230
[37] Liu Y, Chen L, Wang T, Zhang Q, Wang C, Yan J, and Ma L 2015 *ACS Sustain Chem Eng* **3** 1745
[38] Chen L, Li Y, Zhang X, Liu Y, Zhang Q, Wang C, and Ma L 2019 *Energy Procedia* **158** 160
[39] Rezaei P S, Shafaghat H, and Daud WMAW 2015 *RSC Adv* **5** 65408
[40] Dai G, Wang, Zou Q, and Huang S 2018 *Fuel Process Technol* **179** 319-323
[41] Qiao K Shu X, and Zhou F *Appl Catal A, Gen* 2017 **547** 274-282
[42] Zheng Y, Wang F, and Yang X 2017 *J Anal Appl Pyrol* **126** 169-179
[43] Wang JX, Cao JP, Zhao XY, Liu SN, Ren XY, Zhao M, and Wei XY 2019 *Bioresour Technol* **278** 116

Shape selective alkylation of benzene with methanol over different zeolite catalysts

A F A Rahman[1], A A Jalil[1, 2]*, N W C Jusoh[3, 4], M Mohamed[1], N A A Fatah[1] and H U Hambali[1]

[1]School of Chemical and Energy Engineering, Faculty of Engineering, Universiti Teknologi Malaysia, 81310 UTM Johor Bahru, Johor, Malaysia
[2]Centre of Hydrogen Energy, Institute of Future Energy, Universiti Teknologi Malaysia, 81310 UTM Johor Bahru, Johor, Malaysia.
[3]Department of Chemical and Environmental Engineering, Malaysia-Japan International Institute of Technology (MJIIT), Universiti Teknologi Malaysia, Kuala Lumpur, Jalan Sultan Yahya Petra, 54100 Kuala Lumpur, Malaysia.
[4]Centre of Hydrogen Energy, Universiti Teknologi Malaysia, 54100 Kuala Lumpur, Malaysia

*aishahaj@utm.my

Abstract. Benzene methylation is an important process for removing excess benzene in petrochemical industries to produce value-added products such as toluene and xylene. In this study, the performance of three types of zeolites, which is ZSM-5 (HZ), zeolite Y (HY) and zeolite β (Hβ) were investigated as catalyst in the benzene methylation reaction. The catalysts were characterized by N_2 adsorption-desorption and FTIR. The N_2 adsorption-desorption verified that the mesopores volume of the catalysts was in the following order: HB > HY > HZ. While, the FTIR-lutidine revealed that the HB possessed the highest amount of Brönsted acid sites followed by HZ and HY. The catalytic testing at 573 K showed that HZ gave good performance in benzene methylation with 55.2% and 22.1% yield of toluene and xylene, respectively. It is suggested that HZ catalyst with moderate amount of Brönsted acid sites and smallest mesopores volume appear to be effective for shape selective synthesis of toluene and xylene. In addition, pore structure of HZ also contributed to the high catalytic activity of benzene methylation.

1. Introduction

The aromatic hydrocarbons alkylation with methanol is widely used in the chemical industry for production of toluene and xylene [1,2]. Toluene and xylene are widely used as intermediates in the chemical industry to obtain textile fabric, plastics and detergents [3,4]. These chemicals are generally obtained via naphtha pyrolysis and catalytic reforming, along with the production of benzene [5]. However, these processes are greatly depended on the consumption of petroleum and have been called into question because of the lack of petroleum resources. Therefore, benzene methylation is regarded as a potential approach to produce toluene and xylene [6,7].

Previously, these alkylation reaction are performed with strong mineral acids catalysts such as H_2SO_4, HF, and $AlCl_3$, which are highly corrosive and toxic [8,9]. Generally, the acid need to be separated from the alkylation products, which considered as an energy consuming process. Presently, zeolites catalysts such as BEA, H-Y, H-mordenite and H-ZSM-5 are considered as a potential catalyst for the benzene alkylation with methanol because of their advantages such as highly unique shape selectivity for acid catalyzed reaction, environmentally friendly, and less toxic [10–12]. Usually, the alkylation reaction to produce aromatic compounds was influenced by the shape selectivity. The shape selectivity are greatly depends on the acidity, the porosity, the size of channels and the dimensionality of zeolites [13,14].

Thus, the objective of this present work is to examine the effect of the different zeolite towards the catalytic activity of benzene alkylation with. The acidity and porosity properties were comprehensively characterized and their effect on the catalytic activity in the reaction was investigated thoroughly.

2. Experimental

2.1. Catalyst preparation

The ZSM-5, zeolite Y and zeolite β was obtained from Zeolyst International. All zeolites were ion exchange with ammonium nitrate (NH_3NO_4) to form protonated zeolite followed by drying overnight at 383 K and calcined for 3 h under an air atmosphere at 823 K.

2.2. Catalyst characterization

BET method was used to evaluate the textural properties of the catalysts. This was achieved from N_2 adsorption desorption at -196 °C by a Beckman Coulter SA 3100. Analysis were carried out after the samples was degassed at 300 °C for 1 h. The catalysts acidity was obtained by 2,6-lutidine acid probe molecule. The measurements were achieved on FTIR Agilent Cary 640 FTIR spectrometer assembled with a stainless steel cell and CaF_2 windows. All catalysts were pre-treated at 400 °C for 1 h before measurements. Next, the 4 Torr of 2,6-lutidine was adsorbed on pre-treated catalysts for 15 min at 25 °C, followed by degassing at 50 °C change to K for 15 min.

2.3. Catalytic reaction

The benzene methylation was performed in a microcatalytic pulse reactor at atmospheric pressure and 400 °C. Before the reaction, 200 mg of catalyst was pre-treated at 450 °C for 2 h in air stream, then cooled to 400 °C in a nitrogen stream. Then, a 1:1 molar ratio of benzene to methanol passed by to the pre-treated catalyst, and the products were analyzed on a gas chromatograph (7820A) equipped with FID/MS detector. The benzene conversion (X_B), yield (Y_i) and product selectivity (S_i) were calculated by the following equation:

$$X_B = \frac{\sum C_i - C_B}{\sum C_i} \quad (1)$$

$$S_i = \frac{C_i}{\sum C_i - C_B} \times 100\% \quad (2)$$

$$Y_i = \frac{X_B \times S_i}{100} \quad (3)$$

where C_i and C_B are mole number of products and residual benzene.

3. Results and Discussion

3.1. Strcutural properties of catalysts

N_2 adsorption-desorption isotherms of all catalysts are demonstrated in Figure 1. The HZ and HY showed a type I isotherm with H4 hysteresis loop [6,15–17]. In addition, HZ and HY catalysts displayed high nitrogen uptake at $P/P_o < 0.2$, confirming the microporous characteristic of the catalysts. A large portion of the micropore was found in HZ catalyst. Meanwhile, Hβ showed a H3 hysteresis loop with type IV isotherms, signifying the existence of mesopores materials and classification of slit-like pores [18].

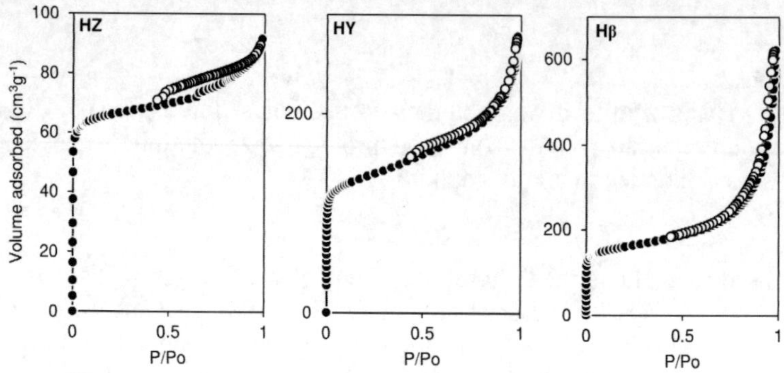

Figure 1. N_2 adsorption-desorption isotherm of catalysts.

From Table 1, Hβ shows the highest surface area (555 m^2g^{-1}) followed by HY (475 m^2g^{-1}) and HZ (225 m^2g^{-1}). The total pore volume of HZ, HY and Hβ are 0.139, 0.427 and 0.943 cm^3g^{-1}, respectively. In addition, Hβ possessed higher mesopore volume as compare to HZ and HY. This result is in agreement with above isotherm result, in which the Hβ showed high nitrogen uptake at $P/P_o > 0.9$.

Table 1. Textural properties of catalysts.

Catalyst	BET surface area (m^2/g)	Pore volume (cm^3/g)	Micropore volume (cm^3/g)
HZ	225	0.139	0.089
HY	475	0.427	0.137
Hβ	555	0.943	0.153

3.2. Acidic properties of catalysts

In this study, 2,6-lutidine was utilized as a probe molecule for the evaluation of the catalyst acidity. Figure 2 shows the FTIR spectra of adsorbed 2,6-lutidine on the zeolite catalyst activated at 350 ºC. It was showed that all the catalysts showed a two doublets bands at 1703 +1677 cm-1 and 1652 + 1628 cm-1, which corresponded to the Brönsted acid sites. Additionally, another peak was observed at 1607 + 1586 cm-1, suggesting the existence of Lewis acid sites [19]. Figure 2B summarizes the variation of band intensity changes in Figure 2A as a function of degassed temperature. The results demonstrated that all catalysts exhibited a strong interaction between BAS and 2,6-lutidine at 50ºC, but the intensities decreased considerably when the temperature was 150ºC, representing the presence of a low amount of moderate to strong BAS It clearly seen that HZ has a moderate amount of BAS and LAS for facilitated benzene akylation reaction [18,20].

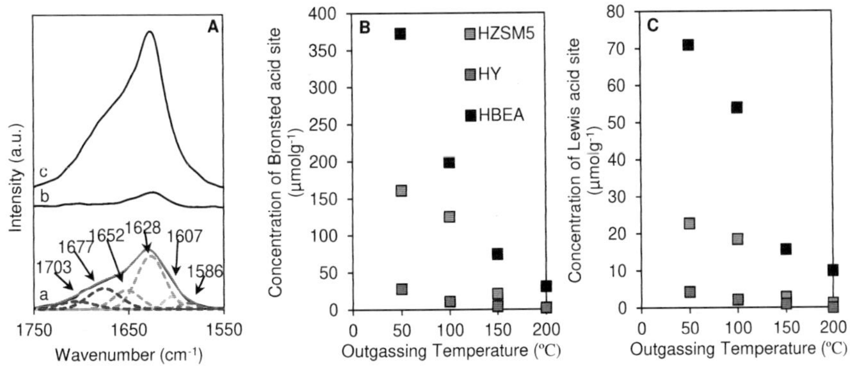

Figure 2. FTIR spectra of 2,6-lutidine for (a) HZ (b) HY (c) Hβ and variation of peak intensity as a function of outgassing temperature 2,6-lutidine adsorption for (B) Brönsted acid site (C) Lewis acid site.

3.3 Catalytic activity
The catalytic activities of the catalysts were performed with respect to benzene methylation at 400 °C. The conversion of benzene and yield of products over all of the catalysts is shown in **Figure 3**. It was showed that the conversion of HZ, HY and Hβ were 89.6, 87.5 and 92.8%, respectively. In addition, the result showed that HZ gave the highest yield of toluene with 61.6 %, followed by HY and Hβ, which may be related to porosity and acidity of catalyst. The acidity in HZ provide an appropriate amount of BAS to initiate the alkylation of benzene for production of toluene[6,18]. Besides acidity, HZ also possesses a small amount of micropore that make these products easy to pass through the pore and inhibited the occurrence of undesired products.

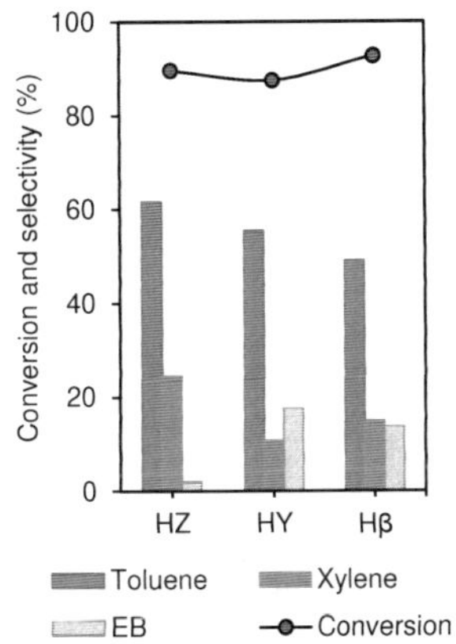

Figure 3. Conversion of benzene and Selectivity of products.

4. Conclusion

In this study, the activity of the HZ, HY and Hβ was studied for the catalytic benzene methylation at 400 °C. It was observed that the different zeolites that have different pore structure, acidity and porosity can influenced the types of products obtained by the catalytic alkylation. Besides, this study also proven that the appropriate amount of Brönsted acid sites and low micropore volume are important in the reaction as the high and low of Brönsted acid sites in HY and Hβ has lowered the toluene yield. While, HZ with low micropore volume and mild acidity exhibited the best catalytic performance with 61.6% yield of toluene.

Acknowledgments
This work was financially supported by the Universiti Teknologi Malaysia through UTM Transdiciplinary Research Grant (Grant No. 06G53), Collaborative Research Grant (07G66) and the awards of UTM Zamalah Scholarship (Anis Farhana Abdul Rahman).

References
[1] Dong P, Li Z, Wang X, Yun H and Li G 2018 *Green Chem. Lett. Rev.* **11** 158–64
[2] Rui J, Lyu J, Hu H, Zhang Q, Wang Q and Li X 2019 *Chinese Chem. Lett.* **30** 757–61
[3] Lyu J, Hu H, Tait C, Rui J, Lou C, Wang Q, Han W, Zhang Q, Pan Z and Li X 2017 *Chinese J. Chem. Eng.* **25** 1187–94
[4] Wang X, Xu J, Qi G, Li B, Wang C and Deng F 2013 *J. Phys. Chem. C* **117** 4018–23
[5] Alabi W, Atanda L, Jermy R and Al-Khattaf S 2012 *Chem. Eng. J.* **195–196** 276–88
[6] Abdul Jalil A, Zolkifli A S, Triwahyono S, Abdul Rahman A F, Mohd Ghani N N, Shahul Hamid M Y, Mustapha F H, Izan S M, Nabgan B and Ripin A 2019 *Ind. Eng. Chem. Res.* **58** 553–62
[7] Lyu J H, Hu H L, Rui J Y, Zhang Q F, Cen J, Han W W, Wang Q T, Chen X K, Pan Z Y and Li X N 2017 *Chinese Chem. Lett.* **28** 482–6
[8] Vernaya O I, Krotova I N, Maksimov Y V and Rostovshchikova T N 2017 *Pet. Chem.* **57** 93–9
[9] Shen Z, Ma C, He J, Wang D, Sun H, Zhu Z and Yang W 2019 *Appl. Catal. A Gen.* **577** 20–7
[10] Ahn J H, Kolvenbach R, Neudeck C, Al-Khattaf S S, Jentys A and Lercher J A 2014 *J. Catal.* **311** 271–80
[11] Wichterlová B, Čejka J and Žilková N 1996 *Microporous Mater.* **6** 405–14
[12] H. U. H, A. A. J, T. J. S, A. A. A, N. A. A. F, I. H and M. S. A 2019 *J. Energy Saf. Technol.* **2** 9–13
[13] Liu K, Xie S, Liu S, Xu G, Gao N and Xu L 2011 *J. Catal.* **283** 68–74
[14] Li Y, Xue B and Yang Y 2009 *Fuel Process. Technol.* **90** 1220–5
[15] Teh L P, Triwahyono S, Jalil A A, Mamat C R, Sidik S M, Fatah N A A, Mukti R R and Shishido T 2015 *RSC Adv.* **5** 64651–60
[16] Jalil A A, Gambo Y, Ibrahim M, Abdulrasheed A A, Hassan N S, Nawawi M G M, Asli U A, Hassim M H and Ahmad A 2019 *Int. J. Energy Res.* **0**
[17] Triwahyono S, Jalil A A, Izan S M, Jamari N S and Fatah N A A 2019 *J. Energy Chem.* **37** 163–71
[18] Ghani N N M, Jalil A A, Triwahyono S, Aziz M A A, Rahman A F A, Hamid M Y S, Izan S M and Nawawi M G M 2019 *Chem. Eng. Sci.* **193** 217–29
[19] Izan S M, Triwahyono S, Jalil A A, Majid Z A, Fatah N A A, Hamid M Y S and Ibrahim M 2019 *Appl. Catal. A Gen.* **570** 228–37
[20] Firmansyah M L, Jalil A A, Triwahyono S, Hamdan H, Salleh M M, Ahmad W F W and Kadja G T M 2016 *Catal. Sci. Technol.* **6** 5178–82

IOP Conf. Series: Materials Science and Engineering **808** (2020) 012004 doi:10.1088/1757-899X/808/1/012004

Toluene side chain alkylation with methanol over silica catalyst

M A H Aziz[1], A A Jalil [2, 3*], A F A Rahman [2], N A A Fatah[2], S M Izan[1], I Hussain[1], H U Hambali[2]

[1]Department of Chemistry, Faculty of Science, Universiti Teknologi Malaysia, 81310 Johor Bahru, Malaysia.
[2]School of Chemical and Energy Engineering, Faculty of Engineering, Universiti Teknologi Malaysia, 81310 Johor Bahru, Johor, Malaysia.
[3]Centre of Hydrogen Energy, Institute of Future Energy, Universiti Teknologi Malaysia, 81310 Johor Bahru, Johor, Malaysia.

* aishahaj@utm.my

Abstract. MSN and SiO_2 catalyst were investigated on side chain toluene alkylation with methanol reaction. Characterization of the catalyst were carried out by XRD, N_2 physisorption analysis, FTIR spectroscopy. A pyrrole adsorption FTIR study reveals shifting of perturbed NH stretching increasing slightly in MSN compared to SiO_2 catalyst revealed that MSN possessed higher basic sites than SiO_2. N_2 adsorption desorption isotherm analysis showed that MSN possessed higher surface area than SiO_2 as well as increased the amount of mesopores in catalyst. The catalytic side chain toluene alkylation with methanol reaction was conducted in the range of 523K-673K under atmospheric pressure. MSN exhibits the highest catalytic performance compared to SiO_2 catalyst.

1. Introduction

Production of polymer materials like styrene-butadiene-styrene (SBS) rubber, polystyrene and resin has gained much courtesy from community nowadays [1]. Styrene monomer serve as the starting material for the production of commercial polymers and commercially derived from the Friedel Craft Alkylation of benzene with ethylene. However, multistep of styrene creation in this process and higher consumption of energy due to difficulty of obligation of ethylbenzene products ratio. Plus, operation of this reaction occurred at higher temperature which more than 550 °C limits its applications [2]. To address these obliges, current researcher moves to alternative side chain toluene alkylation with methanol as it provide reliable way to obtain styrene products. Unfortunately, attempts to design robust catalyst for this reaction still remained being an obstacle which until now the yield of styrene obtained still not reached satisfaction level (< 20%) due to ineffectively catalyst used and rapid deactivation of the catalyst [1–3].

Side chain toluene alkylation with methanol was extensively known required basic sites and Lewis acid sites to accelerate the reaction [4]. Owning the basic sites in the catalyst may proceed the styrene product to convert into ethylbenzene consequently from hydrogen carrier gas or originated from resulted hydrogen molecule formation from decomposition of formaldehyde. Thus, ethylbenzene product formation in this process is inevitable. Additionally, Bronsted acid sites in catalyst will lead to the occurrence of side reaction and emerging of xylene and hydrocarbon product [5]. Hence, it is crucial

IOP Conf. Series: Materials Science and Engineering **808** (2020) 012004 doi:10.1088/1757-899X/808/1/012004

to design a catalyst which possessed the criteria to be good catalyst for toluene side chain alkylation to attain higher selectivity of desired products.

Based on the literature, many efforts of evolving catalyst have been done to evaluate the catalytic performance of side chain alkylation. The effort involving alteration of the Cs-X with variety of elements together with metal or non-metal besides changing the preparation method such as ion exchange, impregnation ad physical mixing [6]. Up to now, no documentation reported on the use of silica catalyst to evaluate side chain alkylation. Herein, the metal free of SiO_2 and MSN catalyst were applied in order to inspect the catalytic performance from both catalyst towards toluene side chain alkylation with methanol.

2. Experimental

Mesoporous MSN were prepared referring to the previous work [7]. The cetyltrimethylammmonium bromide, CTAB (surfactant), ammonium hydroxide solution, NH_4OH and ethylene glycol (EG) were mixed in water (700mL). The molar composition of them were 0.0032:0.2:0.2:01 respectively. The mixture was mixed for 30 minutes together with heating until clear and homogenous solution observed. After that, 1 mmol of 3-aminopropyltriethoxysilane (APTES) and 1.2 mmol tetraethylorthosilicate (TEOS) were slowly added under vigorous stirring for 2 hours until whitely solution observed. The mixture underwent centrifugation and washed with distilled water. The product underwent drying process overnight at 383K followed by calcination process at 823K for 3 hours to eliminate the surfactant.

Catalytic side chain toluene alkylation with methanol was conducted in reactor known as micro catalytic pulse within temperature 523–673 K. Initially, air stream was flew on 200 mg of catalyst for 1 hour at 723K and cooling to desired temperature after catalyst treatment. As soon as temperature was stable, toluene and methanol mixture was injected into the reactor at a flow rate 40 mL/min at reactant ratio of 1. The gas chromatograph equipped with a FID (7820N Agilent Gas Chromatograph) was used to analyze the outlet gas. The conversion of methanol and toluene ($X_{methanol}$ and $X_{toluene}$), selectivity of styrene, ethylbenzene, and other product as well as yield of styrene and ethylbenzene were calculated using equations:

$$X_{\text{methanol}} = \frac{\sum \text{all products}}{\sum \text{all products} + \text{methanol}} \quad (1)$$

$$X_{\text{toluene}} = \frac{\sum \text{all products}}{\sum \text{all products} + \text{toluene}} \quad (2)$$

$$S_{\text{styrene}} = \frac{\sum \text{styrene}}{\sum \text{all products}} \times 100\% \quad (3)$$

$$S_{\text{ethylbenzene}} = \frac{\sum \text{ethylbenzene}}{\sum \text{all products}} \times 100\% \quad (4)$$

$$Y_{\text{styrene}} = \frac{S_{\text{styrene}}}{100} \times X_{toluene} \quad (5)$$

$$Y_{\text{ethylbenzene}} = \frac{S_{\text{ethylbenzene}}}{100} \times X_{\text{toluene}} \quad (6)$$

3. Results and discussions

3.1 Crystallinity and textural properties of catalysts

Low angle XRD pattern was shown in Figure 1. MSN catalyst exhibited three respective peaks at 2θ = 2.4, 4.05, and 4.7° indicate the p6mm hexagonal symmetry in the mesostructured silica of the MSN [8–10]. While, no diffraction peak below than 2θ=10 was observed in SiO_2.

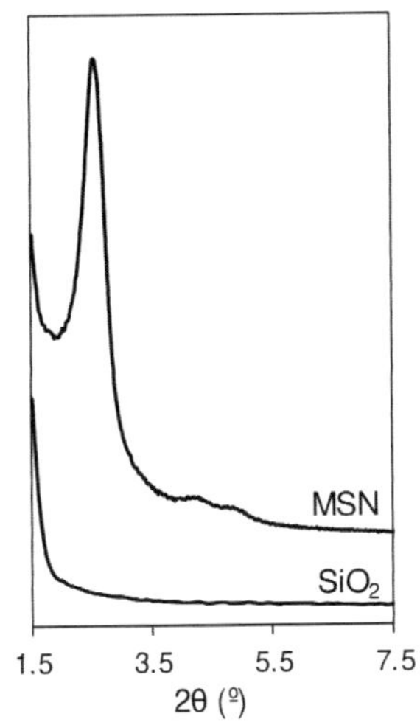

Figure 1. Low angle XRD diffractogram of MSN and SiO_2.

Isotherm with type II was revealed in SiO_2 with H3 hysteresis loop corresponded to macroporous with irregular system of pores material [11]. Meanwhile, MSN catalyst showed isotherm type IV with an hysteresis loop H1 reflected to the attributes of mesoporous materials with exceptionally uniform round and hollow pores [8,9]. Capillary condensation with two steps at P/Po = 0.3 was found in MSN catalyst due to mesoporous (intra- particles) in MSN and (P/Po = 0.9) which at high partial pressure attributed to interparticle textural porosity [9]. Besides, SiO_2 catalyst demonstrated a little narrow capillary condensation with steady slant at high partial pressure which revealed to the nearness of pores with multidimensional tube-shaped channels [7–9].

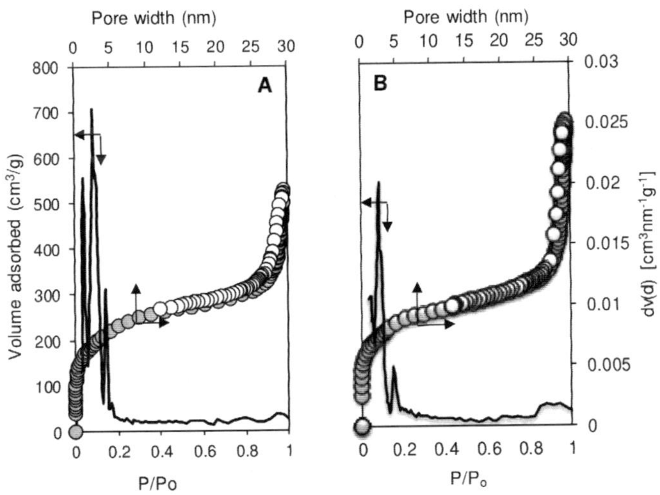

Figure 2. Isotherm data and pore size distribution of catalyst (A) SiO_2 (B) MSN.

As per the isotherm results, the (NLDFT) method was applied for pore distribution identification. It was proved the existence of mesopore in both catalysts. A narrow peak distribution with 4-5nm in MSN attributed to the pores which originated from self-assembly of surfactant during the synthesis procedure. The lessened of micropore volume in MSN due to the increment of mesopores volume in that catalyst. SiO_2 catalyst pore distribution was much lower compared to MSN likely because of amorphous structure of SiO_2 [7]. MSN catalyst possessing high pore volume as well as surface area (851 m^2/g) than SiO_2 which have 473 m^2/g surface area as analyzed in Table 1.

Table 1 Properties of catalysts.

Catalyst	Surface area (m^2/g)	Mesopore volume (cm^3/g)	Micropore volume (cm^3/g)	Total pore volume (cm^3/g)
SiO_2	473	0.75	0.06	0.81
MSN	851	0.94	0.01	1.04

3.2 Chemical properties of catalysts

Highest peak intensity in SiO_2 observed in Figure 2A clarified SiO_2 catalyst owned high basicity. The band at 3800–3400 cm^{-1} which was indexed to the vibration of the NH band stretching in pyrrole which binding with the oxygen structure through hydrogen bonding and connection by via aromatic system with the non-framework cations which happened concurrently and impact one another [11,12]. The shifting of peak intensity of each catalyst was identified in order to determine the basic sites of each catalysts. The intensity peak at 3436 and 3409 cm^{-1} for SiO_2 and 3459 cm^{-1} for MSN catalyst. According to the former study, the strength of the catalysts was observed by the interaction of H atom with basic sites in pyrrole [12,13]. MSN catalyst has revealed the deviation of stretching of NH vibration, ΔV (NH) = −71 cm^{-1} and ΔV (NH) = −94 and -121 cm^{-1} for SiO_2. As observed, SiO_2 has two type of basic sites indicated that presence of two difference nature of basic sites. Thus, it proved that SiO_2 have strength basic sites rather than MSN.

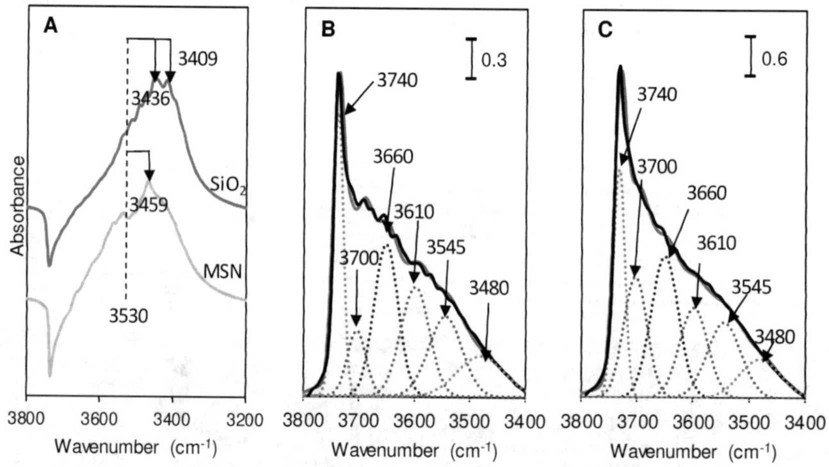

Figure 3. (A) FTIR pyrrole for both catalyst (B and C) FTIR spectra in evacuated system for SiO_2 and MSN catalyst.

FTIR results in evacuated system in both MSN and SiO_2 catalysts were shown in Figure 3B and 3C. Commonly, spectra of both catalysts demonstrated the presence of five bands at 3740, 3700, 3660, 3610, 3545 and broad band at 3480 cm^{-1}. It were corresponded to the terminal silanol group, perturbation of OH through lattice defects, hydroxyl group species, bridging hydroxyl, perturbation of H-bond interaction with the mesoporous channels and silanol group respectively [8–11,14,15]. According to the result observed, it showed that the intensity of every single peak in FTIR spectra in evacuated system augmented two times from SiO_2 to MSN catalyst.

3.3 Catalytic testing

Evaluation catalytic performance of side chain toluene alkylation with methanol was performed at temperature range of 523–673 K using micro catalytic pulse reactor. The methanol conversion, toluene conversion and the yield of the ethylbenzene with styrene product was portrayed in Figure 4. As the results, Figure 4A illustrates that the methanol conversion over silica catalyst which SiO_2 and MSN catalyst. As temperature increased, it was found that the methanol conversion was kept increasing until over MSN catalyst which shows declining of methanol conversion at 673K. It might be due to the lower basic sites possessed in the MSN catalyst. It verified that the basic sites was consumed for the formaldehyde formation via aldol-type condensation as the temperature increase [3,4,16]. As depicted in the pyrrole probed FTIR spectroscopy (Fig 2A), it revealed that the SiO_2 have more basic sites rather than MSN catalyst. It could be the cause of fully converted of methanol.

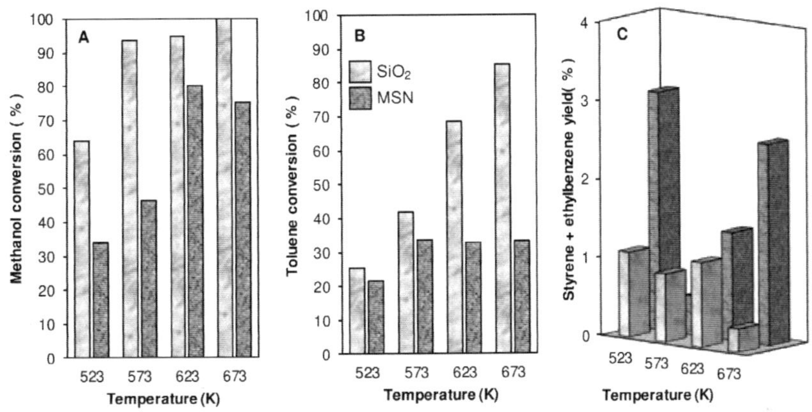

Figure 4. Catalytic toluene side chain alkylation performance (A) Methanol conversion (B) Toluene conversion (C) Styrene and ethylbenzene yield.

Furthermore, Fig 4B demonstrated the toluene conversion over SiO_2 and MSN catalyst within temperature range 523-673K. It was found that toluene conversion was slightly increased from 523K to 673K. It was due adequate amount of basic sites in SiO_2 resulted in higher formation of benzyl anion and 2-phenylethanol intermediates [6,17]. However, MSN catalyst showed lower conversion of toluene at each temperature compared to SiO_2 catalyst. It might be due to fewer basic sites in MSN as evidenced by FTIR adsorbed pyrrole. Preceding work stated that occurrence of less basic sites suppressed the ability of the basic sites to convert toluene into the benzyl anion intermediates [3,5].

Table 2 Distribution of products in side chain toluene alkylation at 523-673K.

SiO_2 catalyst:				
Temperature (K)	523	573	623	673
Methanol conversion (%)	63.9	93.8	94.9	100
Toluene conversion (%)	25.4	41.8	68.5	85.5
Selectivity of products:				
Hydrocarbon	0.224	28.4	90.7	95.7
Benzene	95.5	69.6	7.78	3.98
Ethylbenzene	0.00	0.00	0.00	0.00
Styrene	4.24	2.08	1.57	0.335
Yield of products:				
Hydrocarbon	0.057	11.9	62.1	81.8
Benzene	24.3	29.1	4.92	3.40
Ethylbenzene	0.00	0.00	0.00	0.00
Styrene	1.08	0.869	1.08	0.304
MSN catalyst:				
Temperature (K)	523	573	623	673
Methanol conversion (%)	34.0	46.5	80.3	75.2
Toluene conversion (%)	21.7	33.5	32.9	33.3
Selectivity of products:				
Hydrocarbon	4.83	29.3	35.3	42.4
Benzene	81.1	66.5	60.5	49.9
Ethylbenzene	3.91	1.20	1.19	2.85
Styrene	10.1	3.02	2.98	4.87
Yield of products:				
Hydrocarbon	1.05	9.82	11.6	14.1
Benzene	17.6	22.3	19.9	16.6
Ethylbenzene	0.848	0.402	0.392	0.949
Styrene	2.19	0.10	0.980	1.62

Fig. 4C illustrated the products of styrene and ethylbenzene which considered to be desired product in the reaction process. From the result above, it was discovered that MSN formed more styrene and ethylbenzene product. It was due to the controllable of basic sites in selectivity of styrene and ethylbenzene product in MSN. It can be proved in Table 2 which showed increase of higher desired product selectivity from 0.335 % to 7.72% styrene and ethylbenzene selectivity. From Table 2, SiO_2 produced more hydrocarbon product and decreasing selectivity about 44.3% at 673K in MSN catalyst.

The presence of additional basic sites would lead to the side reaction resulted in higher production of hydrocarbon and benzene product along with quenching the xylene product [5]. Remarkably, no production of styrene product observed in SiO_2 catalyst compared to MSN catalyst suggesting the basic sites possessed in SiO_2 has no ability to transfer hydrogen to form ethylbenzene. Higher hydrocarbon product by SiO_2 might be due to the micropores in the catalyst that will cause diffusion limitations of products. Thus, the formation of hydrocarbon product due to favourable of thermal cracking process. These outcome shows coinciding with N_2 adsorption-desorption result which revealed that desired product favour in high mesopore catalyst.

4. Conclusions

Catalytic performance of toluene side chain alkylation with methanol over silica catalyst (SiO_2 and MSN) were successfully evaluated at atmospheric pressure. The results revealed that a significant enhancement of MSN catalyst towards yield of styrene and ethylbenzene product compared to SiO_2. Fascinatingly, it can be summarized that basicity plays a noteworthy part in dehydrogenation of methanol to formaldehyde and hydrogenation of styrene to form ethylbenzene in order to improve the catalytic performance. Despite of basicity, presence of mesopores also contribute to the augmentation of the selectivity styrene and ethylbenzene product. From the results obtained, it also can be clinched the basic sites in the catalyst also will lead to the formation of the hydrocarbon product and fast decomposition of formaldehyde into the carbon monoxide and hydrogen.

Acknowledgement
The authors are grateful to the financial support for this research from the Ministry of Higher Education Malaysia through UTM Transdisciplinary Research Grant (grant no: 06G53) and MyBrainSc Scholarship (Mohamad Afiq Hakimie Aziz).

References

[1] Lee H, Lee S, Ryoo R and Choi M 2019 *J. Catal.* **373** 25–36
[2] Han Q, Li P, Zhang Y, Lu P, Xu L, Guo H and Xu L 2019 *ChemCatChem* **11** 1610–4
[3] Alabi W O, Tope B B, Al-Khattaf S S, Aitani A M and Hattori H 2012 *Appl. Catal. A Gen.* **443–444** 214–20
[4] Han H, Liu M, Nie X, Ding F, Wang Y, Li J, Guo X and Song C 2016 *Microporous Mesoporous Mater.* **234** 61–72
[5] Li P, Han Q, Zhang X, Yuan Y, Zhang Y, Guo H, Xu L and Xu L 2018 *Catal. Sci. Technol.* **8** 3346–56
[6] Hattori H, Amusa A A, Jermy R B, Aitani A M and Al-Khattaf S S 2016 *J. Mol. Catal. A Chem.* **424** 98–105
[7] Fatah N A A, Triwahyono S, Jalil A A, Ahmad A and Abdullah T A T 2016 *Appl. Catal. A Gen.* **516** 135–43
[8] Rahman A F A, Jalil A A, Triwahyono S, Ripin A, Aziz F F A, Fatah N A A, Jaafar N F, Hitam C N C, Salleh N F M and Hassan N S 2017 *J. Clean. Prod.* **143** 948–59
[9] Jusoh R, Jalil A A, Triwahyono S, Idris A and Noordin M Y 2015 *Sep. Purif. Technol.* **149** 55–64
[10] Fatah N A A, Triwahyono S, Jalil A A, Ahmad A and Abdullah T A T 2016 *Appl. Catal. A Gen.* **516** 135–43
[11] Hamid M Y S, Firmansyah M L, Triwahyono S, Jalil A A, Mukti R R, Febriyanti E, Suendo V, Setiabudi H D, Mohamed M and Nabgan W 2017 *Appl. Catal. A Gen.* **532** 86–94
[12] Förster H, Fuess H, Geidel E, Hunger B, Jobic H, Kirschhock C, Klepel O and Krause K 1999 *Phys. Chem. Chem. Phys.* **1** 593–603
[13] Aziz M A A, Jalil A A, Triwahyono S and Ahmad A 2015 *Green Chem.* **17** 2647–63
[14] Aziz M A A, Jalil A A, Triwahyono S and Sidik S M 2014 *Appl. Catal. A Gen.* **486** 115–22
[15] Ghani N N M, Jalil A A, Triwahyono S, Aziz M A A, Rahman A F A, Hamid M Y S, Izan S M and Nawawi M G M 2019 *Chem. Eng. Sci.* **193** 217–29
[16] Wang B, Huang W and Wen Y 2011 *Energy Sources, Part A Recover. Util. Environ. Eff.* **33** 1933–9
[17] Hattori H, Alabi W O, Jermy B R, Aitani A M and Al-Khattaf S S 2013 *Catal. Letters* **143** 1025–9

IOP Conf. Series: Materials Science and Engineering **808** (2020) 012005 doi:10.1088/1757-899X/808/1/012005

Effect of transition metals (Mo, Mn and Co) on mesoporous ZSM-5 catalyst activity in carbon dioxide reforming of methane

H U Hambali[1,3], A A Jalil[1,2*], A A Abdulrasheed[1], T J Siang[1], N A A Fatah[1], A F A Rahman[1], M A H Aziz[4] and I Hussain[4]

[1] School of Chemical and Energy Engineering, Faculty of Engineering, Universiti Teknologi Malaysia, 81310 UTM Johor Bahru, Johor, Malaysia
[2] Centre of Hydrogen Energy, Institute of Future Energy, Universiti Teknologi Malaysia, 81310 UTM Johor Bahru, Johor, Malaysia
[3] Department of Chemical Engineering, University of Ilorin, PMB 1515, Ilorin, Nigeria
[4] Department of Chemistry, Faculty of Science, Universiti Teknologi Malaysia, 81310 Johor Bahru, Malaysia

*aishahaj@utm.my

Abstract. This paper investigates the effect of transition metals (Mo, Mn and Co) on mesoporous ZSM-5 support in carbon dioxide reforming of methane reaction. The mesoporous ZSM-5 support was synthesized by micro emulsion technique and the metals were loaded by wet impregnation method. It was observed that ZSM-5 supported Co catalyst had higher surface area in comparison to other catalysts, which could favour well dispersion and wider utilization of active component. In addition, the ZSM-5 supported Co catalyst exhibited the highest methane and carbon dioxide conversions of 69 % and 65 % respectively at 850°C. The Co species were more active to decompose methane and carbon dioxide compared to Mo and Mn loaded catalysts. Hence, loading Co on mesoporous ZSM-5 produce an active catalyst in carbon dioxide reforming methane reaction.

1. Introduction

The challenge for cleaner energy sources has continued to receive tremendous attention due to lingering issues of climate change, global energy crisis and tropospheric air quality [1,2]. Hence, processes for conversion of harmful molecules into valuable products are being pursed. Syngas (H_2/CO) is a valuable intermediate for production of chemicals via Fischer-Tropsch process, methanol and oxygenates synthesis [3–5]. Carbon dioxide reforming of methane (CRM) involves conversion of natural gas and carbon dioxide into syngas. The process is considered a viable route to produce pure syngas with H_2/CO ratio of unity. During the transformation process, methane reacts with an equimolar amount of carbon dioxide over a catalyst to yield syngas according to: $CH_4 + CO_2 \rightleftharpoons 2CO + 2H_2$. As such, it is an auspicious technology to address the issues of the two greenhouse gases with high global warming potentials [6]. The emission problems associated with carbon dioxide could be sequestrated by CRM process that is cheap as considered to current carbon capture technologies [7]. The critical issue associated with CRM is the fabrication of highly active and stable catalyst. The catalyst system must be robust to impede carbon deposition and high temperature sintering. A variety of support materials has been developed for CRM such as KCC-1 [3], Al_2O_3 [8], SBA-15[9], MSN [10], SiO_2 [7] and zeolite [11].

The nature of support material is pivotal in controlling the dispersion of metal particles and the stability of active metal particles. CRM has been studied with noble metal (Pt, Rh, Pd, Ru, Ir) based catalysts. Noble metal catalysts displayed remarkable coke resistance, stability and activity [6]. However, their high cost is a major demerit for application as an industrial CRM catalyst. In this scenario, other non-noble metals (Ni, Co, Fe) with low cost becomes potential option. Transition metal catalysts show reasonable performance in CRM reaction with the noble metals producing superior activity and high coke resistance [3]. Reasonable methane and carbon dioxide conversion was recorded by Ni/MCM-41 catalyst at reaction temperature of 750°C [12]. ZSM-5 support with 8 wt.% Ni produced reasonable syngas from CRM reaction [13]. Similarly, the incorporation of nickel on hydroxyapatite support displayed high CRM activity [14]. In addition, Zeng et al. prepared 20% Co/γ-Al_2O_3 catalyst, which showed good potential as an active CRM catalyst [15]. Thus, the present study investigates the activity of transition metals (Mo, Mn and Co) supported on mesoporous ZSM-5 catalyst in CRM reaction. In this paper, a mesoporous ZSM-5 support was synthesized via micro emulsion technique and employed as carrier material in CRM reaction. This study focused on the activity of mesoporous ZSM-5 supported Mo, Mn and Co catalysts in production of syngas from CRM reaction.

2. Experimental

2.1. Materials
The materials used in this study are: Tetraethyl orthosilicate (TEOS), Urea, 1-butanol (C_4H_9OH), Cetyltrimethylammonium bromide (CTAB), ammonium nitrate (NH_4NO_3) and ZSM-5 seed (Si/Al= 23) (MERCK Sdn. Bhd., Malaysia). MoO_3, $Mn(NO_3)_2{\cdot}4H_2O$ and $Co(NO_3)_2.6H_2O$ were used as Mo, Mn and Co precursors respectively obtained from Merck Co. Industrial grade gases (CH_4, CO_2, H_2, N_2 and O_2) were used in catalytic testing. H_2 was used as reducing agent, while N_2 and O_2 were used as carrier and pre-treatment gases respectively.

2.2. Catalyst Preparation
In this study, microemulsion synthesis technique coupled with ZSM-5 seeds crystallization was adopted to prepare mesoporous ZSM-5 according to the previous reported procedure [16,17]. Metal loading was fixed at 5 wt% for all samples via wet impregnation method. The catalysts prepared were denoted as MoZ, MnZ and CoZ representing Mo, Mn and Co loaded mesoporous ZSM-5 catalysts respectively.

2.3. Catalyst Characterization
The phase of the prepared catalysts was confirmed by utilizing a Bruker D8 X-ray diffraction (XRD) analyzer. Specific surface area of all synthesized catalysts was determined from N_2 adsorption–desorption (at 77 K) in an analyzer (Beckman Coulter SA 3100). Catalytic tests were carried out in a micro catalytic fixed bed reactor over different temperature range 550-850°C. 20 mg catalyst with 40–60 meshes size was packed into the isothermal reaction zone of the fixed bed reactor.

Pre-treatment was done for an hour at 850°C under O_2 flowrate of 50 mL min^{-1}. Thereafter, reduction was done at 850°C under H_2 flow rate of 50 mL min^{-1} which was maintained for 1 h. Then, the gaseous feed comprising of CH_4:CO_2:N_2 in a ratio of 1:1:3 were feed into the reactor at flow rate of 100 mL min^{-1}. An online gas chromatograph device (Agilent GC, 7820 N) was used to continuously detect and analyze the product gases. Product sampling at the considered temperatures were conducted an hour after commencement of the reaction. The sampling for each reaction run is 20 min. The feed conversion and corresponding products yield were computed by the following equations:

$$CH_4\ conversion = \frac{F(CH_4)\ in - F(CH_4)\ out}{F(CO_2)\ in} \times 100 \quad (1)$$

$$CO_2\ conversion = \frac{F(CO_2)\ in - F(CO_2)\ out}{F(CO_2)\ in} \times 100 \quad (2)$$

$$H_2\ yield\ = \frac{F(H_2\ )\ out}{2[F(CH_4)in]} \times 100 \tag{3}$$

$$CO\ yield\ = \frac{F(CO\ )\ out}{[F(CH_2)in+F(CO_2)\ out]} \times 100 \tag{4}$$

3. Results and discussion

The X-ray diffractogram of the synthesized catalysts are shown in figure 1. From the XRD analysis, reflections were observed corresponding to ZSM-5 structure as indicated by sharp diffraction peaks in 2θ range of 5–10 and 20–26 respectively [16,18]. The intensity of peaks (figure 1) below 10° slightly increased in the order CoZ > MnZ > MoZ. These might be due to Al removal from the framework, which did not invoke significant damage to the lattice [19]. The peak intensities of ZSM-5 supported Co catalyst had obvious increase, which may be attributed to increase of framework Si species [20]. Furthermore, XRD patterns of MoZ and MnZ seem to be identical. There exist no Mo and Mn species peaks in the XRD patterns, while CoZ demonstrated presence of Co_3O_4 cubic phase by diffraction peaks at 2θ= 31.3°, 36.7°, 45.1°, 55.9°, 59.8° and 65.5° (JCPDS 01-076-1802) [21].

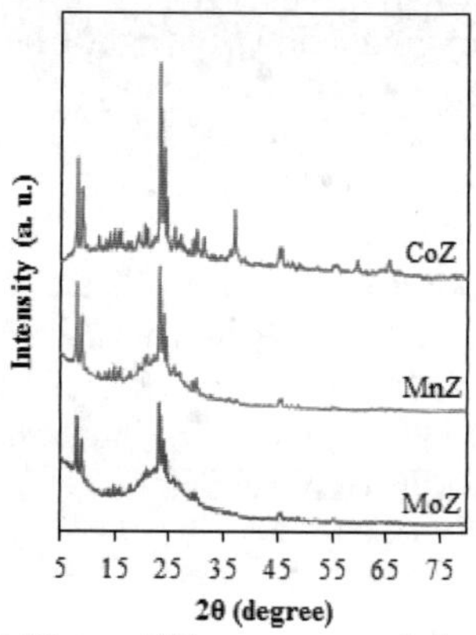

Figure 1. X-ray diffractogram of the catalysts.

Generally, activity of a supported catalyst is significantly related to surface area. Catalyst system with large surface area promote well dispersion and wider utilization of active metals component which result in high CRM performance [22]. Figure 2 illustrates the N_2 physisorption isotherms and the estimated pore volumes of the catalysts are presented in Table 1.

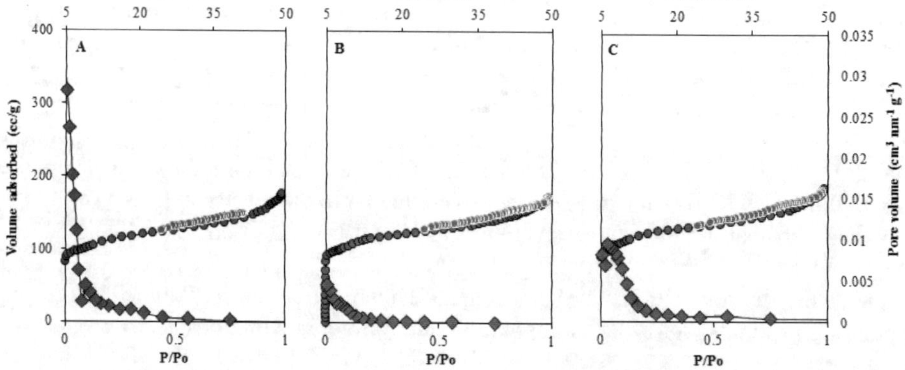

Figure 2. N_2 adsorption–desorption and the pore size distribution (a) MoZ (b) MnZ and (c) CoZ.

The catalysts portrayed type IV isotherms with H3 hysteresis loops. This signals the existence of non-uniform slit-shaped mesopores structure. The surface area was MoZ= 395 m^2/g, Mn= 404 m^2/g and CoZ= 419 m^2/g respectively. The MoZ catalyst had the lowest surface area as compared to other catalysts. The MnZ and CoZ catalysts displayed higher N_2 uptake as compared to MoZ, signifying more porosity. The N_2 uptake at partial pressure (P/Po) are characteristics of micropore presence. Furthermore, the step P/Po of 0.2 and 0.9 are attributed to inter and intra particle pores respectively [23,24]. As reported in literatures [25–27], the microemulsion synthesis condition invoke large amount of interparticle pores, which shows the presence of large surface area materials.

Table 1. Physical catalytic properties.

Sample	Surface area (m^2/g)[a]	Total pore volume (mL/g)[a]	Mesopore volume (mL/g)[b]
MoZ	395	0.2626	0.1722
MnZ	404	0.2838	0.1862
CoZ	419	0.2694	0.1925

[a]Obtain by the BJH method.
[b]Computed by subtracting micropore volumes from total pore volume.

To determine the effect of prepared catalysts in the CRM process, series of tests were conducted in the micro catalytic fixed bed reactor over 550–850°C temperature range. The results are depicted in figure 3. As can be seen, temperature increase led to enhanced conversion and yield due to the endothermicity of CRM reaction. CoZ displayed high reactants conversion of 69 % and 65 % for CH_4 and CO_2 respectively. The CoZ catalyst produced the highest activity than other catalysts. This suggest that ZSM-5 supported Co catalyst is highly active for transformation of CH_4 and CO_2 gases into syngas. These is in accordance with findings in the catalyst characterizations. CoZ catalyst had higher activity as a result of larger amount of pore sizes and higher surface area which lead to better dispersion of active Co species [14]. These observations are affirmed by the XRD and N_2 adsorption–desorption analysis. Moreover, these features aid higher CH_4 and CO_2 adsorption which is pivotal in CRM reaction. Large surface area enhances the adsorption of the reactants. Hence, this observation is directly related to decomposition of CH_4 and CO_2 dissociation. However, the higher conversion of CH_4 than CO_2 suggest suppression of side reactions [11].

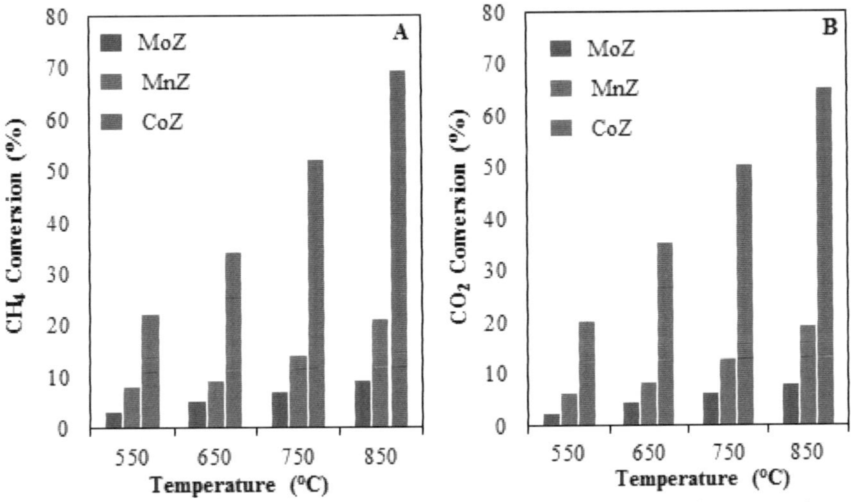

Figure 3. Effect of metal loaded catalysts on (a) CH_4 and (b) CO_2 conversions.

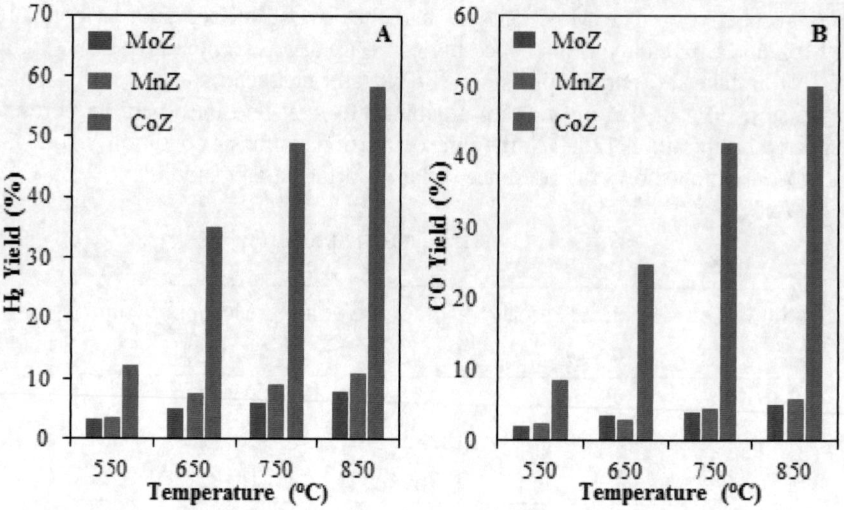

Figure 4. Effect of metal loaded catalysts on (a) H_2 and (b) CO yield.

It is reported that active metal sites play crucial role in CH_4 adsorption to produce CH_X fragments [28,29]. Thus, Co species are more active to decompose CH_4 and CO_2 compared to Mo and Mn loaded catalysts. Fayaz et al. achieved initial H_2 and CO yield of 37.5 % and 39 % respectively over $10Co/Al_2O_3$ catalyst. This is in accordance with findings of Khan et al. [30] and Zeng et al. [15]. Similarly, the catalysts investigated displayed different effect on product yield as shown in figure 4. The H_2 and CO yield for CoZ is 58 % and 50 % respectively. There is significant difference in CRM activity of the CoZ and other catalysts as seen in figure 4. This performance is due to the higher activity of Co, coupled with the greater adsorption and dissociation of reactants which was remarkable on CoZ than other catalysts.

4. Conclusion

Mesoporous ZSM-5 support was synthesized by micro emulsion technique and the metals were loaded by wet impregnation method. This study revealed that ZSM-5 supported Co catalyst is a highly active catalyst system for CRM reaction as compared to MoZ and MnZ catalysts. Higher surface area resulted in higher activity of CoZ catalyst in CRM. The Co species and mesoporous ZSM-5 support had a beneficial interaction which led to the suitable CRM activity. The Co species were more active to decompose CH_4 and CO_2 compared to Mo and Mn loaded catalysts. Thus, CoZ is an efficient catalyst in the quest for a viable CRM catalyst.

Acknowledgments

This work was supported by the Fundamental Research Grant (Scheme Grant No. FRGS/1/2017/STG07/UTM/01/1) and UTM Transdisciplinary Research Grant (Grant No. 06G53).

References

[1] Gambo Y, Jalil A A, Triwahyono S, Abdulrasheed A A 2018 Recent advances and future prospect in catalysts for oxidative coupling of methane to ethylene: A review. *J. Ind. Eng. Chem.* **59** 218–229.

[2] Abdulrasheed A A, Jalil A A, Triwahyono S, Zaini M A A, Gambo Y, Ibrahim M 2018 Surface modification of activated carbon for adsorption of SO_2 and NO_X: A review of existing and emerging technologies. *Renewable Sustainable Energy Rev.* **94** 1067–1085.

[3] Abdulrasheed A A, Jalil A A, Hamid M Y S, Siang T J, Fatah N A A, Izan S M, Hassan N Dry reforming of methane to hydrogen-rich syngas over robust fibrous KCC-1 stabilized nickel

catalyst with high activity and coke resistance. *Int J Hydrogen Energy* (doi:10.1016/j.ijhydene.2019.04.126).

[4] Siang T J, Jalil A A, Hambali H U, Hussain I, Saifulddin M S 2019 Catalytic partial oxidation of methane to syngas over perovskite catalysts. *E3S Web of Conferences*. **6** 1–7.

[5] Hussain I, Jalil A A, Mamat C R, Siang T J, Azami M S, Hambali H U 2019 Role of Promoters in Hoisting the Catalytic Performance for Enhanced CO Methanation. *J. Energy Safety Technology* **02** 15–20.

[6] Abdulrasheed A A, Jalil A A, Gambo Y, Ibrahim M, Hambali H U, Hamid M Y S 2019 A review on catalyst development for dry reforming of methane to syngas : Recent advances. *Renewable Sustainable Energy Rev.* **108** 175–193.

[7] Ayodele O B, Abdullah A Z 2019 Exploring kaolinite as dry methane reforming catalyst support: influences of chemical activation, organic ligand functionalization and calcination temperature. *Appl. Catal. A.* **576** 20–31.

[8] Siang T J, Singh S, Omoregbe O, Bach L G, Phuc N N H, Vo D N 2018 Hydrogen production from CH_4 dry reforming over bimetallic Ni–Co/Al_2O_3 catalyst *Energy Inst.* **91** 683–694.

[9] Siang T J, Pham T L M, Cuong N V, Phuong P T T, Huu N, Phuc H, Duc Q, Vo D N 2018 Combined steam and CO_2 reforming of methane for syngas production over carbon-resistant boron-promoted Ni/SBA-15 catalysts Microporous Mesoporous Mater. **262** 122–132.

[10] Sidik S M, Triwahyono S, Jalil A A, Majid Z A, Salamun N, Talib N B, Abdullah T A T 2016 CO_2 reforming of CH_4 over Ni–Co/MSN for syngas production: Role of Co as a binder and optimization using RSM *Chem. Eng. J.* **295** 1–10.

[11] Bawah A, Malaibari Z O, Muraza O 2018 Syngas production from CO_2 reforming of methane over Ni supported on hierarchical silicalite-1 fabricated by microwave-assisted hydrothermal synthesis *Int. J. Hydrogen Energy* **43** 13177–13189.

[12] Liu D, Lau R, Borgna A, Yang Y 2009 Carbon dioxide reforming of methane to synthesis gas over Ni-MCM-41 catalysts *Appl. Catal. A.* **358** 110–118.

[13] Vafaeian Y, Haghighi M, Aghamohammadi S 2013 Ultrasound assisted dispersion of different amount of Ni over ZSM-5 used as nanostructured catalyst for hydrogen production via CO_2 reforming of methane *Energy Convers. Manag.* **76** 1093–1103.

[14] Boukha Z, Pilar M, Angel M, Gonzalez-velasco J R 2019 Influence of Ca/P ratio on the catalytic performance of Ni/hydroxyapatite samples in dry reforming of methane. *Appl. Catal. A.* **580** 34–45.

[15] Zeng S, Zhang L, Zhang X, Wang Y, Pan H, Su H 2012 Modification effect of natural mixed rare earths on Co/γ-Al_2O_3 catalysts for CH_4/CO_2 reforming to synthesis gas *Int. J. Hydrogen Energy* **37** 9994–10001.

[16] Hambali H U, Jalil A A, Triwahyono S, Jamian S F, Fatah N A A, Abdulrasheed A A., Siang T J 2019 Unique structure of fibrous ZSM-5 catalyst expedited prolonged hydrogen atom restoration for selective production of propylene from methanol *Int. J. Hydrogen Energy*. (doi.org/10.1016/j.ijhydene.2019.11.236).

[17] Teh L P, Triwahyono S, Jalil A A, Firmansyah M L, Mamat C R, Majid Z A 2016 Fibrous silica mesoporous ZSM-5 for carbon monoxide methanation. *Appl. Catal. A.* **523** 200–208.

[18] Hambali H U, Jalil A A, Siang T J, Abdulrasheed A A, Fatah N A A, Hussain I, Azami M S 2019 Effect of ZSM-5 Acidity in Enhancement of Methanol-to-Olefins Process *J. Energy Safety Technology* **4** 9–13.

[19] Rostamizadeh M, Taeb A 2015 Highly selective Me-ZSM-5 catalyst for methanol to propylene (MTP) *J. Ind. Eng. Chem.* **27** 297–306.

[20] Jin Z, Liu S, Qin L, Liu Z, Wang Y, Xie Z, Wang X 2013 Methane dehydro-aromatization by Mo-supported MFI-type zeolite with core–shell structure *Appl. Catal. A.* **453** 295–301.

[21] Abdollahifar M, Haghighi M, Sharifi M 2015 Dry reforming of methane over nanostructured Co/Y catalyst or hydrogen production: Effect of ultrasound irradiation and Co-loading on catalyst properties and performance *Energy Convers. Manag.* **103** 1101–1112.

[22] Pang Y, Dou Y, Zhong A, Jiang W, Gu L, Feng X, Ji W 2018 Nanostructured Ru-Co@SiO_2: Highly efficient yet durable for CO_2 reforming of methane with a desirable H_2/CO ratio *Appl. Catal. A.* **555** 27–35.

[23] Jalil A A, Gambo Y, Ibrahim M, Abdulrasheed A A, Hassan N S, Nawawi M G M, Asli U A, Hassim M H, Ahamd A 2019 Platinum-promoted fibrous silica Y zeolite with enhanced mass transfer as a highly selective catalyst for n-dodecane hydroisomerization *Int. J. Energy Research* 1–16 (doi: 10.1002/er.4545).

[24] Fatah N A A, Triwahyono S, Jalil A A, Salamun N, Mamat C R, Majid Z A 2017 n-Heptane isomerization over molybdenum supported on bicontinuous concentric lamellar silica KCC-1: Influence of phosphorus and optimization using response surface methodology (RSM) *Chem. Eng. J.* **314** 650–659.

[25] Hussain I, Jalil A A, Mamat C R, Siang T J, Rahman A F A, Azami M S, Adnan R H 2019 New insights on the effect of the H_2/CO ratio for enhancement of CO methanation over metal-free fibrous silica ZSM-5: Thermodynamic and mechanistic studies *Energy Convers. Manag.* **199** 1120562.

[26] Hamid M Y S, Jalil A A, Rahman A F A, Abdullah T A T 2019 Enhanced reactive CO_2 species formation via V_2O_5-promoted Ni/KCC-1 for low temperature activation of CO_2 methanation. *React. Chem. Eng.* **4** 1126–1135.

[27] Azami M S, Jalil A A, Hitam C N C, Mamat C R, Siang T J, Hussain I, Hambali H U 2019 A Contemporary Assessment on Composite Titania onto Graphitic Carbon Nitride-Based Catalyst as Photocatalyst *J. Energy Safety Technology* **02** 21–25.

[28] Hambali H U, Jalil A A, Abdulrasheed A A, Siang T J, Vo D V N 2020 Enhanced dry reforming of methane over mesostructured fibrous Ni/MFI zeolite: Influence of preparation method. *J Energy Inst.* 2020 (doi:10.1016/j.joei.2020.01.016).

[29] Abdulrasheed A A, Jalil A A, Hamid M Y S, Siang T J, Abdullah T A T 2020 Dry reforming of CH4 over stabilized Ni-La@KCC-1 catalyst: Effects of La promoter and optimization studies using RSM *J CO_2 Util.* **37** 230–239 (doi:10.1016/j.jcou.2019.12.018).

[30] Khan N A, Kennedy E M, Dlugogorski B Z, Adesina A A, Stockenhuber M 2018 A proposed reaction mechanism for the selective oxidation of methane with nitrous oxide over Co-ZSM-5 catalyst forming synthesis gas (CO+H_2) *Int. J. Hydrogen Energy* **43** 13133–13144.

IOP Conf. Series: Materials Science and Engineering **808** (2020) 012006 doi:10.1088/1757-899X/808/1/012006

Dendritic Mesoporous Ni/KCC-1 for Partial Oxidation of Methane to Syngas

T J Siang[1], A A Jalil[1,2*], H U Hambali[1], A A Abdulrasheedand[1] and M S Azami[3]

[1]School of Chemical and Energy Engineering, Faculty of Engineering, Universiti Teknologi Malaysia, 81310 UTM Johor Bahru, Johor, Malaysia
[2]Centre of Hydrogen Energy, Institute of Future Energy, Universiti Teknologi Malaysia, 81310 UTM Johor Bahru, Johor, Malaysia
[3]Department of Chemistry, Faculty of Science, Universiti Teknologi Malaysia, 81310 UTM Johor Bahru, Johor, Malaysia

*aishahaj@utm.my

Abstract. Depletion of fossil fuel for global energy system and increasing concern on global warming have driven the exploration of alternative and sustainable energy source in realms of academia and industry. This study aims to investigate the physicochemical features of KAUST Catalysis Center-1 supported catalyst (i.e., Ni/KCC-1) and evaluate its catalytic performance for Partial Oxidation of Methane (POM) reaction. N_2 physisorption and XRD analyses confirmed the structural integrity of KCC-1 framework after NiO addition while the growth of Si–O–Ni bonds in KCC-1 structure was corroborated by the FTIR results. The FESEM and TEM images for KCC-1 not only affirm the successful formation of bicontinuous lamellar morphology but also reveal that the three-dimensional spherical structure was originally developed from the centre of microsphere into all axial. The combustion-reforming pathway was determined during reaction run and the H_2/CO ratio ranging of 1.48 to 2.14 was appropriate for synthetic fuel production via Fischer-Tropsch synthesis (FTS).

1. Introduction

The excessive emission of anthropogenic greenhouse gas due to industrial and residential combustion and the depletion of fossil fuel for global energy system have driven the exploration of alternative and sustainable energy source in realms of academia and industry [1]. Syngas emerges as a promising energy source for the global petroleum-based energy system as it is the feedstock for generating synfuel through Fischer-Tropsch (FT) approach [2]. In recent years, syngas produced via partial oxidation of methane (POM, cf. Eq. (1)) has gained substantial attention from researchers as it is capable to provide a practical ratio of H_2/CO around 2 for downstream FT synthesis [2]. However, POM consists of a complex reaction network [3] and simultaneously encounters catalyst deactivation problem caused by deposited carbon on active sites of catalyst [4]. Consequently, designing a new catalyst system is crucial to address the aforementioned problems.

$$CH_4 + 0.5O_2 \rightarrow 2H_2 + CO \quad \Delta H_{298K} = -126 \text{ kJ mol}^{-1} \tag{1}$$

Although the cost-efficiency and excellent cleavage ability of Ni towards hydrocarbon groups make it be widely employed in the reforming processes, the severe deposition of carbonaceous species lead to deactivation is always the biggest drawback for its industrial application [2]. From the above reason, the design of the support system for Ni catalyst has always been paid much attention to overcome the deposition of coke on catalyst surface meanwhile promoting the catalyst performance. In recent years, mesoporous KAUST Catalysis Center-1 (KCC-1) reportedly possesses remarkable physical features, including high surface area, high mechanical stability and thermal stability due to its dendrimeric silica fibers morphologic structure [5]. Additionally, KCC-1 provides oxygen vacancy sites that not only reduce the barriers for CH_4 cleavage but also remove the deposited carbon via gasification by oxidizing gases. Hence, the objective of this study was to assess the physicochemical features of Ni/KCC-1 catalyst and its catalyst performance for POM.

2. Experimental

2.1 Catalyst synthesis

A microwave-assisted hydrothermal method was employed for synthesizing the KCC-1 and Ni/KCC-1 catalyst. A well-stirred mixture solution consists of 737.7 mL of distilled water, 23.8 g of hexadecyltrimethylammonium bromide (CTAB) and 14.9 g of urea (CH_4N_2O) was added into another aqueous mixture solution containing 731.8 mL of toluene (C_7H_8), 31.2 mL of 1-butanol ($CH_3(CH_2)_3OH$) and 63.4 g of tetraethyl orthosilicate (TEOS). The resulting mixture was continuously stirring for 30 min at room temperature before subjected into an oven for 4 h at 120 °C. The white solution was allowed to cool down to room temperature before subjected to oven at 110 °C overnight, followed by calcination in a furnace for 4 h at 550 °C to obtain dendritic mesoporous KCC-1 support.

The 0.5 wt% Ni/KCC-1 catalyst was synthesized by using conventional wetness impregnation approach. A 0.07 g of nickel (II) nitrate hexahydrate ($Ni(NO_3)_2.6H_2O$) was added into 30 mL of distilled water and was homogeneously stirred to form Ni precursor solution. Then, a 2.985 g of as-synthesized KCC-1 was added into solution and the resultant mixture was thoroughly mixed at 60 °C for 2 h, followed by drying at 110 °C overnight and calcination at 800 °C for 5 h for producing Ni/KCC-1 catalyst.

2.2 Characterization methods

The Beckman Coulter SA 3100 unit was employed for evaluating physical properties (*viz.*, average pore diameter, total pore volume and Brunauer-Emmett-Teller (BET) surface area) of specimen. The specimen was pre-treated with N_2 gas flow at 300 °C for 1 h to remove volatile and moist contaminant before measurement. The pore size distribution was estimated via Non-Local Density Functional Theory (NL-DFT) approach. The low-angle and wide-angle X-ray diffractograms of specimen was collected using Bruker Advance D8 diffractometer equipped with Cu K_α radiation (λ=1.544 Å) operating at 40 mA and 45 kV. The diffractograms were recorded in 2θ ranging of 2 to 80° at scanning rate of 0.1° min^{-1} with step size of 0.05°. The Agilent FTIR spectrometer (Cary 640) was used for recording the fourier-transform infrared (FTIR) spectroscopy of specimen at room temperature with a 5 cm^{-1} resolution. The field emission scanning electron microscopy (FESEM) and transmission electron microscopy (TEM) measurement were conducted by the corresponding Zeiss Supra VP35 and JEOL JEM-2100F unit.

2.3 Partial oxidation of methane reaction

The evaluation for partial oxidation of methane was performed at atmospheric pressure and temperature of 800 °C with stoichiometric feedstock ratio ($F_{CH_4}:F_{O_2}=2:1$ with F is molar flow rate (mol s^{-1})). A 0.2 g of pelletized catalyst was sandwiched by quartz wool and placed vertically in the reactor center. The catalyst was pre-treated in H_2 stream of 20 mL min^{-1} for 3 h at 800 °C, followed by the N_2 gas of 60 mL min^{-1}. In order to minimize the external and internal transport resistances, a gas hourly space velocity (GHSV) of 18 L g_{cat}^{-1} h^{-1} was employed during reaction. The gaseous product composition in the effluent outlet was examined by an on-line Agilent 7820N GC unit connected with Carboxen 1010 column and thermal conductivity detector (TCD). The catalytic performance of reaction was assessed with respect to reactant consumption rate (r_j , j: CH_4 or O_2), product formation rate (r_i, i: H_2, CO or CO_2) and H_2/CO ratio, as provided by Eqs. (2) to (4).

$$-r_j\left(mol\ g_{cat}^{-1}\ s^{-1}\right)=\frac{F_j^{In}-F_j^{Out}}{W_{Cat.}} \tag{2}$$

$$r_i\,(mol\ g_{cat}^{-1}\ s^{-1})=\frac{F_i^{Out}}{W_{Cat.}} \tag{3}$$

$$H_2\,/\,CO=\frac{F_{H_2}^{Out}}{F_{CO}^{Out}} \tag{4}$$

where F^{In} and F^{Out} are the corresponding inlet and outlet molar flow rate ($mol\ s^{-1}$) whereas $W_{cat.}$ is the catalyst weight (g).

3. Results & discussion

3.1 Physicochemical attributes of catalyst

3.1.1 N_2 physisorption studies. The N_2 adsorption-desorption isotherms of bare dendritic KCC-1 support and Ni loaded KCC-1 catalyst are presented in Figure. 1A. The curves of both KCC-1 support and Ni/KCC-1 catalyst were determined as type IV isotherm curves associated with H3-type hysteresis loop, indicating the successful formation of lamellar structure of KCC-1 particles [5, 6]. The similar isotherm curve of Ni/KCC-1 compared to that of KCC-1 suggesting the structural integrity of dendritic lamellar KCC-1 morphology after NiO addition. However, the incorporation of NiO particles on KCC-1 surface led to partial blockage of mesopores and thus, resulting in a decline in peak intensity of pore size ranging of 20 to 50 nm, as demonstrated in Figure. 1B.

The embedment of NiO particles on KCC-1 support surface substantially caused a decrease in the both average BET surface area of KCC-1 from 594 to 467 $m^2\ g^{-1}$ and total pore volume from 2.4 to 1.5 $cm^3\ g^{-1}$. Conversely, the average pore diameter raised from 3.6 (KCC-1) to 5.1 nm (Ni/KCC-1) with the NiO embedment reasonably due to the minor collapse of silica framework induced by metal sintering during calcination at high temperature [7]. Although high-temperature calcination process unavoidably results in formation of metal cluster with a bigger crystallite size, it produces a heat-stable structured catalyst [7, 8].

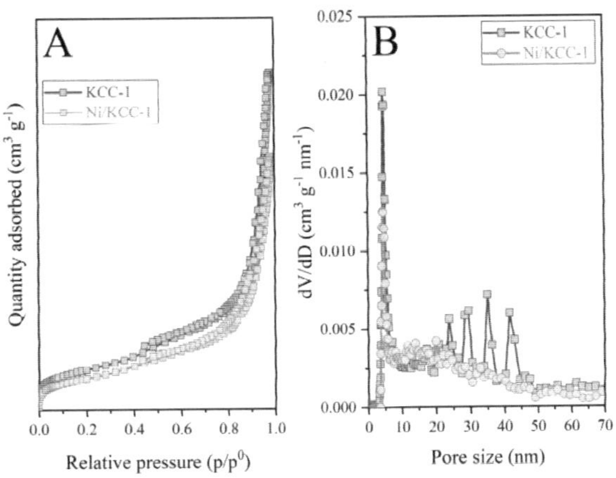

Figure. 1. (A) N_2 physisorption isotherm and (B) NLDFT pore size distribution of KCC-1 support and Ni/KCC-1 catalyst.

IOP Conf. Series: Materials Science and Engineering **808** (2020) 012006 doi:10.1088/1757-899X/808/1/012006

3.1.2 Crystallinity and structural phase studies. Figure. 2A depicts the low-angle XRD patterns of KCC-1 support and Ni/KCC-1 catalyst. Database of Joint Committee on Power Diffraction Standards (JCPDS) was employed for analysing the XRD patterns in this study [9]. An intensive peak at 2θ of 3.29° with a distinctive shoulder at 2θ of 2.34° was discerned, typical characteristic of wormhole-type mesoporous material structure [5]. Notably, the peak intensity of Ni/KCC-1 catalyst at 2θ of 2.34° decreased after NiO incorporation, indicative of the successful embedment of NiO particles. This evolution could explain the observed reduction in the peak intensity attributed to the amorphous SiO_2 framework at 2θ of 20 to 30° for XRD pattern of Ni/KCC-1 catalyst (cf. Figure. 2B). This was in good accordance with the results of N_2 physisorption analysis. Additionally, five peaks located at 2θ of 39.6°, 43.3°, 65.7°, 74.9° and 78.7° were assigned to the NiO crystal planes (1 1 1), (2 0 0), (2 2 0) and (3 1 1), respectively with JCPDS card No. 47-1049 [10].

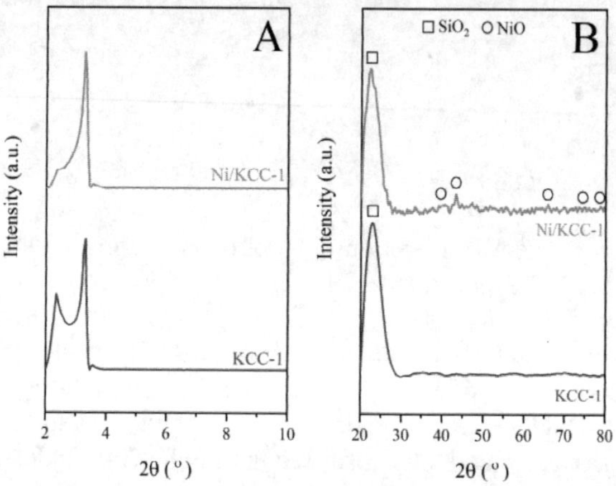

Figure. 2. (A) Low-angle and (B) Wide-angle X-ray diffractograms of KCC-1 support and Ni/KCC-1 catalyst.

Figure. 3A displays the KBr-FTIR spectra of KCC-1 support and Ni/KCC-1 catalyst. The analogous pattern for all spectra suggested the KCC-1 framework was free from structural distortion after NiO addition in agreement with XRD and N_2 physisorption measurement. Both spectra exhibited a broad transmittance band assigned to Si–O–Si bonds in KCC-1 phase from 1300 to 1000 cm^{-1} [11]. Particularly, two noticeable peaks corresponded to longitudinal-optic (LO) and transverse-optic (TO) vibrations for Si–O–Si bonds were discerned at 1240 and 1091 cm^{-1} [12]. The peak detected at 968 cm^{-1} was ascribed to the external Si–OH phase whilst the Si–O vibrations accounted for the peaks observed at 806 and 462 cm^{-1} [13, 14].

From Figure. 3B, the relative peak area percent associated with Si–O–Si groups of Ni/KCC-1 was comparable to that of KCC-1 spectra, corroborating the structural integrity of KCC-1 framework. Notably, the peak intensity of Si–OH for Ni/KCC-1 catalyst was three-fold lesser than that of bare KCC-1. This behavior was probably due to the Si–O–Ni bond formation in the KCC-1 framework by substituting the O–H bonds with Ni ions [10, 11].

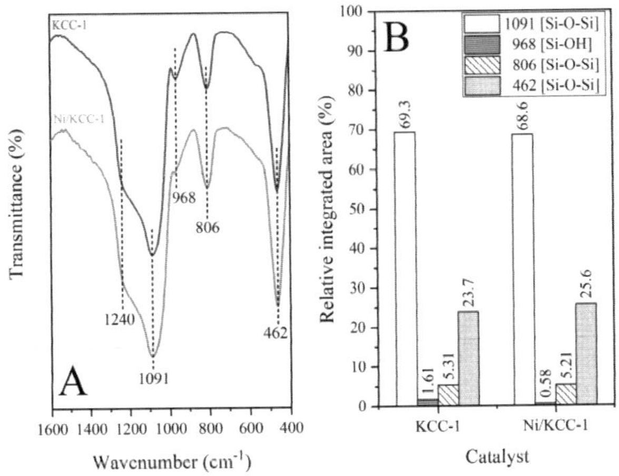

Figure. 3. (A) KBr-FTIR spectra of KCC-1 support and Ni/KCC-1 catalyst. (B) Summary of their band intensities.

3.1.3 Surface morphology. The FESEM and TEM measurement of KCC-1 were conducted as shown in Figure. 4. As seen in Figure. 4A, KCC-1 possessed a bicontinuous lamellar morphology with uniform three-dimensional spherical structure [6]. Additionally, the TEM image (cf. Figure. 4B) reveals that the three-dimensional spherical structure was originally developed from the centre of microsphere into all axial and eventually forming the dendritic KCC-1 particle.

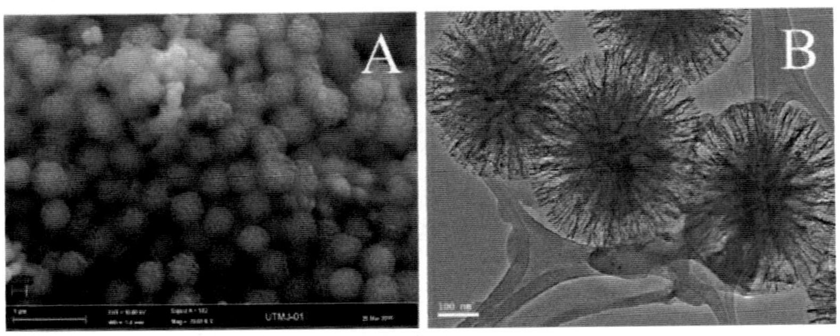

Figure. 4. (A) FESEM and (B) TEM images of fresh fibrous silica KCC-1 support.

3.2 Catalytic partial oxidation of methane evaluation

As demonstrated in Figs. 5A and 5B, the catalytic performance of both pristine KCC-1 support and Ni/KCC-1 was stable within 6 h on-stream. Regardless of catalyst, a total consumption rate achieved in O_2 reactant (7.03×10^{-3} mol g_{cat}^{-1} s^{-1}) and growth in CO_2 formation rate indicates the existing pathway of combustion-reforming, of which CH_4 total oxidation produce oxidizing agents (k_1), followed by reforming processes (k_2) to yield syngas as shown in Eq. (5) [3]. Notably, in comparison with the bare KCC-1, Ni/KCC-1 exhibited around two-fold higher in both overall CH_4 consumption rate and H_2 formation rate. This observation could suggest that the rate of reforming route (k_2) was substantially facilitated than that of combustion route (k_1) over Ni/KCC-1 due to the excellent cleavage ability of Ni towards C–H bonds in CH_4 compounds [10, 11]. However, as seen in Figure. 5B, the Ni/KCC-1 encountered a loss in reaction activity with time on-stream due to deactivated active site resulted from carbon deposition [4] which leads to the slight decline in both H_2 and CO formation rate while CO_2 formation rate increased owing to the excessive unconverted CO_2 from combustion route.

$$CH_4 \xrightarrow[+\,O_2]{k_1} CO_2 + H_2O \xrightarrow[+\,CH_4]{k_2} H_2 + CO \qquad (5)$$

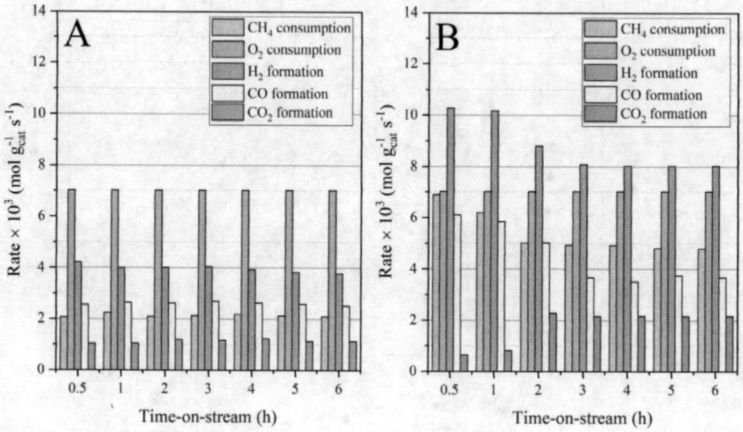

Figure. 5. Gaseous reactants' consumption and products' formation rates of (A) bare KCC-1 support and (B) Ni/KCC-1 catalyst for MPO at temperature of 800 °C and atmospheric pressure with stoichiometric feed composition.

In order to validate the excellent catalytic performance of Ni/KCC-1, a comparison study regarding the catalytic performance of Ni-based catalysts recently employed for POM reaction is provided in Table 1. The H_2/CO ratio obtained over pristine KCC-1 and Ni/KCC-1 in this study falls in a range of 1.48 to 2.14 (cf. Table 1) could be a result of simultaneous steam reforming of methane (SRM), dry reforming of methane (DRM) and reverse water gas shift (RWGS) reaction. Since SRM yields a stoichiometric ratio of H_2 to CO around 3 whereas DRM provides a ratio of H_2/CO lower than unity, with the presence of RWGS, the H_2/CO ratio predictably fell in a range of 1 to 3 appropriate for synfuels generation through FTS [15]. Particularly, a practical H_2/CO ratio for FTS is well known as a value of 2. As seen in Table 1, Ni/KCC-1 not only exhibited a comparable H_2/CO ratio but also provided the highest H_2 yield amongst the Ni-based catalysts, confirming the enhancement in rate of H_2 generation over Ni/KCC-1. This could be a result of the uniqueness of dendrimeric silica fibers morphologic structure providing the remarkable reactant accessibility to the active sites and hence, enhancement in reaction rate and product formation rate [5, 6, 11, 14].

Table 1. Catalytic performance listing of Ni-based catalysts recently employed for POM reaction.

Catalyst	CH_4/O_2 ratio	T (°C)	GHSV (L g_{cat}^{-1} h^{-1})	H_2 yield[a] (%)	H_2/CO ratio	Ref.
KCC-1	2/1	800	54	43	1.48	This study
Ni/KCC-1	2/1	800	54	65	2.14	This study
Ni/SiO_2	2/1	800	25	49	2.02	[16]
Ni/ZrO_2-SiO_2	2/1	800	25	40	2.09	[16]
Ni/Al_2O_4	2/1	650	60	42	2.29	[17]
$NiAl_2O_4$-CeO_2	2.5/1	700	60	58	-	[17]

[a]H_2 yield (%) = $\frac{F_{H_2}^{Out}}{2F_{CH_4}^{In}} \times 100\%$

4. Conclusions

The Ni/KCC-1 catalyst was synthesized via microwave-assisted hydrothermal approach and was scrutinized for POM at 800 °C and atmospheric pressure with stoichiometric feed composition. The N_2 physisorption and XRD analyses confirmed the structural integrity of KCC-1 after NiO addition and the successful incorporation of NiO particles into mesoporous KCC-1 framework. The KBr-FTIR results reveal the formation of Si–O–Ni bonds in the Ni/KCC-1 catalyst. Both bare KCC-1 support and Ni/KCC-1 catalyst exhibited a total consumption rate achieved in O_2 reactant and increasing CO_2 formation rate, indicating the existing combustion-reforming pathway for syngas production. The obtained H_2/CO ratio over KCC-1 supported catalysts in a range from 1.48 to 2.14 was appropriate for synthetic fuel production via FT synthesis.

Acknowledgments
The authors are appreciatively acknowledging the financial support sponsored by the Universiti Teknologi Malaysia via UTM Transdisciplinary Research Grant (Grant No. 06G53) and Collaborative Research Grant (Grant No. 07G62) as well as the Universiti Teknologi Malaysia Zamalah Scholarship for this work.

References

[1] Energy Information Administration, *Annual energy outlook 2018: With projections to 2050* (Government Printing Office, 2018).

[2] Siang T J, Minh D P, Singh S, Setiabudi H D, Vo D V N, Recent advances in hydrogen production through bi-reforming of biogas *Fuel Processing and Energy Utilization* vol 1, ed S Nanda *et al* (London: Chapman and Hall) chapter 5 pp 71–89.

[3] Nguyen T H, Łamacz A, Krztoń A, Liszka B and Djéga-Mariadassou G 2016 *Appl. Catal. B–Environ.* **182** 385–91

[4] Argyle M and Bartholomew C 2015 *Catal.* **5** 145–269

[5] Febriyanti E, Suendo V, Mukti R R, Prasetyo A, Arifin A F, Akbar M A, Triwahyono S and Marsih I N 2016 *Langmuir* **32** 5802–811

[6] Hamid M Y S, Triwahyono S, Jalil A A, Jusoh N W C, Izan S M and Abdullah T A T 2018 *Inorg. Chem.* **57** 5859–69

[7] Usman M, Daud W W and Abbas H F 2015 *Renew. Sust. Energ. Rev.* **45** 710–44

[8] Pan C J, Tsai M C, Su W N, Rick J, Akalework N G, Agegnehu A K, Cheng S Y and Hwang B J 2017 *J. Taiwan Inst Chem Eng.* **74** 154–86

[9] JCPDS Powder Diffraction File, International centre for diffraction data, PA: Swarthmore, 2000

[10] Siang T J, Bach L G, Singh S, Truong Q D, Phuc N H H, Alenazey F and Vo D V N 2019 *Int. J. Hydrogen Energy* **44** 20839–50.

[11] Sidik S M, Triwahyono S, Jalil A A, Aziz M A A, Fatah N A A and Teh L P 2016 *J. CO_2 Util.* **13** 71–80

[12] Azizi S N, Ghasemi S and Chiani E 2013 *Electrochimica Acta* **88** 463–72

[13] Pudukudy M, Yaakob Z and Akmal Z S 2015 *Appl. Surf. Sci.* **330** 418–30

[14] Sidik S M, Jalil A A, Triwahyono S, Abdullah T A T and Ripin A 2015 *RSC Adv.* **5** 37405–14

[15] Özkara-Aydınoğlu Ş 2010 *Int. J. Hydrogen Energy* **35** 12821–28

[16] Guo S, Wang J, Ding C, Duan Q, Ma Q, Zhan, K and Liu P 2018 *Int. J. Hydrogen Energy* **43** 6603–13

[17] Gil-Calvo M, Jiménez-González C, de Rivas B, Gutiérrez-Ortiz J I and López-Fonseca R 2017 *Appl. Catal. B–Environ.* **209** 128–38

IOP Conf. Series: Materials Science and Engineering **808** (2020) 012007 doi:10.1088/1757-899X/808/1/012007

Hydroconversion of *n*-hexane over Pt-supported on fibrous silica mordenite catalysts: effect of transition metals on acidity and activity

M Ibrahim[1,4], A A Jalil1[1,2*], W F W Zakaria[1], M Izan[3], A A Fauzi[1] and I Hussain[3]

[1]School of Chemical and Energy Engineering, Faculty of Engineering, Universiti Teknologi Malaysia, 81310 UTM Johor Bahru, Johor, Malaysia
[2]Centre of Hydrogen Energy, Institute of Future Energy, Universiti Teknologi Malaysia, 81310 UTM Johor Bahru, Johor, Malaysia
[3]Department of Chemistry, Faculty of Science, Universiti Teknologi Malaysia, 81310 UTM Johor Bahru, Johor, Malaysia
[4]Department of Chemical Engineering, Abubakar Tafawa Balewa University, PMB 0248, Bauchi, Bauchi State Nigeria

*aishahaj@utm.my

Abstract. The effect of incorporation of either Zr, Co, Mo or Zn on Pt/HFSM was investigated for the hydroconversion of *n*-hexane in a pulsed micro-reactor at a temperature range 423-623 K. The acidity of all the studied catalysts were investigated using pyridine adsorption and Fourier Transform Infrared Spectroscopy. The FTIR-pyridine study showed that the catalysts possessed different concentration and strength of both Lewis and Brønsted acid sites. It was revealed that the moderate acid sites distribution on PtZr/HFSM and PtZn/HFSM facilitated the isomerization of the intermediate hexene while the strong acid sites were selective towards cracking reaction. The trend for the bimetallic catalysts studied in *n*-hexane hydroisomerization is in the order: PtZr/HFSM>PtZn/HFSM>PtMo/HFSM>PtCo/HFSM. The highest activity for hydrocracking and hydrogenolysis were exhibited by PtMo/HFSM and PtCo/HFSM respectively. The PtZr/HFSM exhibited the highest hydroisomerization activity due to the moderate acidity and abundant Lewis acid sites capable of generating protonic acid sites by the hydrogen-spill over phenomena.

1. Introduction.

Generally, hydroconversion of *n*-alkanes is done over a bifunctional catalyst consisting of metal function and acid sites for the rearrangement of the intermediate species [1]. Typical example of such catalysts are platinum supported on zeolites. Zeolites are widely utilized support materials that can be modified to accomplish optimum efficiency in a various applications [2]. In recent years, extensive research on zeolites has been devoted to a detailed description of the active site in the zeolite structure. Several reports indicated that zeolites suffer low molecular diffusion efficiency, quick catalyst deactivation and consequently, weak activity. The application of Pt/HMOR has received considerable attention in hydroconversion of *n*-hexane, several strategies for modification of the properties of modernite were reported with both improved mesoporosity and appropriate acidic properties. Improving the catalytic performance is a vital issue for enhancing the selectivity of the process [3].

Herein, infrared spectroscopy of pyridine was used to study the acid sites concentration and strength in metal promoted Pt supported fibrous silica mordenite catalysts, prepared by microemulsion system assisted seed-recrystalization. The synthesized catalysts were loaded with Pt and second metals as promoters and tested in the hydroconversion of *n*-hexane.

2. Experimental

2.1 Preparation of parent fibrous silica mordenite
Fibrous silica mordenite (FSM) was synthesized using a microemulsion system via self-assembly process. Briefly, *n*-butanol and toluene were mixed, tetraethyl orthosilicate (TEOS) was added to the mixture and stirred vigorously in a Teflon. Subsequently, a solution prepared from CTAB, urea, and water as the template, mordenite as seed was quickly added into the mixture. After vigorous stirring to obtain homogeneity in the solution, the mixture was exposed to 4 h hydrothermal treatment at 393 K. The solid products were centrifuged, washed with deionized water and dried in oven for 24 h at 383 K. Finally, the synthesized material was calcined at 823 K for 6 h. The H-form was obtained by ion exchange with ammonium nitrate solution and calcined at 823K denoted as HFSM.

2.2 Preparation of metal loaded HFSM
The metal precursors were used to load 5 wt% of each on to HFSM support via the incipient wetness impregnation technique. The respective precursors are zirconium ($ZrOCl.6H_2O$), cobalt ($Co(NO_3)_2{\cdot}6H_2O$), molybdenum (MoO_3) and zinc ($Zn(NO_3)_2{\cdot}6H_2O$). The solid catalysts obtained were dried overnight in an oven at 383 K and calcined at 823 K for 3 h. All the synthesized catalysts were loaded with 0.5 wt% platinum with an aqueous solution of chloroplatinic acid (H_2PtCl_6) as Pt precursor denoted as PtCo/HFSM, PtZn/HFSM, PtZr/HFSM and PtMo/HFSM.

2.3 Characterization
The XRD pattern were obtained with Bruker Advance D8, 40 kV, 40 mA recorded using Cu kα radiation of 0.15418 nm at a 2Θ angle ranging from 5° to 40° with scanning rate of 0.02°. The morphological structure of the catalyst was examined using FESEM-EDX JEOL JSM-6701F. Pyridine was used as adsorbed probe molecule to study the acidity of the catalyst. Pressure of 2 Torr was established for the adsorption of pyridine at 423 K for 30 mins. The strength of the Brønsted acid sites and Lewis acid sites were analysed by pyridine desorption at Low (423 K), medium (523 K) and high (623 K) temperatures for 30 mins.

2.4 Catalytic testing
The catalytic performance of synthesized catalysts was tested for *n*-hexane hydroconversion in a micro-catalytic pulse reactor at 423-623 K and ambient pressure under hydrogen stream. Briefly, 0.2 g of catalyst was pretreated with flowing oxygen stream for 1 h and hydrogen stream for another 3 h, both at a flowrate of 100 ml/min and 673 K, it was subsequently cooled down to 423 K in the hydrogen stream. The temperature was allowed to stabilized, then 2 μmol dose of *n*-hexane was passed over the activated catalyst. The reactor effluent was trapped at 77 K, followed by flash-evaporation and the products were analyzed by an online 6090N Agilent Gas Chromatograph equipped with an HP- 5 Capillary Column and an FID detector. The selectivity (Si) of particular product were calculated according to Equations 1.

$$Si(\%) = \frac{[C]i}{\sum[C]i - [C]\ \text{residual}\ n\text{-hexane}} \times 100 \tag{1}$$

Where the molar concentration are represented by $[C]_i$ for a particular product and $[C]_{\text{residual } n\text{-hexane}}$ for residual *n*-hexane.

3. Results and discussion

3.1 Synthesis mechanism

The formation mechanism for fibrous silica mordenite zeolite (FSM) is illustrated in figure 1. Firstly, a typical synthesis involves the formation of microemulsion system consisting of water-toluene-1-butanol by using CTAB as template and urea [4]. Mordenite seed act as primer, the hydrophilic interaction of the CTAB head with the mordenite zeolite forming reverse micelles while the organic phase interacts with the hydrophobic tail. The most important step in the resultant mesoporous structure being the microspheres formed from CTAB template, the negatively-charged surface of mordenite served as the nucleation center. Butanol facilitates the CTAB incorporation on mordenite zeolite at the initial stage while urea which at as a catalyst, provides the basic condition for the hydrolysis of TEOS with TEOS as the precursor [5]. Furthermore, this well oriented arrangement of the CTAB formed a template for hydrolyzed silica growth on the mordenite. Finally, the silanol groups and ethanol obtained as hydrolysis proceed condense to form repeated unit of Si-O-Si (polysiloxane) by cross-linking process. Then, the calcination process removes all the organic matter and surfactants in the template ultimately giving rise to the core-shell structured FSM with radially oriented fibrous silica on the mordenite core.

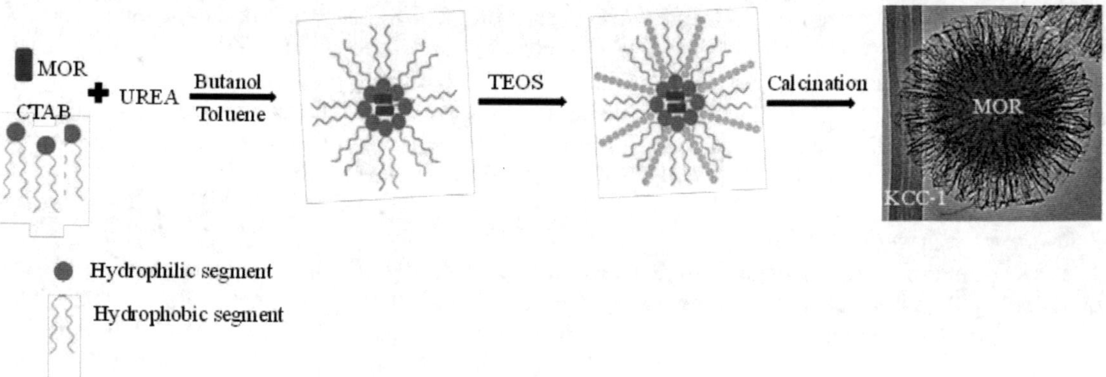

Figure 1. Schematic illustration of the formation process of fibrous silica mordenite (FSM).

3.2 Morphology and crystallinity study

The surface study carried out by field emission scanning electron microscopy (FESEM) revealed that the FSM consist of microspheres with diameter in the range of 650-750 nm with fibrous morphology as depicted in figure 2. The crystalline structure of FSM as determined by XRD analysis is presented. The diffraction peaks at 2θ = 6.55°, 8.85°, 9.76°, 13.52°, 15.40°, 19.70°, 22.42°, 25.73°, 26.44° and 27.64 corresponds to the characteristics peaks for mordenite zeolite as identified by JCPDS data card no. 43-0171[6]. These results confirmed the incorporation of fibrous silica onto mordenite.

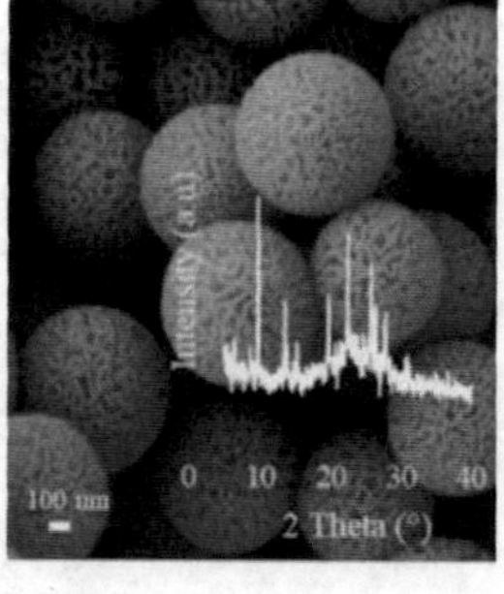

Figure 2. FESEM image and XRD pattern of fibrous silica mordenite.

3.3 Acidity study

The type and strength of acid sites have a strong influence on the catalytic performance in *n*-hexane hydroconversion. In order to investigate acidic properties of the studied catalysts, pyridine as probe molecule was absorbed on the samples. The IR spectra of pyridine absorbed on metal promoted Pt supported protonated FSM (HFSM) catalysts and then desorbed at different temperatures are shown in figure 3. The adsorption band at 1545 cm^{-1} corresponds to pyridinium ions adsorbed on Brønsted acid sites while the adsorption band at 1454 cm^{-1} corresponds to coordinately adsorbed pyridine on Lewis acid sites [7]. It is well known that the framework of mordenite consist of silica (SiO_4) and alumina (AlO_4) tetrahedrally connected. The bridging OH (Si-OH-Al) correspond to the Brønsted acid sites. Besides, the aluminum species not in the zeolite framework described as extra-frame work aluminum (EFAl) as well as the metal modification account for the Lewis acid sites. As shown in figure 3 , the acid strength is determined after desorption of pyridine at 423 K, 523 K and 623 K and the amount of pyridine on the catalyst corresponds to weak, medium and strong acid sites respectively. PtZr/HFSM has the highest amount of Lewis acid sites. Both PtZr/HFSM and PtZn/HFSM displayed a moderate acidity as both catalysts retained about 50% of the pyridine after desorption at 623 K. It is obvious that PtCo/HFSM has the highest amount of Brønsted acid sites and new acidic centers for Lewis acid sites. In addition, it loses small amount of the adsorbed pyridine even at high reaction temperature of 623 K. This is an indication of high concentration of strong acid sites possessed by the catalyst. The PtMo/HFSM also retained about 75% of the adsorbed pyridine at 623 K.

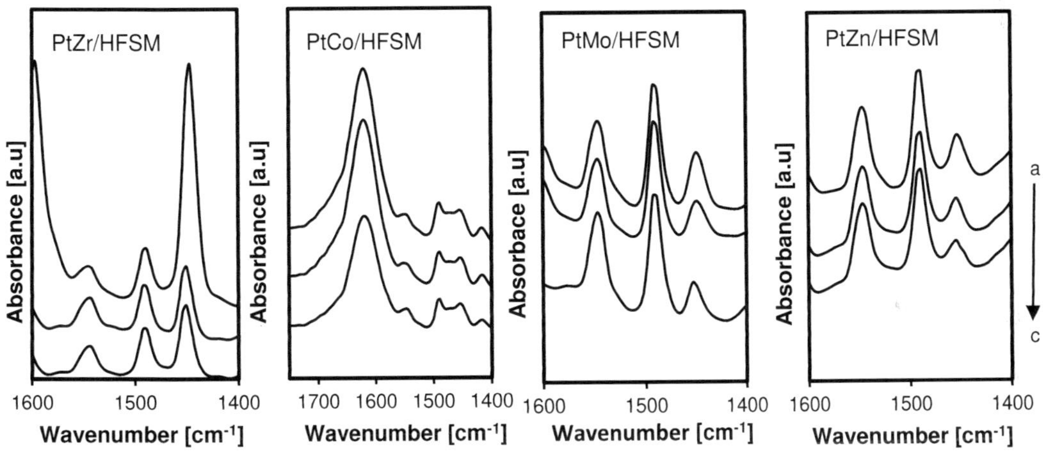

Figure 3. FT-IR spectra of pyridine desorbed at (a) 423 K (b) 523 K and (c) 623 K.

3.4 Catalytic performance

The catalytic activities of bimetallic catalysts promoted with different promoters as a function of reaction temperatures in terms of rate of conversion and selectivity are presented in figure 4 and figure 5 respectively. The activitity in *n*-hexane hydroconversion is in the following order: PtMo > PtZr > PtZn > PtCo. The significant increase in activity for all catalysts was observed with reaction temperature (figure 4). The rate of conversion for PtMo/HFSM was higher by a factor of 10 at 623 K for *n*-hexane hydroconversion.

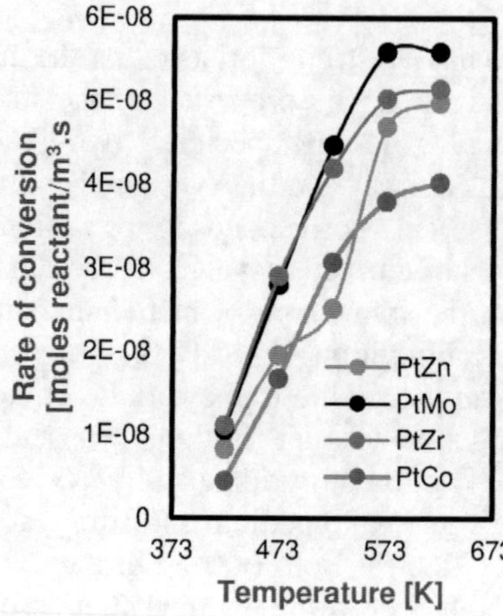

Figure 4. Catalytic performance of all studied catalysts.

For hydroisomerization, the selectivity is in the following order: PtZr >PtZn > PtMo > PtCo for *n*-hexane at all the temperature range investigated. The moderate acidity especially in Lewis acid sites for the PtZr/HFSM catalyst favors hydroisomerization. For hydrocracking activity, the following order was observed PtMo > PtCo > PtZn> PtZr. Meanwhile, for hydrogenolysis, PtCo/HFSM showed the highest selectivity probably due to the strong and abundant Brønsted acid sites (figure 5).

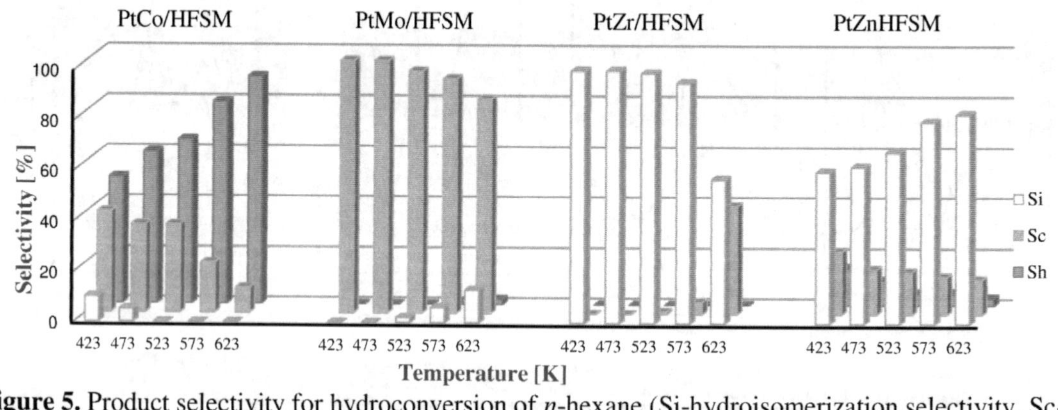

Figure 5. Product selectivity for hydroconversion of *n*-hexane (Si-hydroisomerization selectivity, Sc-cracking selectivity and Sh-hydrogenolysis selectivity).

Scheme 1. depicts the plausible mechanisms of *n*-hexane hydroconversion over metal promoted Pt-HFSM catalysts with different metals. All catalysts were active in the hydroconversion, hydroisomerization proceeds over Zr and Zn promoted catalysts while the cracking reaction was predominate over Mo promoted catalyst. Hydroisomerization process involves the bifunctional mechanism by which the Pt dehydrogenate the *n*-hexane and that is subsequently rearranged by acid catalyzed reaction in which the isohexene were formed [8]. The presence of medium strength acid sites and abundant Lewis acidity was significant in *n*-hexane isomerization, hence the increase in formation of iso-hexane products as the intermediates easily desorbed from the acid sites. The hydroisomerization products (iso-hexane) were generated by the hydrogenation of the iso-hexene intermediates. For the Mo promoted catalyst, cracking reaction proceeds via a dimerization-cracking pathway as observed in the product distribution. The hydroconversion process involves the dimerization of *n*-C6 to form *n*-C12 and subsequent formation of higher hydrocarbons and cracked products. However, for Co-promoted catalyst, the hydrogenolysis of *n*-hexane overwhelmed the reaction chemistry, as the absence of higher hydrocarbon products exclude the possible dimerization-cracking pathway [7]. Most likely, hydrogenolysis occurs on the surface of PtCo/HFSM catalyst with very strong acid sites enabling longer residence time to create the environment for such process to occur.

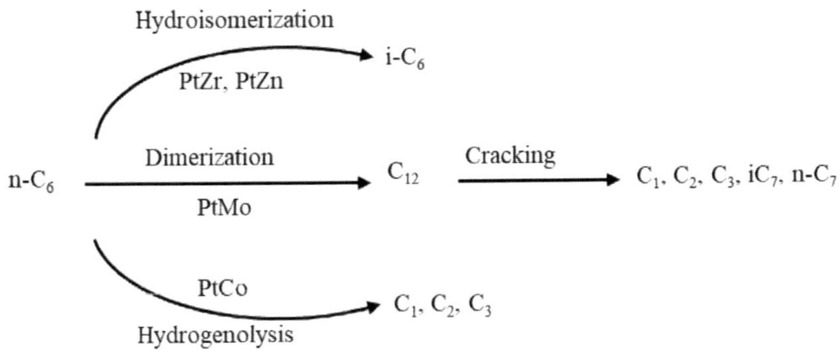

Scheme 1. Plausible reaction mechanisms for *n*-hexane hydroconversion over PtM/HFSM catalysts with different metals (Zr, Zn, Mo and Co).

Acknowledgments
This research study was sponsored by Fundamental Research Grant Scheme (Grant No. FRGS/1/2017/STG01/UTM/01/1) from Ministry of Higher Education Malaysia and also the staff intervention by Tertiary Education Trust Fund through Abubakar Tafawa Balewa University Bauchi, Nigeria.

References

[1] Radlik M, Matus K and Karpiński Z 2019 n-Hexane Hydrogenolysis Behavior of Alumina - Supported Palladium – Platinum Alloys *Catal. Letters* **149** 3176–8
[2] Teh L P, Triwahyono S, Jalil A A, Firmansyah M L, Mamat C R and Majid Z A 2016 Fibrous silica mesoporous ZSM-5 for carbon monoxide methanation *Appl. Catal. A Gen.* **523** 200–8
[3] Jafari M, Rafiei R, Amiri S, Karimi M, Iranshahi D, Rahimpour M R and Mahdiyar H 2013 Combining continuous catalytic regenerative naphtha reformer with thermally coupled concept for improving the process yield *Int. J. Hydrogen Energy* **38** 10327–44
[4] Firmansyah M L, Jalil A A, Triwahyono S, Hamdan H, Salleh M M, Ahmad W F W and Kadja 2016 Synthesis and characterization of fibrous silica ZSM-5 for cumene hydrocracking *Catal. Sci. Technol.* **6** 5178–82
[5] Jalil A A, Gambo Y, Ibrahim M, Abdulrasheed A A, Hassan N S, Nawawi M G M, Asli U A, Hassim M H and Ahmad A 2019 Platinum-promoted fibrous silica Y zeolite with enhanced mass transfer as a highly selective catalyst for *n*-dodecane hydroisomerization *Int. J. Energy Res.* 1–16
[6] Zhang W, Xie J, Hou W, Liu Y, Zhou Y and Wang J 2016 One-pot template-free synthesis of Cu-MOR zeolite toward efficient catalyst support for aerobic oxidation of 5 hydroxymethylfurfural under ambient pressure *ACS Appl. Mater. Interfaces* **8** 23122–32
[7] Setiabudi H D, Triwahyono S, Jalil A A, Kamarudin N H N and Aziz M A A 2011 Effect of iridium loading on HZSM-5 for isomerization of *n*-heptane *J. Nat. Gas Chem.* **20** 477–82
[8] Izan S M, Triwahyono S, Jalil A A, Majid Z A, Fatah N A A, Hamid M Y S and Ibrahim M 2019 Applied Catalysis A , General Additional Lewis acid sites of protonated fi brous silica @ BEA zeolite (HSi @ BEA) improving the generation of protonic acid sites in the isomerization of C 6 alkane and cycloalkanes *Appl. Catal. A, Gen.* **570** 228–37

IOP Conf. Series: Materials Science and Engineering **808** (2020) 012008 doi:10.1088/1757-899X/808/1/012008

Metal Removal from Industrial Waste by Hydrochloric Acid

W C Siaw, M R Musram Rakunman, N Kamaruzaman, T Tsuji and N Abdul Manaf*

Department of Chemical Process Engineering, Malaysia-Japan International Institute of Technology (MJIIT), Universiti Teknologi Malaysia, Kuala Lumpur, Malaysia

*norhuda.kl@utm.my

Abstract. Leaching of metals from industrial solid waste in hydrochloric acid, HCl solution was studied. In this study, the significant of parameters such as HCl concentration, temperature, liquid to solid ratio and leaching time on the removal percentage of metal impurities were investigated using response surface methodology (RSM) with central composite design (CCD). The ANOVA study concluded that the quadratic model was fitted well to the 27 experimental runs based on p-value (<0.0001), R^2 (0.9876) and Adj-R2 (0.9731). This study also revealed that the liquid to solid ratio was the most significant factor for the leaching of metal impurities compared to HCl concentration for this specific industrial solid waste. However, temperature and leaching time showed no significant impact on the leaching rate of metal impurities. Under suitable conditions, a high removal percentage of metal (88 %) was readily achieved at condition of 6 M HCl at 70 °C with liquid to solid ratio of 10 over a leaching period of 70 mins.

1. Introduction

As a highly developing country, Malaysia produces approximately 3.2 million tonnes of industrial solid waste which consist of hazardous materials and heavy metals. This value shows that the rapid industrialization has contributed to the waste generation in which these wastes are non-biodegradable and tend to accumulate in soil. Moreover, these wastes have a long half-life that increases its probability to essentially form a cumulative toxin. This will lead to soil and water pollutions. Ojo et al. [1] stated that industrial solid waste was a serious issue in soil pollution due to its capability to leach through the soil and contaminate the underground water when it comes into contact with acid rain. The long-term exposure of metal dust will lead to Parkinson's-like disease and neurological effects. Furthermore, long term exposure to metal oxide will cause respiratory problems and reproductive effects [2]. Recycling initiatives have been receiving much attention lately. However, in Malaysia, less than 5 % of solid waste is recycled and only 42 % from the total amount of waste is incinerated or chemically treated before dispose to the environment while the remaining 43 % of waste is just dumped in landfill [3]. This kind to disposal attitude or activities has led to the excretion of leachate and methane which can pollute the waterway under the ground. According to the waste statistic generated by the World Bank in year 2010, Malaysia had produced 42.2 % of methane, 25.0 % of carbon dioxide and 10.4 % of nitrous oxide [2]. This harmful greenhouse gaseous generated from solid waste has taking concern from various Malaysian federal and states government and they started to focus on reducing the amount of greenhouse gaseous generation. Therefore, a recent challenge is to find efficient waste utilization processes that can be used to extract the precious elements or metals from industrial solid waste.

Pyrometallurgical and hydrometallurgical methods are the common and conventional methods that use to remove metal impurities or mineral from industrial waste [4]. Pyrometallurgical method is an extraction method that involves thermal treatment that results a physical and chemical transformation in minerals which enable to remove the precious metals [5]. Whereas, hydrometallurgical method offers an interesting alternative for the recovery of metals and minerals by using solvent extraction or leaching method [6]. The most widely medium that a use in hydrometallurgical method is HCl due to it is a low-cost leaching agent and has fast dissolution between metal with Cl^- solution. Halli et al. [7] had investigated removal of heavy metal from the electric arc furnace (EAF) steel slag via HCl leaching. According to their research, 75 % of Fe was able to leach from steel slag with 3.0 M HCl at temperature of 90 °C with 1.18 mm particle size [7]. Besides that, Alafara et al. [8] also studied the leaching kinetic of Nigerian ilmenite ore by HCl acid. In their study, 77% of iron-ore was dissolved within 120 min at a stirring speed of 360 rpm. Additionally, Hernández et al [9] was studied the effect of HCL concentration towards the leaching process. Their analysis indicated that, 80 % of Ti and 85 % of Fe are dissolved as the concentration of HCl increased from 1 M to 6 M. On the other hand, an addition of NH_4Cl in HCl leach solution during leaching of laterite nickel ore was done by Li et al. [10]. In their process, it demonstrated a good leaching rate of Ni (88 %), Co (75 %), Mn (96 %) and Fe (21 %) with 2 M NH_4Cl and 3 M HCl at leaching temperature of 90 °C for 90 mins. They also proved that NH_4Cl exhibited a promising ability to activate H^+, as evident in the simulation result [11]. Ruijing et al. [12] investigated the leaching effects of metals from electroplating sludge by HCl. They indicated that 80.6 % of Cu was leached under 1.5 mol/L HCl, liquid to solid ratio of 10:1 at 40 °C. Besides, a mixed acid solution of HCl and H_3PO_4 also provided good efficient in extracting more valuable metals such as Cu (80.6 %) and Fe (39 %) [12].

Response surface methodology, RSM is the most common and convenient method for design of experiment which can be used to optimize a leaching or extraction process. Sushanta et. al [13] optimized the operating parameters on ash reduction from low-grade coal by using central composite design, CCD. A quadratic model was proposed to correlate the independent variables for maximum ash reduction at optimum condition. The linear effects, square effects and interactive effects of parameters were analyzed by a model plot ANOVA. Their study revealed that HF concentration was the most significant parameters with F-statistic of 933.04 which affect the large extend of ash reduction [13]. Based on their studies, second degree polynomial model was well-fitted to the experimental data at 0.003 p-value and 0.97 R^2 value. Another method from RSM, box Behnken design, BBD method was implemented by Hong et al. [14] with 5 repetition experimental runs to estimate the optimum condition of V extraction from stone-coal. From the ANOVA plot, it displayed that concentration of leach solvent was the most significant parameter with F-value of 84.21 and the model was well-fitted to the experimental data at R^2 of 0.97 [14]. D-optimal design was also one of the RSM method with 4 factors and 3 levels that employed to analyse the response on the leaching of CuS by H_2SO_4 [15]. This design technique had been applied by Soghra et al. [15] to evaluate the effect of particle size, temperature, liquid to solid ratio and redox potential parameters on the efficiency of copper extraction. Their test results indicated that the model was significant to the experimental data at a correlation coefficient, R^2 of 0.96.

In this work, the significant of parameters on the response for instance removal percentage of metal impurities was investigated. The parameter such as HCl concentration, temperature, liquid to solid ratio and time were considered as affecting parameters on the leaching process. RSM based on CCD method is performed to optimize the process parameters. An empirical model is also developed by using the design of experiment to determine the optimum leaching condition and the significance of each parameter towards the response.

2. Material and methods

2.1. Material

An untreated industrial waste was used in this study. A 106 μm of powder slag from industrial waste was prepared through grinding with a mechanical grinder (Vigor mix 700 Swing Type Dry Grinder) and sieved by using GILSON 106 μm stainless steel test sieve. The metal composition in industrial waste

IOP Conf. Series: Materials Science and Engineering **808** (2020) 012008 doi:10.1088/1757-899X/808/1/012008

was then determined by wet chemical analysis using HCl acid digestion followed by inductively coupled plasma optical emission spectrometry, ICP-OES (PerkinElmer NexION 2000 B).

2.2. Leaching experiment
In this study, leaching process with HCl acid was performed by using a round-bottomed flask (750 mL nominal capacity) equipped with a rotary evaporator and a water bath (Heidolph Hei-VAP Platinum 2 Rotary Evaporator). The agitation of the leaching content was carried out at a rotation speed of 200 rpm. The experimental leaching set-up was shown in Figure 1. After leaching process, the solid residue and liquid leachate were separated by vacuum filtration. The elemental composition of metal in liquid leachate was analyzed by ICP-OES. While, the solid residue was washed with distilled water to remove excess Cl^- and filtered in glass funnel by using Whatman filter paper (125 mmΦ) and dried in the hot air oven at 60 °C for 24 hr. Finally, the percentage removal of metal was calculated from Equation. (1).

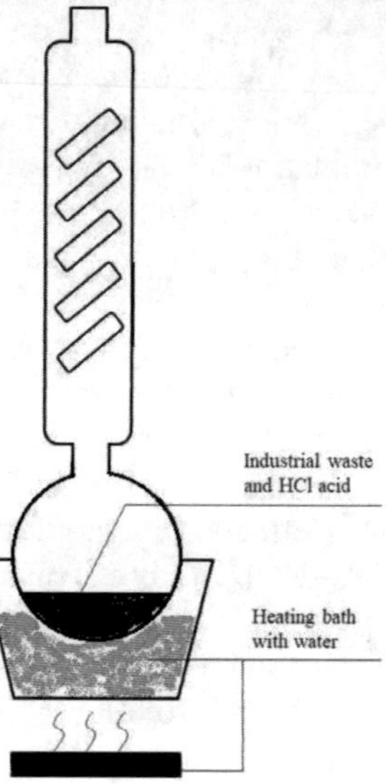

Figure 1. Experimental setup for leaching process

$$Removal\ (\%) = [L] / [R] \times 100\ \% \qquad (1)$$

where *[L]* is the elemental concentration of metal in liquid leachate in ppm and *[R]* is the elemental concentration of metal in raw material (industrial waste) also in ppm.

2.3. Experimental design and statistical analysis
The RSM method is a combination of numerical and statistical technique, which has been commonly used for modelling and optimization study. CCD, one of the RSM methods with 3 levels was adopted to study the main interaction of factors on the response in leaching test. In this study, 4 factors: HCl concentration, M (A), temperature, °C (B), liquid to solid ratio (C) and leaching time, mins (D) were concerned in the leaching tests. The minimum and maximum levels of each input variable were selected in accordance with previous studies [8,12,16]. The coded levels and actual values of input factors were listed in Table 1. The design generated 27 experimental runs which were comprised of 16 factorials points, 8 axial points and 3 centered points as shown in Figure 3. Statistical analysis of the experimental data was performed using Design-Expert 6.0.6 software in which response surface plots were also generated. P-values were used for the significance test of parameters and the accuracy of polynomial

model was determined from R^2 coefficient. When the P-value is smaller than the selected confidence level (e.g. 95%, α = 0.05), it indicated that the parameter tested is statistically significant [17]. The functional relation which also known as mathematical model can be estimated by second-order polynomial (quadratic function) which generally expressed as Equation (2) [18]:

$$Y = \beta_0 + \sum \beta_i X_i + \sum_{i\neq j}^{n} \beta_{i\neq j} X_i X_j + \sum \beta_{ii} X_i^2 + \varepsilon \quad (2)$$

For 4 input variables, the above model can be written as:

$$Y = \beta_0 + \beta_1 X_1 + \beta_2 X_2 + \beta_3 X_3 + \beta_4 X_4 + \beta_{12} X_1 X_2 + \beta_{13} X_1 X_3 + \beta_{14} X_1 X_4 + \beta_{23} X_2 X_3 + \beta_{24} X_2 X_4 + \beta_{34} X_3 X_4 + \beta_{11} X_1^2 + \beta_{22} X_2^2 + \beta_{33} X_3^2 + \beta_{44} X_4^2 + \varepsilon \quad (3)$$

where Y is response variable, X_1, X_2, X_3 and X_4 were manipulated variables.
While β_0 is a constant, β_1, β_2, β_3 and β_4 are linear coefficients. While β_{12}, β_{13}, β_{14}, β_{23}, β_{24} and β_{34} are cross-product coefficients, β_{11}, β_{22}, β_{33} and β_{44} are quadratic coefficients.

Table 1. The coded levels and actual values of input factors.

Factor		Levels		
		Low	center	High
		-α	0	A
A	[HCl], M	2.0	4.0	6.0
B	Temperature, °C	30.0	50.0	70.0
C	Liquid to solid ratio (unit here)	10.0	20.0	30.0
D	Leaching time, mins	60.0	180.0	300.0

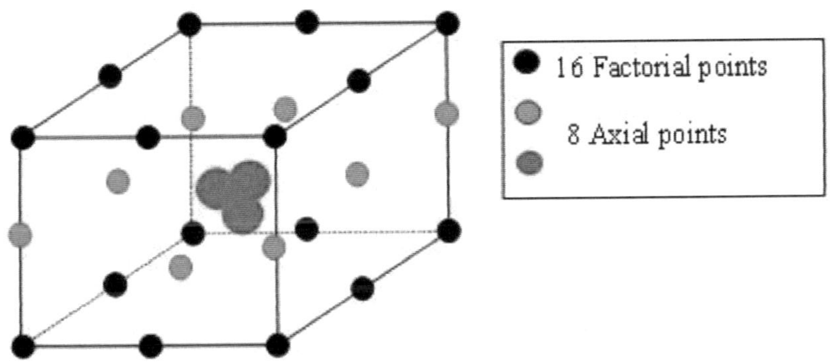

Figure 2. Location of points in the 2^4 factorials design [13]

3. Results and discussion

3.1. Leaching mechanism of HCl acid leaching on metal impurities

The element analysis of industrial waste revealed that more than 90 % of Fe was found in the industrial waste as refer to Table 2. The significant quantity of Fe was presented in the oxide form of the compound such as wuestite (FeO) and magnetite (Fe_3O_4) as refer to Figure 3. The results showed more than 90 % of Fe existed in industrial waste as shown in Table 2.

Table 2. Metal composition of industrial waste sample used in the present study.

Element	Cr	Mn	Ni	B	Fe
Content (%)	2.44	2.72	2.21	1.41	91.22

X-ray diffraction, XRD analysis was used for mineralogy study of industrial waste. The major mineral of the Fe element was reported as magnetite (Fe_3O_4) and wuestite (FeO) as illustrated in Figure 3.

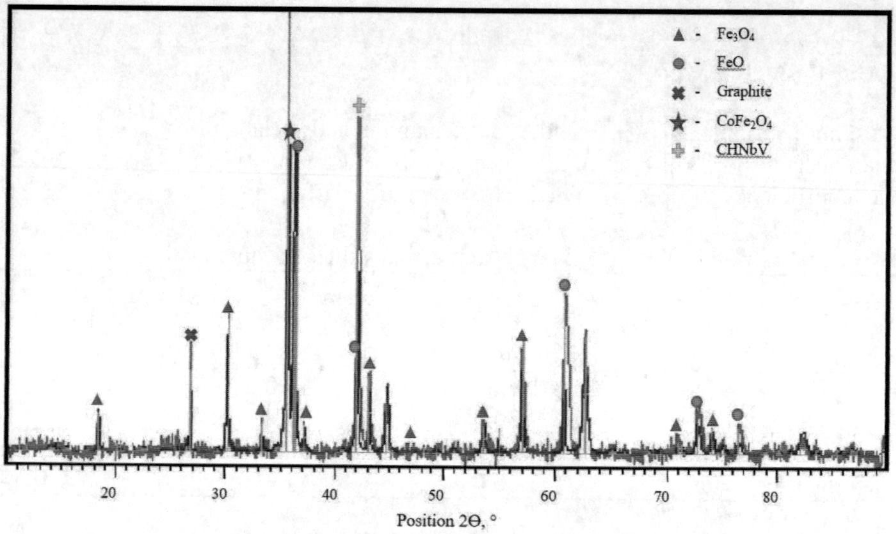

Figure 3. XRD pattern of industrial waste

The leaching process mechanism occurred by the HCl acid diffusion, which acid reacted to the metal impurities and dissolves almost the entire Cr, Mn, Ni, B and Fe components. Acid concentration, temperature, liquid to solid ratio and time were the main parameters that influenced the dissolution behavior of metal during the leaching process. Compared to other chemical solvents (i.e. H_2SO_4, HNO_3), HCl acid was one of the most efficient and well-known reagents that provided a greater potential of elution in industrial applications [19]. This is because HCl provides a high degree of dissolution where high affinity of Cl^- towards the metal in the industrial waste [11]. The major metal (Fe) in waste was reacted with HCl acid during the leaching and formed soluble chloride compounds such as $FeCl_3$ as shown in Equations. (4), (5) and (6) [20].

$$Fe_2O_3\,(s) + HCl\,(aq) \rightarrow 2\,FeCl_3\,(aq) + 3\,H_2O\,(l) \tag{4}$$

$$Fe_2O_3\,(s) + 6\,H^+ + 6\,Cl^- \rightarrow 2\,Fe^{3+} + 6\,Cl^- + 3H_2O\,(l) \tag{5}$$

$$Fe_2O_3\,(s) + 6\,H^+ \rightarrow 2\,Fe^{3+} + 3H_2O\,(l) \tag{6}$$

3.2. Development of model

The leaching experimental data through RSM was evaluated via ANOVA to determine the significance of model [21]. For this leaching study, the required experimental runs with coded level and actual level are given in Table 3, which showed the design of experiment and experimental result. As the output proposed from RSM, the quadratic model was aliased. The empirical model in terms of code factor for removal of metal impurities, (*Y*, %) was shown in Eq. (7):

$$Y = 62.22 + 6.63A + 0.69B - 15.44C + 0.041D - 0.19A^2 + 2.29B^2 - 0.19C^2 - 0.19D^2 - 1.67AB - 0.64AC - 2.24AD - 1.55BC - 1.32BD - 2.35CD \quad (7)$$

Table 3. Experimental factors in coded and actual units and experimental response.

Exp run	Independent factor in coded form				Independent factor in actual form				Metal removal (%)
	A	B	C	D	A (M)	B (°C)	C	D (mins)	
1	+1	-1	+1	-1	6	30	30	60	60.60
2	+1	0	0	0	6	50	20	180	70.35
3	-1	-1	+1	+1	2	30	30	300	42.10
4	-1	-1	-1	+1	2	30	10	300	76.10
5	+1	-1	-1	-1	6	30	10	60	85.04
6	0	0	+1	0	4	50	30	180	48.28
7	0	0	0	0	4	50	20	180	60.30
8	+1	+1	-1	+1	6	70	10	300	85.94
9	+1	+1	-1	-1	6	70	10	60	88.20
10	0	+1	0	0	4	70	20	180	69.10
11	-1	+1	-1	-1	2	70	10	60	72.59
12	-1	-1	+1	-1	2	30	30	60	40.89
13	+1	+1	+1	-1	6	70	30	60	56.19
14	-1	0	0	0	2	50	20	180	56.48
15	0	-1	0	0	4	30	20	180	62.70
16	+1	-1	+1	+1	6	30	30	300	54.24
17	0	0	0	+1	4	50	20	300	63.15
18	0	0	0	0	4	50	20	180	60.30
19	-1	-1	-1	-1	2	30	10	60	61.40
20	0	0	0	0	4	50	20	180	57.70
21	0	0	0	-1	4	50	20	60	63.68
22	+1	+1	+1	+1	6	70	30	300	45.90
23	+1	-1	-1	+1	6	30	10	300	86.70
24	-1	+1	+1	+1	2	70	30	300	41.79
25	0	0	-1	0	4	50	10	180	78.55
26	-1	+1	-1	+1	2	70	10	300	77.90
27	-1	+1	+1	-1	2	70	30	60	44.50

In order to fit the experimental data to the model, the significance of the regression model with individual model coefficient and lack of fit was tested through ANOVA analysis. These significant factors were normally ranked based on the F-values, p-values, R^2 and Adj-R^2 [21]. In Table 4, the ANOVA for the polynomial model is presented. High significant of the corresponding coefficient was shown as in high F-value and low p-value (Prob. > F) [22]. In the leaching experiment, the F-value of 68.17 and p-value of <0.0001 indicated that the model was significant since both values were at a confidence level of 99.99 %.

In the analysis, the contribution of each factor and their interactions are importance. It helps to understand the role and impact of different process conditions on output response. The individual model terms are significant when the p-values were less than 0.05 [18]. Among the individual terms, acid concentration (A) and liquid to solid ratio (C) were ranked as the highest significant with the p-value of

IOP Conf. Series: Materials Science and Engineering **808** (2020) 012008 doi:10.1088/1757-899X/808/1/012008

<0.0001 respectively. Whereas, leaching temperature and time were considered as the non-significant terms with the p-value of 0.2449 and 0.9435 respectively. This analysis indicated that the metal removal rate in leaching test was mainly influenced by acid concentration and liquid to solid ratio. Meanwhile, the interaction terms such as AB, AD, BC, BD and CD were also highly significant with the p-values less than 0.05 and provided a large contribution in the F-value and the total sum of squares.

The fitting of the model is further verified by R^2 and Adj-R^2. The value of R^2 beyond 0.90 defined the model has good agreement between the experimental and the predicted value (metal removal)[16]. The values of R^2 and Adj-R^2 for this fitted model were 0.98 and 0.97, respectively. The value of R^2 indicated that 98.8 % of the total variation of each factor in leaching test can be explained by these linear, interaction and quadratic effects [20,23]. Therefore, it concluded that the model was well fitted to experiment data and highly significant as shown in Figure 4. Adequate Precision is also the main analysis which used to measure the signal-to-noise ratio [13,18]. An adequate precision ratio greater than 4 indicates the model is highly precise [24,25]. From the analysis, the adequate precision ratio was displayed 27.04 in which indicated an adequate signal. Finally, the diagnostic plot illustrated in Fig. 4 showed that there was no outlier presented in the data and the selected model provided good prediction of the response variables.

Table 4. ANOVA results for response surface quadratic model for removal of metal impurities.

Source	Sum of square	Degree of freedom	Mean square	F value	Prob. > F (P value)	Remarks
Model	5400.05	14	385.72	68.17	< 0.0001	Significant
A	792.15	1	792.15	139.99	< 0.0001	Significant
B	8.16	1	8.16	1.50	0.2449	Not significant
C	4291.39	1	4291.39	758.40	< 0.0001	Significant
D	0.030	1	0.030	5.232E-03	0.9435	Not significant
A^2	0.096	1	0.096	0.017	0.8986	Not significant
B^2	13.51	1	13.51	2.39	0.1483	Not significant
C^2	0.096	1	0.096	0.017	0.8986	Not significant
D^2	0.096	1	0.096	0.017	0.8986	Not significant
AB	44.36	1	44.36	7.884	0.0161	Significant
AC	6.55	1	6.55	1.16	0.3030	Not significant
AD	79.92	1	79.92	14.12	0.0027	Significant
BC	38.56	1	38.56	6.82	0.0228	Significant
BD	27.98	1	27.98	4.95	0.0461	Significant
CD	88.17	1	88.17	15.58	0.0019	Significant
Residual	67.90	12	5.66			
Lack of fit	63.39	10	6.345	2.81	0.2906	Not significant
Pure error	4.51	2	2.25			
Total	5467.95					
R^2	0.9876					
Adj-R^2	0.9731					
Adequate precision	27.039					

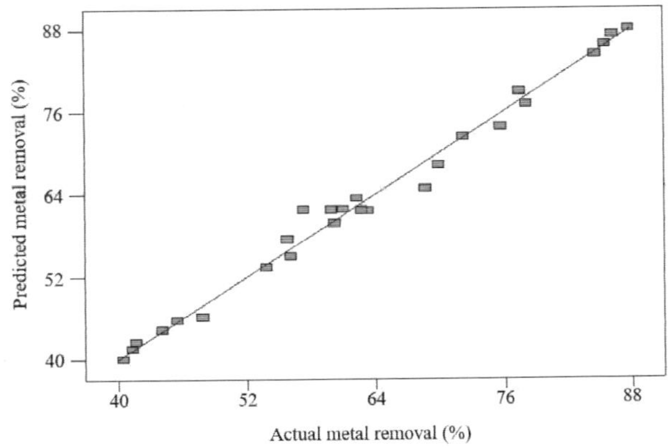

Figure 4. Comparison between the model prediction and experimental data.

3.3. Individual effect of variables

Figure 5 shows the effect of HCl concentration on metal removal in the range of 2 to 6 M for 150 mins. The effect of HCl concentration against metal removal percentage was monitored every 30 mins at the constant temperature of 70 °C and solid to liquid ratio of 1:10. As can be seen, there is a significant increase of metal removal percentage from 30 to 60 mins which the increment of HCl concentration from 2 to 6 M. This prevision can be discussed according to the reaction mechanism in Eqs. (4) and (5) in which Cl^- is generated from HCl acid that acts as a strong activator to dissolve the metals from the industrial waste and it is strongly affecting the dissolution rate of metal [26]. Therefore, it can be stated that high concentration of HCl will increase the amount of Cl^- as activator and react with metal impurities to form soluble compounds, Fe_2Cl_3. These soluble compounds can be easily removed through filtration.

Figure 6 shows the effect of liquid to solid ratio on metal removal in the range of 1:10 to 1:30 M for 150 mins. The effect of liquid to solid ratio against metal removal percentage was monitored every 30 mins at the constant temperature of 70 °C and 6M HCl. As can be seen, there is an increment of metal removal percentage with the reduction of liquid to solid ratio from 1:30 to 1:10. As the liquid to solid ratio reduces, the metal impurities can be dispersed effectively and reacted with acid thereby enhancing the percentage of metal removal. Additionally, the high concentration of HCl assists in reducing the viscosity of the solution which enhances the availability of Cl^- to react with metal impurities since more surface contact between metal impurities and acid are provided. Similar profile of metal removal percentage was found in the studies conducted by [12,27].

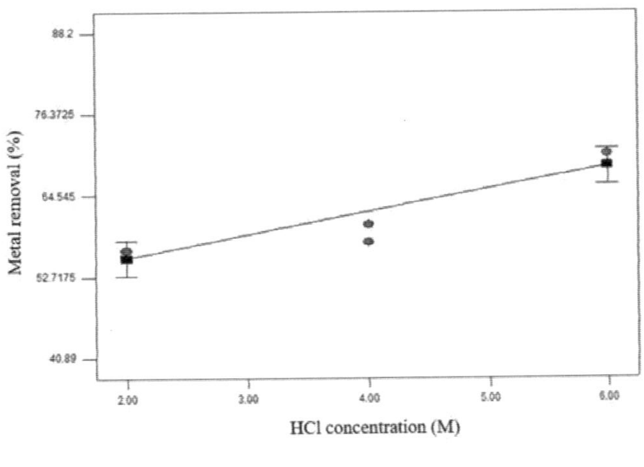

Figure 5. Effect on HCl concentration on metal removal.

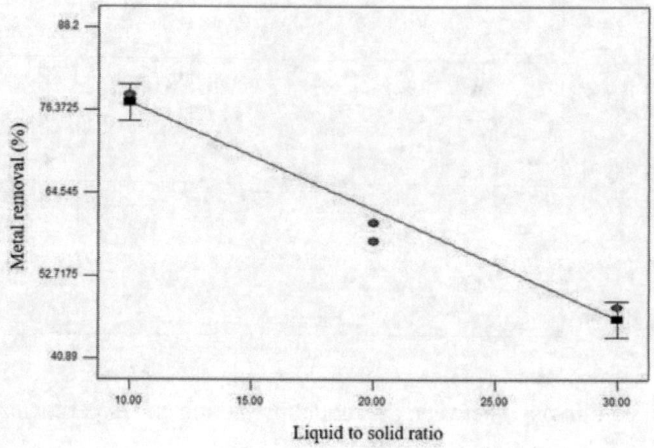

Figure 6. Effect on liquid to solid ratio on metal removal.

3.4. Interaction on parameters

Figure 7 shows the mutual effect of HCl concentration and temperature on metal removal. As can be seen, the metal removal increases with increasing of HCl concentration. In the meanwhile, there is no noticeable change for the metal removal at the entire temperature range (30 °C to 70 °C). At 6 M HCl, higher metal removal has been reported at higher HCl concentrations since more Cl^- ion is generated which enhance the reaction rate.

Figure 8 shows the combined effect of HCl concentration and time on metal removal. As illustrated, leaching time results in increasing the metal removal at 2 M HCl while an inverse trend is observed at 6 M HCl. This is due to less activator (Cl^-) is generated at low acid concentration and more time is required for Cl^- to fully dissolved with metals from industrial waste. However, at high HCl concentration, high amount of Cl^- is formed and this ion can dissolve the metals in short period. As the leaching time goes on, the soluble metal chloride compounds will start precipitate and form solid compound which lead to decrease in the metal removal.

Figure 9 shows the interaction effect between temperature and time on metal removal. As can be seen the leaching behavior illustrated in Fig. 9 is similar to Fig. 8. The metal removal is increasing with time at 30 °C and vice versa at 70 °C. At constant leaching time (60 mins), temperature enhances the rate of reaction since the Fe complexation is decreased at higher temperature at refer to reaction Eq. (4). However, as the leaching time goes on, some bubbles will be found in the solution and start forming a yellow precipitation. These bubbles and precipitations reduce the penetration between acid with metal and hence the leaching rate is reduced.

Figure 10 shows the mutual effect of temperature and liquid to solid ratio metal removal. As can be easily seen, at higher liquid to solid ratio, the metal removal is reduced with increased in amount of solid (industrial waste). Generally, the metal removal increases with reducing the solid densities and viscosity since high amount of leaching acid is added to low content of solid. Bedsides, no noticeable change is observed between temperature and metal removal with constant liquid to solid ratio.

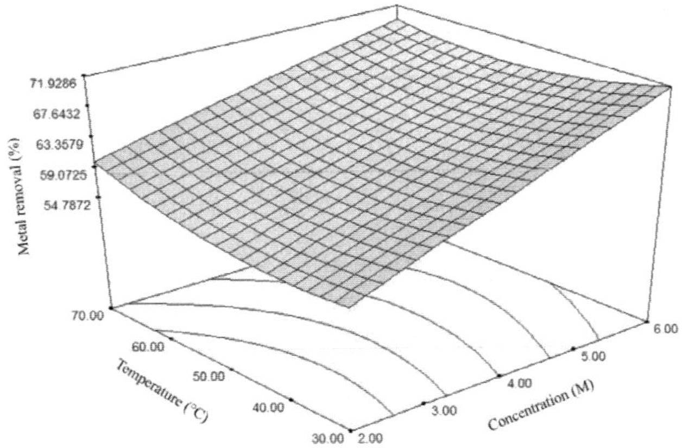

Figure 7. Combined effect of concentration and temperature on metal removal

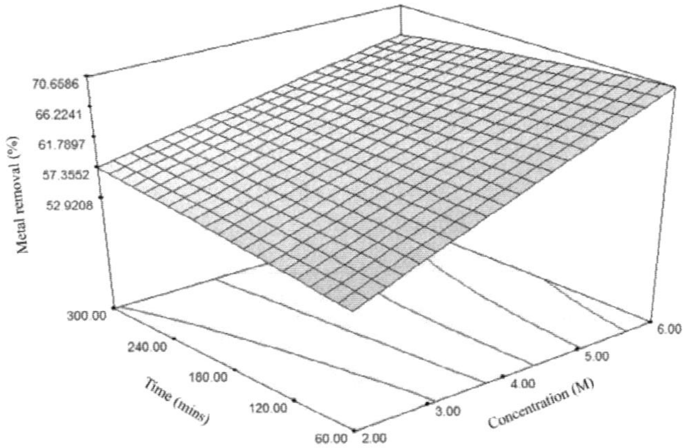

Figure 8. Combined effect of concentration and time on metal removal.

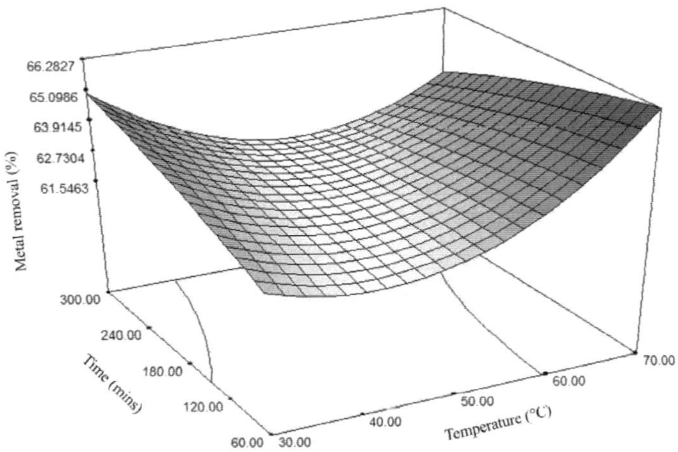

Figure 9. Combined effect of temperature and time on metal removal.

IOP Conf. Series: Materials Science and Engineering **808** (2020) 012008 doi:10.1088/1757-899X/808/1/012008

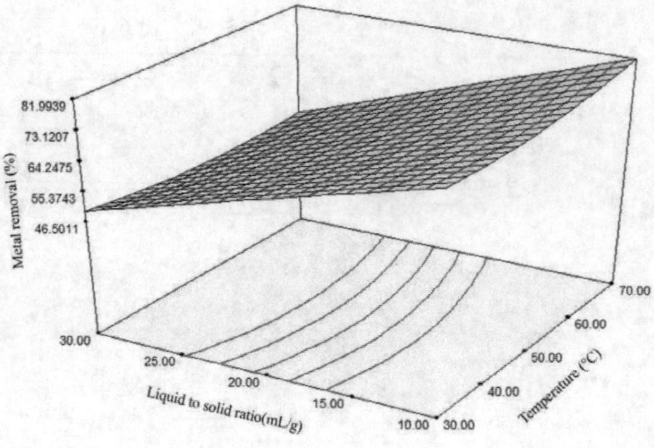

Figure 10. Combined effect of temperature and liquid to solid ratio on metal removal.

3.5. Optimization study

In this study, numerical optimization in CCD design is adopted to determine the optimum leaching condition at maximum of metal removal. In the numerical optimization, the desired goal for each factor and response was chosen. In optimization goals, in range level for each factors and maximum level for response were selected are summarized in Figure 11. The selected optimum operating condition for metal removal from industrial waste via leaching method is at 5.94 M HCl, reaction temperature of 69.47 °C, liquid to solid ratio of 1:10 and leaching time 152.30 mins. This optimum condition was able to obtain 85.4 %. A validation between predicted response and experimental data was calculated by relative standard error, RSE delineated in Eq. (8). The result indicated the error was 2.9 % as shown in Table 6. Thus, it can be concluded that, most of the metal impurities were effectively removed under suggested optimum condition.

$$RSE = \frac{P-E}{P} \times 100\ \% \quad (8)$$

Where P and E is the predicted value and experimental value respectively.

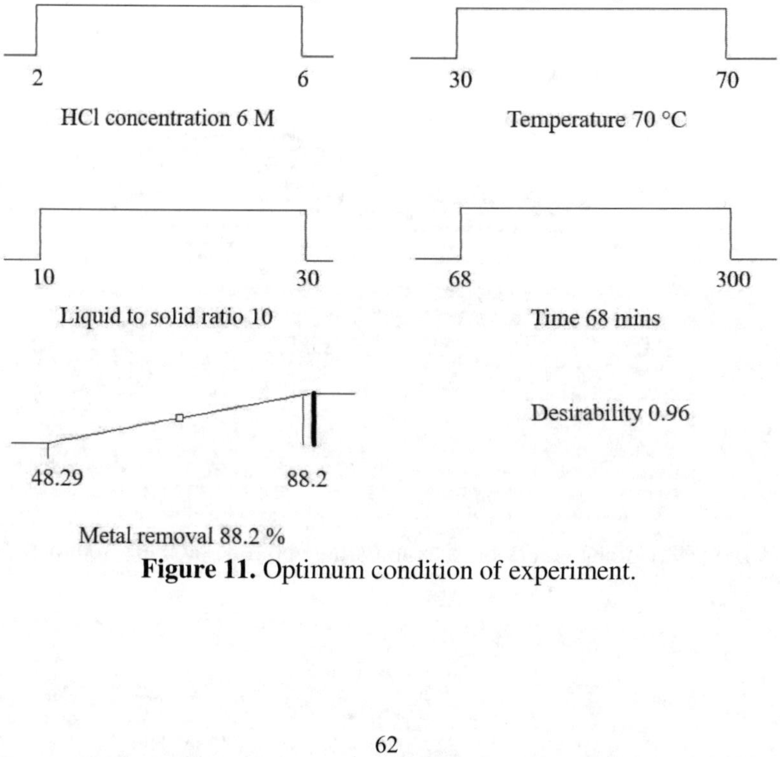

Figure 11. Optimum condition of experiment.

Table 5. Reliability results of optimum conditions.

Predicted value of metal removal (%)	Actual value of metal removal (%)
88.2	85.4

4. Conclusion
Conclusively, HCl leaching method is a efficeincy method for metal removal from industrial waste. RSM with CCD method was carried out to investigate the significant parameters with their interaction on metal removal percentage from industrial waste. Based on the ANOVA analysis the model were found as statistically significant as well with the R^2 and Adj-R^2 values were observed to be 0.98 and 0.97 respectively. Besides, liquid to solid ratio and HCl concentration were identified to be the most significant parameters for the leaching process with p-value of <0.0001. The metal removal linearly increased with the HCl acid concentration and reduction of liquid to solid ratio. The optimum condition was found to be 6 M HCl, 70.00 °C reaction temperature, liquid to solid ratio of 1:10 and 68 mins of leaching time. The predicted and experimental percentages of metal removal were 88.32 % and 85.0 % respectively with the RSE error of 2.9 %.

References
[1] Adekeya E A, Ojo M A and Ajayi O O 2011 Contributions of metal welding workshops to environmental to environmental pollution *J. Environ.* Issues Agric. Dev. Ctries **3** 1–7.

[2] WHO, Hazard prevention and control in the work environment: Airborne dust, *WHO (World Heal. Organ). (1999)* 96. http://www.who.int/occupational_health/publications/en/oehairbornedust.pdf%0Ahttp://www.who.int/occupational_health/publications/airdust/en/.

[3] P.C. Corporation, Less than 1 % of waste out of 3 million tonnes get recycled, *STAR. (2019)* 2018–2020. https://www.thestar.com.my/news/nation/2019/01/24/less-than-1-of-waste-out-of-3-million-tonnes-get-recycled.

[4] Sadri F, Nazari A M and Ghahreman A 2017 A review on the cracking, baking and leaching processes of rare earth element concentrates *J. Rare Earths* **35** 739–752.

[5] Jadhav U U and Hocheng H 2012 A review of recovery of metals from industrial waste Industrial management and organisation *J. Achiev. Mater. Manuf. Eng.* **54** 159–167.

[6] Zhang P, Guo Q, Wei G, Meng L, Han L, Qu J and Qi T 2016 Leaching metals from saprolitic laterite ore using a ferric chloride solution *J. Clean. Prod.* **112** 3531–3539.

[7] Halli P, Hamuyuni J, Revitzer H, and Lundström M 2017 Selection of leaching media for metal dissolution from electric arc furnace dust *J. Clean. Prod.* **164** 265–276.

[8] Alafara A B , Adekola F A, and Folashade AO ***2005*** Quantitative Leaching of a Nigerian Iron Ore in Hydrochloric Acid *Jasem.* **9** 15–20.

[9] Hernández R A H, García F L, Cruz L E H and Luévanos A M 2013 Iron removal from a kaolinitic clay by leaching to obtain high whiteness index *IOP Conf. Ser. Mater. Sci. Eng.* **45**

[10] Li H, Eksteen J, and Oraby E 2018 Hydrometallurgical recovery of metals from waste printed circuit boards (WPCBs): Current status and perspectives – A review *Resour. Conserv. Recycl.* **139** 122–139.

[11] Li J, Li D, Xu Z, Liao C, Liu Y and Zhong B 2018 Selective leaching of valuable metals from laterite nickel ore with ammonium chloride-hydrochloric acid solution *J. Clean. Prod.* **179** 24–30.

[12] Su R, Liang B and Guan J 2016 Leaching Effects of Metal from Electroplating Sludge under Phosphate Participation in Hydrochloric Acid Medium *Procedia Environ. Sci.* **31** 361–365.

[13] Behera S K, Meena H, Chakraborty S and Meikap B C 2018 Application of response surface

methodology (RSM) for optimization of leaching parameters for ash reduction from low-grade coal *Int. J. Min. Sci. Technol.* **28** 621–629.

[14] Liu H, Min Zhang Y, Huang J, Liu T, Nan Xue N and Hua Shi Q 2017 Optimization of vanadium (IV) extraction from stone coal leaching solution by emulsion liquid membrane using response surface methodology *Chem. Eng. Res. Des.* **123** 111–119.

[15] Salehi S, Noaparast M and Shafaei S Z 2016 Response surface methodology (RSM) for optimization of chalcopyrite concentrate leaching with silver-coated pyrite *Physicochem. Probl. Miner. Process.* **52** 1023–1035.

[16] Qiu T, Yan H, Li J, Liu Q and Ai G 2018 Response surface method for optimization of leaching of a low-grade ionic rare earth ore *Powder Technol.* **330** 330–338.

[17] Kumar K A, Sharma G K, Khan M A and Semwal A D 2015 Optimization of Multigrain Premix for High Protein and Dietary Fibre Biscuits Using Response Surface Methodology (RSM) *Food and Nutrition Science* **2015** 747–756.

[18] Javed U, Farooq R, Shehzad F and Khan Z 2018 Optimization of HNO3 leaching of copper from old AMD Athlon processors using response surface methodology *J. Environ. Manage.* **211** 22–27.

[19] Senanayake G 2011 Acid leaching of metals from deep-sea manganese nodules - A critical review of fundamentals and applications *Miner. Eng.* **24** 1379–1396.

[20] Golpayegani M H and Abdollahzadeh A A 2017 Optimization of operating parameters and kinetics for chloride leaching of lead from melting furnace slag *Trans. Nonferrous Met. Soc. China (English Ed. 27 (2017)* 2704–2714.

[21] Zhao Q, Liu C, Li B, Zevenhoven R, Saxén H and Jiang M 2018 Recovery of chromium from residue of sulfuric acid leaching of chromite *Process Saf. Environ. Prot.* **113** 8–87.

[22] Yazici E Y and Deveci H 2014 Ferric sulphate leaching of metals from waste printed circuit boards *Int. J. Miner. Process.* **133** 39–45.

[23] Bohlouli A, Afshar M R, Aboutalebi M R and Seyedein S H 2016 Optimization of tungsten leaching from low manganese wolframite concentrate using Response Surface Methodology (RSM) *Int. J. Refract. Met. Hard Mater.* **61** 107–114.

[24] Rauf M A 2019, Determination of optimum operating conditions of carmine decoloration by UV / H_2O_2 using response surface methodology, *J. Harzad. Mater.* **161** 281–286.

[25] Zhang Z, Peng J, Srinivasakannan C, Zhang Z and Zhang L 2010 Leaching zinc from spent catalyst : Process optimization using response surface methodology *J. Harzad. Mater.* **176** 1113–1117.

[26] Teo Y Y, Lee H S, Low Y C and Choong S W 2018 Hydrometallurgical Extraction of Zinc and Iron from Electric Arc Furnace Dust (EAFD) using Hydrochloric Acid *J. Phy. Sci.* **29** 49–54.

[27] Agatzini-Leonardou S, Oustadakis P, Tsakiridis P E and Markopoulos C 2008 Titanium leaching from red mud by diluted sulfuric acid at atmospheric pressure *J. Hazard. Mater.* **157** 579–586.

IOP Conf. Series: Materials Science and Engineering **808** (2020) 012009 doi:10.1088/1757-899X/808/1/012009

Effect of ligand type on CO_2 adsorption over amine functionalized fibrous adsorbents

Noor Ashikin Mohamad[1,2,*], Nur Afifah Zubair[1,2], Ebrahim Abouzari Lotf[1,3], Mohamed M. Nasef,[1,4,] Arshad Ahmad[1,3], Roshafima Rasit Ali[1,2], TAT Abdullah [1,3], Teo Ming Ting[5]**

[1]Advanced Materials Research Group, Center of Hydrogen Energy, Universiti Teknologi Malaysia, Jalan Semarak 54100, Kuala Lumpur, Malaysia.
[2]Malaysia-Japan International Institute of Technology (MJIIT), Universiti Teknologi Malaysia, Jalan Sultan Yahya Petra, 54100 Kuala Lumpur, Malaysia.
[3]Department of Chemical Engineering, Faculty of Chemical and Energy Engineering, Universiti Teknologi Malaysia, Johor Bahru, 81310, Johor, Malaysia.
[4]Chemical Engineering Department, Universiti Teknologi Petronas, 32610 Seri Iskandar, Perak, Malaysia.
[5]Radiation Processing Technology Division, Malaysian Nuclear Agency, 43000 Kajang, Selangor, Malaysia

*nashikinmohamad@gmail.com, **mohamed.nasef@utp.edu.my

Abstract. In this study, ligands originated from glycidyl methacrylate (GMA) and vinyl benzyl chloride (VBC) monomers were grafted onto fibrous polyethylene/polypropylene (PE/PP) substrate and the resulting grafted copolymers were functionalized with ethylenediamine (EDA). The changes in the chemical structure and crystallinity of the aminated adsorbents were evaluated using Fourier transform infrared spectroscopy (FTIR) and X-ray diffraction (XRD), respectively. The PE/PP-*g*-PGMA-EDA adsorbent showed a CO_2 adsorption capacity of 1.73 mmol/g from pure gas, which is 87% lower than that of PE/PP-*g*-PVBC-EDA (3.24 mmol/g) at 30 bars despite the higher density in the former adsorbent. The effect of temperature, selectivity and CO_2 gas composition was further investigated for the PE/PP-*g*-PVBC-EDA adsorbent.

1. Introduction

Natural gas (NG) is a mixture of a few gas components and it is rich in methane with impurities such as carbon dioxide (CO_2), nitrogen (N_2) and hydrogen-sulphide (H_2S). It can be classified into five categories, such as dry gas, wet gas, sweet gas, sour gas and acid gas. Sour and sweet gas is described as NG with and without significant percentage of H_2S [1]. Sweet NG reservoirs are depleting worldwide and the remaining sour gas reservoirs has been attracting the interest for exploitation to fulfil the increasing demand on NG [2,3]. However, NG comes with up to 80% of CO_2, which is highly corrosive in the presence of moisture causing a rapid corrosion in the pipelines and equipment [4]. Hence, removal of CO_2 from NG is mandatory in order to improve the calorific value and meet the typical pipeline specifications of 2-5% of CO_2 [5]. The highest CO_2 content in NG was found in the Platong and Erawan fields in Thailand with CO_2 content of 90%. Moreover, up to 71% to 87% of CO_2 contents have left an estimated 46 to 13 trillion cubic feet of NG reserves in Indonesia and Malaysia undeveloped due to presence of high CO_2 content. This represents a significant CO_2 capture demand and subsequent storage opportunity in these countries [6].

The large amount of CO_2 from NG should be removed during the offshore NG exploitation especially during the pretreatment to prevent the corrosion to the floating liquefied NG production storage and off-loading platform (LNG-FPSO) equipment. To date, the main technologies for CO_2 removal are absorption, membrane separation and adsorption. Absorption is the best technology for removal of CO_2 from NG [4]. However, it has a number of disadvantages including onshore central plant operation of high cost and corrosions challenges [7]. Moreover, the limited space of LNG-FPSO and the swing caused by waves and typhoons implies that the CO_2 removal plant should be compact, safe and reliable. The limitation in LNG-FPSO has divert the attention towards membrane separation technology due to their modular installation, smaller footprint, lower weight, easy maintenance and minimum utility requirements [8]. However, membranes cannot cope with NG with high CO_2 contents [9]. On the other hand, adsorption is a promising alternative technology due to their simplicity, low operational requirements, low cost, ease of control and high efficiency [10,11]. Hence, this study explores the development of fibrous adsorbent for CO_2 capture based polyethylene coated polypropylene (PE/PP) non-woven sheets.

The fibrous sheets are converted to adsorbent by imparting chemical ligands using the radiation-induced grafting (RIG) technique and subsequent functionalization [12–15]. The inert polymer substrates were grafted with vinyl or acrylic monomers such as glycidyl methacrylate (GMA) and vinyl benzyl chloride (VBC) [16]. The incorporated epoxy group from GMA and methyl chloride group from VBC provide fibrous substrate the ability to host amine group such as ethylenediamine (EDA) in a post-grafting reactions. Although both final adsorbents have same functional groups, but the impact of having aliphatic and aromatic structure of PGMA and PVBC on the adsorption capacity in not known. Thus, the aims of this study are to investigate the effect of the type of ligands incorporated in PE/PP structure by RIG technique and followed by amination with EDA on the CO_2 adsorption performance.

2. Materials and methods

2.1 Chemicals and Reagents

The PE/PP nonwoven sheet was grafted with GMA and VBC monomers according to the reported procedure [17]. The synthesis variables were tuned to confer PE/PP substrates with similar degree of grafting in the range of 156 -166%. Ethanol (99.6%) was purchased from Chemiz and used for washing of samples. EDA (≥99%) was purchased from Sigma-Aldric. Pure CO_2 gas (99.8%) and Pure CH_4 gas (99.995%) were supplied by Linde Malaysia (Sdn Bhd). Deionized water (DI) was obtained from Barnstead Nanopure Diamond Lab Water Purification System (ThermoFisher, Waltham, USA)

2.2 Apparatus and measurements

The changes in the chemical composition after grafting and amination were observed by Fourier transform infrared spectroscopy using a Nicolet iS50 FT-IR spectrometer with spectra obtained in a frequency range of 500-4000 cm^{-1} using 32 scans and 4 cm^{-1} resolutions. The crystalline structural changes of the samples were investigated by X-ray diffraction (XRD) using a PANalytical Empyrean analyzer at a Bragg's angle in the range of 5-70°. All analysis was carried out using samples of pristine PE/PP, PGMA grafted PE/PP (PE/PP-*g*-PGMA), PVBC grafted PE/PP (PE/PP-*g*-PVBC), aminated PE/PP (PE/PP-*g*-PGMA-EDA) and (PE/PP-*g*-PVBC-EDA). CO_2 adsorption capacity of the aminated adsorbent were evaluated using a gravimetric sorption analyzers-isoSORP®, Rubotherm-TA instruments operating at a temperature of 30 °C and a pressure range of 5-30 bar.

2.3 Amine Functionalization

A 0.1 g of PE/PP-*g*-X-EDA, where X stands for PGMA and PVBC substrates were placed in a 100 ml round bottom flask containing 30 ml of pure EDA solution. The reaction was carried out under reflux for 6 h. The EDA-functionalized substrates were removed and washed repeatedly with water and ethanol

IOP Conf. Series: Materials Science and Engineering **808** (2020) 012009 doi:10.1088/1757-899X/808/1/012009

then dried at 60 °C for 5 h. The weight changes of PE/PP-*g*-X were recorded before and after the reaction. The amination percentage was calculated using eqn 1:

$$\text{Percent of amination } (\%) = \frac{(\mathrm{Wa} - \mathrm{Wg})/W_{EDA}}{(\mathrm{Wa} - \mathrm{Wo})/W_X}\ x100\% \qquad (1)$$

Where, W_X and W_{EDA} are the molecular weights of GMA/VBC and EDA (g/mol), respectively. W_a and W_g are the weights (g) of the PE/PP-*g*-X substrates before and after amination of both type of ligands, respectively. The W_o is the weight of PE/PP substrate before grafting.

2.4 CO_2 adsorption tests

0.1 g of PE/PP-*g*-PGMA-EDA and PE/PP-*g*-PVBC-EDA substrates were placed inside the sample container and the adsorption equilibrium of carbon dioxide (CO_2) and methane (CH_4) were performed in a magnetic suspension balance (MSB) of gravimetric sorption analyser-isoSORP® from Rubotherm-TA instruments. Details of the basic principles, components and operational procedure of MSB was described elsewhere [18–20]. Prior to the adsorption, the first step in the experiment procedure is the pre-treatment process which carried out to remove moisture content at temperature 80°C for 2 hours under vacuum condition and after that, buoyancy measurements were performed at 30°C from vacuum up to 30 bar using nitrogen gas with 500 ml/min of flow rate. The last procedure was the adsorption equilibrium measurement which measured at 30°C from vacuum to 30 bar with flowrate 500 ml/min for both pure CO_2 and pure CH_4. The effect of temperature and selectivity study were both conducted at three different temperatures (30°C, 40°C and 50°C) by using the same flowrate and operating pressure. Hence, the best operating temperature was selected to further examine on the performance using different CO_2/CH_4 composition.

3. Results and discussion

Figure 1 shows reaction scheme for preparation of CO_2 fibrous adsorbent by grafting of GMA or VBC onto PE/PP and subsequent functionalization of PE/PP-*g*-X with EDA, where X referred to PGMA and PVBC ligands which were capable of hosting EDA. The presence of amine groups conferred the basicity needed to capture CO_2 molecules (chemisorption) which is categorized as Lewis acid [21]. The percent of amination obtained for PE/PP sample treated with two different ligands such as PGMA and PVBC was 60% and 47.37%, respectively. The values obtained were calculated based on equation (1). The results show that, PGMA sample contains 21% higher concentration of amine compared to PVBC.

IOP Conf. Series: Materials Science and Engineering **808** (2020) 012009 doi:10.1088/1757-899X/808/1/012009

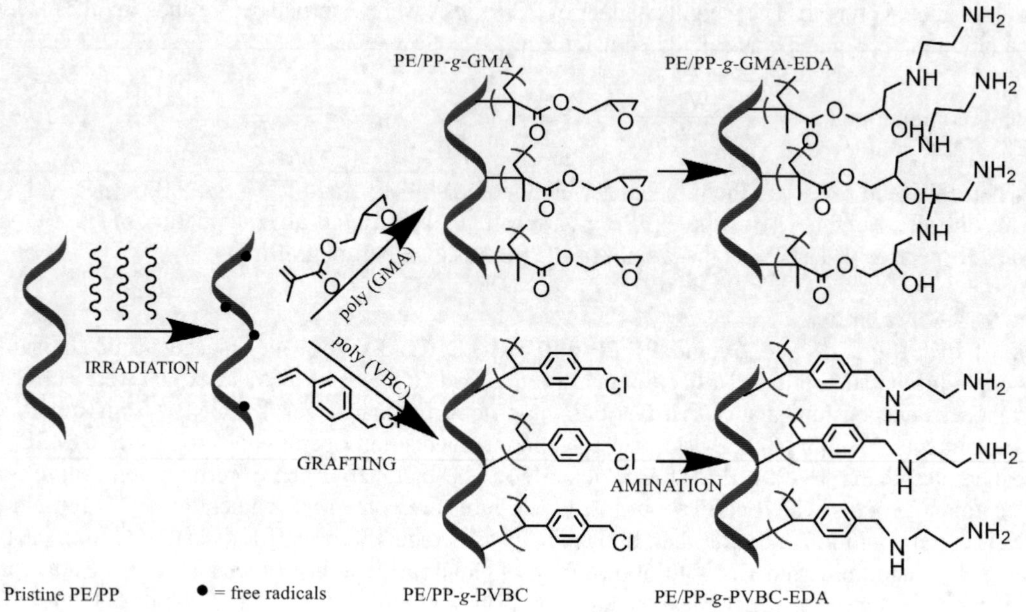

Figure 1. Reaction scheme for grafting of GMA/VBC onto PE/PP and subsequent amination with EDA.

3.1 Chemical and structural changes

Figure 2a shows FTIR spectra of pristine PE/PP and the corresponding PE/PP grafted with PGMA and the aminated adsorbent. The emergence of new peaks at 840 cm^{-1}, 905 cm^{-1} and 1740 cm^{-1} confirmed the presence of epoxy group and –C=O stretching vibration originated from PGMA chains attached to the PE/PP backbone. This proves that PGMA was covalently bonded to PE/PP substrate [12]. The disappearance of the peaks at 840 cm^{-1} and 905 cm^{-1} which is parallel with the appearance of peak at 3500 cm^{-1} assigned for –OH confirm the opening of epoxy ring and the introduction of EDA. This was further evidence from the appearance new peaks 1658 cm^{-1}, 1527 cm^{-1} and a broad band at 3300 cm^{-1} is due to –NH and $-NH_2$ [13,22]. Figure 2b shows FTIR spectra of pristine PE/PP, grafted PE/PP (PE/PP-*g*-PVBC) and aminated PE/PP (PE/PP-*g*-PVBC-EDA). The bands at 838, 702 and 676 cm^{-1} are originated from the *p*-substituents on the aromatic ring and the peak at 1265 cm^{-1} is from -C-Cl stretching vibration, which provide a strong evidence of the successful grafting of VBC onto PE/PP substrate [23,24]. The successful incorporation of EDA in PE/PP-*g*-PVBC was confirmed by the presence of bending vibrations of $-NH_2$ at 1490 cm^{-1} and the bending mode of -CN at 1330 cm^{-1} [25].

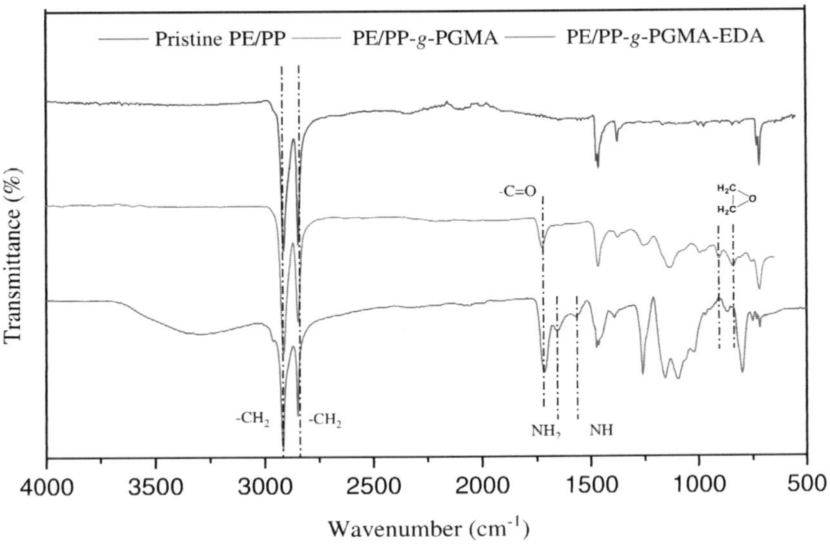

Figure 2(a). FTIR spectra of Pristine PE/PP (black), PE/PP-g-PGMA (red) and PE/PP-g-PGMA-EDA (blue).

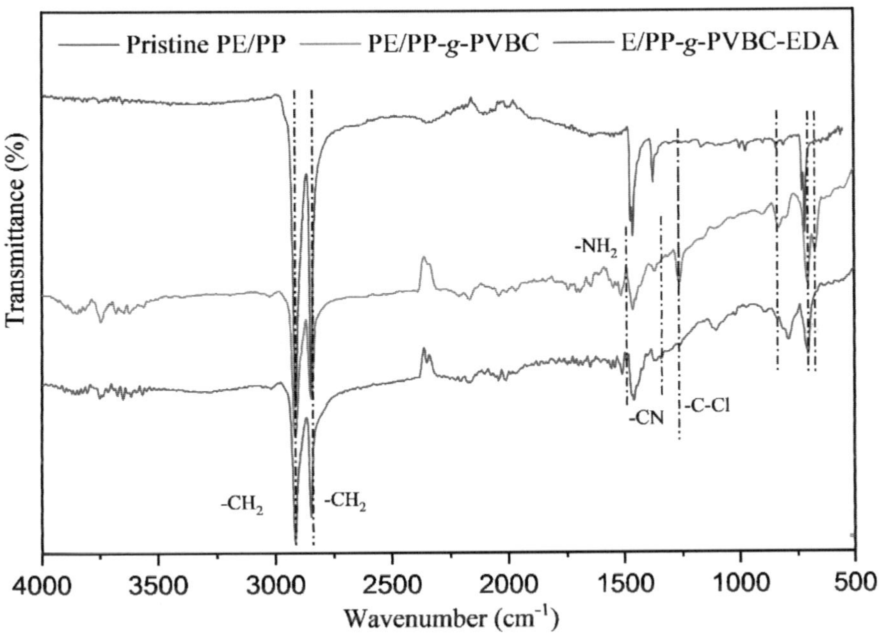

Figure 2(b). FTIR spectra of pristine PE/PP (black), PE/PP-g-PVBC (green) and PE/PP-g-PVBC-EDA (purple)

Figure 3 shows XRD diffractograms of pristine PE/PP, PE/PP-*g*-PGMA, PE/PP-*g*-PGMA-EDA, PE/PP-*g*-PVBC and PE/PP-*g*-PVBCA-EDA samples. The Bragg's angle at each diffraction peaks occurred at the same positions (2θ in the range of 10-30°), indicating that the crystalline structure of pristine PE/PP was well preserved after grafting and subsequent amination. This suggests that grafting and functionalization took place in the amorphous region and is a complete agreement with literature [26,27].

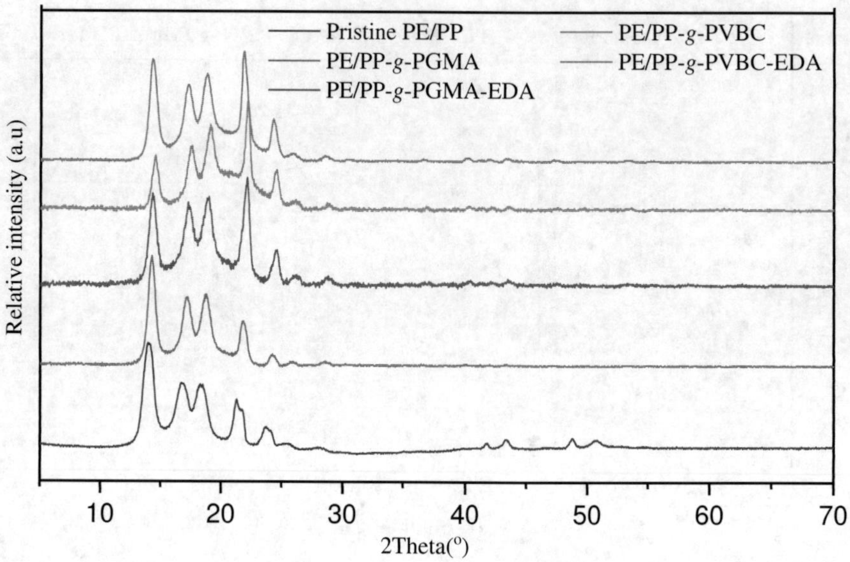

Figure 3. XRD diffractograms of pristine PE/PP (black) PE/PP-g-PGMA (red), PE/PP-g-PGMA-EDA (blue), PE/PP-g-PVBC (green) and PE/PP-g-PVBC-EDA (purple).

3.2 CO_2 adsorption by two adsorbents

Figure 4 shows the Variation of CO_2 adsorption capacity with pressure for pristine PE/PP, PE/PP-g-PGMA, PE/PP-g-PVBC, PE/PP-g-PGMA-EDA and PE/PP-g-PVBC-EDA. The CO_2 adsorption capacities for pristine PE/PP, PE/PP-g-PGMA and PE/PP-g-PVBC samples were included as references. As can be obviously seen, the adsorbent having aminated PVBC ligand showed an adsorption capacity of 3.24 mmol/g whereas that for the adsorbent containing aminated PGMA ligand was 1.73 mmol/g at 30 bar and 30 °C. Such unexpected result shows that the former adsorbent has 87% higher CO_2 adsorption capacity than the latter despite the early observation revealing that the PGMA ligand hosted 21% EDA more than that of PVBC. This likely to be due to the low degree of flexibility of PGMA, which is highly amorphous in nature and tested below it glass transition temperature compared to that of PVBC, which has a rubbery nature. This allowed more CO_2 adsorption in the latter adsorbent than the former.

Additionally, the high amine content of PGMA-containing adsorbent may have caused a negative impact on the CO_2 adsorption performance by blocking and/or reducing the free volume in the adsorbent, thus decreased the gas diffusion rate which leading to lower CO_2 adsorption capacity [28].

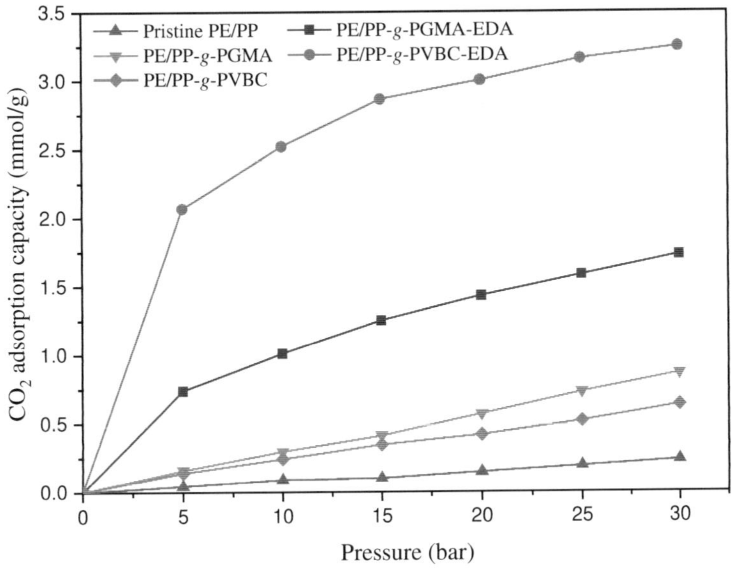

Figure 4. Variation of CO_2 adsorption capacity with pressure for pristine PE/PP (blue), PE/PP-g-PGMA (green), PE/PP-g-PVBC (purple), PE/PP-g-PGMA-EDA(black) and PE/PP-g-PVBC-EDA(red).

The effect of temperature the adsorption capacity of the adsorbent containing aminated was carried out and the data is presented in Figure 5. The adsorption performance was reduced by only 7.56% when operating temperature changed from 30°C to 40°C with adsorption capacity dropping from 3.24 to 3.00 mmol/g at 30 bar. Further rise in the temperature to 50°C showed no impact on the adsorption performance. This trend suggests that the adsorbent is stable up to 50°C and can be operated at such temperature despite the minor decrease in the capacity. Figure 5 also shows the adsorption behavior of the adsorbent toward pure CH_4 gas. Obviously, the CH_4 adsorption is extremely low or negligible and this confirms that the adsorbent is highly selective to CO_2. This is due to the presence of basic group (amine), which is covalently bonded the PE/PP substrate which has a strong affinity to Lewis acids of CO_2 molecules [29,30].

Table 1 provides the data for the selectivity of CO_2 over CH_4 at different pressures and temperatures. The pressure varied from vacuum to 30 bar and the temperature from 30°C to 50°C. Overall, the highest equilibrium selectivity of CO_2 over CH_4 was at 40°C with a huge gap as compared with 30°C and 50°C operating temperatures. Moreover, most of the measured data shows a decrease in the selectivity with the increase in the pressure. This is because at higher pressure, there is only less strong adsorption sites to adsorb CO_2 molecules whereas, at low pressures, more CO_2 molecules occupied the strong adsorption sites [31]. Despite the low CO_2 selectivity at 30°C compared to 40°C, the measured capacity of the adsorbent when pure CH_4 was feed to the system showed negligible capacity at all temperatures. Hence, 30°C was selected as an operating temperature to study the effect of different CO_2 composition on CO_2 adsorption capacity.

IOP Conf. Series: Materials Science and Engineering **808** (2020) 012009 doi:10.1088/1757-899X/808/1/012009

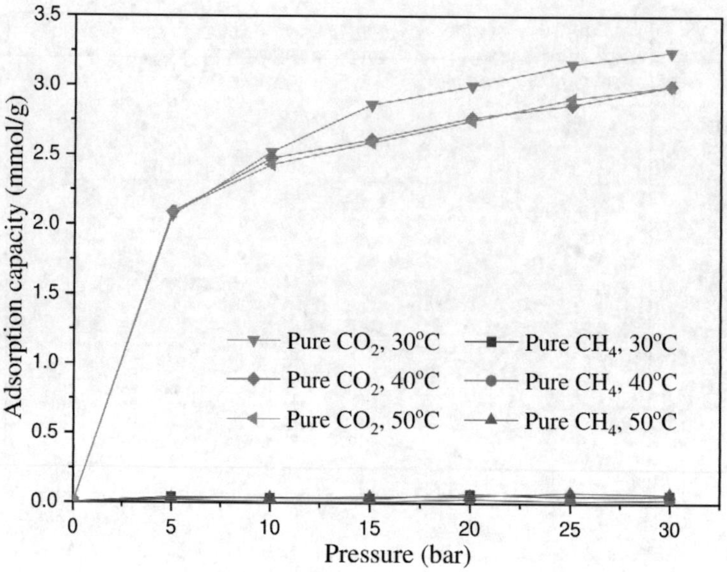

Figure 5. Variation CO_2 adsorption capacity with pressure for PE/PP-g-PVBC-EDA adsorbent with pure CO_2 and pure CH_4 at different temperatures

Table 1. Selectivity of CO_2 over CH_4 at different pressures and temperatures on PE/PP-g-PVBC-EDA

Sample	Pressure (bar)	Adsorption capacity (mmol/g)(CO_2)			Adsorption capacity (mmol/g)(CH_4)			Equilibrium selectivity (CO_2/CH_4)		
		Temperature (°C)			Temperature (°C)			Temperature (°C)		
		30	40	50	30	40	50	30	40	50
PE/PP-*g*-PVBC-EDA	5	2.068	2.090	2.090	0.037	0.018	0.027	55.152	115.000	77.966
	10	2.522	2.477	2.431	0.032	0.005	0.035	79.286	545.000	69.032
	15	2.863	2.613	2.590	0.033	0.007	0.042	86.897	383.333	61.622
	20	2.999	2.772	2.749	0.062	0.030	0.051	48.000	93.846	53.778
	25	3.158	2.863	2.908	0.049	0.014	0.075	64.651	210.000	38.788
	30	3.245	2.999	2.999	0.055	0.026	0.066	59.500	114.783	45.517

Figure 6 shows the variation of CO_2 adsorption capacity with pressure for PE/PP-g-PGMA-EDA with different CO_2/CH_4 compositions. As can be seen, the adsorption capacity increased when CO_2 composition increased from 20-100% at 30°C and 30 bar. This is mainly due to strong interaction and high selectivity between the CO_2 and the amine anchored in the substrate compared to CH_4. As the pressure increases, the diffusion of CO_2 through substrate also increased, hence increased the adsorption capacity [8]. From pure CO_2 to 20% of CO_2 gas composition, the performance was degraded in an increasing order of 4.06% < 5.11% < 8.46% < 16.81%. This might be due to the decreasing of chances for CO_2 to reach the amine active site which might be subjected to physical attachment from CH_4 molecules. This observation reveals that the present adsorbent performance shows a potential for the removal of CO_2 from NG stream containing CO_2 up to 80%.

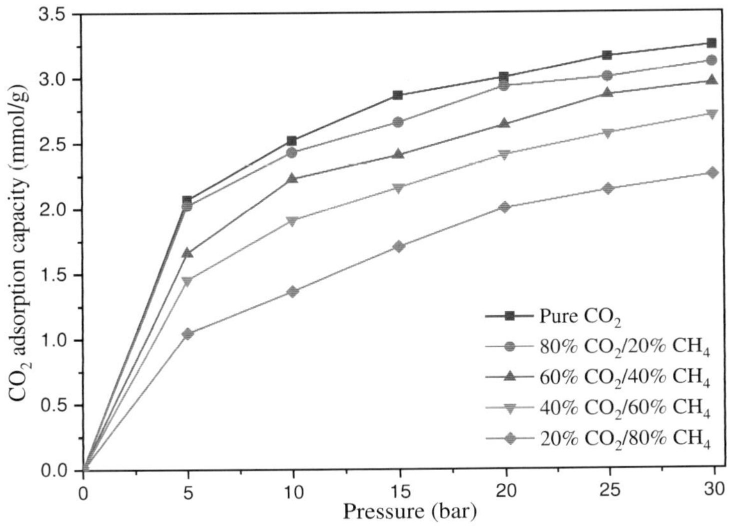

Figure 6. Variation of CO_2 adsorption capacity with pressure for PE/PP-g-PGMA-EDA with different CO_2/CH_4 compositions.

4. Conclusion

The effect of the type of ligands (aminated PGMA and PVBC) incorporated to PE/PP fibrous substrates on the CO_2 adsorption capacity was investigated. The percent of amination showed that the PE/PP-*g*-PGMA-EDA has higher amination value with 60% compared to PE/PP-*g*-PVBC-EDA with 47.37%. However, the CO_2 adsorption capacity performance displayed an opposite trend in which the PE/PP-*g*-PGMA-EDA adsorbent showed a capacity of 1.73 mmol/g compared to3.24 mmol/g for PE/PP-*g*-PVBC-EDA with pure CO_2 gas at 30 bar. Moreover, the equilibrium selectivity (CO_2/CH_4) analysis showed a great value of 114.78 at 40°C and 30 bar. The effect of temperature showed that the optimum CO_2 adsorption capacity was obtained at temperature 30°C. The effect of different CO_2 compositions on the adsorption revealed that the adsorption capacity increased with the increase in the percent of CO_2 in the gas mixture. It can be concluded that the fibrous adsorbent containing aminated PVBC ligands has a strong potential for future application in removal of CO_2 from natural gas.

Acknowledgment

This work was supported by the Universiti Teknologi Malaysia (UTM) and Malaysia Thailand Joint Association (MTJA) under grant (Vot:4C172).

References

[1] Oyeneyin B 2015 *Introduction to the Hydrocarbon Composite Production System* vol 63

[2] Tan L S, Lau K K, Bustam M A and Shariff A M 2012 Removal of high concentration CO_2 from natural gas at elevated pressure via absorption process in packed column *J. Nat. Gas Chem.* **21** 7–10

[3] Tan L S, Shariff A M, Lau K K and Bustam M A 2012 Journal of Industrial and Engineering Chemistry Factors affecting CO_2 absorption efficiency in packed column : A review *J. Ind. Eng. Chem.* **18** 1874–83

[4] Tay W H, Lau K K, Lai L S, Shariff A M and Wang T 2019 Current development and challenges in the intensified absorption technology for natural gas purification at offshore condition *J. Nat. Gas Sci. Eng.* **71** 102977

[5] Safari M, Ghanizadeh A and Montazer-rahmati M M 2009 Optimization of membrane-based CO_2-removal from natural gas using simple models considering both pressure and temperature effects *Int. J. Greenh. gas Control* **3** 3–10

[6] Tan L S, Shariff A M, Lau K K and Bustam M A 2012 Factors affecting CO_2 absorption efficiency in packed column: A review *J. Ind. Eng. Chem.* **18** 1874–83
[7] Jameh A A, Mohammadi T, Bakhtiari O and Mahdyarfar M 2019 Synthesis and modification of Zeolitic Imidazolate Framework (ZIF-8) nanoparticles as highly efficient adsorbent for H_2S and CO_2 removal from natural gas *J. Environ. Chem. Eng.* **7** 103058
[8] Jusoh N, Lau K K, Ying F Y and Shariff A M 2016 Bulk CO_2/CH_4 Separation for Offshore Operating Conditions using Membrane Process *sains Malaysiana* **45** 1707–14
[9] Hatab F A, Abdullatif N, Marzouk S A M and Al-Marzouqi M H 2019 Experimental and modeling of CO_2 removal from gas mixtures using membrane contactors packed with glass beads *Sep. Purif. Technol.* **217** 240–6
[10] Chen S, Tian M, Tao Z, Fu Y, Wang Y and Liu Y 2019 Effect of swing on removing CO_2 from offshore natural gas by adsorption *Chem. Eng. J.*
[11] Abu-Zahra M R M, Abbas Z, Singh P and Feron P 2013 Carbon Dioxide Post-Combustion Capture : Solvent Technologies Overview, Status and Future Directions *Mater. Process. energy Commun. Curr. Res. Technol. Dev.* 923–34
[12] Kavaklı C, Barsbay M, Tilki S, Güven O and Kavaklı P A 2016 Activation of Polyethylene/Polypropylene Nonwoven Fabric by Radiation-Induced Grafting for the Removal of Cr(VI) from Aqueous Solutions *Water. Air. Soil Pollut.* **227**
[13] Mahmoud Nasef M, Abbasi A and Ting T M 2014 New CO_2 adsorbent containing aminated poly(glycidyl methacrylate) grafted onto irradiated PE-PP nonwoven sheet *Radiat. Phys. Chem.* **103** 72–4
[14] Abbasi A, Nasef M M, Faridi-Majidi R, Etesami M, Takeshi M and Abouzari-Lotf E 2018 Highly flexible method for fabrication of poly (Glycidyl Methacrylate) grafted polyolefin nanofiber *Radiat. Phys. Chem.* **151** 283–91
[15] Rojek T, Gubler L, Nasef M M, Abouzari-lotf E and Accepted J 2017 Polyvinylamine containing adsorbent by radiation induced grafting of N-vinylformamide onto UHMWPE films and hydrolysis for CO_2 capture *Ind. Eng. Chem. Res.* **56** 5925–34
[16] Nasef M M and Güven O 2012 Radiation-grafted copolymers for separation and purification purposes: Status, challenges and future directions *Prog. Polym. Sci.* **37** 1597–656
[17] Mahmoud Nasef M, Abbasi A and Ting T M 2014 New CO_2 adsorbent containing aminated poly(glycidyl methacrylate) grafted onto irradiated PE-PP nonwoven sheet *Radiat. Phys. Chem.* **103** 72–4
[18] Reiser S and Türk M 2019 Influence of temperature and high-pressure on the adsorption behavior of CO_2 on MCM-41 and SBA-15 *J. Supercrit. Fluids* **144** 122–33
[19] Chen M, Kang Y, Zhang T, Li X, Wu K and Chen Z 2018 Methane adsorption behavior on shale matrix at in-situ pressure and temperature conditions: Measurement and modeling *Fuel* **228** 39–49
[20] Lee S P, Mellon N, Shariff A M and Leveque J M 2018 High-pressure CO_2-CH_4 selective adsorption on covalent organic polymer *J. Nat. Gas Sci. Eng.* **50** 139–46
[21] Mohamad N A, Lotf E A, Nasef M M and Ahmad A 2019 A comparison of CO_2 adsorption behaviour of mono- and diamine-functionalised adsorbents *E3S Web Conf.* **90** 8
[22] Yu J, Le Y and Cheng B 2012 Fabrication and CO_2 adsorption performance of bimodal porous silica hollow spheres with amine-modified surfaces *RSC Adv.* **2** 6784
[23] Jeong S K, Lee J S, Woo S H, Seo J A and Min B R 2015 Characterization of Anion Exchange Membrane Containing Epoxy Ring and C–Cl Bond Quaternized by Various Amine Groups for Application in Fuel Cells *Energies* **1** 7084–99
[24] Ting T M, Nasef M M and Aw E E L 2019 Kinetics of radiation grafting of glycidyl methacrylate and vinylbenzyl chloride onto polymer fibers *J. Eng. Sci. Technol.* **14** 646–58
[25] Chen Y and Lu D 2014 Journal of Colloid and Interface Science Amine modification on kaolinites to enhance CO_2 adsorption *J. Colloid Interface Sci.* **436** 47–51
[26] Teng Y, Li L, Xu G, Zhang K and Li K 2016 Promoting effect of inorganic alkali on carbon

dioxide adsorption in amine-modified MCM-41 *Energies* **9**

[27] Hashim K 1998 Cation Exchange Membranes by Radiation-Induced Graft Copolymerization of Styrene onto PFA Copolymer Films . I . Preparation and Characterization of the Graft Copolymer *J. Appl. Polym. Sci.* **73** 2095–102

[28] Jahandar Lashaki M, Khiavi S and Sayari A 2019 Stability of amine-functionalized CO_2 adsorbents: A multifaceted puzzle *Chem. Soc. Rev.* **48** 3320–405

[29] Chiang Y, Wu C and Chen Y 2020 Separation and Purification Technology Effects of activation on the properties of electrospun carbon nano fibers and their adsorption performance for carbon dioxide *Sep. Purif. Technol.* **233** 116040

[30] Peng A, Qi S, Liu X, Xue D, Peng S and Yu G 2019 N-doped porous carbons derived from a polymer precursor with a record- high N content : E ffi cient adsorbents for CO_2 capture *Chem. Eng. J.* **372** 656–64

[31] Hong J, Yoon T, Kim S, Kim M, Kim H, Yang H and Bae Y 2019 Microporous and Mesoporous Materials Extraordinarily selective adsorption of CO_2 over N_2 in a polyethyleneimine-impregnated NU-1000 material *Microporous Mesoporous Mater.* **281** 84–91

IOP Conf. Series: Materials Science and Engineering **808** (2020) 012010 doi:10.1088/1757-899X/808/1/012010

Gold extraction via cyanide leaching using alkaline-based empty fruit bunch activated carbon

Nurul Shahiera Shafie*, Nabilah Zaini, and Nurul Farahanim Ali

Shizen Conversion & Separation Technology (SHIZEN), Department of Chemical Process Engineering, Malaysia-Japan International Institute of Technology, Universiti Teknologi Malaysia, Kuala Lumpur, Malaysia

*nshahiera29@gmail.com

Abstract. This research is carried out to extend the knowledge on the modernized method of modifying the physical and structural properties of empty fruit bunches (EFB), a low-cost, highly accessible and abundant agricultural waste in Malaysia, for gold adsorption process. To complete this aim, EFB was modified by using an alkaline-based chemical to improve the surface area, porosity volume and surface chemistry of sorbents. The structure of the empty fruit bunch is characterized by using characterization techniques such as Thermogravimetric analysis (TGA) and Energy-Dispersive X-ray spectroscopy. The gold solution was prepared by mixing gold with a sodium cyanide solution to undergo a leaching process for 24 hours. The solution was then used for the adsorption experiment. The performance of the alkaline-based EFB-AC to separate gold from the cyanide leaching solution was further studied on the effect of contact time, pH and agitation rate. The gold adsorption capacity achieved by the alkaline-based EFB-AC was measured by calculating the difference between the initial and final gold concentrations using Atomic Absorption Spectrometry (AAS). The maximum gold adsorption capacity findings of contact time, pH, and agitation rate were found to be 2.5 hours, pH of 10 and 60 rpm, respectively. This study reveals that the alkaline-based EFB-AC has the potential to extract gold as an alternative activated carbon in the gold adsorption process.

1. Introduction

Gold is a valuable metal that has been commonly used in various fields such as the electrical and electronic industries, jewellery, and medicine, as well as a catalyst in many chemical processes [1]. Gold is an asset that has several attractive properties such as chemical stability, high redox potential and a good conductor for heat and electricity [2]. According to an executive at the Malaysian Chamber of Mines, the gold industry in Malaysia has grown by three-fold in the past decade from RM221 million in 2006 to RM780 million in 2015. Gold mining in Malaysia can be found in several regions including Pahang, Johor, Sabah, Terengganu, and Kelantan [3]. Malaysia exports around 50 tonnes of gold jewellery annually to Middle Eastern countries like the United Arab Emirates and Turkey, an amount which is worth RM6 to RM7 billion [4].

There are several techniques and processes for gold extraction from the mineral mixture. The techniques include amalgamation, sluicing, magnetic, centrifuge and hydrometallurgy [5]. Cyanide leaching treats gold in a hydrometallurgical process from its ores where the processes rely on the recovery of gold from the enriched gold leach pulp by a suitable activated carbon [6]. This process has high sensitivity, a high percentage of gold extraction and low operational cost [7]. There are three types of carbon adsorption which are carbon-in-pulp (CIP), carbon-in-leach (CIL) and carbon-in-column (CIC).

Activated carbon derived from the coconut shell has been the preferred adsorbent due to its selectivity for gold [8]. However, an alternative activated carbon should be explored from other abundantly available and cost-effective materials that can be found in Malaysia. The Malaysian palm oil industry is one of the largest in the world and produces large amounts of by-products, mainly empty fruit bunches [9]. The empty fruit bunch has a high carbon content and hydroxyl groups which can be modified to improve its adsorption potential. In this study, gold adsorption is done using empty fruit bunches to analyse whether they can be used to be an alternative activated carbon for gold adsorption in the gold mining industry.

2. Materials and methods

2.1. Materials

An empty fruit bunch was selected as a precursor in this study and this material underwent physical modifications via carbonization and activation steps as previously reported [10]. Potassium hydroxide (KOH) was used to impregnate the empty fruit bunch to produce alkaline-based EFB-AC. Sodium cyanide (NaCN), AR, 97%, was purchased from Evergreen Engineering & Resources and was used for cyanide leaching. Gold samples were collected from a gold mine in Pahang.

2.2. Experimental procedures

Gold was mixed with sodium hydroxide (NaOH) to adjust the pH, then sodium cyanide with a concentration of 1 g/L was prepared before being added to the mixture. The solution was left to undergo the leaching process for 24 hours and filtered to obtain a gold cyanide solution. Then, the gold concentration was analyzed by using Atomic Absorption Spectrometry (AAS).

The empty fruit bunch was characterized structurally by using a thermogravimetric analyzer (TGA), and electron dispersive x-ray (EDX). Thermogravimetric analysis (TGA) is an analysis technique used to determine the changes in the weight of each component of the sample as the temperature is increased. The electron dispersive x-ray (EDX) is used by detecting the x-ray reflected from the sample during the injection of an electron beam to determine the elemental compositions.

Afterwards, the adsorption experiments were performed by placing the prepared alkaline-based EFB-AC in 50mL of the gold cyanide solution for two hours. The effects of pH, agitation rate and adsorption time were studied and evaluated in this experiment. The solution was then filtered, and AAS was used to analyze the gold concentration.

3. Results and discussion

3.1. Thermogravimetric analysis (TGA)

Thermogravimetric analysis (TGA) was used to determine the changes in the weight of the samples with respect to the increasing temperature [11]. The x-axis indicates the temperature (°C), while the y-axis represents the percentage of weight loss. The temperature range for this analysis was between 20 to 1000°C. Figure 1 shows the TGA curve of the unmodified empty fruit bunch (EFB).

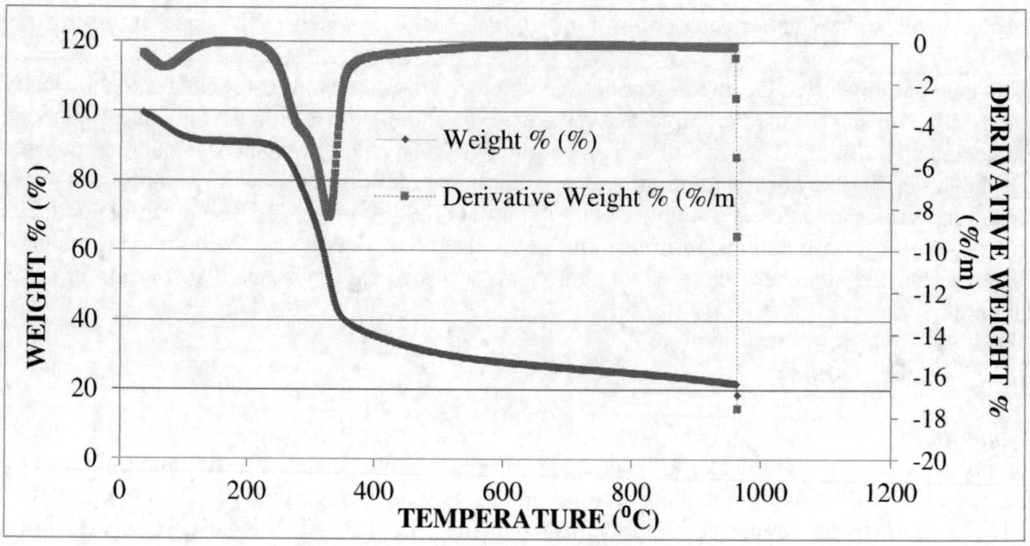

Figure 1. TGA curve for unmodified empty fruit bunch.

Three degradation steps are observed which signify the weight loss due to the removal of moisture content, removal of cellulose and hemicellulose contents, and the formation of char, respectively [12]. The first degradation observed at a temperature range between 20 to 100°C is smooth and slightly bent downward. This TGA result indicates the removal process of moisture and extractives. The second degradation range for the empty fruit bunch occurs at 250 to 350°C where the weight loss can be explained by the degradation of hemicelluloses, lignin, and celluloses. The production of char empty fruit bunch takes place in the third stage which occurs at a temperature range of 390 to 1000°C.

3.2. Energy dispersive X-ray spectroscopy (EDX)

Energy Dispersive X-Ray Spectroscopy (EDX) serves to detect the elemental compositions of the prepared sample [13]. Table 1 shows the elemental compositions of an empty fruit bunch in different states which are raw, carbonized, and activated carbon.

Table 1. Elemental compositions of empty fruit bunch.

Sample	Empty Fruit Bunch (EFB)		
Elemental Composition (%)	Raw	Char	Activated Carbon
Carbon (C)	54.85	51.03	63.88
Oxygen (O)	41.27	9.78	10.24
Sodium (Na)	-	2.54	0.85
Magnesium (Mg)	0.94	1.91	2.46
Potassium (K)	2.40	32.21	18.45
Calcium (Ca)	0.54	2.54	2.02

Table 1 indicates that an empty fruit bunch is categorized as a natural material as the elemental composition results indicates the presence of cellulose, hemicellulose and lignin. Table 1 shows that the elemental composition for carbon (C) and oxygen (O) for the raw empty fruit bunch are 54.855 and 41.27%, respectively. The increment of carbon content in the empty fruit bunch activated carbon correlates with the presence of inert gas which improves the carbon composition [13]. The high composition of carbon affects the adsorption of gold as it is influenced by the exchange with the charged groups that exist in the surface of activated carbon [14].

3.3 Gold adsorption study

Gold adsorption via the cyanide leaching process was determined through the sorption of gold onto alkaline-based EFB-AC. Figure 2, Figure 3 and Figure 4 show the percentage of gold adsorption capacity in terms of the effect of time adsorption, effect of pH and effect of agitation rate, respectively.

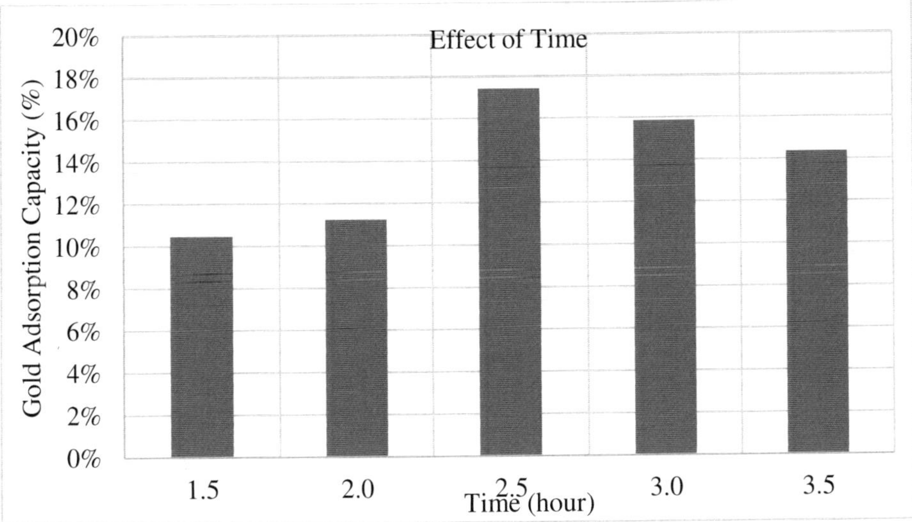

Figure 2. Gold adsorption capacity from effect of time.

Figure 2 presents the effect of contact time on the gold adsorption capacity between the range of 1.5 to 3.5 hours. It can be observed that the activated carbon empty fruit bunch adsorbent has a much higher adsorption capacity at 2.5 hours. The percentage of adsorption increases from 1.5 to 2.5 hours then decreases slightly. This is due to the vacant adsorption sites on the sorbents at the beginning and as the sites are filled, the sorption decreases. As the adsorption sites become filled after a while, there are repulsive forces between the solute molecules on the solid and bulk phases which makes the sorption decreases [15]. 2.5 hours is the highest gold adsorption capacity recorded, which is why the time for study on the effect of pH and the effect of agitation rate conducted for 2.5 hours.

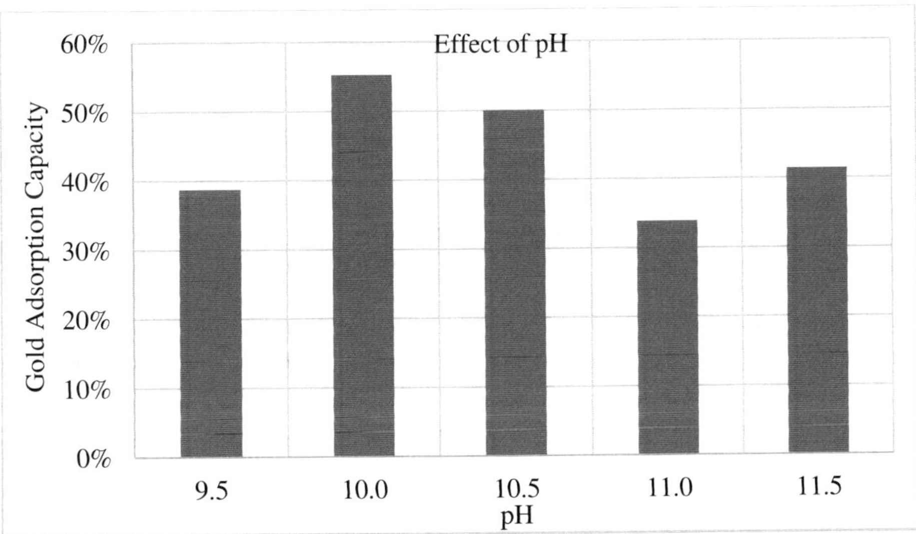

Figure 3. Gold adsorption capacity from effect of pH.

In this study, the performance of the alkaline-based EFB-AC was studied in the pH between 9.5 to 11.5 to prevent hydrogen cyanide gas formation. The gold adsorption capacity of the alkaline-based EFB-AC affects the surface charges of the adsorbent and the speciation of the metal adsorbate is significantly affected by the pH as can be seen in Figure 3. The results suggest that gold adsorption is optimal at pH 10 and decreased from pH of 10.5. Based on the pH analysis, the gold adsorption capacity of the alkaline-based EFB-AC was affected greatly due to the surface charges of the adsorbent and the speciation of the metal adsorbate [16]. At a very basic pH solution, gold cyanide ion may be deprotonated, which can cause repulsive interaction between the charge of alkaline-based EFB-AC and gold cyanide.

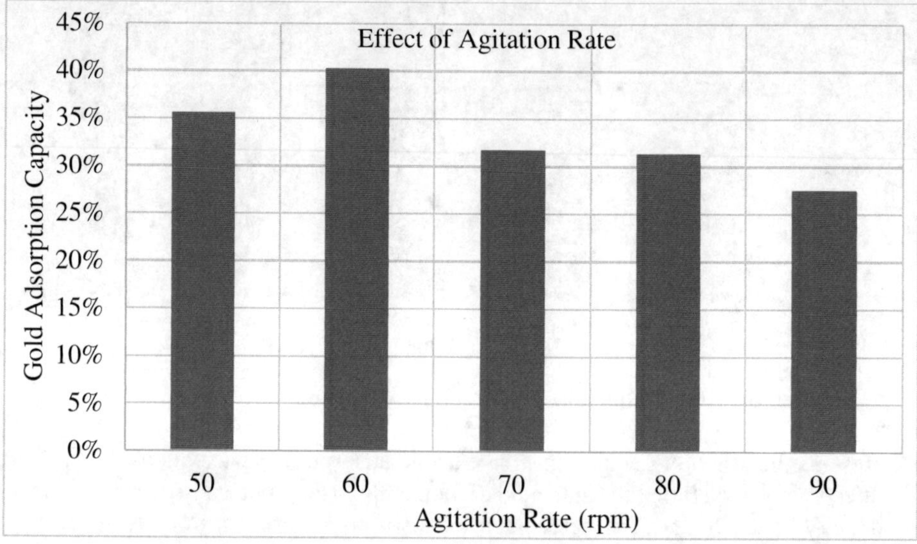

Figure 4. Gold adsorption capacity from effect of agitation rate.

Figure 4 shows the effect of agitation rate of 50 to 90 rpm towards gold sorption onto alkaline-based EFB-AC. The maximum gold adsorption is achieved at 60 rpm. The percentage capacity of gold adsorption increases until 60 rpm and decreases from 70 to 90 rpm. This is due to the bond breaking between the gold trapped onto the alkaline-based EFB-AC as there is a decrease in the boundary layer thickness around the adsorbent particle as a higher agitation rate is used [17]. Besides, it is possible that with less of the CN- being available for reactions with the activated carbon surface groups ultimately resulted in a decrease of gold adsorption.

4. Conclusions

The empty fruit bunch was chosen as a precursor for this research due to its high potential. Through carbonization, the empty fruit bunch was physically modified which enhanced the adsorption process. Structural analysis was used to characterize the samples through energy dispersive x-ray spectroscopy (EDX) and thermogravimetric analysis (TGA). Based on the TGA and EDX results, the transformation of carbonized samples to activated carbon is proven. The percentage of gold adsorption towards the empty fruit bunch activated carbon in terms of the effect of pH, agitation rate and adsorption time has been studied. The maximum gold adsorption capacity for effect of pH, agitation rate and adsorption time were found to be 10, 60 rpm and 2.5 hours, respectively.

References

[1] Khosravi R, Azizi A, Ghaedrahmati R, Gupta V K and Agarwal S 2017 *J. Industrial & Eng. Chem.* **54** 464-471
[2] Pita M, Gutierrez-Sanchez C, Olea D, Velez M, Garcia-Diego C, Shleev S and De Lacey A L 2011 *J. Phys. Chem.* **115** 13420-13428
[3] Malaysia Delegation 2008 *Thailand: Coordinating Committee for Geoscience Programmes in East and Southeast Asia (CCOP)*
[4] Siow E 2018 *Singapore Bullion Market Association*
[5] Eugene W W L and Mujumdar A S 2009 *Minerals, Metals, and Materials Technology Centre, National University of Singapore.*
[6] Gupta C K 2017 *Hydrometallurgy in Extraction Processes* Volume II Routledge
[7] Yanuar E 2015 *Procedia Chemistry* **17** 59-65
[8] Soleimani M and Kaghazchi T 2008 *Chinese J. Chem. Eng.* **16** 112-118
[9] Chang S H 2014 *Biomass and Bioenergy* **62** 174-181
[10] Nasri N S, Sazali N A, Hamza U D and Anirman, N L *Advances in Environmental Biology* **9** 62-66
[11] Shahedifar V and Rezadoust A M 2013 *J. Reinforced Plastics and Composites* **32** 681-688
[12] Sulaiman, F., & Abdullah, N. (2014). Pyrolytic product of washed and unwashed oil palm wastes by slow thermal conversion process. Journal of Physical Science, 25(2), 73.
[13] Polini A and Yang F 2017 *Nanofiber Composites for Biomedical Applications* 97-115 Woodhead Publishing
[14] Adhoum N and Monser L 2002 *Chemical Engineering and Processing: Process Intensification* **41** 17-21
[15] Morcali M H Zeytuncu B Ozlem E and Aktas S 2015 *Materials Research* **18** 660-667
[16] Srivastava V C Mall I D and Mishra I M 2006 *Journal of hazardous materials* **134** 257-267
[17] Zahoor M 2011 *Journal of the Chemical Society of Pakistan* **33** 305-312

IOP Conf. Series: Materials Science and Engineering **808** (2020) 012011 doi:10.1088/1757-899X/808/1/012011

Renewable biogas from anaerobic digestion of biomass: influence factors in life cycle assessment

R Abu[1], M A A Aziz[1, 2,*], C H C Hassan[3] and Z Z Noor[1, 3]

[1]School of Chemical and Energy Engineering, Faculty of Engineering, Universiti Teknologi Malaysia (UTM), 81310 UTM Johor Bahru, Johor, Malaysia
[2]Centre of Hydrogen Energy, Institute of Future Energy, Universiti Teknologi Malaysia (UTM), 81310 UTM Johor Bahru, Johor, Malaysia
[3]Centre for Environmental Sustainability and Water Security (IPASA), Research Institute of Sustainable Environment (RISE), Universiti Teknologi Malaysia (UTM), 81310 UTM Johor Bahru, Johor, Malaysia

*m.arif@utm.my

Abstract. Based on the current generation rate of 1 kg^{-1} $person^{-1}$, the production of organic waste in Malaysia is estimated to reach 9 metric tons per year by the year 2020. Components of these wastes, however, can be used to generate biogas, not only to decrease waste-related issues, but also to produce renewable energy. There is a growing interest in resource recovery and waste/energy integration through biogas generation from organic waste through anaerobic digestion method. However, due to the anaerobic digestion process varies in different facilities, thus proactive assessment on the status of biogas production and its effect on the environment through life cycle assessment is vital. The objective of this review is to assess factors that affect environmental performance results such as the system boundaries setting, the databases used and the life cycle impact assessment methods applied. This review underscores the fact that goal definition and scope, functional units, system boundaries setting, characterization and life cycle impact assessment methodology, as well as types of software and databases used influence and affects the life cycle assessment results. It suggests that for future cross study comparisons, all assessment guidelines should be addressed to avoid biased comparisons on climate performance between different alternatives.

1. Introduction

Biomass is classified as a renewable source that comprises of water and carbon dioxide, utilizing daylight as a source of energy. This contains food, harvested crop, trees, farm and forestry residues, aquatic plants, animal and municipal waste [1, 2]. The environmental benefits of using biomass are derived from its cycle of life. Output source on biomass could, in theory, biodegrade into carbon dioxide and water toward their finish cycle, preferably generating no "additional" carbon dioxide. Renewable biomass feed stocks are grains, oils, fats, proteins, and lignin's from a chemical perspective [3].

Food waste (FW) is a source of biomass. It is the largest element of municipal solid waste (MSW) that comprises restaurants and canteen waste, households, as well as food processing plants. These large generations of FW brings about a wide range of environmental and sanitary problems, and this should not be ignored. It is expected that by 2025, FW volume in Asian countries will increase to 4.16 billion tons compared to the current 2.78 billion tons [4]. Every year, as stated by Xiao et al. [5], the foods produced are lost at about one-third over the whole food supply chain, and this is estimated to be around 1.3 billion tons of food. In 2018, FW in Malaysia reached nearly 16,700 metric tons on a daily basis. This includes 4000 metric tons that are still suitable for consumption, and should not be disposed [6]. Generally, MSW in Malaysia comprises 50% FW, of which, 70% is dumped in a landfill site [7].

Currently, several municipalities in Malaysia, like for example Petaling Jaya city council, have been utilizing anaerobic digestion (AD) dry continuous type (Cowtech technology) which is capable of processing approximately 500 kg-1ton/day of FW [8]. Other city councils such as Subang Jaya, Sepang, and Putrajaya are utilizing anaerobic waste digesting system facility to treat FW. The resulting biogas is used for direct heating in the kitchen, as well as generating electricity for lamps. According to Hoo et al. [9] production of biogas from FW is estimated at almost 60 Mm^3 of CH_4 that can be produced annually. This is equivalent to 16.3 MW of electricity, based on FW generated in 2010. The estimation is based on methane gas emissions derived from the Intergovernmental Panel on Climate Change (IPCC) biological treatment formula. The sale of electricity through the Feed-in-Tariff (F.i.T) scheme is expected to generate revenue of nearly Ringgit Malaysia 42 million. Nonetheless, due to the lack of a research facility, its real potential is not optimized.

FW has increased tremendously in line with the steady growth of the economy. It is an extremely unpleasant material that however, can be treated as a valuable potential source of bio methane, which can play a significant role in renewable and sustainable energy development. Hence, increasing consideration has been made towards technology advancement in this field. Apparently, AD has become the most effective way to treat energy-rich and high humidity organic substantial's [10]. It is relatively cost affordable and sustainable for long-term energy production that consistently becomes the most popular investigation trend. AD will be of value to both nutrient recovery and energy renewable. Hence, it is being investigated extensively because it offers favourable alternatives and has a high wastes valorisations value. Intensive explanations of the treatment methods, including landfilling, composting, incineration, integrated treatment, and anaerobic digestion can be concluded comprehensively in Table 1. These contents are used here to assist in evaluating the advantages, and disadvantages of each FW treatment method. Compared to landfill, AD requires less input energy for operation, fewer spaces and there is still potential for efficiency improvement [10-12].

Table 1. Results of the comprehensive evaluation of the different current methods FW Management.

		Landfill[a]	**Composting**[a]	**Incineration**[a]	**Integrated**[a] **Treatment**	**Anaerobic**[b] **Digestion**
Economy	Investment	Moderate	Moderate	Large	Large	Large
	Operating cost	Cheap	Medium	Expensive	Expensive	Medium
	Land Occupation	Large scale	Medium scale	Small scale	Small scale	Small scale
Ecology	Discharge gas	Biogas	Foul odour	Acid gas	Acid gas	Biogas
	Waste Water	Leachate	Little leachate	-	Little leachate	-
	Soil	Leachate Contamination	Heavy Metal Contamination	Gasses Pollutant Residues	Gasses Pollutant Residues	-
Technology	Requirement	All	Putrescible	High calorific waste	All	Putrescible
	Dependability	High	Medium	Medium	Medium	Medium
	Security	High	High	Medium	Medium	Medium
	Standard of control	Low	Medium	High	High	High
Community	Convenience	Moderate	Significant	Significant	Significant	Significant
	Resource	Little	Medium	Large	Large	Large
	Reduction	Nil	20% reduction	80% reduction	90% reduction	50% reduction

a [12][13] b [11]

In correlation with different biofuels, biogas is progressively adaptable, since it tends to be created from hydrocarbons, proteins, and fats, including wet and secondary materials [14, 15]. Generally, biogas is generated from FW, sewage sludge and manure, but the biogas production method and arrangement is differed between regions, with various feedstocks, production, and use [16]. Thailand, Indonesia, Brazil, China, the United States of America and European countries for instance, deployed their accessible local timber and biomass resources to substitute coal and oil [17]. In Malaysia, biomass energy sources are acquired from timber waste and several cultivations i.e. palm oil, rubber, coconut, sugarcane, cocoa, rice paddy, animal waste and urban waste [18]. However, the biggest biomass energy is contributed by the palm oil industry due to its enormous plantation in Malaysia. Bruijstens et al. [19] stated that FW has higher methane content, which can reach nearly 85% compared to other wastes. A study conducted by Naesko Environment's Association [20], has shown percentage variation on the biogas yielded based on four types of waste. Compared to household waste (50-60%), all other wastes have a higher methane percentage (food waste 68%; agricultural waste 60-75%; wastewater treatment sludge 60-75%).

AD has numerous processes involving methanogenic archaea and bacterial actions. To assist the degradation of organic matter, four stages would be involved sequentially such as hydrolysis, acidogenesis, acetogenisis, and methanogenisis. During the treatment, macromolecular organic component is degraded into liquefy monomers which then to tinier molecules of short group fatty acid, pyruvic acid, amino acid, acetic acid and others. These eventually ingested into hydrogen and acetic acids and converted by methanogenic archaea into methane [4]. FW is highly considered suitable for AD bio methane potentiality approximated at range 200 to 670ml CH_4/g added Volatile Solids (VS).

During the AD process, organic waste might be altered into burnable biogas (principally methane, as well as gases variation, such as, hydrogen sulphide and hydrogen [5]. Parameters, for instance, volatile solids reduction and methane yield or so were employed to evaluate investigation efficiency. In the procedure to improve the quality of methane production, various aspects of AD are investigated such as additional supplementation, functional alternations, optimization of parameters and microbial behaviour. Besides biogas, the anaerobic digestion process transfers secondary products- bio fertilizer also known as digestate. As a large share of the source of nourishment content of the feedstock ends up in digestate, it can be used as bio compost contributing to improved soil nutrient management [21, 22].

However, as biomass transformation to bioenergy increases ecological issues, the subject on ecological sustainability assessment of bioenergy pathways has been extensively conducted. Consequently, a life cycle assessment (LCA) as a comprehensive method is used to determine and guarantee the environmental sustainability of the biogas. LCA is a holistic approach for surveying a system's environmental burdens by characterizing the fuel, materials utilized, and contamination discharged into the ecological system. As stated by Aziz et al. [23], particularly in Malaysia's biogas production there are less LCA studies have been conducted. This study aims to assess the biogas generation from biomass towards environmental impacts through LCA. This review highlights the influence factors that affect the life cycle assessment results, for example, goal determination and scope, functional units, boundary systems setting, characterization, assessment methodology on life cycle impact, impacts categories, finally the selection of software and databases.

2. LCA for biogas generation from anaerobic digestion of biomass

There is expanding attention in the assessment of the ecological impacts of biogas as a cleaner and more sustainable power source utilising approach of LCA. This reviewed paper discussed the investigations on LCA biogas generation from several countries with various characteristic points of view. The studies depended on the described LCA approach that stated in ISO 14040 and 14044. The consideration of the exhaustion of non-renewable energy source and the ecological issue, for instance, an a worldwide temperature alteration identified with ozone harming substance releases has incited an extension in examining at organic wastes biogas creation for a more feasible energy production.

Biogas processing from biomass products is highly cost saving for the creation of sustainable energy. This is because the abundance of ready feedstocks such as MSW, wastewater discharges, horticultural and animal waste can be exploited to generate energy. Reusing of natural wastes is likewise a better and more practical ecological expenditure that be able profit residents and the environment simultaneously. However, effect of feedstocks to lower ecological burden need to be defined; for example the use of straw, corn silage, or co generation between energy crop-agricultural residues with MSW food residue feedstock mixture [24]. Table 2 indicates 20 of the studies evaluated, 13 of which involved European countries such as Luxembourg, Germany, Italy, the United Kingdom, Switzerland, and Sweden. One study was from the African nation, Ethiopia and the other six were conducted in Asian nations (three in China, and one each in Vietnam, Singapore, Indonesia). Recently, the LCA research on biogas production has increases especially in Asian and North America regions. The developing concern for ecological assessment related to the generation and application of biogas utilizing LCA was to evaluate the environmental performance and the sustainability of the energy pathways [25].

Table 2 shows the summarized of 20 LCA biogas studies with different study context, the methodology used, and feedstock type, use of biogas and bio fertilizer, and the categories of environmental impacts. Referring to Table 2, the studies on life cycle assessment for AD plants in this review could have conducted out quite differently. Researchers found a number of importance influences in the way biogas by AD plants conducted. For example, the life cycle impact assessment method applied still play an important role in affecting the LCA results. The LCA results can be obtained much better, but there is stillroom for improvement especially complex systems that include assumptions and plenty of records. Many LCA results can be improved with appropriate system boundaries setting, the used of measurable functional unit (FU) as well as the robust databases selected. Difficulties can arise when there is more than one crucial element to be perceived. Thus, it is important to identify which elements have a larger influence on some of the other viable elements; and these may be described qualitatively. The following section addresses components and aspects such as goal definition and scopes, functional unit, system boundaries, use of biogas, use of bio fertilizers and categories of impacts that have been covered in the reviewed studies. As the LCA usage is likely to increase in the future, relevant LCA establishments were needed with the detailed guidelines provided. This is important for future potentiality on cross study comparisons and increasing the general quality in assessment.

2.1. Defining the Goal and scope of reviewed studies

The conducted LCA study is intended for its certain utility, the target audiences or for info dissemination. The utmost intention of using LCA on the biogas from AD is to assess on the environmental existence cycle effects of managing organic waste as they are achieved in those cases. In this segment, the plans of the LCA researches need to be described as clearly as possible. The classified scopes were ecological performance, ecological and economic performance, ecological and energy performance as well as resource efficiency. Fiez et al. [22] has assessed on environmental performance of FW biogas digesters that has carried out in Xiaoguwei Island, Guangzhou, China. The study was comparing the impacts on environmental of different waste treatment such as FW to biogas production (with upgrading to fuel, and with electricity production), incineration, landfill, and pig farms. The results indicated that the FW to biogas plant provides about 0.3MNm^3 bio methane and a reduction of greenhouse gas emissions of about 1780 tons of CO_2-eq each year.

Table 2. Goals, System Boundary, Functional Unit and Method LCIA of the Biogas from AD Treatment

Reference [a]	Country	Type of Waste	Goals	Aspects/System Boundary	Functional Unit	System of Reference	Database/software for LCA	Method LCIA	Midpoint	Endpoint	Biogas application	Bio fertilizer application
[22]	China	Treated Food Waste	Biogas potential, Climate performance	Cradle to gate	1 ton of Treated Food Waste	China condition	Monte Carlo Simulation	Calculation GWP_{100}			Biogas, electricity, vehicle fuel	Bio Fertilizer
[24]	Germany	Animal slurry, plant biomass, industrial waste/other	Environmental performance	Biogas production and utilization pathways	1 ton of feedstock mixture	German condition	SimaPro 7.2, Eco invent v2.1	ReCiPe	x	x	Bio methane, CHP	Fertilizer
[26]	Luxembourg	Plant biomass	Energy and environmental performance	Cradle to gate	1MJ injected into the natural gas grid	The imported natural gas system	Umberto 5.5, Eco invent 2.0	Impact2002, Eco-indicator 99	x	x	Injected into the natural gas grid	NA
[27]	Switzerland	Plant biomass, industrial waste/other	Environmental performance	Cradle to gate	1 ton of feedstock	Swiss condition	Eco invent v2.0,	Eco indicator 99, Ecological Scarcity 2006 method, IPCC 2007	x	x	Fuel, CHP	
[28]	Italy	Treated Food Waste	Environmental performance	Cradle to gate	1 ton treated Food Waste	Italy condition	SimaPro 7	CED, CML 2001	x		Bio methane upgrading into natural gas, CHP, vehicle fuel	Fertilizer
[29]	Sweden	Food Waste, Sewage Sludge	Environmental and economic performance	Cradle to gate	NA	NA	ORWARE Model	CML 2001 baseline	x		Vehicle purpose	Soil fertilizer
[30]	Vietnam	Animal slurry	Resource efficiency	Cradle to farm gate	Sum of products (the sum of 1 kg of each product delivered at the farm gate)	NA	ExFA, Eco invent v2.2 database	CEENE method	x		Household cooking	Base Fertilizer
[31]	China	Plant biomass	Environmental performance	Cradle to gate	1 ton of pre dried straw	NA	SimaPro, Eco invent database,	Eco-indicator 99 (H), IPCC 2007 GWP	x	x	Compressed LNG, CHP	Fertilizer
[32]	United Kingdom	Animal slurry, plant biomass, industrial waste/other	Environmental balance	Cradle to gate	1 Mg of dry matter feedstock input	NA	LCAD Eco screen tool, Eco invent v3.1	CML 2010	x		Bio methane, CHP	Fertilizer
[33]	Ethiopia	Animal slurry	Environmental performance	Gate to gate	The amount of energy needed to produce 1MJ heat	Dung combustion for cooking	GaBi Software, Eco invent v2.2	CML	x		Household cooking	Fertilizer, pastures
[34]	China	Animal slurry	Energy and environmental performance	Cradle to gate	2136 ton of manure	NA	Weighting method	Weighting method	x		Injected to the biogas grid, CHP	Fertilizer
[35]	Italy	Industrial waste	Environmental performance	Cradle to industry gate	1kg of tomato puree	NA	Sima Pro, Eco invent database	ILCD method	x		CHP	Organic fertilizer

[36]	Germany	Plant biomass	The impact of regional factors	Cradle to grave	1 kg of fresh matter of maize, 1 kWh of electricity	NA	GaBi 4.4	GEMIS	x		CHP	Fertilizer
[37]	Italy	Animal slurry, plant biomass, industrial waste/other	Environmental performance	Cradle to grave	1.02 kWh of electrical energy, 10.92MJ of thermal energy, 1.86kg of compost	A real pilot plant system	Sima Pro, Eco invent 2.1 database	Eco Indicator 99	x		CHP	Soil Fertilizer
[38]	Germany	Animal slurry, plant biomass	Environmental and economic performance	Cradle to gate	1kWh$_{el}$ emission	Fossil reference system	Umberto 5.6, Eco invent 2.2 database	NA	x		CHP	Fertilizer
[39]	Norway	Industrial waste/other	Environmental performance	Cradle to gate	1MJ of electricity produced	Synthetic fertilizers and electricity from natural gas	ARDA v.1.8, Eco invent database	NA	x		CHP	Fertilizer, soil amendment
[40]	United Kingdom	Industrial waste/other	Environmental performance	Cradle to gate	1MWh of heat and electricity	AD-CHP system with electricity and heat geneartion	GaBi v4.4,	CML 2011	x		CHP	Fertilizer
[41]	Singapore	Food Waste	Environmental performance	Gate to gate	1000 ton of FW	NA	NA	NA	x		Electricity	Compost
[42]	Norway	Food Waste	Resource efficiency and environmental performance	Cradle to grave	1 ton Dry Matter mixed organic waste entering RBA	Value chain of Romerike biogas plant (RBA)	ARDA	ReCiPe	x	x	Liquefied Biogas	
[43]	Indonesia	Palm Oil Mill Effluent	Energy and environmental performance	Gate to gate	1 ton of fresh fruit bunch	NA	SimaPro, IPCC 2013	Impact 2002+	x		Electricity	Land application

Software

Biogas Application Other

CML: Centrum voor Milieukunde Leiden, , IPCC: Intergovernmental Panel of Climate Change, ReCiPe: RIVM and Radboud University, CML, and PRE' Consultants, CED:Cumulative Energy Demand; GWP100: Global Warming Potential (100years); NA: Not Applicable ; EDIP: Environmental Design of Industrial Products; TRACI 2: Reduction and Assessment of Chemical and Other Environmental Impacts; ILCD: International Reference Life Cycle Data System, CEENE: Cumulative Exergy Extracted from the Natural Environment; GEMIS: Global Emissions Model for integrated Systems; ORWARE: Organic waste research, GaBi:Ganzheitlichen Bilanzierung (German for holistic balancing), ReCiPe: RIVM and Radboud University, CML, and PRE' Consultants
CHP: Combine Heat and Power, LNG: Liquid Natural Gas, LBG: Liquefied Bio Gas,

GWP: Global Warming Potential, CED: Cumulative Energy Demand, NA: Not Applicable

IOP Conf. Series: Materials Science and Engineering **808** (2020) 012011 doi:10.1088/1757-899X/808/1/012011

Meanwhile in another study by Fiez et al [44] a multi-criteria method applied to evaluate the right quality of biogas feedstocks and the production of bio fertilizer was assessed. The study examined aspects of efficient costing, feasible technology, energy and ecological performance, accessibility, policy, competition and other relevant issues. The importance of the finding is to structure information relevancy for facilitating strategic overviews, informed decision making and communicating its result.

2.2. Functional units

FU has to be defined as the type of function or service the products' system delivers. The FU reason is to have a reference to which the inputs and outputs are related. The FU is essential to enable the similar cross study comparability between the LCA studies. Therefore, the FU itself needs to be measurable. Gradually, the options of the FU in biogas structures refer on power, distance, mass, volume or a hectare for land used. As shown in Table 2 the LCA studies of biogas manufacturing were having variation FU, for examples; ton [22,24,27-28,31,34,41-43], kg [30,35,36], Mg [32], m^3, MJ [26,33,39], kWh [37-38], MWh [40], ha and km.

2.3. Boundaries Setting

According to Fiez et al. [22] the effects of environmental and energy assessment are very dependent on the choice of system boundaries, assumptions and include effect categories. The boundaries setting indicate the system or process as well as the inputs and outputs of the system that to be included under the consideration for LCA evaluation purposes. There are 13 studies from total 20 reviewed LCA-studies took into consideration the system boundaries from cradle-to-gate [22,26-32,34-35,38-39,40], 3 studies from cradle to grave [36-37,42], 3 studies from gate to gate [33,41,43], and a biogas production and usage respectively [24]. The cradle-to-gate method consists of the complement of feedstock until the biogas production, and in a certain case focused effectively on unique stages within the system boundaries. In the general system, as illustrated in Figure 1, the dotted traces are the flows of energy and the solid traces are the flows of material. Definition system in Figure 1 is inclusion of feedstocks transportation, different processes waste, and the bio fertilizer product. In an extra correct system, biofertilizer products must also be covered in addition to the transportation of these flows and in a few cases, the fuel that is produced. Dependent on the substrate and/or the intended first-class of the products the need for different processes will vary. Five principal processes, AE, have been defined: - A: sorting- B: pre-treatment- C: anaerobic digestion- D: bio residual treatment — E: biogas usage and upgrading.

Bernstad et al. [45] discussed how variation in process parameters and boundary setting may influence the result of the LCA and made suggestions to improve interior equivalence within treatment options in a comparable LCA as well as among various LCAs. For example as reported by Chen [12] AD requires huge energy consumption (diesel provision) using larger trucks due to the process of collection, transporting collection to the waste treatment facility (electricity provision) and labourers' cost. The journey from the collection sites or small transfer station will release emission to air. Emission to air in AD process especially in collecting and transporting even in low concentration is considered as environmental burden.

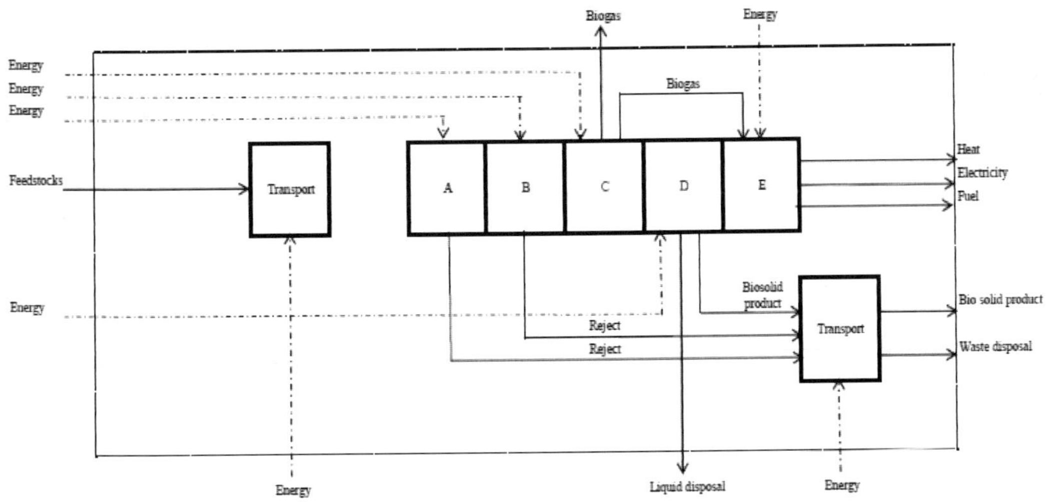

Figure 1. General value chains for biogas system definition [42].

2.4. Characterization method and life cycle impact assessment

Life cycle impact assessments (LCIA) transform the life cycle inventory (LCI) results in a limited number of indicator scores, which are useful in defining the scores of indicators that show relative magnitude of impacts on the environment. As stated in Table 3, the LCA practitioners in biogas plants might have conducted the LCIA in a different way. The selection of the LCIA approach used can have a major impact on the outcome; however, no clear information or recommendations are available for general users to choose reasonable LCIA methods. Therefore, there is a need to understand each LCIA method or combination of methods used for purposes of comparisons. Furthermore, the findings can be used to elaborate the literature in future studies. These could be very useful when it is used in biogas plants with similar features. In addition, this analysis may also be part of future research on comparing biogas plants with other organic fractions treatments of AD from LCA perspectives.

The LCIA method is consists of two elements, midpoint and endpoint. Midpoint has lower uncertainties because the impacts that can be modelled are considered. Midpoint impact assessment has indicators that concentrate on a single environmental issue e.g. climate change or acidification. This will give a more concise overview of the system's impacts on the environment. The environmental impact results obtained using different LCIA midpoint approaches are identical for comparable categories. Moreover, the LCIA techniques such as ReCiPe, EDIP and Impact 2002+ offer greater assessment and comparative categories than CML and TRACI (as shown in Table 3) [46].

With respect to the definitions of global warming (called "Global Warming" in CML, TRACI, EDIP, and Impact 2002+ and "Climate Change" in ReCiPe Midpoint), expressed as kilograms of carbon dioxide equivalent (kg CO2 eq); it can be noted that there is almost the same relative result because the same reference was derived from the IPCC. Referring to Table 3, the midpoint studies using CML approaches conducted by Grosso et al; Eriksson et al; Styles et al; Lansche et al. and Whiting et al.[28-29,32-33,40], the results showed the best indication related to the potential for global warming reduction and improvement. Another type of environmental impact that can be correlated with different methods is the degradation of the ozone layer (called "Ozone Depletion" in ReciPe, TRACI and EDIP and "Ozone Layer Depletion" in CML and Impact 2002+, expressed as kilograms of chlorofluorocarbons equivalent (kg CFC-11 eq). Such quantitative analysis is not feasible for impacts such as eutrophication and toxicity that have different indicators of specific ecological compartments depending on the method selection. For acidifications, the findings are not convergent due to differences in spatially similar characterization variables and different models characteristics adopted according to each approach. An example of the differences in CML (acidification, as kg SO_2 eq), in ReCiPe (terrestrial acidification, as kg SO_2eq), EDIP (acidification as square meter) in TRACI (acidification as H^+ moles eq) and in Impact

IOP Conf. Series: Materials Science and Engineering **808** (2020) 012011 doi:10.1088/1757-899X/808/1/012011

2002+ looking at terrestrial acidification (as kg SO_2eq) and the aquatic acidification as (kgSO_2eq) is chosen [24, 26, 28]. Meanwhile, the GEMIS (Global Emissions Model for Integrated Systems) calculations have established the standard for contemporary assessments of the life cycle. GEMIS is a free lifecycle and material flow analysis tool for the public domain, which also provides a comprehensive database on energy carriers, materials energy, and material processes.

Midpoint have robust model and data, which is recognized as a problem oriented LCIA method. It may lead to reliable interpretation and open possibilities for comparisons between different studies. However midpoint can also be harder to interpret because it has too many abstract meanings e.g. radiative forcing - acidification. Besides, such features are not the only environmental effects of the emissions listed on LCI, but there are measures of the potential impact. Therefore, endpoint as damage oriented LCIA method is an alternative for simplification on the LCIA interpretation. For instance endpoint (e.g. ReCiPe 3 points and Eco Indicator 99) were indicators that showed the impact of the environment on three higher aggregations level on the effect on human health, biodiversity, and finally resource scarcity [24, 26, 27].

With regard to the single score, the relative contribution of the respiratory categories in organics in IMPACT 2002+ and Eco-Indicator 99 (H) methods and particulate matter formation in the ReCiPe endpoint (H) method can be compared [26,24,31]. Both categories are related to emissions of particulate, sulphur oxides, ammonia, and nitrogen oxides with some little differences in characterization steps and all of them are expressed in Disability Adjusted Life Years (DALY), at endpoint level. Another comparison can be made between the relative weight of land use in Eco indicator 99 (H) and Impact 2002+ methods [26, 27]. The Ecological Scarcity method presents no specific environmental impact category related to land [27]. The categories of Energy Resources in Ecological Scarcity, Fossil Fuels in Eco Indicator 99 (H) and Non-renewable energy in Impact 2002+ generally showed similar results, as exemplified by a comparison study conducted by Jury et al. and Stucki et al. [26, 27].

Table 3. Life cycle impact assessment (LCIA) methods [46].

LCIA methods	Midpoint/ Endpoint	Number of impact categories evaluated	Single score damage categories	Regional validity	REF.
CML 2001	Midpoint	10	NA	World, excluding acidification and photo oxidants (Europe) production.	[28], [29], [32],[33], [40]
Impact 2002+	Midpoint	15	NA	The simplest version of Europe. Calculations were performed for the intake fraction (toxicity effect category) for a European spatial model based on a grid of 200x250 km. A multi-continental version of this model was made available to test pollution inventories across all continents.	[26],[43]
	Endpoint	13	i-Human health ii-Ecosystem quality iii-Climate Change Resources		
EDIP 2003	Midpoint	19	NA	Europe (factors with a European average value for up to 44 regions or countries within Europe). Global for the categories of global impact.	[NA]

Eco Indicator 99	Endpoint	11	i-Human health ii-Ecosystem quality iii-Resources	Global environmental climate, ozone depletion and resource affect categories. European model for other types of effects: It is presumed that all emissions occur in Europe. Damage that happens outside Europe is also taken into account when using the European impact situation, if the lifetime of the atmosphere is long (some toxic substances, radioactive substances, etc.). Dutch model-based acidification/eutrophication, land use is driven on the Swiss model.	[26],[27],[31],[37]
TRACI 2	Midpoint	9	NA	Emissions in the United States, impacts for acidification, eutrophication, and smog development in North America, and ozone depletion and global warming around the world. Human and eco-toxicity is not a common location in TRACI, but in the United States. EPA standards are used for human health factors and recommendations for risk assessment	[NA]
ReCiPe	Midpoint	18	NA	Europe. Global climate change, degradation of atmospheric layers and resource.	[24],[42]
	Endpoint	17	i-Human health ii-Ecosystem iii-Resources	Europe. Global climate change, degradation of atmospheric layers and resource.	
Ecological Scarcity	Endpoint	7	NA	The original method for Switzerland has been established. Different versions of the Ecological Scarcity approach for other countries or regions of the world were established.	[27]
Others					
Calculation GWP_{100}	Midpoint				[22]
ILCD	Midpoint				[35]
IPCC 2007	Midpoint				[27],[31]
CED	Midpoint				[28]
CEENE	Midpoint				[30]
Weighting					[34]
GEMIS					[36]

CML: Centrum voor Milieukunde Leiden, , IPCC: Intergovernmental Panel of Climate Change, ReCiPe: RIVM and Radboud University, CML, and PRE' Consultants, CEENE: Cumulative Exergy Extracted from the Natural Environment; CED:Cumulative Energy Demand; GWP100: Global Warming Potential (100years); NA: Not Applicable ; EDIP: Environmental Design of Industrial Products; TRACI 2: Reduction and Assessment of Chemical and Other Environmental Impacts; ILCD: International Reference Life Cycle Data System, GEMIS: Global Emissions Model for Integrated Systems

Poeschl et al. [24] indicated that single scores using ReCiPe methods could show the most significant outcomes when comparing variation on technologies of small scale and large-scale biogas plant, and identifying the 'hot spot' potential where it is possible to achieve further reduction of environmental impacts. The outcome indicates preference for small plants. Tri-generation would be the most promising path for sustainable biogas use; compared to the generation of electricity only. Tri-generation combined heat and power (CHP) could be reduced the overall impact on the environment by almost 200%. Cavalett et al. [46] extended the debate whether and to what degree the efficiency of the system is influenced by using of the various Life Cycle Impact Assessment (LCIA) presented throughout the research study of the Brazilian environmental impact analysis of bio refinery. The assessment considered the following LCIA methods: CML 2001, Impact 2002+, EDIP 2003, Eco-indicator 99, TRACI 2, ReCiPe, and Ecological Scarcity 2006. Although there is a consensus on the outcomes of similar categories of environmental impact using different midpoint LCIA measures, when considering single-score metrics, the use of various techniques of LCIA could results in different interpretations. However, ReCiPe was indicated as the best endpoint method to evaluate environmental impact as shown in the summarized framework in Figure 2.

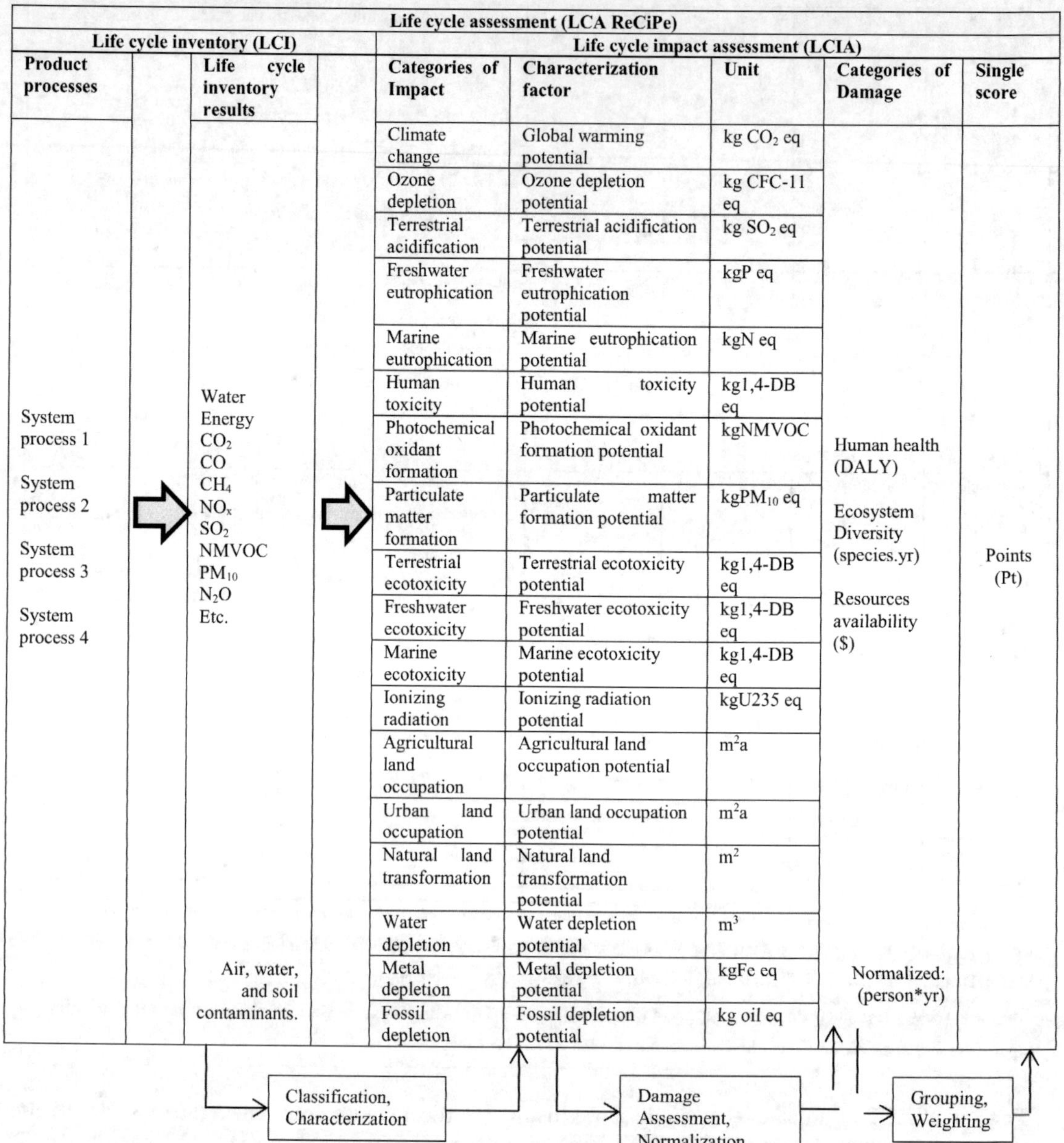

Figure 2: Summarized LCIA ReCiPe framework [24].

2.5. Impact categories

Some research that focused mainly on greenhouse gases and energy balance assessed greenhouse gas emissions, potential for global warming (GWP), and greenhouse gas mitigation potential (GMP), Cumulative energy demand (CED), cumulative exergy demand (CExD), net energy gain (NEG) and primary energy demand (PED). In several studies, these metrics were calculated to assess the energy efficiency of biogas production. Based on Bosch et al. [47], CED and CExD offered a reliable energy efficiency evaluation of the production of biogas system compared to the method of natural gas. Many biogas production studies have shown that biogas-generated energy contributes to greenhouse emissions reductions compared to fossil fuel [26, 48, and 49].

Bacenetti et al. [50] explained that the using of cogenerated heat effectively and reduce methane emissions at the side of strength self-consumption have been the primary factors to enhance the results of the CED and GHG emissions. In the LCA conducted by Seldal et al. [42], the GWP had the point of interest. It was observed that biogas plant from the AD value chain had an overall impact on GWP of 455kg CO_2-eq/FU. By the defined processes, the household waste collection had the greatest effect. When adopting the value chain compared to using alternative options such as chemical fertilizer and diesel, a burden of 747 396 kg CO_2-eq can be averted.

AD treatment releases emissions that direct from the digestive process or indirect from transient emissions of valve malfunction or transitory and this concept has developed by Gentil et al. [51] occurs during the upstream-operation-downstream, which all relevant processes were accounted for. Besides, the produced biogas is necessary to be enhanced, compressed/liquefied and transferred. The fugitive emissions from anaerobic digestion plants may emit CH_4 during treatment and upgrading which may largely affect the general Global Warming Potential (GWP) from the treatment chain. Such discharges change between different technologies and, as stated by Eggleston et al. [52], range from 0 to 10 vol. percent of the biogas generated, but close to 0 percent in plants with unexpected discharges. According to Börjesson and Berglund [53] and Møller et al. [54], both studies noted that the outlawed discharge of methane is a key parameter in connection with Greenhouse Gas (GHG)-outflows from AD systems (in both cases allowing discharges equal to 0–3 vol. percent of methane delivered). While others [55-58] do not consider these discharges at all in their assessments, Bjarnadottir et al. [59] and Ia Cour Jansen et al. [60] consider fugitive methane discharges compulsory.

In the meantime, Lantz et al. [61] and Pertl et al. [62] indicate that during the technology upgrade decision and the consequent fear of CH_4-outflows, the total treatment chain GWP may be significantly affected. As stated in the IPCC rules, the ignition of recovered biogas is minimum in connection to a worldwide temperature alteration since CO_2 emissions are of biogenic starting point (and along these lines GWP neutral) and the outflows of CH_4 and N_2O are insignificant (IPCC, 2006), whereas Ia Cour Jansen et al. [60] states that biogas discharges is necessary be considered. Biogas outflows from ignition rely upon the kind of gas engine, and lean-burn gas engines, commonly utilized at AD plants, often produce higher CH_4 in contrast with different gas engines [63]. While the use of LCA is likely to increase in this area in the near future, it is rather important to develop up more standards that are comprehensive in order to improve the overall consistency of assessments as well as the possibility of cross-study comparisons.

2.6. Software's/databases

Software is also one of the influencing factors in the LCA results. The AD system may look simple at one glance, of reality; however, any technology must be carefully modelled. The version of the waste management program requires a substantial amount of information that includes characteristics of waste, system of collection, different choices for treatment, background structures, and so forth. Such changes were introduced for future trends update, offering future portions of waste, fuel expenses, electricity mixes, and lots of others. Since used software varies for example SimaPro, Gabi, ORWARE, EaseTech etc. the most important of software selection is its databases. GaBi for instance has more than 20,000 accessible databases and this will provide very good visualization to represent assessment results compared to other software which only used common database provider i.e. Eco Invent (13,300 databases) [64].

There is an extensive division among popular LCA applications as with SimaPro, GaBi, Team, Umberto and the programs specifically designed for use in a particular area. In waste treatment schemes, for example: ORWARE (organic waste research), IWM-2 (integrated waste management II), WISARD (waste – integrated systems for assessment of recovery and disposal), WRATE (waste resources assessment tool for the environment), and EASEWASTE (environmental assessment of solid waste systems and technologies). GEMIS (Global Emissions Model for Integrated Systems) is a free lifecycle and material flow analysis tool for the public domain, which also provides a comprehensive database on energy carriers, resources, energy and material processes. Such software will be used to assess the present and further to version-new waste management systems. Figure 3 showed the databases used for LCA reviewed studies in biogas plant by AD.

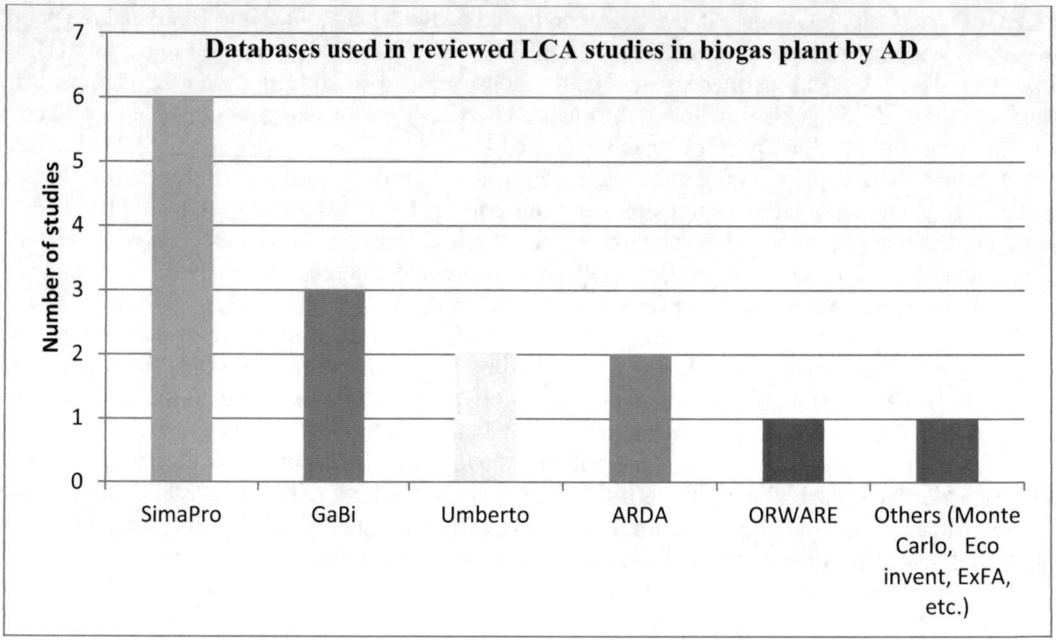

Figure 3: Databases for LCA reviewed studies in biogas plant by AD.

Kulczycka et al. [65] stated that differences among the research in the assessment are associated to technical assumptions with input data, form of technology, used inventories and data production. It is also determined that the models are subject to special assumptions are especially associated with various times after the information system has been developed (e.g. assumptions for the time horizons for landfill emissions). Further optimization of these tools to geographical situations will have an extra impact on the consequences. Nevertheless, each of the LCA programs can display a few benefits and downsides if assessed strictly from the waste management perspectives. The common LCA programs frequently encompass complete databases; however, they are now not continually specific for the waste treatment systems and suitable for national/regional situations [65].

Nonetheless, compared to a standard LCA model, the concept of ORWARE is designed to capture the waste characteristics control applying an SFA method and additionally include estimates of the economy. The use of a version unlike ORWARE, the concept of the district heating system has been popular and is in addition defined by Eriksson et al. [29], ORWARE is to an excessive volume feasible to adapt to the system studied precisely. ORWARE is beneficial due to (1) its potential to each treatment of environmental and biological wastes efficiently and (2) the potential of the new model procedures.

3. Recommendations

It is recommended that the modelling performed on to biogas plants from AD treatment be improved to give better results such as by defining a system with more processes to produce more equal results, thus more processes and more reliable inputs and outputs are needed to be included. So, the result with much lesser uncertainty will be generated by a system. Secondly, is by comparison of the value chain with relevant literature studies. It is strongly suggested to LCA practitioners to perform both a Material Flow Analysis and LCA for robust analysis [42]. Finally, a selection of suitable database to model the study system on, for example by using GaBi which has 20,000 accessible data plus Eco invent databases 13,300 data for the foreground processes to reduce uncertainties. This means the impacts that can be modelled are will be considered to have with robust model and data, and finally could lead to accurate analysis and open comparability between different studies [64, 65]. Regarding the AD alternatives for organic waste treatment the single stage dry continuous as well as single stage wet high rate digestion system can effectively handle organic waste. Both technologies are capable of significantly reducing the water consumption and energy expense. However, in high generation of organic waste in urban area with fewer spaces available, dry continuous system is more preferable because it has a small reactor, cheap pre-treatment, very little wastewater production, composting waste well applied and less heat needed. This is supported by Poeschl et al. [24] which indicated that small scale AD technologies is the promising approach to reduce impact on the environmental. The two-stage AD systems with separation of hydrolysis and methanogenesis in separate reactors could have shorter processing times and higher organic loading levels (up to 50 kg-COD/m3/d) than the single-stage systems, but their Total Solid > 20%, the dilution of the waste stream causes a substantial increase in the energy needed for heating, pumping and expanding reactors. Meanwhile, the performance of the three-stage digestion systems has more complex operation, more costly investment, more maintenance and operating resources compared to two-stage AD systems, and the large-scale three-stage system is not currently a good option [66].

4. Conclusions

LCA has much to offer regarding decision-making aid for a preference towards a cleaner AD process. This is beneficial because the identification of the GWP contributing inventory of AD is of crucial importance. Such determination is helpful to lead to the mitigation procedure to decrease the GWP of AD. Since direct energy use in AD plants changes generally, it ought to as per all surveyed rules, be tended to especially during the treatment, biogas utilization and upgrading. Despite different facilities used, the analysis of the same functional unit of biogas, operation mode, system boundaries setting, databases used and life cycle impacts assessment method are also very relevant to avoid biased comparisons on climate performance between different alternatives. Extra investigation and study on various AD scales, waste streams, and process systems are fundamental to better understand their impacts on explicit and GWP in general.

References

[1] Abeliotis K Lasaridi K and Chroni C 2015 *J. Toxic. and Envir. Chem.* **98**
[2] Sheldon R 2011 *Catalysis Today.* **16**
[3] Metzger J O and Eissen M 2004 *J. Comptes Rendus Chimie.* **7**
[4] Ren Y, Yu M, Wu C, Wang Q, Gao M, Huang Q and Liu Y 2018 *J. Biores. Techno.* **247**
[5] Xiao B, Qin Y, Zhang W, Wu J, Qiang H, Liu J and Li Y Y 2018 *J. Biores. Techno.* **249**
[6] Suzannah P 2018. *https://www.nst.com.my/news/nation/2018/06/380793/ugly-culture-food-wastage-growing-among-malaysians*. Retrieved 5/10/2019
[7] Nadzri Y 2013. *In 10th Annual Waste Management Conference & Exhibition Malaysia*
[8] The Star 2017. *https://www.thestar.com.my/metro/metro-news/2017/09/09/smart-waste-solution-lab-helps-ease-dumping-problem-in-ss*. Retrieved 5/10/2019
[9] Hoo P Y, Hashim H, Ho W S and Tan S T 2017 *J. Chemi. Engine. Transac.* **56**
[10] Xu F, Li Y, Ge X, Yang L and Li Y 2018 *J. Biores. Techno.* **247**
[11] Zhu S, Gao H and Duan L 2018. *J. IOP Confer. Series: Earth and Environ. Science.* Vol **153**
[12] Chen X 2012 MSc Thesis

[13] Bong C P C, Lim L Y, Ho W S, Lim J S, Klemeš J J, Towprayoon S, , ... and Lee C T 2017 *J. of Clean. Prod.* **146**
[14] Weiland P 2003 *J. Appl. Biochem. Biotechno.* **109**
[15] Angelidaki I, Karakashev D, Batstone D J, Plugge C M and Stams A J M 2011. Vol **494**
[16] Svensson M, Baxter D, Drosg B, Bochmann G, Mc Cabe B, Toyoma J, Al Seadi T, Rasi S, Theobald O and Bastide G 2015 *France IEA: Paris*
[17] Chuah T, Wan Azlina A, Robiah Y, Omar R 2006 *Int. J. Green Energy*. **3** 346-323
[18] Aziz N I H A and Hanafiah M M 2017 *J Green Energy.* **1** 323–346
[19] Bruijstens A J, Beuman W P H, Molen M V D, Rijke J D, Cloudt R P M, Kadijk G, Camp O O D and Bleuanus 2008 *J. Biores. Techno.*
[20] Naesko Environment's Association 2009. http://biogasrenewableenergy.info/biogas_composition.html. Retrieved 5 October 2019
[21] Ammenberg J and Feiz R 2017 *J. Resour. Conserv. Recycl.* **122**
[22] Feiz, R, Ammenberg J, Björn A, Guo Y, Karlsson M, Liu Y and Trygg K 2019 *J. Sustain.* **11** 1556
[23] Aziz N I H A, Hanafiah M M and Gheewala S H 2019 *J. Biomass and Bioenergy.* **122**
[24] Poeschl M, Ward S and Owende P 2012 *J. Clean. Prod.* **24**
[25] Cherubini F, Bargigli S and Ulgiati S 2009 *J. Energy.* **34**
[26] Jury C, Benetto E, Koster D, Schmitt B and Welfring J 2010 *J. Biomass Bioenergy.* **34** 66-54
[27] Stucki M, Jungbluth N and Leuenberger M 2011 Switzerland. *ESU Services Ltd Commissioned.*
[28] Grosso M, Nava C, Testori R, Rigamonti L and Viganò F 2012 *J. Waste Manage. and Research.* **30**
[29] Eriksson O, Bisaillon M, Haraldsson M and Sundberg J 2016 *J. Environ. Manage.* **175**
[30] Nhu T T, Dewulf J, Serruys P, Huysveld S, Nguyen C V and Sorgeloos P 2015 *J. Resour. Conserv. Recycl.* **102**
[31] Wang Q L, Li W, Gao X, and Li S J 2016 *J. Bioresour. Technol.* **201**
[32] Styles D, Dominguez E M and Chadwick D 2016 *J. Sci. Total Environ.* **560**
[33] Lansche J and Müller J 2017 *J. Clean. Prod.* **165**
[34] Wang Y, Wu X, Tong X, Li T, and Wu F 2018 *J. Clean. Prod.* **192**
[35] Bacenetti J, Duca D, Negri M, Fusi A and Fiala M 2015 *J. Sci. Total Environ.* **526** 97-88
[36] Dressler D, Loewen A and Nelles M 2012 *Int. J. Life Cycle Assess.* **17** 1115-1104
[37] Cappelli A, Gigli E, Romagnoli F, Simoni S, Blumberga D and Palerno M 2015 *J. Energy Procedia.* **72** 10–13
[38] Lauer M, Dotzauer M, Hennig C, Lehmann M, E Nebel and Postel J 2017 *J. Int. J. Energy Res.* **41** 80-63
[39] Iordan C, Lausselet C and Cherubini F 2016 *J. Environ. Manag.* **184** 527-517
[40] Whiting A and Azapagic A 2014 *J. Energy.* **70** 193-181
[41] Tong H, Shen Y, Zhang J, Wang CH, Ge T S and Tong Y W 2018 *J. Appl. Energy.* **225** 1143–1157
[42] Seldal T J 2014 Master Thesis. *Norwegian University of Science and Technology.*
[43] Nasution M A, Wibawa D S, Ahamed T, Noguchi R 2018 *J. Clean. Prod.* **184** 1028–1040
[44] Feiz R and Ammenberg J 2017 *J. Res. Conserv. and Recyc.* **122** 373-387
[45] Bernstad A and Ia Cour Jansen J 2012 *J. Waste management.* **32**
[46] Cavalett O, Chagas M F, Seabra J E and Bonomi A 2013 *Int. J. Life Cycle Assess.* **18**
[47] Bösch M E, Hellweg S, Huijbregts M A, Frischknecht R 2007 *Int. J. Life Cycle Assess*. 1**2** 190-181
[48] Börjesson P and Berglund M 2006 *J. Biomass Bioenergy*. **30** 485-469
[49] Edelmann W, Baier U, and Engeli H 2005 *J. Water Sci. Technol.* **52** 208-203
[50] Bacenetti J, Sala C, Fusi A, and Fiala M 2016 *J. App. Energy.* **179** 669-686
[51] Gentil EC, Damgaard A, Hauschild M and Finnveden G 2010 *J. Waste Management*. **30**
[52] Eggleston S, Buendia L, Miwa K, Ngara T and Tanabe K 2006 *J. Waste Management*. **5**
[53] Börjesson P and Berglund M 2007 *J. Biomass and Bioenergy***. 31**
[54] Møller J, Boldrin A and Christensen T H 2009 *J. Waste Management and Research.* **27**

IOP Conf. Series: Materials Science and Engineering **808** (2020) 012011 doi:10.1088/1757-899X/808/1/012011

[55] Aye L and Widjaya E R 2005 *J. Waste Management.* **26**
[56] Diggelman C and Ham R 2003 *J. Waste Management and Research.* **21**
[57] Chaya W and Gheewala S H 2007 *J. Clean. Prod.* **15**
[58] Güereca L P, Gassó S, Baldasano J M and Jiménez-Guerrero P 2006 *J. Res. Conserv. and Recyc.* **49**
[59] Bjarnadottir H J, Fridriksson G B, Johnsen T, Sletnes H 2002 Finland. *Nordtest Report TR 517.*
[60] la Cour Jansen J, Christensen T, Davidsson Å, Lund Hansen T, Jönsson H and Kirkeby J 20017 602. *Nordic Council of Ministers, Copenhagen*
[61] Lantz M, Ekman A and Börjesson P 2009 Report 69. *Environ. and Energy Sys. Studies, Lund University, Lund Sweden*
[62] Pertl A, Mostbauer P and Obersteiner G 2010 *J. Waste Management.* **30**
[63] Nielsen O K, Lyck E, Mikkelsen M H, Hoffmann L, Gyldenkærne S and Winther M 2008 *Denmark.* http://www.dmu.dk/Pub/FR667.pdf Retrieved 5/10/2019
[64] http://www.gabi-software.com/international/index/ Retrieved 5/10/2019
[65] Kulczycka J, Lelek L, Lewandowska A and Zarebska J 2015 *Polish J. of Envir. Studies.* **24**
[66] Van D P, Fujiwara T, Leu Tho B, Toan S, Phu P, Hoang Minh G and Minh G H 2019 *J. Envir. Eng. Research.* **25** 1-17

IOP Conf. Series: Materials Science and Engineering **808** (2020) 012012 doi:10.1088/1757-899X/808/1/012012

Preparation of porous membrane with graphene oxide for vanadium redox flow battery

Noor Fatina Emelin Nor Fadzil[1,2], Ebrahim Abouzari-Lotf [2,3,5*], Saidatul Sophia Md Sha'rani [1,2], Teo Ming Ting [4], Nurfatehah Wahyuny Che Jusoh [1,2] and Arshad Ahmad [2,3]

[1] Malaysia–Japan International Institute of Technology, Universiti Teknologi Malaysia, Malaysia
[2] Center of Hydrogen Energy, Universiti Teknologi Malaysia, Malaysia
[3] Department of Chemical Engineering, Universiti Teknologi Malaysia, Malaysia
[4] Malaysian Nuclear Agency (MNA), 43000, Kajang, Selangor, Malaysia

*ebrahim@utm.my

Abstract. Graphene oxide (GO) was functionalized onto both sides of grafted porous polyethylene/polypropylene (PEPP) membrane sheet through sonication and simple immersion. Graphene oxide is a trending material in the field due to its excellent properties such as highly resistant to alkaline/acidic and strong mechanical strength, low cost and easily accessible. PEPP was successfully modified and the fiber gap of PEPP was reduced by grafting with GMA functionalized with GO. With addition of GO, modified PEPP membrane became hydrophilic with the addition of hydrogen bond from the GO. The FTIR results also showed traces of GMA on the surface of PEPP and after functionalized the membrane, it has shown prominent new peak which correspond to the C bond from GO from ultrasonication method.

1. Introduction

1.1. Vanadium Redox Flow Battery

Vanadium redox flow battery (VRFB) has attracted significant attention for its great potential as electrochemical energy storage (EES) due to its scalability and versatility in comparison to other redox flow battery options such as Zinc-Bromide or Iron- Chromium. Other properties of VRFB are low electrolytes contamination as VRFB uses all vanadium states (V^{2+}, V^{3+}, V^{4+}, and V^{5+}) in both positive and negative electrolytes, good electrochemical reversibility, great performance and long life cycle [1–4]. VRFB typically consists of set containers for electrolytes, a pump and a single cell unit as it basic setups.

The main piece of VRFB is the membrane as the membrane act as a divider to inhibit cross-mixing of electrolytes from both sides as well as to complete the circuit as it allowing ion transfer from anode to cathode [5-6]. An ideal characteristics look for VRFB are low vanadium permeability, high mechanical and chemical stability, high ion conductivity and low cost [7–9]. The favorable choice of membrane used in VRFB is the Dupont Nafion series including N115, N117, N112, N1135 as well as N1100 which preferred for their high proton conductivity at an average of 100mS cm-1 and high mechanical strength and chemical stability. Even though Nafion is commercially used for VRFB, it had a significant setback as it has high vanadium permeability which causing a drop in the performance of

VRFB and also low ion selectivity [2,7,10,11]. Nafion membrane cost also a concern to be considered as it is highly expensive costing around \$500 to \$1000 per m2 [2,12,13].

1.2. Porous Membrane

Porous membrane has been in the spotlight as an alternative to replace Nafion. The porous membrane is a non-ionic membrane which has many potentials such as high stability, low costs and good proton conductivity [7,13,14]. However pristine porous membrane has been found to have high vanadium permeability despite its plenty advantages; thus further modifications are needed. Surface modification is a well-known method that can be divided into two, surface grafting and surface coating. Surface coating involved deposition of a thin layer onto the membrane surface however, it is unstable and the coated layers can be washed away during washing or while in an application. This is mostly due to the weak physical adsorption bond between coated layers and the membrane. Surface grafting does not face this problem by totally inserted the monomers onto the surface of the membrane by covalent bonding interaction.

Grafting polymerization for chemical modification and various techniques can be used such as UV photo-irradiation, plasma and high energy irradiation that can also cause recovery in water flux and hydrophilicity of the membrane [15]. However, UV irradiation have disadvantages such as poor penetration ability compared to gamma rays. The plasma method also has drawbacks as the chemical and physical alteration can only occur on the surface of the membrane. The best method to modify the surface of a membrane is by radiation-induced graft polymerization and has advantages such as homogenous reaction, no chemical inhibitors, high penetration rate and low amount of monomers needed to modify a membrane surface [15]. Previous studies reported even with grafting, the porous membrane is still unable to surpass Nafion. Thus, extra work required to make a porous membrane to meet the ideal criteria for VRFB application.

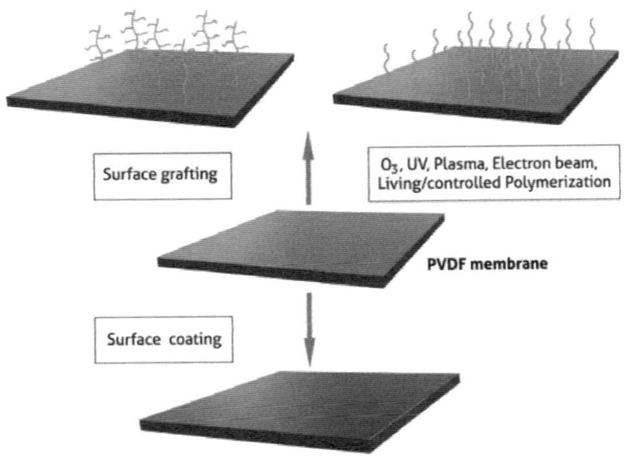

Figure 1. Route of surface modification of surface grafting and surface coating for porous membrane according to previous studies. [16]

1.3. Graphene Oxide

Carbon-based materials such as graphene and graphene oxide (GO)-based membrane have garnered a huge interest as substitute materials for VRFB applications in recent years. The versatile potentials for various applications such as separations and purifications, high selectivity for both gasses and liquids and antifouling are one of the attractions of GO-based membrane. Other properties of GO that attract researchers are strong mechanical strength, easily accessible, high resistance to strong acids/alkaline and organic solvents and low cost [17–19]. The potentials of GO are similar to the ideal characteristic properties for the VRFB membrane. The integration of GO with other types of nanomaterials will increase the potential of GO and making it an excellent potential for conversion and energy storage devices [20]. Past studies indicated the addition of GO is helping to reduce the permeability issues of porous membranes as the hydrophilicity of porous membrane increased with the formation of hydrogen bonds on the surface of the membrane [19].

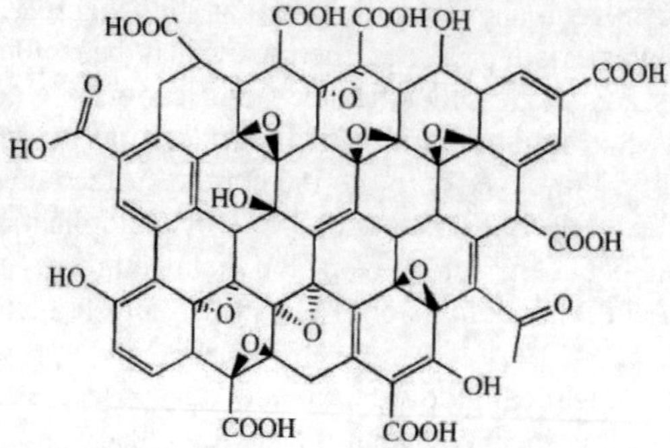

Figure 2. Molecular structure of Graphene Oxide.

In this study, pristine porous membrane polyethylene/polypropylene (PEPP) is modified using surface grafting with glycidyl methacrylate (GMA) and functionalized with graphene oxide (GO). The morphology and microstructure of the membrane were characterized by Scanning Electron Microscope (SEM) and Fourier-transform infrared (FTIR) spectroscopy.

2. Experimental

2.1. Material
Graphene oxide with monolayer content of 95% purchased from Graphenea, USA. Grafted PEPP with GMA was received from Nuclear Malaysia with degree of grafting of 180°. All reagent and membrane were used as received without further purification or treatment.

2.2. Characterization
The composite membranes SEM images were attained using the Philips XL30 field emission scanning electron microscope (FESEM) after coating with 5 nm Au. Fourier transform-infrared (FTIR) was executed using Agilent Cary 660 spectrometer.

2.3. Grafting
In the preparation of PEPP-g-GMA, all irradiation grafting will be done by Nuclear Malaysia Agency and the PEPP-g-GMA will be used as received. The reaction was performed following past work published somewhere else [21,22]. Usually, the PEPP membrane was placed in a polyethylene (PE) zip lock bags which were purged using purified nitrogen gas, sealed and irradiated using an electron beam accelerator (EPS 3000) to a total dose of 50 kGy, 100 kGy, and 300 kGy. Degree of grafting (dg) was determined by gravimetric method from the following equation:

$$d_g(\%) = \frac{W_g - W_s}{W_s} \times 100 \quad (1)$$

Where W_s and W_g are the weights of the original substrate and final grafted samples, respectively.

2.4. Functionalizing membrane
The PEPP-g-GMA will be functionalized with 5.0 % GO using sonication and a simple immersion method to determine which method is better. In both methods, PEPP-g-GMA will be functionalized for

IOP Conf. Series: Materials Science and Engineering **808** (2020) 012012 doi:10.1088/1757-899X/808/1/012012

24hr to ensure complete functionalization. In simple immersion, grafted PEPP was immersed inside a 25ml vials with 5.0 % GO solution. The vial was then secured using a retort stand and put inside a water bath with silicon oil and placed on top of a heating plate. Temperature estimated around 60°C was applied. Even though no heat was applied in the ultrasonication process, the temperature still rises and to ensure the same condition on both methods, the heat was applied to the simple immersion method. Similar to simple immersion, grafted PEPP was placed inside a 25ml vial with a 5.0 % GO solution as well for sonication. The vial was then put in a 50ml beaker filled with water before placed inside an ultrasonicator and left for 24hours with no heat applied. After 24h hours, both membranes were washed with deionized water for 5.0 hours to remove any excess GO and dried in an oven overnight.

3. Results and Discussion

3.1. Microscopy analysis of PEPP-g-GMA and proton conductivity

Images from Figure 3 showing a morphology of pristine PEPP membrane (a) and modified PEPP membrane (b). It is clearly observable in (b) modified PEPP has changed with its morphology compared to as it pristine state in (a). The structure of PEPP has been altered and thicker than it was before which reduces the gap between one fibres to another. The pore size of pristine PEPP has been reduced and thus help to reduce the high vanadium permeability of pristine PEPP. High vanadium permeability will cause the battery to self-discharge and results in low coulombic efficiency of VRFB. With reduced pore size, lower vanadium permeability was expected but this will impact low proton transfer as well. The role of graphene oxide was to promote the proton transfer to ensure modified PEPP will still have a good proton conductivity. Upon functionalization of GO, PEPP membrane change its hydrophilicity to a hydrophilic membrane from the hydrogen bond from GO. Hydrophilic properties of PEPP helps in boosting the proton transfer across PEPP membrane. Proton conductivity of modified PEPP was conducted under a temperature of 30°C and 100% relative humidity and was recorded to be 22.7 $mS.cm^{-1}$. **Figure** is an illustration of the mechanism of the functionalization of porous PEPP membranes with graphene oxide.

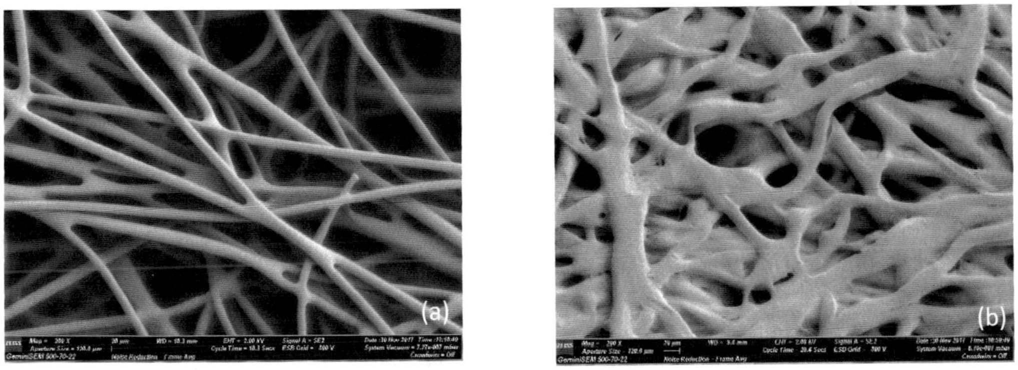

Figure 3. SEM images displaying morphology of pristine PEPP (a) and modified PEPP membrane (b).

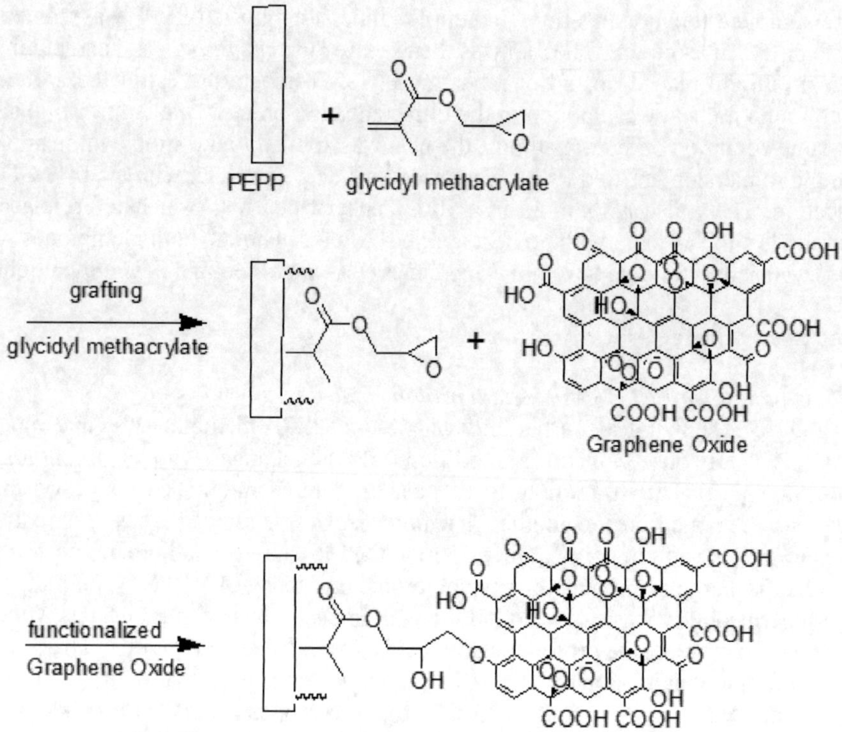

Figure 4. Mechanism of functionalization of porous PEPP membrane with graphene oxide.

3.2. FTIR analysis of PEPP-g-GMA and functionalized membrane

The IR spectrum of PEPP-g-GMA in **Figure** showed that there was a significant peak at 1724 cm^{-1} band which corresponds to the C=O bond from the structure of GMA. The peak around 3000-2950 cm^{-1} was the vibration of the CH_2 group. Upon functionalized with GO the peak around 1724 cm^{-1} for immersion reduced while sonication seems the same. A new peak also appeared around 1649 cm^{-1} at immersion and sonication indicated a sp^2-hybridized C bond from the GO. The peak at 905 cm^{-1} was associated with epoxy of GMA however in immersion the peak seems reduced more than sonication. These peaks would indicate the reaction between GMA and GO, however, GO also had epoxy on its structure. Thus the slight reduction for sonication might indicate that the epoxy from GMA was reacted with GO but the signal was overlapped with its own epoxy reading while in immersion there was less reaction of GMA and GO. From this, we can conclude that the best technique for functionalizing GO between the two methods was ultrasonication. During the ultrasound process, the high-intensity ultrasound enables chemical reaction which functionalized GO with grafted PEPP membrane.

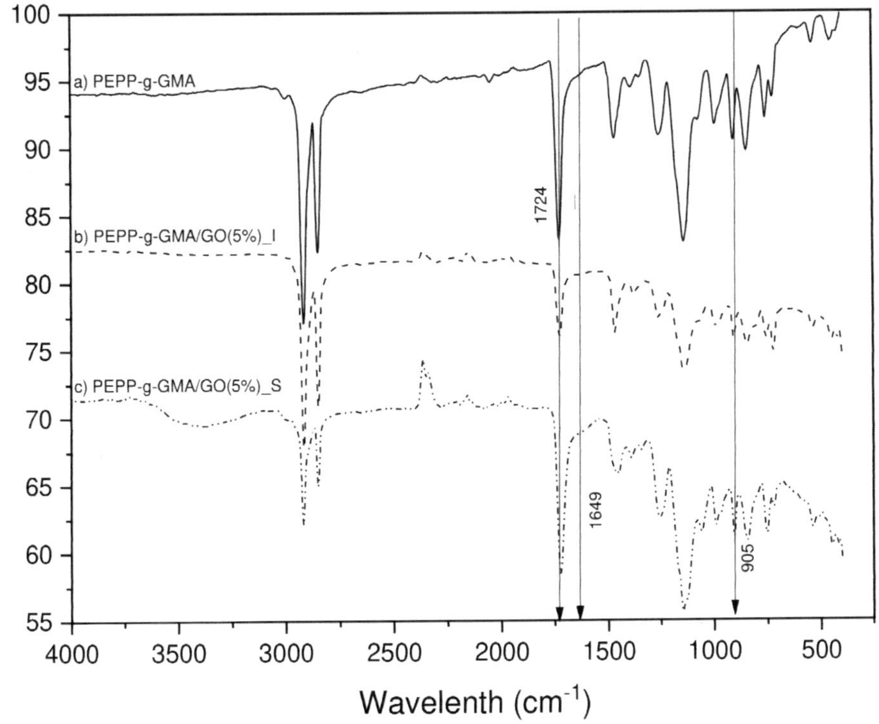

Figure 5. FTIR spectra of combined of PEPP-g-GMA (a) and functionalized PEPP-g-GMA with 5% GO content through simple immersion (b) and sonication (c).

4. Conclusion

The increasing of PEPP thickness showed that PEPP was successfully modified and thus reducing the porous size of PEPP. This will theoretically helps to reduce the vanadium permeability of the porous membrane. Further test is required to test the vanadium permeability of the modified PEPP. The FTIR results also showed traces of GMA on the surface of PEPP. After functionalized the membrane, it was found that sonication was the best method out of the two option as it has shown more prominent new peak at 1649 cm^{-1} which correspond to the C bond from GO. The proton conductivity of the modified PEPP was recorded at 22.7 mS cm^{-1}. In compared with N117 which the commercial membrane for VRFB, proton conductivity of modified PEPP was lower by 69.9%. However, if compared with other modified porous membranes such as PES, modified PEPP has a similar value which was recorded to be around 22.4 mS cm^{-1}. The graphene oxide content may be varied and optimized in future work to find the best concentration to improve the performance of the modified membrane.

Acknowledgments

The research was primarily supported by the joint partnership research program of MTJA #4C172 and Research Univesity Grant UTM #16J49.

5. References

[1] Parasuraman A, Lim T M, Menictas C and Skyllas-kazacos M 2012 Review of material research and development for vanadium redox flow battery applications *Electrochim. Acta*

[2] Shi Y, Eze C, Xiong B, He W, Zhang H, Lim T M, Ukil A and Zhao J 2020 Recent development of membrane for vanadium redox fl ow battery applications : A review **238** 202–24

[3] Weber A Z, Mench M M, Meyers J P, Ross P N, Gostick J T and Liu Q 2011 Redox flow batteries: a review *J. Appl. Electrochem.* **41** 1137–64

[4] Chen H, Cong T N, Yang W, Tan C, Li Y and Ding Y 2009 Progress in electrical energy storage system: A critical review *Prog. Nat. Sci.* **19** 291–312

[5] Wang N, Peng S, Li Y, Wang H, Liu S and Liu Y 2012 Sulfonated poly(phthalazinone ether sulfone) membrane as a separator of vanadium redox flow battery *J. Solid State Electrochem.* **16** 2169–77

[6] Li X, Zhang H H, Mai Z, Zhang H H and Vankelecom I 2011 Ion exchange membranes for vanadium redox flow battery (VRB) applications *Energy Environ. Sci.* **4** 1147–60

[7] Qiao L, Zhang H, Lu W, Xiao C, Fu Q, Li X and Vankelecom I F J 2018 Advanced porous membranes with slit-like selective layer for flow battery *Nano Energy* **54** 73–81

[8] Choi E Y, Strathmann H, Park J M and Moon S H 2006 Characterization of non-uniformly charged ion-exchange membranes prepared by plasma-induced graft polymerization *J. Memb. Sci.* **268** 165–74

[9] Sizov V E, Kondratenko M S, Gallyamov M O and Stevenson K J 2018 Advanced porous polybenzimidazole membranes for vanadium redox batteries synthesized via a supercritical phase-inversion method *J. Supercrit. Fluids* **137** 111–7

[10] Jiang B, Wu L, Yu L, Qiu X and Xi J 2016 A comparative study of Nafion series membranes for vanadium redox flow batteries *J. Memb. Sci.* **510** 18–26

[11] Cha S H 2015 Recent Development of Nanocomposite Membranes for Vanadium Redox Flow Batteries *J. Nanomater.* **2015**

[12] Yuan Z, Duan Y, Zhang H, Li X, Zhang H and Vankelecom I 2016 Advanced porous membranes with ultra-high selectivity and stability for vanadium flow batteries *Energy Environ. Sci.* **9** 441–7

[13] Lu W, Yuan Z, Zhao Y, Zhang H, Zhang H and Li X 2017 Porous membranes in secondary battery technologies *Chem. Soc. Rev.* **46** 2199–236

[14] Yuan Z, Duan Y, Zhang H, Li X, Zhang H and Vankelecom I 2016 Advanced porous membranes with ultra-high selectivity and stability for vanadium flow batteries *Energy Environ. Sci.* **9** 441–7

[15] Shin I H, Hong S, Lim S J, Son Y S and Kim T H 2017 Surface modification of PVDF membrane by radiation-induced graft polymerization for novel membrane bioreactor *J. Ind. Eng. Chem.* **46** 103–10

[16] Subasi Y and Cicek B 2017 Recent advances in hydrophilic modification of PVDF ultrafiltration membranes – a review: part I *Membr. Technol.* **2017** 7–12

[17] Ma J, Ping D and Dong X 2017 Recent developments of graphene oxide-based membranes: A review *Membranes (Basel).* **7**

[18] Abouzari-Lotf E, Etesami M and Nasef M M 2018 *Carbon-Based Nanocomposite Proton Exchange Membranes for Fuel Cells* (Elsevier Inc.)

[19] Bystrov V S, Bdikin I K, Silibin M V., Karpinsky D V., Kopyl S A, Goncalves G, Sapronova A V., Kuznetsova T and Bystrova V V. 2017 Graphene/graphene oxide and polyvinylidene fluoride polymer ferroelectric composites for multifunctional applications *Ferroelectrics* **509** 124–42

[20] Mahmood N, Zhang C, Yin H and Hou Y 2014 Graphene-based nanocomposites for energy storage and conversion in lithium batteries, supercapacitors and fuel cells *J. Mater. Chem. A* **2** 15–32

[21] Abouzari-Lotf E, Ghassemi H, Nasef M M, Ahmad A, Zakeri M, Ting T M, Abbasi A and Mehdipour-Ataei S 2017 Phase separated nanofibrous anion exchange membranes with polycationic side chains *J. Mater. Chem. A* **5** 15326–41

[22] Mohamed Mahmoud Nasef, Masao Tamada, Noriaki Seko E A-L 2014 Advances in the development of functional polymers using radiation induced emulsion polymerization *Recent Res. Devel. Polym. Sci.* **12** 107–28

IOP Conf. Series: Materials Science and Engineering **808** (2020) 012013 doi:10.1088/1757-899X/808/1/012013

Heterogeneous catalyst application in biodiesel production: Needs to focus on cost effective and reusable catalysts

Peter Abiodun Jeremiah[1,2,4], Aishah Abdul Jalil[2,3,*], and Moses Aderemi Olutoye[4]

[1] School of Engineering Technology, Department of Chemical Engineering, Federal Polytechnic, Bida, Nigeria
[2] School of Chemical and Energy Engineering, Faculty of Engineering, Universiti Teknologi Malaysia, 81310UTM Johor Bahru, Johor, Malaysia.
[3] Center of Hydrogen Energy, Institute of Future Energy, Universiti Teknologi Malaysia,81310 UTM Johor Bahru, Johor, Malaysia.
[4] School of Infrastructure, Process Engineering and Technology, Department of Chemical Engineering, Minna, Niger State, Nigeria

*aishahaj@utm.my

Abstract. Start your abstract here… Uses of heterogeneous catalyst in bio-energy production also refer to as green energy has been in existence and well researched. Majority of recent heterogeneous catalysts produced focus on optimizing yield of biodiesel from a single feedstock without concerted efforts been made to consider the cost of production. They are mostly developed and produced from synthetic chemicals with their attendants high cost of production. The present review summarizes the needs to produce heterogeneous solid catalyst from wastes and natural resources like clay which is available in all parts of the world.

1. Introduction

The uses of heterogeneous catalyst in fatty acid methyl esters (FAME) production was introduced to solve the problems encountered in applying homogeneous catalysts. Some of such problems include: soap formation, difficulty in separation of the final products leading to generation of large waste water, poor recoverability of products etc among others. Heterogeneous catalyst essentially describes processes that involve different phases,solid, ,liquid and gases. The reactants are mostly liquid or gaseous states while catalyst are solid. Some merits of heterogeneous catalyst that favours its general acceptability in (FAME)production includes: ease separation from the final products, reusability, non-toxic and is environmentally friendly. The process can easily be operated whether batch, continuous or semi-batch mode [1–4]. Helwani et al., [5] asserted that heterogeneous catalyst are mixtures of metallic oxides mostly from non-noble metals. The author said that the catalyst contains both basic and acidic components which are essential properties in converting triglycerides in oil to biodiesel in the presence of alcohol which is similar to those of homogeneous catalyst. [3]. The reaction mechanism also has been reported showing some similarities to those of homogeneous catalyst [6, 7]. Solid acid catalyst is widely employed, if the oil contains high free fatty acid (FFA) and water content. Also, solid acid catalysts require high operating temperature to achieve reasonable conversion to desired products [4 8]. Despite all these advantages of heterogeneous catalyst, mass transfer across interphase boundary remain a major challenge. This is common with most heterogeneous catalyzed process. The reaction rate is reduced resulting in low yield. Hillion et-al.; Mbaraka and Shanks, [8, 9] have suggested a way of overcoming interphase diffusion limitation in heterogeneous catalysts. These authors suggested the use of promoters or supports which can provide large surface area and pore sizes for active components of the catalyst in order to overcome this limitation and enhance mass transfer conversion. Aluminum oxide, Al_2O_3 (alumina) had been recognized as the suitable promoter for the transesterification reaction without catalyst loss. Heterogeneous catalyst can also be manipulated via understanding of its reaction mechanism to include desired catalyst properties so that the presence of FFA's or water does not adversely affect reaction steps during transesterification. Other advantages of the heterogeneous catalyst include its prolonged lifetime for FAME production and the fact that it is relatively cheap in comparison to homogeneous catalyst [2].

2. Clay as a source of heterogeneous catalyst

Clay is composed of particles of hydrous aluminum silicates and other mixture of metallic oxides and minerals. Clay is available naturally and in modified forms and has several applications as catalytic materials to carry out some typical industrial reactions [10, 11]. Clay minerals are important constituents of the earth crust not only because of their abundance but merely because of their chemical activity. It is found almost everywhere but differs greatly in purity. Clay in its pure condition is comparatively rarely found, the larger portion of the deposit being composed of impure clay. Pure clay is a silicate of aluminum and when found in large quantities it is called Kaolin. Clay material has a wide application for example; it is used in paper, cement and porcelain, sand casting, bricks, dishware, and glass industries and in the removal of heavy metals from waste water and air purification. It is inexpensive, abundant and widely fond in every areas and region of the earth crust. [12]. Clay contains several metallic oxides: amongst these is silicon oxide (SiO_2), aluminum oxide, calcium oxide (CaO), iron oxide and so on. The presence of these metallic oxides justifies its usage for biodiesel production. The CaO and Al_2O_3 serve as the basis and acidic component while SiO_2 serves as support for the catalyst which makes it to act like a bi-metallic catalyst.

3. Recent review of some heterogeneous solid catalyst

Ghiaci et al., developed a highly active novel heterogeneous catalyst from biontshell by a tri-step procedure: in complete carbonization- KF impregnation activation and produced biodiesel from rape seed oil [10]. The study revealed that the synthesized catalyst was active for catalyzing transesterification and also recyclable. Transesterification of rape seed oil using 3 wt% of biontshell catalyst, 9: 1 molar ratio of methanol to oil and reaction temperature of 70°C yielded 97.5 % biodiesel within 3 h.

Wei et al. [13], studied waste egg shell as low-cost solid catalyst for biodiesel production from soybean oil and showed high activity. The experimental result showed that transesterification of soybean oil with the waste eggshell catalyst calcined at 1000°C under the reaction condition of 3 wt% catalyst, 9:1 methanol to oil molar ratio at 65°C for 3 h exhibited the best catalytic activity yielding 95% biodiesel. Table 1 below contains some basic heterogeneous solid catalyst for transesterification reaction stating type of catalyst, feedstock used, optimum reaction conditions, biodiesel yield and reference. Also, Table 2 shows the catalyst obtained from naturally occurring and waste materials used for biodiesel production.

IOP Conf. Series: Materials Science and Engineering **808** (2020) 012013 doi:10.1088/1757-899X/808/1/012013

Table 1: Some basic heterogeneous solid catalyst used for biodiesel production .

Catalyst	Feedstock	Optimum reaction conditions	Biodiesel FAME yield (%)	Reference
KF/CaO	Chinese tallow seed oil	T= 65°C, t= 2.5h, M:O =12:1	96.8	[14]
KNO_3/CaO	Rape oil	T= 65°C, t= 3h M:O=6:1, W = 1 wt.%	98.0	[15]
CaO/ZnO	Ethyl Butyrate	T= 60°C, t= 2h, M:O=12:1,W= 1.3 wt.%	90.0	[16]
Atta pulgite (ATP)/ K_2CO_3	Soybean oil	T= 60°C, t= 3h, M:O = 22:1, W = 8 wt.%	93.4	[17]
MnZr/ AlO_3	Waste cooking oil	T= 150°C, t= 5h, M:O=14:1 W = 2.5 wt.%	93.0	[18]
Fe_2O_3-MnO-SO_4^{2-} /ZrO_2 nanoparticle solid catalyst	Waste cooking oil	T= 180°C, t= 5h, M:O = 20:1 W = 3 wt.%	96.5	[19]
Na_2ZrO_3	Ricinus communis oil	T= 65°C, t= 3h, M:O = 15:1 W = 5 wt.%	99.9	[20]
SrO	Palm oil	T= 80°C, t= 3h, M:O = 12:1 W = 5 wt.%	98.2	[21]
KF-Atta-2M-5	Silybum Marianum oil	T= 65°C, t= 5h, M:O = 12:1 , W = 12 wt.%	93.5	[22]
20-CeO_2/Li/SBA-15	Cotton seed oil	T= 65°C, t= 4h, M:O = 40:1 W = 10 wt.%	98	[23]
Magnetic Fe_3O_4/MCM-41 composite-supported sodium silicate	Soybean oil	T= 78°C, t= 8h, M:O = 25:1 W = 3 wt.%	99.2	[24]
MgO/y-Al_2O_3	Soybean oil	T= 60°C, t= 6h, M:O = 6:1 W = 5 wt.%	85.0	[25]
ZnO/y-Al_2O_3	Castor oil	T= 80°C, t= 6h, M:O = 6:1 W = 5 wt.%	97.0	[25]
Calcium glyceroxide	Rape seed oil	T= 60°C, t= 3h, M:O = 8:1 W = 1 wt.%	96.0	[26]

Tungsten (W) supported TiO_2/SiO_2	Waste cooking oil	T= 65°C, t= 4h, M:O = 30:1, W= 5 wt.%	98.0	[27]
K-pumice	Microalgae oil	T= 62°C, t= 10mins, M:O = 12:1 W= 10 wt.%	85.0	[28]

Note: T – reaction temperature; t – reaction time; M:O – methanol to oil; W – catalyst loading

Table 2: Some of the catalyst obtained from naturally occurring and waste materials used for biodiesel production.

Catalytic Materials	Feedstock	Reaction/ Operating condition	Biodiesel yield (%)	Reference
Clamshell	Waste frying oil	C.C. – 3.0% R.T – 60 °C Rt – 180min C.T – 900 °C	89.0	[21]
Waste animal bone	Palm oil	C.C. – 5-25% R.T – 65 °C Rt – NM C.T – 200 – 1000 °C	96.78	[29]
Snail shell	Waste frying oil	C.C. – 1.4% R.T – 50 – 65 °C Rt – 5 – 8 h C.T – 900 °C	87.28	[30]

Note: C.C. – catalyst concentration; RT – reaction temperature; Rt – reaction time; C.T – calcinations temperature; NM-not mentioned

4. General trend in recent catalyst synthesis in biodiesel production

The high performance at the optimum conditions given earlier, is due to relatively high surface area and the net negative charge on the clay [23]. Also the Attapulgate clay used in this work has needlelike shaped silicate with particle size approximately 20 nm in diameter with the length of varying range spanning hundreds of nm to μm [17]. It also contains hydrated magnesium aluminum silicates that contains ribbon of 2:1 phyllosilicate structure. This unique morphology and structure provides reasons for its relatively high yield amidst other operating conditions. Also in the work of high yield recorded by [23] was due to combined effect of Li and Ce in a single catalyst. The bi-metal catalysts possesses greater thermal stability and Li have higher activity. In this work, both metal ions were immobilized over mesoporous silica surface which enhanced its activity. The yield is due to higher surface area from calcined fusion of waste chicken and fish bones catalyst ($34.73 m^2 g^{-1}$) which is better than the uncalcined fusion of waste chicken and fish bone catalyst ($0.87 m^2 g^{-1}$) [29]. Increase in surface area and pore volume of the calcined fusion waste chicken and fish bones after calcination at 1000°C for 4h was responsible for better performance in FAME yield.

In general, majority of recent research in heterogeneous catalysts from wastes and natural resources (clay) for biodiesel production uses interplay of these variables to produce cost effective and reusable catalysts.

5. Conclusion

The synthesis of clay-Bi-Zn composite catalyst for transesterification of waste frying oil is one of the efforts made at producing cost effective and reusable catalyst. Various types of clay supported catalysts can be developed for biodiesel production. In this review, the necessity to focus on wastes and naturally occurring materials to develop heterogeneous catalyst to produced biodiesel from reaction between alcohol and oil (triglycerides) has been emphasized.

Acknowledgments

This research study was sponsored by the Universiti Teknologi Malaysia through Transdiciplinary Research Grant (Grant No. 06G53) and Collaborative Research Grant (Grant No. 07G62) and Federal Polytechnic, Bida, Nigeria.

References

[1] Zabeti M, Wan Daud W M A, and Aroua M K, 2009 Activity of solid catalysts for biodiesel production: A review *Fuel Process. Technol.* **90** 770–777
[2] Yan S, Dimaggio C, Mohan S, Kim M, Salley S O, and Ng K Y S 2010 Advancements in Heterogeneous Catalysis for Biodiesel Synthesis *Top. Catal.* **53** 721–736
[3] Endalew A K, Kiros Y, and Zanzi R 2011 Heterogeneous catalysis for biodiesel production from Jatropha curcas oil (JCO) *Energy* **36** 2693–2700
[4] Yu X, Wen Z, Li H, Tu S, and Yan J 2011 Transesterification of Pistacia chinensis oil for biodiesel catalyzed by CaO – CeO_2 mixed oxides *Fuel* **90** 1868–1874
[5] Helwani Z, Othman M R, Aziz N, Fernando W J N, and Kim J 2009 Technologies for production of biodiesel focusing on green catalytic techniques : A review *Fuel Process. Technol.* **90** 1502–1514
[6] Lotero E, Goodwin J, Bruce J A, Suwannakarm K, Luu Y and Lopez D 2006 The catalyst of biodisiel systhesis *Catalysis* **19** 41–83
[7] Di-Serio M, Tesser R, Pengmei L, and E. Santacesaria 2008 Heterogeneous Catalysts for Biodiesel Production *Energy Fuels* **9** 207–217
[8] Hillion G, Delfort B, Pennec D, Bournay L and Chodorge J A 2003 Biodiesel production by a continuous process using a heterogeneous catalyst *ACS Division of Fuel Chemistry, Preprints* **48** 636-8 vol. 48, no. 1, pp. 636–638, 2003.
[9] Mbaraka I K, Shanks B H 2006 Conversion of oils and fats using advanced mesoporous heterogeneous catalysts *J Amer Oil Chem Soc* **83** 79–91
[10] Ghiaci M, Aghabarari B, and Gil A 2011 Production of biodiesel by esterification of natural fatty acids over modified organoclay catalysts *Fuel* **90** 3382–3389
[11] Kallo D 2001 Applications of Natural Zeolites in Water and Wastewater Treatment *Geochem.* **45** 519–550
[12] Olutoye M A and Hameed B H 2013 A highly active clay-based catalyst for the synthesis of fatty acid methyl ester from waste cooking palm oil *Applied Catal. A, Gen.* **450** 57–62
[13] Wei Z, Xu C and Li B 2009 Bioresource Technology Application of waste eggshell as low-cost solid catalyst for biodiesel production *Bioresour. Technol.* **100** 2883–2885
[14] Wen Z, Yu X, Tu S, Yan J and Dahlquist E 2010 Synthesis of biodiesel from vegetable oil with methanol catalyzed by Li-doped magnesium oxide catalysts *Appl. Energy* **87** 743–748
[15] Encinar J M, González J, Pardal A and Martínez G 2010 Rape oil transesterification over heterogeneous catalysts *Fuel Process. Technol.* **91** 1530–1536

[16] Alba-Rubio A C, Santamaría-González J, Mérida-Robles J M, Moreno-Tost R, Martín-Alonso D, Jiménez-López A and Maireles-Torres P 2010 Heterogeneous transesterification processes by using CaO supported on zinc oxide as basic catalysts *Catal. Today* **149** 281-287
[17] Ye B, Qiu F, Sun C, Li Y and Yang D 2014 Biodiesel production from soybean oil using heterogeneous solid base catalyst *J. Chem. Technol. Biotechnol.* **89** 988-997
[18] Amani H, Ahmad Z, Asif M and Hameed B 2014 Transesterification of waste cooking palm oil by MnZr with supported alumina as a potential heterogeneous catalyst *Journal of Industrial and Engineering Chemistry* **20** 4437-4442
[19] Elhassan F, Rashid U and Taufiq-Yap Y H 2014 Synthesis of waste cooking oil-based biodiesel via effectual recyclable bi-functional Fe2O3MnOSO42−/ZrO2 nanoparticle solid catalyst *Fuel* **142** 38–45
[20] Martínez Ponce A, Mijangos G, Romero I, Hernández-Altamirano R, Mena-Cervantes V and Gutiérrez S 2018 A novel green one-pot synthesis of biodiesel from Ricinus communis seeds by basic heterogeneous catalysis *Journal of Cleaner Production* **196** 340–349
[21] Roschat W, Phewphong S, Kaewpuang T and Promarak V 2018 Synthesis of glycerol carbonate from transesterification of glycerol with dimethyl carbonate catalyzed by CaO from natural sources as green and economical catalyst *Materials Today: Proceedings* **5** 13909-13915
[22] Takase M, Pappoe A N M, Afrifa E A and Miyittah M 2018 High performance heterogeneous catalyst for biodiesel production from non-edible oil *Renewable Energy Focus* **25** 24-30
[23] Malhotra R and Ali A 2018 Lithium-doped ceria supported SBA−15 as mesoporous solid reusable and heterogeneous catalyst for biodiesel production via simultaneous esterification and transesterification of waste cottonseed oil *Renew. Energy* **119** 32-44
[24] Xie W, Han Y and Wang H 2018 Magnetic Fe3O4/MCM-41 composite-supported sodium silicate as heterogeneous catalysts for biodiesel production *Renew. Energy* **125** 675-81
[25] Navas M B, Lick I D, Bolla P A, Casella M L and Ruggera J F 2018 Transesterification of soybean and castor oil with methanol and butanol using heterogeneous basic catalysts to obtain biodiesel *Chem. Eng. Sci.* **187** 444-454
[26] Esipovich A, Rogozhin A, Danov S, Belousov A and Kanakov E 2018 The structure, properties and transesterification catalytic activities of the calcium glyceroxide *Chem. Eng. J.* **339** 303-316
[27] Kaur M, Malhotra R and Ali A 2018 Tungsten supported Ti/SiO2 nanoflowers as reusable heterogeneous catalyst for biodiesel production *Renewable Energy* **116** 109-119
[28] Cercado A P, Ballesteros F and Capareda S 2018 Ultrasound assisted transesterification of microalgae using synthesized novel catalyst *Sustain. Environ. Res.* **28** 234-239
[29] Obadiah A, Swaroopa G A, Kumar S V, Jeganathan K R and Ramasubbu A 2012 Biodiesel production from Palm oil using calcined waste animal bone as catalyst *Bioresour. Technol.* **116** 512-6
[30] Birla A, Singh B, Upadhyay S N and Sharma Y C 2012 Kinetics studies of synthesis of biodiesel from waste frying oil using a heterogeneous catalyst derived from snail shell *Bioresour. Technol.* **106** 95-100

IOP Conf. Series: Materials Science and Engineering **808** (2020) 012014 doi:10.1088/1757-899X/808/1/012014

Enhanced Degradation Rates of Paracetamol in Aqueous Solution using Silver Doped Durio Zibethinus Husk Catalyst

G E Harimisa[1], M H Mustapha[1], A Masudi[1], N. W. C Jusoh [1,2,*] and L S Tan[1]

[1] Department of Chemical Process Engineering, Malaysia Japan International Institute ofTechnology, (MJIIT), Universiti Teknologi Malaysia Kuala Lumpur, Jalan Sultan Yahya Petra, 54100 Kuala Lumpur, Malaysia

[2] Center of Hydrogen Energy, Institute of Future Energy, Universiti Teknologi Malaysia Kuala Lumpur, Jalan Sultan Yahya Petra, 54100 Kuala Lumpur, Malaysia

*nurfatehah@utm.my

Abstract. In the recent years, traces of paracetamol in water body had increases due to the increasing usage and production of it. Since the discharge of wastewater containing paracetamol into water body can cause harm to environment and human health, it is crucial to find a suitable treatment method to overcome this problem. Advanced Oxidation Process (AOPs) with the assist of heterogeneous catalyst received growing interest as one of prospective treatment method. Some of noble catalysts such as platinum (Pt), gold (Au) and Silver (Ag) could serve as alternative to popular semiconductor catalyst. This study report a series of Ag (5, 10, 15 wt.%) loaded to Durio Zibethinus Husk (Ag/DZH) with detailed physicochemical properties using X-ray diffraction (XRD), Fourier Transform Infrared (FTIR), X-ray photoelectron spectroscopy (XPS) and Scanning Electron Microscopy (SEM). Then, the respective catalysts were tested for photodegradation of paracetamol (PCT) in aqueous solution. It was showed that the catalysts have a potential to degrade 5 mg/L of paracetamol solution (pH 9) with the highest degradation rate of 0.01197 mg/L.min using 0.03 g/L of 10 wt.% Ag/DZH catalyst. This result could serve as alternative for DZH utilization and reduce environmental pollution.

1. Introduction

The discharge of Emerging Contaminants (ECs) into water stream may become serious problem to environment and human. As one of ECs, more than 80 pharmaceutical compounds and metabolites were detected in ground and surface water over US and Europe in last decades [1]. Improper medication of these contaminants from hospital, personal care and pharmaceutical industry cause several side effect to human [2, 3]. Hence, these pharmaceutical wastes should be handled carefully to conserve environmental sustainability and human welfare.

Many methods were reported to remove these contaminants such as adsorption, coagulation, membrane, and biological. Biological technique was constrained for antibiotics which are bacteria resistance. Meanwhile, adsorption, coagulation and membrane could only transfer pollutant to other places without destruction [4, 5]. Nowadays, the increasing interest was centralized to Advanced Oxidation Process (AOPs), especially on photo-catalysis, which involve heterogeneous catalyst in assistance of external light, such UV irradiation [6, 7]. This method could degrade the contaminants and achieve complete mineralization.

The common photo-catalysts were focus on semiconductor type such as TiO_2, ZnO, CdS, ZrO_2 and SnO_2. However, the latest research shown that noble metal such as platinum (Pt), gold (Au) and silver (Ag) poses as the promising photocatalyst [8, 9]. Despite their high activity, the detailed exploration on silver as photo-catalyst was still scanty. Previously, some support material such as zeolite, silica, carbon and alumina were successfully enhanced their photocatalytic activity [10-13]. In response to this, an abundant agriculture waste, *Durio Zibethinus Husk* (DZH) that consist of cellulose will gave an advantage as a support material. According to previous report, DZH comprise of 13.1% of hemi-cellulose, 60.5% of α-cellulose and 73.5% of hello-cellulose. This characteristic is attractive to assist adsorption which also lead to high degradation of organic pollutants [14] [10]. Therefore, the investigation to modify DZH as catalyst support will also be beneficial for reducing environmental waste.

In this work, simple electrochemical method was utilize to synthesis Ag (5, 10 and 15 wt%) loaded on DZH (Ag/DZH) catalyst. Then, its physicochemical properties were evaluated with X-ray diffraction (XRD) to identify its crystallinity and purity, Scanning Electron Microscopy (SEM) to observe its morphology. Meanwhile, the responsible functional group that affected PCT degradation was investigate with Fourier Transform Infrared Spectroscopy (FTIR). Then, X-ray photoelectron spectroscopy (XPS) was carry out to identify its chemical oxidation. Furthermore, the synthesized catalyst undergo the photocatalytic reaction to degrade Paracetamol (PCT) in aqueous solution.

2. Materials and Methods

2.1. Reagent and Materials
Platinum and silver plates were purchased from Nilaco Metal, Japan. Tetraethylammonium perchlorate (TEAP) was prepared using our previous procedure[15]. Hydrochloric acid (HCl) and Sodium Hydroxide (NaOH) were obtained from Evergreen Engineering and Resources, Malaysia. The paracetamol was obtained from Pusat Kesihatan Kuala Lumpur. The DZH was collected at Setapak Jaya, Kuala Lumpur, Malaysia.

2.2. Preparation of Support Material
The collected DZH sample was washed with deionized water to remove debris and impurities. Prior to cut into small pieces, the sample was dried under the sunlight for 3 days. Finally, the sample was dried at 110 °C overnight and grounded to 0.67 mesh.

2.3. Synthesis of Ag Loaded on DZH
The pristine Ag was prepared by modifying our previous protocol [11]. An open system of electrolysis cell consists of silver plate anode (2 × 2 cm^2) facing a platinum plate cathode (2 × 2 cm^2). The current density was adjusted at 120 mA/cm^2 and 0 °C at ambient pressure. 10 mL of distilled water and 0.1 M TEAP were utilized as supporting electrolyte. After the electrolysis, the sample was dried at 110 °C for 12 h. The same preparation method was carried out for Ag/DZH sample with addition of DZH before the electrolysis. The required weight of Ag supported on DZH was calculated based on the Faraday's law,

$$t = \left(\frac{F}{I}\right)(z \times n) \tag{1}$$

where t is total time (s), F is Faraday's constant of 96,486 C mol^{-1}; I is applied current during electrolysis; z is number of electron transferred or valence number of metal ions, and n is number of mole (n= m/M).

2.4. Characterization of Catalyst
The crystallinity structures of the catalysts were conducted using Bruker Advance D8 X-ray powder diffractometer (XRD). The functional group in DZH assisted for pollutants degradation was elucidated with Fourier Transform Infrared (FTIR) Shimadzu IRT Tracer-100 at wavelength scan of 400-4000 cm^{-1}. The morphologies of the catalysts were observed using Scanning Electron Microscopy (SEM). Finally, the chemical environment exist in the catalysts were identified with X-ray photoelectron spectroscopy (XPS) conducted on a Kratos Ultra spectrometer equipped with an Mg Kα radiation source (10 mA, 15 kV).

2.5. Photocatalytic Testing
The photocatalytic activities of the samples were evaluated towards Paracetamol degradation in aqueous solution. The experiments were carried out in a batch reactor connected to 36 W of fluorescence lamp as visible light sources. 0.003 g of the catalyst was added to 100 mL of Paracetamol solution with desired concentration. This mixture was stirred for 1 h in dark to obtain equilibrium of adsorption desorption between catalyst and solution. Therefore, the sample was irradiated with lamp for another 2 h. During the reaction, aliquots of 2 ml were take periodically at interval of 10 min and centrifuged before analyzing with UV Vis spectrometry Agilent at 325 nm.

3. Result and Discussion

3.1. Characterization of the Catalyst
The XRD pattern of the pristine Ag and various loading percentage of Ag onto DZH were shown in Figure 1. The peak characteristics of Ag was appeared at 44.40° (200), 64.45° (220), 77.40° (311) and 81.69° (222). The structure of these nanoparticles was face centered cubic as confirmed with JCPDS file no. 01-087-0717 as metallic Ag [16]. The principal peak of these spectra increased when increasing the content of Ag from 5 to 15 wt.%. Then, the crystallite size of these catalysts were estimated using Scherer equation by observing the main peak and position of two theta (2θ), using,

$$D = \frac{0.9\lambda}{\beta cos\theta} \tag{2}$$

where λ is XRD wavelength, β as full width at half maximum (FWMH), and θ is bragg's angle.

From this calculation, the crystallite sizes are 46.91, 18.25, 18.26 and 41.04 nm for pristine Ag, 5 wt.% Ag/DZH, 10 wt.% Ag/DZH, 15 wt.% Ag/DZH, respectively. This result verified that DZH could effective supported smaller size of Ag at 5 wt.% and 10 wt.% as a result of good dispersion of Ag over DZH surfaces.

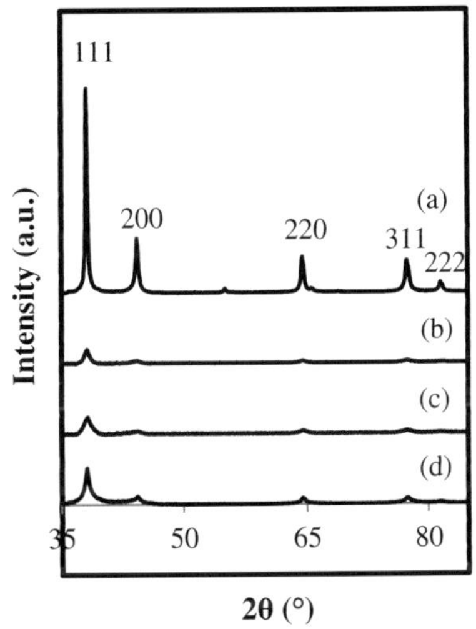

Figure 1. XRD patterns of (a) Ag, (b) 5 wt.% Ag/DZH, (c) 10 wt.% Ag/DZH and (d) 15 wt.% Ag/DZH

The major morphology of Ag as depicted in Figure 2 (a) ws pre-dominantly spherical as well as rectangular [17]. Meanwhile, some of shiny particles of Ag was observed over DZH surface as presented in Figure 2 (b) for 10 wt.% Ag/DZH. This phenomenon is indication that Ag-NPs can be deposited on DZH with certain loading content.

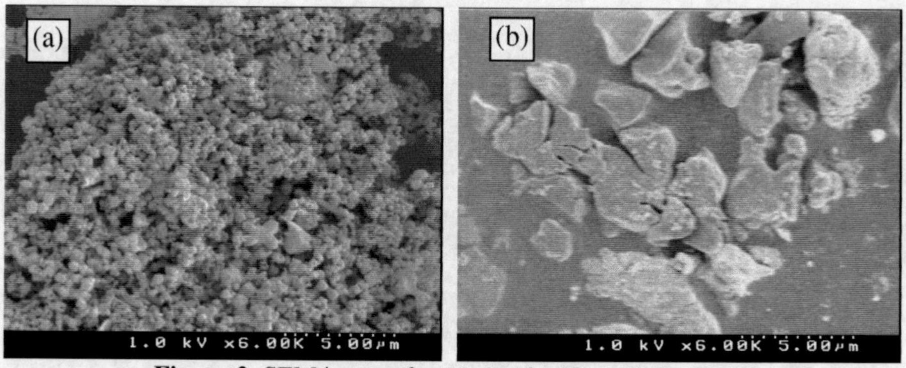

Figure 2. SEM images for (a) Ag (b) 10 wt.% Ag/DZH

Then, FTIR analysis for DZH, Ag and synthesized Ag/DZH catalysts are showed in Figure 3. In DZH spectra as presented in Figure 3 (a), there are three prominent peaks at 1016, 1650 and 3302 cm^{-1}. The intense peak at 1016 cm^{-1} was assigned of C-O bonding of polysaccharide, where 1650 and 3302 cm^{-1} was attributed to hydroxyl group. For Ag catalyst, a broad peak at 631 cm^{-1} can be assigned to Ag peak (Figure 3 (b)). This peak could also be observed when Ag loaded on DZH surface ((Figure 3 (c-e)), with a slightly shift to a lower wavenumber, suggesting the possible interaction of Ag with the DZH support. In addition, the characteristic of C-O bond are shifting of from 1016 cm^{-1} to 1111 cm^{-1}. The C-O in polysaccharide form might helped in the adsorption of paracetamol before the degradation process occurred. In addition, an appearance of new peak at 792 cm^{-1} may be a sign of new bond formation between Ag-NPs and DZH. A broad peak also observed at 3394 cm^{-1} for all the catalyst, attribute to the adsorbed H_2O molecules [18]. This peak tend to decrease with the increasing amount of Ag loading suggesting that the greater the amount of Ag loaded onto the DZH, the fewer hydroxyl groups were adsorbed on the catalyst's surface. The similar trend also observed at peak 1652 cm^{-1} which also assigned to hydroxyl group.

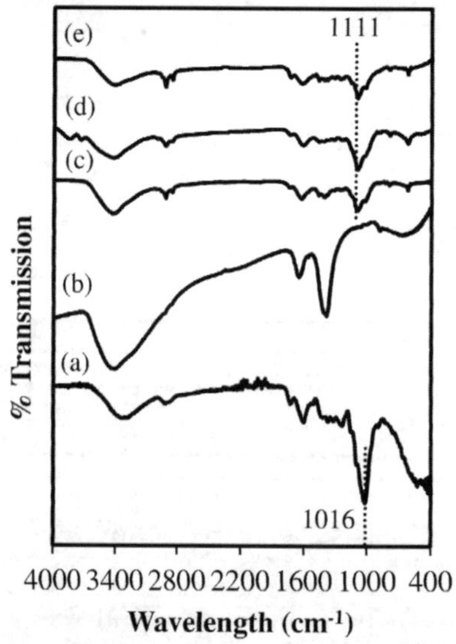

Figure 3. FTIR spectra for (a) DZH, (b) Ag, (c) 5 wt.% Ag/DZH, (d) 10 wt.% Ag/DZH and (e) 15 wt.% Ag/DZH

XPS spectra of Ag and 10 wt.% Ag/DZH were shown in Figure 4 to recognize their chemical oxidation. As expected, the spectra of Ag only exhibit two peak of $3d_{3/2}$ at 374.35 eV and $3d_{5/2}$ at 368.15 eV. However, this peak were shifted to 373.45 eV and 367.75 eV for $3d_{3/2}$ and $3d_{5/2}$, respectively in 10 wt.%

Ag/DZH catalyst. The shifting wavelength and reducing intensity indicate interaction between Ag and DZH, which aligned with the report by Belluso et al. (2019) [19].

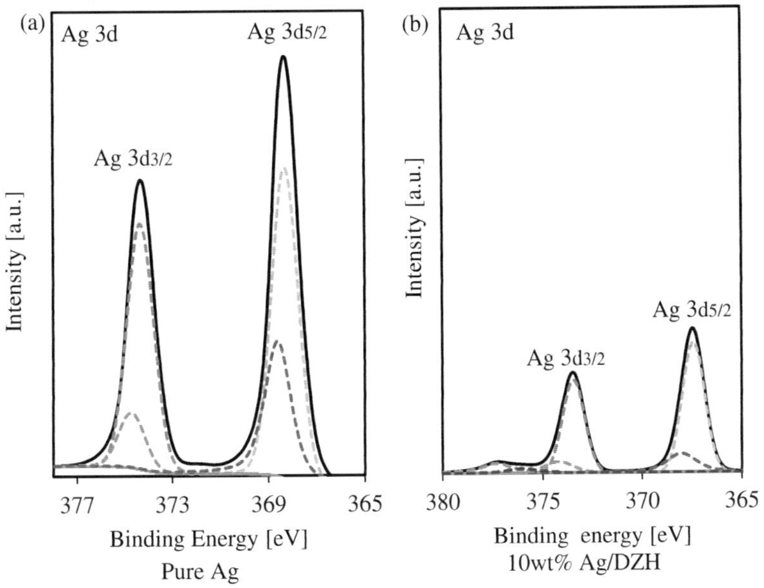

Figure 4. XPS spectra for (a) Ag and (b) 10 wt.% Ag/DZH

3.2. Photodegradation of Paracetamol

3.2.1 Effect of pH. The degradation rates of PCT can be affected by the pH of solution. pH of the solution can determine the surface charge properties of the catalyst, the size of aggregates formed, the charge of PCT and catalyst surface as well as the concentration of hydroxyl radicals. The experiment was run with 0.03 g/L of 5 wt.% Ag/DZH, 5 mg/L of PCT solution at 30°C. With this condition, the photodegradation of PCT on Ag/DZH was studied in the pH range from 5 to 11. Figure 5 showed that the photodegradation of paracetamol is not favoured in an acidic solution (pH $<$ 5). The paracetamol solution has a non-ionic form that caused the water solubility to be minimized and adsorption onto the catalyst was not effective. However, by increasing the solution pH to 9, it enhanced the photodegradation and increased the rate of reaction up to 0.0107 mg/L.min. This phenomenon happened because high pH caused more hydroxide groups available on Ag/DZH surface and form more OH, which it increased the effectiveness and efficiency of PCT degradation [20]. However, when the pH was increasing to pH 11, it produced a poor adsorption of paracetamol, thus decrease the degradation rates.

3.2.2 Effect of Catalyst Dosage. The effect of catalyst dosage was studied between 0.03 g/L to 0.09 g/L using 5 wt.% Ag/DZH at pH 9, 5 mg/L paracetamol solution in 30°C condition. The results were illustrated in Figure 6. As can be seen, the addition of more catalyst resulted to decrease in the degradation rate from 0.01070 mg/L.min to 0.0008 mg/L.min. When the catalyst dosage was increased, the number of active sites that could be absorbed more photons also increased. However, the increment of catalyst dosage seemed to lead to the excess dosage and caused the suspension turbidity in the solution. As a result, the light penetration reduced and the efficiency of the photocatalytic process decreased [21].

IOP Conf. Series: Materials Science and Engineering **808** (2020) 012014 doi:10.1088/1757-899X/808/1/012014

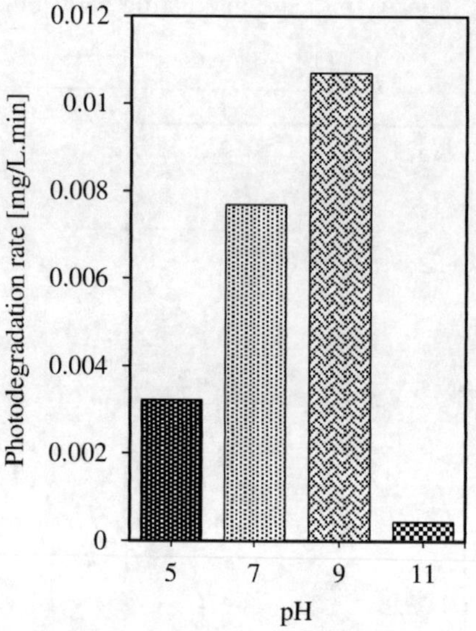

Figure 5. Effect of pH on degradation of paracetamol [Initial Concentration: 5 mg/L; Catalyst: 5 wt.%Ag/DZH; Catalyst Dosage: 0.03 g/L; Temperature: 30 °C]

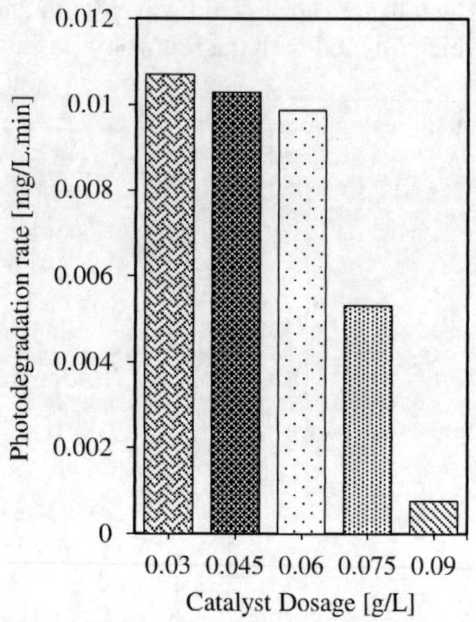

Figure 6. Effect of catalyst dosage on degradation of paracetamol [Initial Concentration: 5 mg/L; Catalyst: 5 wt.%Ag/DZH; pH: 9; Temperature: 30 °C]

3.2.3 Effect of Ag Loading on DZH The photodegradation performance of Ag and various Ag loading on DZH catalyst was evaluated using 0.03 g/L catalyst dosage at pH 9 for 5 mg/L PCT concentration. As demonstrated in Figure 7, 10 wt.% Ag/DZH shows the highest degradation rate (0.01197 mg/L.min). The different photocatalytic efficiencies were achieved for each catalyst, which most likely due to the effect of the Ag dispersion and composition in the DZH surface. Similar trend was reported by Shen et al.(2008) during deposition of ZnO on the silica nanoparticles where well dispersion of ZnO resulted in high photodegradation of methylene blue [22]. Although a high degradation rate was also observed for pure Ag, these results also illustrate the important role of DZH as a support material in reducing the amount of Ag used, which is beneficial from an economic point of view.

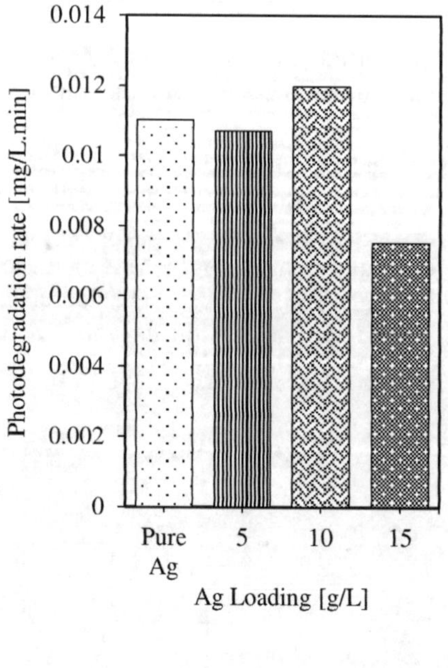

Figure 7. Effect of Ag loading on degradation of paracetamol [Initial Concentration: 5 mg/L; Catalyst dosage: 0.03 g/L; pH: 9; Temperature: 30 °C]

4. Conclusion

This study focused on simple electrochemical approach to synthesis Ag and series of Ag loaded on ZH (Ag/DZH). Their physicochemical properties were studied in detail with XRD, SEM, FTIR and XPS. This report verified that smaller particle size was obtained when Ag loaded to DZH catalyst support. The synergistic effect between particle size and nature of DZH enhanced the photodegradation towards Paracetamol. In summary, 10 wt.% Ag/DZH poses highest degradation rate compared to pure Ag and other Ag/DZH catalysts.

Acknowledgment

This research was supported by Ministry of Education (MOE) through Fundamental Research Grant Scheme (FRGS/1/2018/STG07/UTM/02/11) and MJIIT UTM.

References

[1] Bataduwaarachchi V R and Weeraratne C L 2016 Global medication waste management practices: challenges and opportunities in developing countries *2016* **5** 5

[2] Wilkinson J, Hooda P S, Barker J, Barton S and Swinden J 2017 Occurrence, fate and transformation of emerging contaminants in water: An overarching review of the field *Environ. Pollut.* **231** 954-70

[3] Xu J, Xu Y, Wang H, Guo C, Qiu H, He Y, Zhang Y, Li X and Meng W 2015 Occurrence of antibiotics and antibiotic resistance genes in a sewage treatment plant and its effluent-receiving river *Chemosphere* **119** 1379-85

[4] de Andrade J R, Oliveira M F, da Silva M G C and Vieira M G A 2018 Adsorption of Pharmaceuticals from Water and Wastewater Using Nonconventional Low-Cost Materials: A Review *Ind. Eng. Chem. Res.* **57** 3103-27

[5] Menkiti M C, Okoani A O and Ejimofor M I 2018 Adsorptive study of coagulation treatment of paint wastewater using novel Brachystegia eurycoma extract *Appl Water Sci* **8** 189

[6] Mirzaei A, Chen Z, Haghighat F and Yerushalmi L 2016 Removal of pharmaceuticals and endocrine disrupting compounds from water by zinc oxide-based photocatalytic degradation: A review *Sustainable Cities and Society* **27** 407-18

[7] Bethi B, Sonawane S H, Bhanvase B A and Gumfekar S P 2016 Nanomaterials-based advanced oxidation processes for wastewater treatment: A review *Chemical Engineering and Processing: Process Intensification* **109** 178-89

[8] Gomes J F, Lopes A, Bednarczyk K, Gmurek M, Stelmachowski M, Zaleska-Medynska A, Quinta-Ferreira M E, Costa R, Quinta-Ferreira R M and Martins R C 2018 Effect of Noble Metals (Ag, Pd, Pt) Loading over the Efficiency of TiO2 during Photocatalytic Ozonation on the Toxicity of Parabens *ChemEngineering* **2** 4

[9] Qing W, Chen K, Wang Y, Liu X and Lu M 2017 Green synthesis of silver nanoparticles by waste tea extract and degradation of organic dye in the absence and presence of H2O2 *Appl. Surf. Sci.* **423** 1019-24

[10] Rusu C N and Yates J T 1997 Defect Sites on TiO2(110). Detection by O2 Photodesorption *Langmuir* **13** 4311-6

[11] Jusoh N W C, Jalil A A, Triwahyono S, Karim A H, Salleh N F, Annuar N H R, Jaafar N F, Firmansyah M L, Mukti R R and Ali M W 2015 Structural rearrangement of mesostructured silica nanoparticles incorporated with ZnO catalyst and its photoactivity: Effect of alkaline aqueous electrolyte concentration *Appl. Surf. Sci.* **330** 10-9

[12] Khusnun N F, Jalil A A, Triwahyono S, Jusoh N W C, Johari A and Kidam K 2016 Interaction between copper and carbon nanotubes triggers their mutual role in the enhanced photodegradation of p-chloroaniline *Phys. Chem. Chem. Phys.* **18** 12323-31

[13] Sapawe N, Jalil A A, Triwahyono S, Sah R N R A, Jusoh N W C, Hairom N H H and Efendi J 2013 Electrochemical strategy for grown ZnO nanoparticles deposited onto HY zeolite with enhanced photodecolorization of methylene blue: Effect of the formation of SiOZn bonds *Appl. Catal. A: Gen* **456** 144-58

[14] Abidin M A Z, Jalil A A, Triwahyono S, Adam S H and Nazirah Kamarudin N H 2011 Recovery of gold(III) from an aqueous solution onto a durio zibethinus husk *Biochemical Engineering Journal* **54** 124-31

[15] Jusoh N W C, Jalil A A, Triwahyono S and Mamat C R 2015 Tailoring the metal introduction sequence onto mesostructured silica nanoparticles framework: Effect on physicochemical properties and photoactivity *Appl. Catal. A: Gen* **492** 169-76

[16] Jaafar N F, Jalil A A, Triwahyono S, Efendi J, Mukti R R, Jusoh R, Jusoh N W C, Karim A H, Salleh N F M and Suendo V 2015 Direct in situ activation of Ag0 nanoparticles in synthesis of Ag/TiO2 and its photoactivity *Appl. Surf. Sci.* **338** 75-84

[17] Mehmood A, Murtaza G, Bhatti T M and Kausar R 2014 Enviro-Friendly Synthesis of Silver Nanoparticles Using Berberis lycium Leaf Extract and Their Antibacterial Efficacy *Acta Metallurgica Sinica (English Letters)* **27** 75-80

[18] Janardhanan R, Karuppaiah M, Hebalkar N and Rao T N 2009 Synthesis and surface chemistry of nano silver particles *Polyhedron* **28** 2522-30

[19] Belusso L C S, Lenz G F, Fiorini E E, Pereira A J, Sequinel R, Bini R A, Felix J F and Schneider R 2019 Synthesis of silver nanoparticles from bottom up approach on borophosphate glass and their applications as SERS, antibacterial and glass-based catalyst *Appl. Surf. Sci.* **473** 303-12

[20] Khezrianjoo S and Revanasiddappa H D 2015 Effect of operational parameters and kinetic study on the photocatalytic degradation of m-cresol purple using irradiated ZnO in aqueous medium *Water Quality Research Journal* **51** 69-78

[21] Konstantinou I K and Albanis T A 2004 TiO2-assisted photocatalytic degradation of azo dyes in aqueous solution: kinetic and mechanistic investigations: A review *Appl. Catal. B: Environ* **49** 1-14

[22] Shen W, Li Z, Wang H, Liu Y, Guo Q and Zhang Y 2008 Photocatalytic degradation for methylene blue using zinc oxide prepared by codeposition and sol–gel methods *J. Hazard. Mater.* **152** 172-5

IOP Conf. Series: Materials Science and Engineering **808** (2020) 012015 doi:10.1088/1757-899X/808/1/012015

Bayesian Analysis for Assessing Risks of Rotating Equipment and Its Financial Loss

Rifky Oktarinanda[1] and Norafneeza Norazahar[1,2,*]

[1] School of Chemical and Energy Engineering, Faculty of Engineering, Universiti Teknologi Malaysia, 81310 Johor Bahru, Johor, Malaysia
[2] Centre of Hydrogen Energy, Institute of Future Energy, Universiti Teknologi Malaysia, 81310 Johor Bahru, Johor, Malaysia

*norafneeza@utm.my

Abstract. The petrochemical industry is equipped with major and rotating equipment, such as storage tanks and pumps, to produce the required outputs. The equipment could pose a high level of risks should the organization failed to provide control and mitigation measures. Risk analysis is a conventional and practical technique in identifying hazards and assessing risks associated with equipment. The risk analysis can be integrated with the Bayes's theorem to assess risks of major and rotating equipment. Therefore, this paper presents the use of Bayesian analysis for assessing risks of rotating equipment and its financial loss, specifically in the petrochemical industry. The study begins with developing a cause-and-effect relationship using a Bayesian network. The probability of failure of equipment and its financial loss can be analyzed based on the Bayesian network. The cause-and-effect relationship could also reveal human factors contributing to the probability of failure of equipment. This paper focuses on an accident involving pumps during maintenance activity in the petrochemical plant. The human factors issues contributing to the accident and the financial loss are inadequate procedures and poor work processes.

1. Introduction

Organizations or companies make decisions based on the risk analysis. In engineering field, risk analysis could help the organization or management to develop or design the engineering system. Understanding the risks and the process for analysing the risks have a significant impact to the outcome of engineering design.

The process on identifying hazards and risk factors that have potential to cause harm is known by risk assessment. But, the term not only for this. Risk assessment also focusing on analyse and evaluate the risk associated with the hazard. Find appropriate ways to eliminate the hazard, and to control the risk if the hazard cannot be eliminated is another term where risk assessment take part [1].

Risk assessment have two different perspective. Engineering perspective and insurance company perspective. Engineering here refers to the engineer and all equipment that involve in a chemical process plant. All the hazard material and dangerous equipment will be the main consideration for the risk assessment due to the high potential for causing harm. Risk assessment on engineering perspective focusing on risk management before, during, and after the events of hazards.

Insurance company focus on relationship between risk assessment and financial. Financial for insurance company related to payment for equipment loss and for injury worker. Because of this, insurance company will consider about risk assessment. Because it can help them to manage their financial and to reduce payment for the injury of the worker and equipment loss [2], [3].

Risk assessment in the insurance company is the important to be consider by the company. Since the insurance companies are selling what numerous people consider to be a risk mitigation. The insurance companies find an assortment of risk they have to mitigate.

1.1. Risk assessment for engineering and insurance companies

The petrochemical industry and the insurance company may have different perspectives on risks. The petrochemical industry focuses on the consideration of the risk assessment about harm factors and hazardous materials [4], [5]. They concern the safety of workers, safe working with equipment, and preventing workers from accidents [6]. This perspective also considering the procedure on how to treat the injury when it already happened. With the number of equipment and hazardous materials on a plant, the probability of getting injured during work is high. The risk assessment also helps the engineer to control the maintenance of the equipment. The petrochemical industry can prepare for scheduled maintenance and its costs when they know the failure probability of the equipment.

The risk assessment helps insurance companies to balance the profitability with the probability of injury sustained by the workers, as well as the probability of equipment failure. For well-established insurance company companies, the risks around investment and asset or liability matching far exceed the risk of morbidity or mortality. Risk is made up of two parts; the probability of the equipment going wrong and the negative consequences.

The insurance company focuses on getting benefit from the risk assessment. If the company has a lower number of accidents, the insurance company gets the benefit as the engineering company keeps paying for the insurance following the contract. The insurance company should understand the workplace and the equipment in the industry to provide a suitable coverage premium.

In case the accident happened, the insurance coverage does not state either on the premium or the deal. Such situation could cause both the engineering and insurance companies to blame each other. The insurance company may not pay for the accident cost because it not stated on the premium. On the other hand, the engineering company asks for responsibility because they paid a certain amount of money for insurance. The insurance company does not cover equipment with a high probability of failure or errors. Thus, the workers need to work carefully with the equipment, and they need to get full knowledge about the procedure to operate the equipment.

1.2. Objectives

The objectives of research work are as follows:

i) To identify the presence of human factors during the maintenance activities of rotating equipment.
ii) To develop cause-and-effect relationship that can contribute to financial loss to insurance company.
iii) To analyse risk of equipment considering the financial loss.

2. Methodology

The study focuses on the centrifugal pump, which is commonly used the petrochemical industry. Centrifugal pumps handle most types of liquid and design in two configurations; vertical and horizontal.

2.1. Risk Analysis Sequences

In order to achieve the objectives of research, a method have been used to get the cause-and-effects relationship for the risk assessment [7]. The method is known as the Bayesian Network. And there is a flow as shown in Figure 1 that need to follow to complete the research.

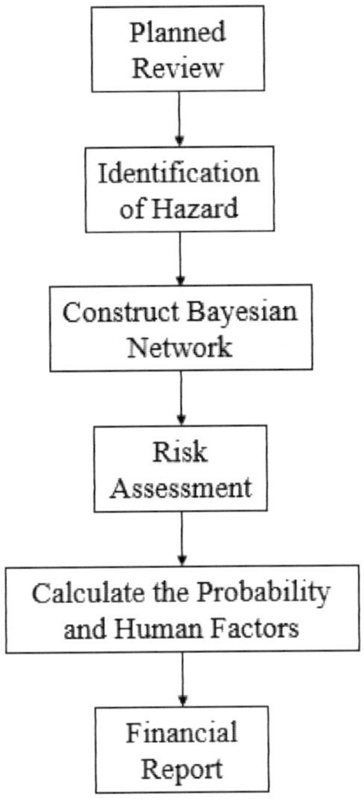

Figure 1. The sequences of risk analysis.

2.2. Bayesian Network

Bayesian Network is a method to combine prior information about a parameter with information, for this research is risk assessment from sample to guide the statistical process. Bayesian Network are a type of Probabilistic Graphical Model that can be used to make models from data/or expert opinion [8]. Bayesian Network expect to demonstrate prohibitive dependence, and along in this way causation, by addressing unforeseen dependence by edges in a planned outline. Through these connections, one can successfully lead inference on the unpredictable factors in the diagram utilizing factors.

2.3. Four Major Analytics Disciplines

A Bayesian Network represent the simple probabilistic relationship between a set of variables, their contingent conditions, and it gives a minimal representation of a joint probability distribution.

In term of probability distribution, Bayesian Network can be categorized into four major analytics disciplines as shown in Figure 2. Descriptive analytics, diagnostic analytics, predictive analytics, and perspective analytics

Bayesian Network are probabilistic because they are made from probability distribution and also use the rules of probability for prediction and anomaly detection [8]. The purpose is to think and diagnostics, decision making under vulnerability and time arrangement expectation.

Any nodes in a Bayesian Network is in every case restrictively autonomous of its everything non-descendants given that node's parents. In this way, the joint probability distribution of every irregular variable in the chart put into a series of conditional probability distribution of random variables given by the parents [9]. Consequently, we can make a full probability model by centering just at the conditional probability distribution in each node.

The Bayesian Network is more perplexing than other models, utilizing tens or even several nodes. It is likewise essential to take note of that each node in a diagram ought to be associated with no less than one edge to another node. Something else, the isolated node is autonomous to every single residual node (additionally to the result variable), and in this manner there is no compelling reason to consider this node [10].

The main weakness is that Bayesian Networks require prior probability distributions. The other weakness is, Bayesian Network need a fully parameter probability model that generally rules out the use of procedures. Where this is not optimal for several model assumptions that strong against a wide range of true situations.

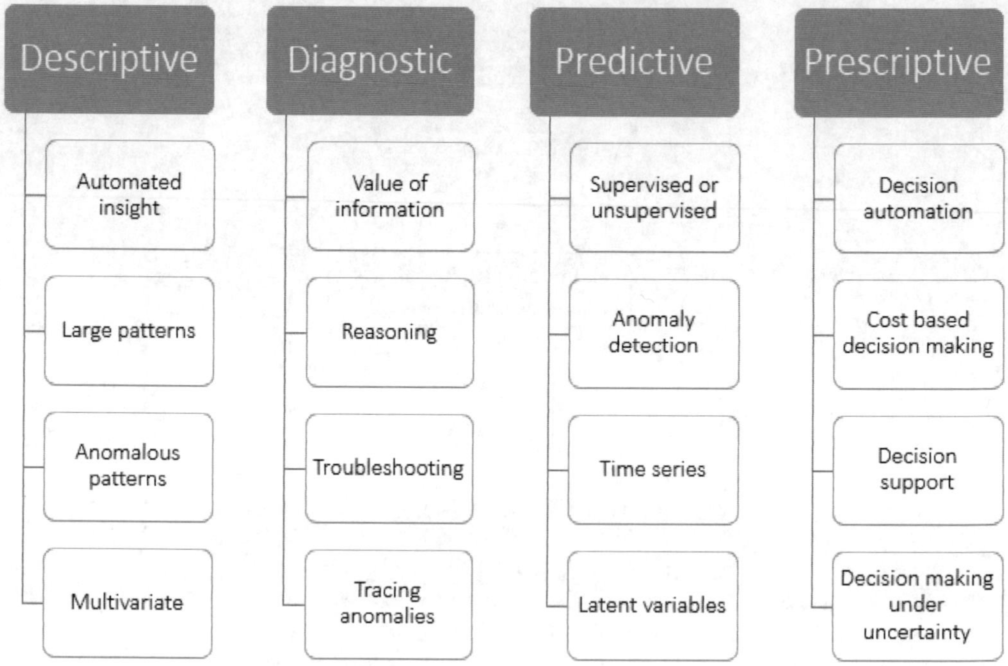

Figure 2. Four areas analysing using Bayesian.

2.4. Develop a Bayesian Network

There are two ways to build a Bayesian network. Either manual construction or automatic construction using the databases. Each of this method have their own advantages and disadvantages.

Manual construction of a Bayesian Network accepts earlier expert knowledge of the hidden space. The initial step is to assemble a directed acyclic graph, trailed continuously venture to survey the conditional probability distribution in every node [8].

Not quite the same as the manual construction, automatic learning did not require expert information on the fundamental space. Bayesian network may gain naturally coordinate from the databases utilizing experience-based calculations that previously implicit proper programming. In any case, the inconvenience is that automatic construction puts more necessities on the data. For the most part automatic construction calculations require no missing data in the dataset, which is much of the time an extraordinarily strong doubt eventually.

Bayesian analysis can include estimation or guess probability due to scarcity of data or data not available. It is also required to have enough data to fit the algorithm's requirements for dependable estimates of the conditional probability distribution. For manual construction, the conditional probability distribution is assumed to be a prior known. Automatic learning then involves in both of network structure creation and conditional probability distribution prediction.

2.5. Software

There are several types of useful software that deal with graphical models. The most common software is the GeNie, HUGIN, BUGS, and R Software. For this research, we use the Genie software to build the Bayesian network. Graphical models use a conditional probability distribution at each node in the graph. If the conditional probability distribution unknown, it can find from the data using estimation of conditional frequencies of probability distribution.

The calculation and Bayesian Network can be done using the HUGIN software. The software provides graphical user interface allowing for interactive model building and learning [11]. Calculation in the HUGIN software can be done by generate all of the probabilities in percentages. The probability can be obtained by dividing the percentages by 100 percent.

2.6. Data Collection

To collect the data related to the accidents during maintenance activity of pump, time limitations have been set to limit the newest accident cases taken. The limit starting from the 1990 until now. The reason is because this time range is the suitable one to get the data and convert it into the probability of failure and human error factors.

The data taken from the report paper that include all the information about the accident. The important part that must include on the data is the overview about the accident, what cause the accident, sequence, background of the accident, the number of victims, and the financial loss.

3. Bayesian Network

In order to complete the research, the Bayesian Network has been developed to show the most human factor that cause the accident on the workplace [12], [13]. The relation of the human factors is illustrated in Figs. 3, 4 and 5. The arrow on the Bayesian Network shown the cause-and-effect relation between each factor.

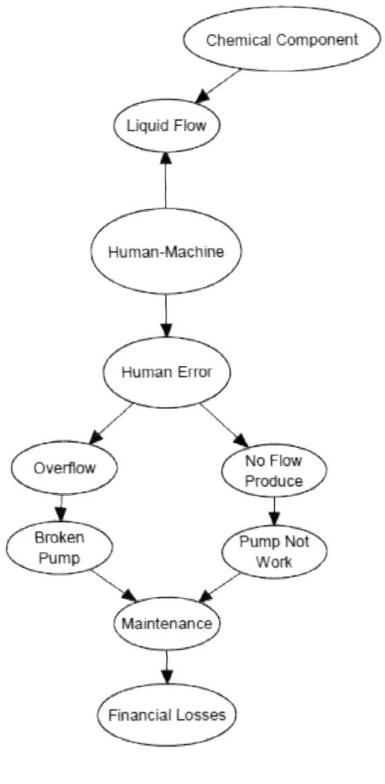

Figure 3. Bayesian network for liquid flow

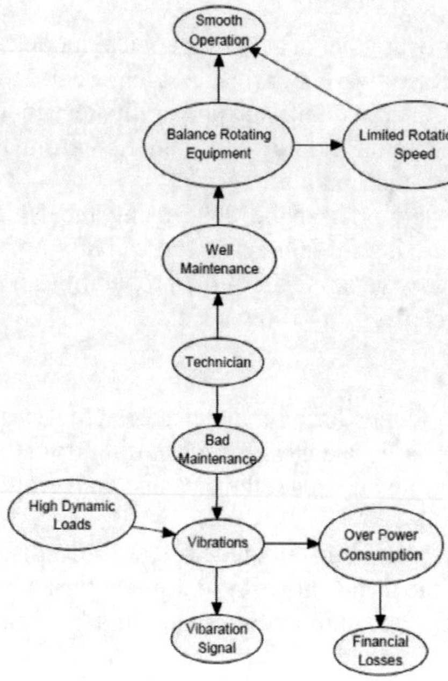

Figure 4. Bayesian network for smooth operation

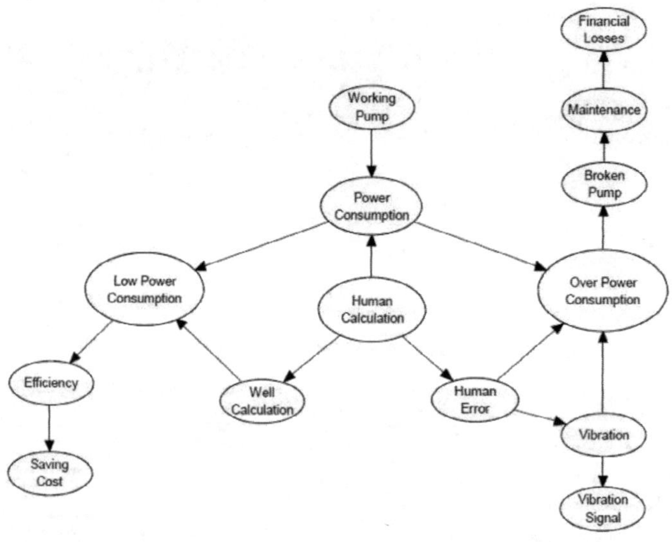

Figure 5. Bayesian network for power consumption

4. Conclusion

Risk assessment is an important procedure in the petrochemical industry. It helps to reduce the number of accidents occur at the workplace. Risk assessment also shown the procedure to handle the accident if the accident already happened. Based on the cause-and-effect relationship Bayesian Network, the probability of failure of equipment has been calculated. Because there is no data yet about the factors of failure, the probability will be initially set to 0.5. The probability will be updated based on data obtained from the investigation report or previous accidents. Based on the updated probability, the financial loss could be estimated for the failure of rotating equipment in the petrochemical industry.

Acknowledgments
The authors thank to Universiti Teknologi Malaysia (UTM) for the financial support. The cost center number of the research funding is Q.J130000.2646.15J46 and Q.J130000.3551.06G00.

References
[1] Ido M, Peter C, Liliane P, and Mohammad S 2018 Identifying the Cause of Human Error in Maintenance Work in Developing Countries *International Journal of Industrial Ergonomics* **68** 222-230
[2] Peter M 2016 *January Financial Risk Management – Sources of Financial Risk and Risk Assessment* (United Kingdom: Heriot-Watt University)
[3] Kartashova O I, Molchanova O V, Axana T 2018 Insurance Risks Management Methodology *J. Risk Financial Manag.* **11** 75
[4] Junaid H and Faisal K 2012 Risk-based asset integrity indicators *Journal of Loss Prevention in the Process Industries* **25** 544–554
[5] Faisal K and Mahmoud H 2004 Risk-based maintenance of ethylene oxide production facilities *Journal of Hazardous Materials* ***A108*** 147–159
[6] Alireza N, Faisal K, Scott M, Paul A, and Travis D 2014 Determination of Human Error Probabilities in Maintenance Procedures of a Pump *Process Safety and Environmental Protection* **92** 131-141
[7] Ari A, Clare P, and Leon S 2014 Identification of the human factors contributing to maintenance failures in a petroleum operation *Human Factors* **56 (2)** 306–321
[8] Michal H 2014 *Bayesian Networks* (United State: Boston University)
[9] The National Institute of Standards and Technology (NIST) 2012 *Guide for Conducting Risk Assessments* (United State: National Institute of Standards and Technology)
[10] Michael H F 2008 *Risk Assessment in Engineering – Principles, System Representation & Risk Criteria* (ISBN 978-3-909386-78-9)
[11] Ayele Y Z and Barabadi A 2017 *IOP Conf. Ser.: Mater. Sci. Eng.* **276** 012019
[12] Alireza N, Nima K, Faisal K, Scott M, and Rouzbeh A 2013 The role of human error in risk analysis: Application to pre- and post-maintenance procedures of process facilities *Reliability Engineering and System Safety* **119** 251–258
[13] Kate R, James B, Sarah G C, and Alexandra H 2016 *Human and Organisational Factors in Major Accident Prevention: A Snapshot of the Academic Landscape* (United Kingdom: RAND Europe)

IOP Conf. Series: Materials Science and Engineering **808** (2020) 012016 doi:10.1088/1757-899X/808/1/012016

Fibrous silica induced narrow band gap TiO_2 catalyst for enhanced visible light-driven photodegradation of methylene blue

A A Fauzi[1], A A Jalil [1,2*], M Mohamed[1], N A Naseri[1], C N C Hitam[1], N F Khusnun[1], N S Hassan[1], A F A Rahman[1], F F A Aziz[1] and M S M Azmi[3]

[1] School of Chemical and Energy Engineering, Faculty of Engineering, Universiti Teknologi Malaysia, 81310 UTM Johor Bahru, Johor, Malaysia
[2] Centre of Hydrogen Energy, Institute of Future Energy, 81310 UTM Johor Bahru, Johor, Malaysia
[3]Department of Chemistry, Faculty of Science,Universiti Teknologi Malaysia, 81310 UTM Johor Bahru, Johor, Malaysia

*aishahaj@utm.my

Abstract. Fibrous silica titania (FST) was synthesized via hydrothermal method and evaluated on photodegradation of methylene blue (MB). The catalyst was characterized using field emission scanning electron microscopy (FESEM), Fourier transform infrared (FTIR) spectroscopy, ultraviolet-visible diffuse reflectance spectroscopy (UV-Vis DRS) and N_2 adsorption–desorption. The photocatalytic activity was performed under different reaction condition namely pH, catalyst dosage and MB initial concentration. FST demonstrated higher performance (99.9%) for 10 mg L^{-1} of MB initial concentration than titania (TiO_2) using 0.25 g L^{-1} of catalyst dosage at pH 5 for 2 h. The superior performance towards photodegradation of MB under visible light demonstrated by synthesized FST was due to unique morphology of fibrous, high surface area, narrow band gap and more active site (Si-O-Ti). Kinetics study indicated that the photodegradation of MB was well fitted with pseudo-first order Langmuir-Hinshelwood model and adsorption was the rate-limiting step. The FST maintained its photocatalytic activities for up to five cycles reaction with slightly catalyst deactivation, suggesting that the FST is suitable to be implied in the photocatalytic reaction.

1. Introduction

In textile dyeing industry, 1-20% of the total world production of dyes is lost during the dyeing process [1]. Methylene blue (MB) is a cationic dye commonly used for printing, dyeing cotton, wool, and leather. Consequently, the discharged MB in wastewater can threat to aquatic life and human being due to its carcinogenic and toxicity [2]. Various alternatives have been introduced such as adsorption, biodegradation and coagulation. However, these methods have proven to be markedly ineffective due to the production of secondary pollutant, which requires high cost and energy consumption [3]. Thus, another alternatives method is urgently needed.

Advanced Oxidation Processes (AOPs) have attracted much attention among researchers for degradation of targeted pollutants due to the simple and low-cost processes [4]. Among AOPs, heterogeneous photocatalysis using semiconductors such as zinc oxide, titanium dioxide, copper oxide

IOP Conf. Series: Materials Science and Engineering **808** (2020) 012016 doi:10.1088/1757-899X/808/1/012016

and zirconium dioxide has been proved efficient for degrading aquatic and atmospheric organic contaminants through the hydroxyl radicals. Among semiconductor photocatalysts, titanium dioxide (TiO_2) is widely used in various areas such as photocatalysis, self-cleaning and dye-sensitized solar cells due to its excellent chemical and physical properties, strong photooxidation power and low cost [4]. However, TiO_2 consists of several shortcomings including low surface area, wide band gap, less electron hole separation and difficulty in isolation and reusability [6,7], which reduce its performance for adsorption and degradation of pollutant in visible light. Hence, the modification of TiO_2 is required in order to improve the performance in visible light.

To overcome these drawbacks, most of researchers have interested on coupling of TiO_2 with porous silica in order to prevent the agglomeration of TiO_2 particles and increase surface area [8]. Previous studies have been reported the synthesis of nanostructured TiO_2 or TiO_2 supported on this supports can enhance surface area and particle size [6]. Moreover, it has been found that the coupling of SiO_2 with TiO_2 able to effectively prevent charge recombination in dye-sensitized solar cell [8]. The synthesis of Ti-containing large pores materials such as mesoporous silica (MS) has been also reported [9]. In spite of this modification enhance the properties of TiO_2 supported on porous material, the accessibility of pollutant towards catalyst seemed decreasing due to tubular structure and high possibility of pores clogging which partially collapse after thermal treatment.

Latterly, fibrous morphology has aroused attention among researches in enhancing the performance due to its unique morphology, high accessibility of active sites and surface area [10]. In addition, this morphology has open pore channels and controllable pore sizes during synthesis such as by altered amount of surfactant, heating and stirring time. Besides, morphological modification of the catalysts also was extensively developed aiming the active catalyst for visible light region. There are several types of morphological modifications previously reported such as nanorod, nonocubes, nanospheres and flower-like [9]. To the best of our knowledge, synthesis of fibrous silica titania (FST) was conducted in this study for photodegradation of MB. The catalysts were characterized by FESEM, UV-vis/DRS, FTIR and N_2 adsorption–desorption. The synthesized FST catalyst has a unique morphology, narrow band gap, high surface area, more interaction of Si-O-Ti that acts as an active site for improving electron-hole separation and high stability that can enhance the photocatalytic performance in visible light.

2. Materials and Methods

2.1. Materials

Cetyltrimethylammonium bromide (CTAB, 99%) and methylene blue (MB, 100%) were purchased from Merck Sdn. Bhd., Malaysia. Urea (99.5%) and toluene (99.5%) were purchased from QRec, Malaysia. 1-Butanol (99.5%) and tetraethyl orthosilicate (TEOS, 95%) were purchased from Sigma Aldrich and Acros Organics, Malaysia respectively. Commercial TiO_2 powder catalyst (JRC-TiO_2) was supplied by the Catalysis Society of Japan. All chemicals were of analytical grade and used without any treatment.

2.2. Catalyst preparation

FST was synthesized by using hydrothermal method which had been established in previous study with some modification using TiO_2 seed (JRC TiO_2-2) [11]. 25 g of CTAB and 14 g of urea were diluted in distilled water and stirred for 5 minutes. 6 g of titania seed, 600 mL of toluene and 30 mL of 1-butanol were added into the solution. After that, 56 mL of TEOS was then added into the solution and stirred for 2 hours. After 2 hours, the resulting solution was kept in microwave for 6 hours at 120°C. The product was left to dry overnight at 110°C before calcined at 550°C for 3 hours.

2.3. Characterization

The physicochemical properties of the synthesized FST were characterized using several instruments. The morphological of the catalyst were analysed by FESEM (JEOL JFC-1600, Japan). The interactions existed in the catalysts were analyzed by FTIR spectroscopy (Perkin Elmer Spectrum GX FTIR Spectrometer, U.S). IR absorbance data with wavenumbers ranging from 400 to 4000 cm^{-1} was obtained. The wavelength of the catalysts was measured using UV-Vis/DRS which recorded over a range of wavelengths from 200 to 800 nm using a Perkin-Elmer Spectrophotometer and the band gap was calculated by equation below as shown below. The band gap was also obtained by plotting kulbelka munk.

$$Band\ gap\ (eV) = \frac{1240}{Wavelength} \quad (1)$$

2.4 *Photocatalytic testing*
The photoactivity of the catalyst was tested via photodegradation of MB. An amount of photocatalyst was dispersed into 250 mL of MB with a desired concentration and pH. The solution was stirred in a dark for 1 hours and then exposed under visible light for another 2 hours. During the process, 1.5 mL of MB solution sample was collected every 15 minutes and centrifuged using Beckman Coulter Microfuge at 14,000 rpm for 10 minutes. Then, the concentration of MB centrifuged liquid was examined by UV-Vis spectrophotometer (Cary 60 UV-vis Agilent, U.S) at the absorption band of 664 nm. The experiment was conducted under different parameter such as catalyst dosage (0.1-0.5 g L^{-1}), MB initial concentration (10-100 mg L^{-1}), pH (3-9) and. While, kinetic study was analysed based on the performance for different initial concentration. The efficiency of photocatalytic degradation was calculated by equation below.

$$Degradation\ (\%) = \frac{Co - Ct}{Co} \times 100 \quad (2)$$

3. Results and Discussion

3.1 *Morphological studies of FST*
The morphological properties of FST catalyst was investigated by field emission scanning electron microscopy (FESEM) as shown in figure 1A and B at magnification 50,000 X and 10,000 X, respectively. In Fig. 1A, it was obvious that a clear spherical shape with dendrimeric fibre was formed, signifying that the FST with fibrous morphology was successfully synthesized by hydrothermal method. Meanwhile, figure 1B illustrates the particle size of FST ranging from 0.46 to 0.77 μm. Similar phenomenon has been reported in previous study where the fibrous silica ZSM5 showed a well-defined uniform spherical structure with cockscomb-like surface and high porosity [12]. This morphology is expected to provide more accessibility of active sites for enhancing the adsorption of pollutants and photodegradation due to its unique structure [8,12]. Besides, the morphology of TiO_2 anatase showed low surface area with agglomeration as reported in previous study [13]. Hence, the unique morphology of FST would also enhance the performance compared to TiO_2.

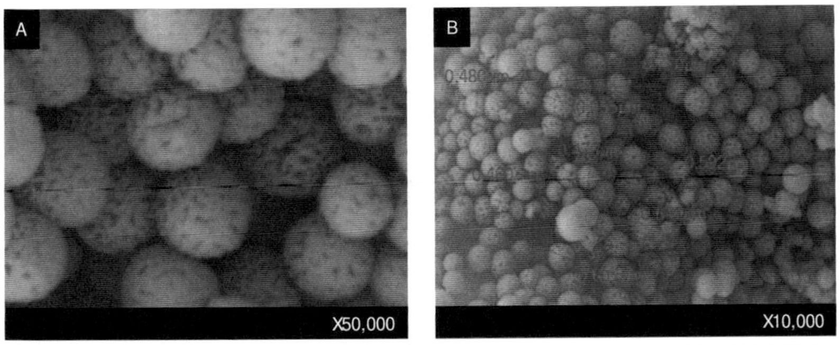

Figure 1. FESEM images of FST catalyst at magnification A) 50,000X and B)10,000X.

3.2 Vibrational spectroscopy analysis

Figure 2 illustrates the FTIR spectra of the FST, SiO_2 and commercial TiO_2 to verify the functional groups present in catalysts. It was observed that all catalysts have two bands at 3452 cm^{-1} and 1630 cm^{-1}, which attributed to a stretching and bending vibrations of adsorbed water molecules, respectively [14]. Synthesized FST resulted in increasing intensity for these bands, which indicates that the introduction of TiO_2 on fibrous silica (FS) has made the catalyst to possess more surface hydroxyl groups. The hydroxyl groups play an important role in the photocatalytic reaction since they can capture photoinduced holes and form hydroxyl radicals with high oxidation capability [15]. The bands at 1088, 796 and 468 cm^{-1} were ascribed to asymmetric and symmetric stretching, and bending vibrations of Si-O-Si, respectively [14]. Compared to the commercial TiO_2 and FS spectra, it was clearly observed that the intensity for FST at these bands were increasing. This trend indicated a major arrangement of silica framework during synthesis [16].

The bands in the range of 900-1000 cm^{-1} illustrated the significant region usually involved in the rearrangement of SiO_2 framework subsequent to modification [17]. Previous study has reported that the band existed in this range was assigned to Si-O-H stretching, non-bridging free broken Si-O stretching and Si-O-Ti bonds [18]. In this study, the band at 966 cm^{-1} was increased for FST showing that silica framework had been embedded by titania. The existence of a band at 680 cm^{-1} was observed for commercial TiO_2 and FST which corresponded to Ti-O-Ti bands signifying the presence of TiO_2 in the catalyst [19]. However, this band was seen reducing for FST, which might be due to the formation of more Si-O-Ti bonds. Thus, FST has proved that the formation of Si-O-Ti which act as an active site for electron hole separation can enhance the degradation performance.

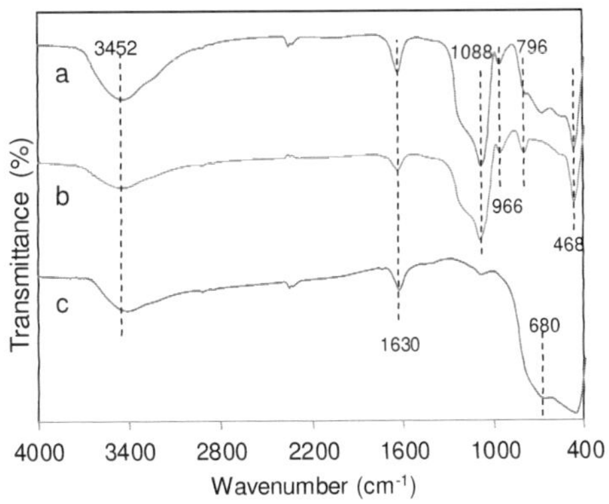

Figure 2. FTIR spectra of catalysts (a) FST, (b) Fibrous silica and (c) commercial TiO_2 at region 400-4000 cm^{-1}.

3.3 Band gap analysis

The UV–vis DRS spectra of FST and TiO_2 are illustrated in figure 3A. It can be seen that the wavelength of FST was shifted to above 400 nm compared to TiO_2 which was below 400 nm, demonstrating a significant improvement of the catalysts towards visible light absorption [17]. Kubelka-Munk plot was used to estimate the band gap energy of the prepared samples as shown in figure 3B. The band gaps were optically obtained as presented in table 1. It was observed that the band gap of FST (2.92 eV) was lower than commercial TiO_2 (3.19 eV), which attributed to easy generation of electrons and holes of FST under visible light irradiation and resulted in higher photocatalytic activity. It has been previously reported that the wide band gap above 3.0 eV contributes to high recombination rate of the photo-induced electrons and holes, which then limits the photocatalytic activity [20]. Hence, it is believed that the FST could result in a better performance in visible light. Table 1 lists the surface area of FST and TiO_2. The synthesized FST showed increasing surface area compared to TiO_2. This would lead to high accessibility of active sites and hence improved the catalytic performance [8].

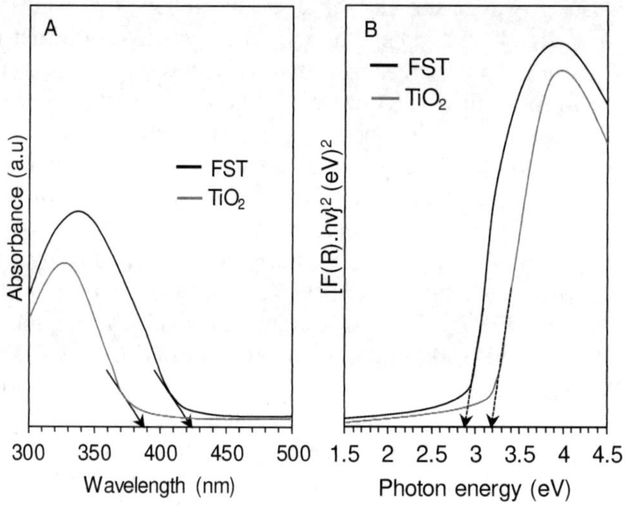

Figure 3. Kulbelka munk plot of the FST and TiO_2.

Table 1. Band gap and surface area analysis.

Catalysts	Surface area, S_{BET} ($m^2\ g^{-1}$)[a]	Pore volume, V ($cm^3\ g^{-1}$)[b]	Average pore size (nm)[b]	Band gap (eV)[c]
Commercial TiO_2	20	0.1	32.6	3.19
FST	616	1.54	45.7	2.92

[a] Calculated from N_2 adsorption measurements using BET method
[b] Calculated from N_2 adsorption measurements using t-plot method
[c] Calculated band gap using kulbelka-munk

3.4 Photocatalytic performance and kinetic study

The photocatalytic performance of FST was studied with respect to the photodegradation of MB and compared with TiO_2 commercial as illustrated in figure 4. It was clearly showed that when the reaction was performed in the dark for 1 hour, the adsorption capacity of FST became much higher than commercial TiO_2. This was probably due to the adsorption property of silica-supported TiO_2, which might further influence the photodegradation of MB [21]. After adsorption step, the reactions continued in visible light for another 2 hours. The photodegradation rate of MB was seen to favour FST compared to commercial TiO_2 with 99.9% and 94.6%, respectively. At 60 minutes, almost all of the MB was degraded by FST catalyst while the commercial TiO_2 required more than 120 minutes. This was due to the superior properties of FST including unique morphology, narrow band gap, high surface area and more Si-O-Ti interaction.

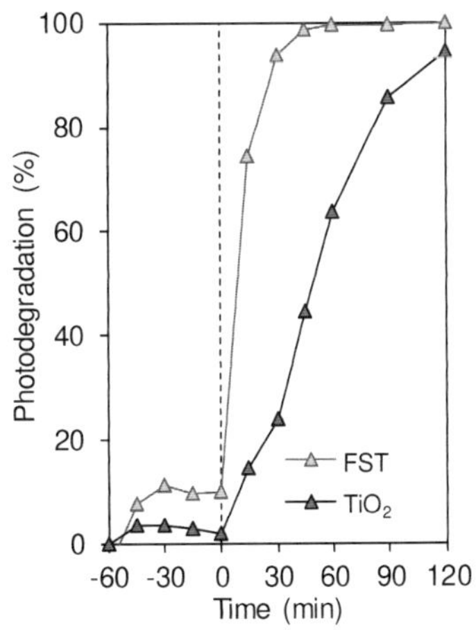

Figure 4. Photodegradation of MB using FST and TiO_2 [pH=5, W=0.25 g L^{-1}, initial concentration =10 mg L^{-1}, t=2 h].

The effects of pH were investigated ranging from of 3-9 in the presence of the FST catalyst with the results shown in figure 5A. It was seen that under various pH, the best performance was at pH 5, indicating that the reaction was favoured in a slightly acidic condition. Fig. 5B shows the effects of catalyst dosage towards photodegradation of MB at pH 5 using 10 mg L^{-1} of MB solution. The catalyst dosage was varied from 0 to 0.5 g L^{-1}. It was noticed that the photodegradation percentage of MB as the catalyst dosage increases. This might be due to the increase in active site by increasing the catalyst dosage, thus enhancing the catalytic reaction [17]. Although the overall performance was similar in the range of 0.125-0.5 g L^{-1}, the best performance was attained by 0.25 g L^{-1} of FST dosage with 99.9% under visible light compared to others, which is more favourable in adsorption process. Meanwhile, further increase in catalyst dosage resulted in a decrease of photodegradation of MB due to the increase in catalyst dosage that induced the turbidity of the system, causing light scattering and hindered the light penetration as well as triggered agglomeration of the catalyst that inhibited the photodegradation process [18]. The excess amount of catalyst dosage might lead to a decreasing production of electron-hole pairs due to inhibition of light penetration, thus producing fewer hydroxyl radicals and making the reaction rate decreased.

The photodegradation of MB at different initial concentrations ranging from 10 to 100 mg L^{-1} using 0.25 g L^{-1} of FST dosage at pH 5 is shown in figure 5C. The result clearly showed that the photodegradation percentage for 10, 25, 40, 55, 70, 85 and 100 mg L^{-1} were 99.9%, 97.9%, 87.8%, 80.7%, 73.6%, 58.5% and 47.6%, respectively. Based on the results, it can be seen that the increase in initial concentration of MB led to the decrease in performance. As the initial concentration of MB increases, more dye molecules were adsorbed on the surface of FST, which resulted in lack of direct contact with the holes or hydroxyl radicals making inhibitive effect of dye degradation [22]. The performance of the catalyst at 10 mg L^{-1} was the highest which was 99.9% possibly due to the less adsorption of dye molecules on the catalyst surface. Hence, more photons could penetrate in the catalyst due to more accessible of light.

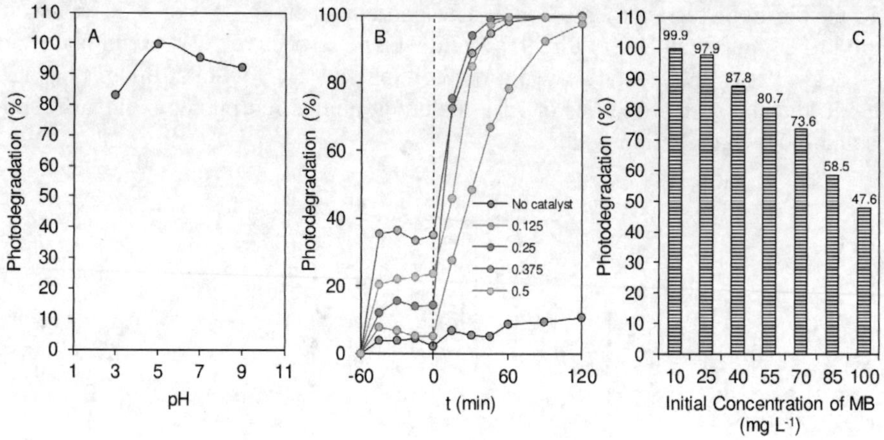

Figure 5. Photodegradation of MB (A) Effect of catalyst dosage [Co = 10 mg L^{-1}; pH = 5; t = 2 h], (B) Effect of initial concentration [pH = 5; W = 0.25 g L^{-1}; t = 2 h].

Based on the initial concentration, the photodegradation kinetics of MB was also studied. A linear fitting of $\ln(C_o/C_t)$ versus irradiation time (t) was plotted in figure 6A and the data was summarized in table 2. The results demonstrated that 10 mg L^{-1} gave the highest photodegradation rate among others concentrations. Lower photodegradation rate was shown by increasing concentrations than 10 mg L^{-1} due to the existence of large amount of adsorbed dye molecules result in lack of direct contact between the catalyst and the holes or hydroxyl radical.

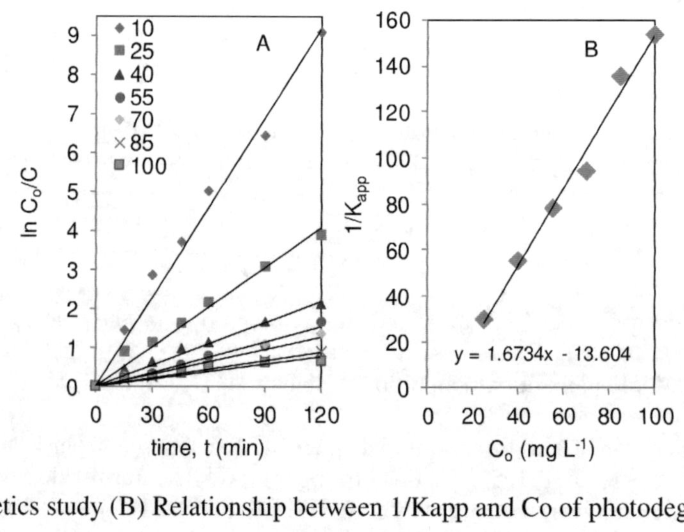

Figure 6. (A) Kinetics study (B) Relationship between 1/Kapp and Co of photodegradation of MB.

Table 2. The kinetics parameter of photodegradation using FST.

Initial concentration ($mg\ L^{-1}$)	Reaction rate constant, k (min^{-1})
10	0.0766
25	0.0339
40	0.0182
55	0.0128
70	0.0098
85	0.0074
100	0.0065

Predominantly, for most of the organic compound, the effect of the photocatalytic degradation was outlined by pseudo-first-order kinetic model of Langmuir-Hinshelwood (L-H) model [19]. The L-H model is commonly used to interpret the kinetic data due to the limitation of mass transfer that occurs on the catalyst surface. At low concentration of MB, the general equation of the rate of photodegradation is given by equation 3.

$$\ln\left(\frac{C_O}{C_t}\right) = k_r\, K_{LH} = K_{app}\, t \tag{3}$$

The straight line plotted on a graph in figure 6A confirmed that photocatalytic degradation of MB follows the pseudo-first-order kinetic model. The slope of each line represents the apparent first-order rate constant [23]. Based on the L-H model, a linear plot was obtained by plotting 1/kapp as a function of C_o in figure 6B.

$$\frac{1}{K_{app}} = \left(\frac{1}{k_r}\right) C_O + \frac{1}{k_r\, K_{LH}} \tag{4}$$

Referring to Fig. 6B, the slope of the graph and the y-intercept were 1.6734 and -13.604 respectively. By simplifying the data obtained with equation 4, the reaction rate constant (k_r) and the adsorption equilibrium constant (K_{LH}) were calculated and the resulted values were 0.5976 mg L^{-1} min^{-1} and 0.123 L mg^{-1} respectively. The value of k_r was larger than K_{LH} which signified that the reaction would occur at the surface of the catalyst and it represents the rate determining step in the photocatalytic process [24,25].

3.5 Stability test

The durability of catalyst in wastewater treatment is also a plays an crucial role in order to reduce the operational cost [26]. Hence, the stability of FST for photodegradation of MB was conducted by repeating the reaction. As shown in figure 7, the FST possess high stability stable with only a slightly deactivation even after five successive recycles for the degradation. This was due to the heat treatment, which caused agglomeration of the catalysts which led to decrease of surface areas [27].

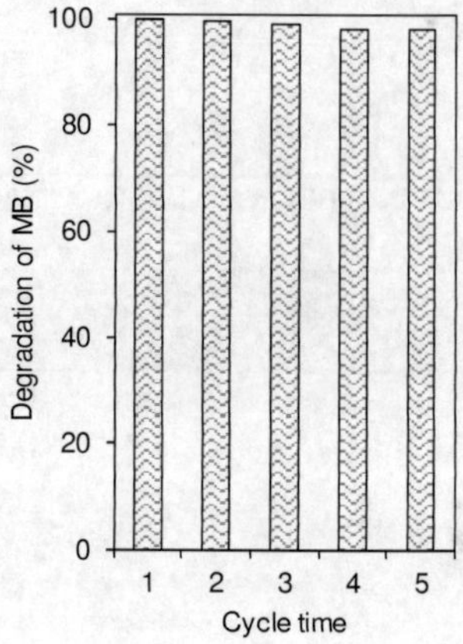

Figure 7. Stability test of FST for photodegradation of MB.

4. Conclusions

As a conclusion, FST was successfully synthesized via hydrothermal method and characterized by field emission scanning electron microscopy (FESEM), fourier transform infrared (FTIR), UV-vis diffuse reflectance spectroscopy (UV-vis/DRS) and N_2 adsorption-desorption. The potential of the synthesized FST and commercial TiO_2 were evaluated towards photocatalytic degradation of MB under visible light for 2 hours. The conditions were varied during reaction such as pH, catalyst dosage and initial concentration. It was proven that FST exhibited higher photodegradation percentage of MB than commercial TiO_2 which were 99.9% and 94.6% respectively. This was due to the unique morphology, narrow band gap, existing of interaction of Si-O-Ti that acts as an active site and large surface area obtained by FST. This study had proved that the most effective parameters for photocatalytic degradation of MB were 0.25 g L^{-1} of catalyst dosage, 10 mg L^{-1} of initial concentration and pH 5 for 2 hours of irradiation time. Based on the Langmuir-Hinshelwood model, k_r showed larger value than K_{LH}, which suggests that the reaction between OH radicals and dye molecules over the catalyst surface represents the rate determining step in photocatalytic process.

Acknowledgments

The work was supported by Fundamental Research Grant No. FRGS/1/2017/STG07/UTM/01/1 (Grant No. 4F969)

References

[1] Elmoubarki R, Mahjoubi F Z, Tounsadi H, Moustadraf J, Abdennouri M, Zouhri A, El Albani A and Barka N 2015 *Water Resour. Ind.* **9** 16-29

[2] El-Yazeed W S A and Ahmed A I 2019 *Inorg. Chem. Commun.* **105** 102-111

[3] Khalil M, Anggraeni E S, Ivandini T A and Budianto E 2019 *Appl. Surf. Sci.* **487** 1376-1384

[4] Mustapha F H, Jalil A A, Mohamed M, Triwahyono S, Hassan N S, Khusnun N F, Hitam C N C, Rahman A F A, Firmanshah L and Zolkifli A S 2017 *J. Clean Prod.* **168** 1150-1162

[5] Chong M N, Jin B, Chow C W and Saint C 2010 *Water Res.* **44** 2997-3027

[6] Fauzi A A, Jalil A A, Mohamed M, Triwahyono S, Jusoh N W C, Rahman A F A, Aziz F F A, Hassan N S, Khusnun N F and Tanaka H, J 2018 *Environ. Econ. Manag.* **227** 34-43

[7] Nguyen C H, Fu C C and Juang R S 2018 *J. Clean Prod.* **202** 413-427

[8] Singh R, Bapat R, Qin L, Feng H and Polshettiwar V 2016 *Catalysis* **6** 2770-2784

[9] Ke X, Xu L, Zeng C, Zhang L and Xu N 2007 *Micropor. Mesopor. Mater.* **106** 68-75

[10] Polshettiwar V, Cha D, Zhang X and Basset J M 2010 *Angew. Chem. Int. Ed.* **49** 9652–9656

[11] Firmansyah M L, Jalil A A, Triwahyono S, Hamdan H, Salleh M M, Ahmad W F W and Kadja G T M 2016 *Catal. Sci. Tech.* **6** 5178

[12] Aziz F F A, Jalil A A, Triwahyono S and Mohamed M 2018 *Appl. Surf. Sci.* **445** 84-95

[13] Nasikhudin, Diantoro M, Kusumaatmaja A and Triyana K 2018 *Journal of Physics: Conf. Series.* **1011** 012069

[14] Karim A, Jalil A, Triwahyono S, Sidik S, Kamarudin N, Jusoh R, Jusoh N and Hameed B J 2012 *Colloid Interface Sci.* **386** 307-314

[15] Huo Y D, Wang X C, Wu L, Chen X F, Ding Z X, Wang X X and Fu X Z 2008 *Chemosphere.* **72** 414-421

[16] Rahman A F A, Jalil A A, Triwahyono S, Ripin A, Aziz F F A, Fatah N A A, Jaafar N F, Hitam C N C, Salleh N F M and Hassan N S 2017 *J. Clean Prod.* **143** 948-959

[17] Hitam C N C, Jalil A A, Triwahyono S, Ahmad A, Jaafar N F, Salamun N, Fatah N A A, Teh L P, Khusnun N F and Ghazali Z 2016 *RSC Adv.* **6** 76859

[18] Vasconcelos D C L, Nunes E H M, Gasparon M and Vasconcelos W L 2011 *Mater. Sci. App.* **2** 1375

[19] Ansari S A and Cho M H 2016 *Sci. Rep.* **6**

[20] Kim H J, Shul Y G and Han H 2005 *Top Catal.* **35** 287-293

[21] Xu C, Rangaiah G and Zhao X 2014 *Ind. Eng. Chem. Res.* **53** 14641-14649

[22] Ranjbar P Z, Ayati B and Ganjidoust H 2019 *J. Environ. Sci.* **79** 213-224

[23] Yang L Y, Dong S Y, Sun J H, Feng J L, Wu Q H and Sun S P 2010 *J. Hazard Mater.* **179** 438-443

[24] Adam F, Muniandy L and Thankappan R 2013 *J. Colloid Interface Sci.* **406**, 209-216

[25] Hassan N S, Jalil A A, Triwahyono S, Hitam C N C, Rahman A F A, Khusnun N F, Mamat C R, Asmadi M, Mohamed M, Ali M W and Prasetyoko D 2018 *J. Taiwan Inst. Chem. Eng.* **82** 322-330

[26] Zhang S, Chen X and Song L 2019 *J. Hazard. mater.* **367** 304-315

[27] Abderrahmane H, Luc B, Pascal R, Ahmed A, Hatem E, Rabah B and Yannick C 2016 *Mater. Des.* **109** 634-643

IOP Conf. Series: Materials Science and Engineering **808** (2020) 012017 doi:10.1088/1757-899X/808/1/012017

Tailoring amount of TiO_2 doped onto fibrous silica ZSM-5 for enhanced photodegradation of paracetamol

F F A Aziz[1], A A Jalil [1,2*], C N C Hitam[1], N S Hassan[1], A F A Rahman[1], and A A Fauzi[1]

[1] School of Chemical and Energy Engineering, Faculty of Engineering, Universiti Teknologi Malaysia, 81310 UTM Johor Bahru, Johor, Malaysia
[2] Centre of Hydrogen Energy, Institute of Future Energy, 81310 UTM Johor Bahru, Johor, Malaysia

*aishahaj@utm.my

Abstract. Photocatalytic degradation of paracetamol has received much attention due to its low energy consumption, high efficiency and safety to environment. A simple electrolysis system was used to synthesis a series of titania (1-5 wt%) supported on fibrous silica ZSM-5 (TiO_2/FZ). The FESEM images showed the cockscomb-like surface of FZ was preserved even after the addition of TiO_2. FTIR results confirmed that TiO_2 is successfully incorporated with FZ through the Si-O-Ti bonds. The adsorption edge of TiO_2 shifted to the higher wavelength after supported on FZ and lowering the band gap of TiO_2. The photodegradation of paracetamol was resulted in the following order: 3 wt% TiO_2/FZ (90%) > 5 wt% TiO_2/FZ (71%) > 1 wt% TiO_2/FZ (65%). 3 wt% TiO_2/FZ exhibits excellent photodegradation of paracetamol owing to the highest number of Si-O-Ti bonds and lowest band gap. The result obtained give new insight for other wastewater treatment processes over photodegradation of various pollutants.

1. Introduction

Nowadays, the accumulation of paracetamol in wastewaters represent a pollution problem that must be solved urgently due to their potential toxic effects to aquatic organisms [1]. Thus, many efforts have been used to develop suitable methods that can remove the paracetamol from water due to its inappropriate conditions for biodegradation [2,3]. Advanced Oxidation Process (AOP) has been broadly used as one of practical way in wastewater treatment technology [4]. The technique uses the light that plays an effective role in producing highly free radical species such as hydroxyl radicals and superoxide radicals by using heterogeneous photocatalyst semiconductors [5]. The semiconductors that commonly used are TiO_2 (titanium dioxide), ZnO (zinc oxide) and CdS (cadmium sulfate) [6-8]. Titanium dioxide (TiO_2) has been established as the most active photocatalyst and semiconductors for environmental remediation [9]. However, unsupported TiO_2 has some limitations in its practical use, such as the difficulties in recycling, low adsorption capacity and high aggregation tendency [10]. Those drawbacks have driven the production of silica zeolite as supported photocatalysts [11].

Zeolite, has been marked as a promising supported material to improve the properties of TiO_2 due to their unique porous network, large surface area and excellent adsorption of reactants [12,13]. However, conventional zeolite has low accessibility of active site due to its morphology. Recently, fibrous silica ZSM-5 (FZ) has been introduced to increase the accessibility of active site that allowed good achievement in catalytic reaction including benzene methylation and cumene hydrocracking [14,15]. Previously, we studied the good potential of in situ electrochemical method for producing

IOP Conf. Series: Materials Science and Engineering **808** (2020) 012017 doi:10.1088/1757-899X/808/1/012017

various metal nanoparticles such as NiO_2, TiO_2 and ZnO doped onto MSN [16-18]. All of the catalysts showed exceptional performances for excellent CO_2 methanation and photodegradation of various organic contaminants. Considering the advantages of fibrous materials, thus we extent the studies on synthesization of TiO_2 catalysts supported on FZ (TiO_2/FZ) by electrolysis [9], where the effect of TiO_2 amount doped on FZ for degradation of paracetamol has not been reported. All of the samples were characterized by FESEM, FTIR and UV-Vis DRS. Kinetics studies of the photodegradation of paracetamol are also reported. The modified catalyst, would be useful as a new catalyst especially for applications in wastewater treatment.

2. Methods

2.1 Materials
ZSM-5 zeolite (98% purity) with Si/Al=23 was obtained from Zeolyst International. Paracetamol (4-acetamidophenol) with 97% purity, was purchased from Fisher Scientific (M) Sdn. Bhd., Malaysia. Cetyltributylammonium bromide (98%, CTAB), urea (99%), butanol (99.5%), toluene (95.5%) and tetraethyl orthosilicate (98%, TEOS), tetraethylammonium bromide (>99%, TEAB) and hydrogen perchloric acid (97%, HP) from Merck Sdn. Bhd., Malaysia were used. The titanium plates were acquired from Nilaco Metal, Japan (99% purity). Tetraethylammonium perchlorate (TEAP) was used as supporting electrolyte and the synthesis method will be explained in the synthesis.

2.2 Synthesis of catalysts
FZ was prepared by microemulsion method [15]. Toluene and butanol were added in the mixture of water (250 mL), CTAB (20 g) and urea (5 g). After 30 min, ZSM-5 seed (1 g) and TEOS (3 mL) was added and stirred for another 2 h. Then, the mixture was exposed to microwave radiation at 393 K (2 h). The sample was dried at 383 K (24 h) and calcined at 823 K (6 h) with heating rate 273 K min^{-1}. The TiO_2/FZ catalysts were prepared according to the literature and the weight percentage (wt%) of Ti doped was varied between 1% to 5% by electrolysis method [19]. Firstly, TEAP was synthesized by adding TEAB and HP into 600 mL of distilled water. Then the solution was retained overnight for crystallization. Next, the solution was filtered and the crystal was collected before being added with 150 mL distilled water. After that, the solution was heated at 353 K to form homogeneous solution and retained for the recrystallization. The TEAP formed as a white needle in 90% yield [20]. TiO_2/FZ was prepared by electrolysis method. the FZ was added in a cell fitted with a Pt plate (2 cm x 2 cm) cathode and a Ti plate (2 cm x 2 cm) anode which containing distilled water (20 mL) and TEAP. The electrolysis was conducted at a constant current 120 mA cm^{-2} and at 273 K under atmospheric pressure and Ti ions were produced to undergo titanation. After electrolysis, the sample was dried overnight at 383 K and calcined at 823 K for 6 h.

2.3 Characterization
Surface morphology of the prepared catalysts were observed by field emission scanning electronmicroscopy (FESEM) (Zeiss supra VP35). The samples' functional group was utilized by PerkinElmer Spectrum GX FTIR Spectrometer (400-4000 cm-1). UV-Vis diffuse reflectance spectra were collected by using PIKE Technologies DiffusIR between (200- 800 nm). The band gap of each samples was determined based on the Eq. 1,

$$E_g(eV) = \frac{1240}{\lambda} \quad (1)$$

where Eg is the band gap (eV) and λ is the wavelength of the absorption edges (nm).

IOP Conf. Series: Materials Science and Engineering **808** (2020) 012017 doi:10.1088/1757-899X/808/1/012017

2.4 Photodegradation of paracetamol
The photocatalytic performances of all samples toward the photodegradation of paracetamol were carried out in a batch reactor, in which 36 W metal halide lamp (400 nm) for visible light source is served with a cooling system. 0.375 g L^{-1} of sample was added in 10 mg L^{-1} of paracetamol solution (100 mL). The mixture was magnetically stirred for 1 h in the dark to get the equilibrium of paracetamol adsorption-desorption on the photocatalysts surfaces. The samples were withdrawn for every 30 mins and the concentration of paracetamol was estimated according to the adsorption band wavelength at 274 nm by UV-Vis spectrophotometry.

3. Results and Discussion

3.1. Structural investigation
The morphological properties of the catalysts were analyzed using FESEM and shown in Fig. 1. The effect of titanium on the FZ structure can be verified from FESEM analysis. It can be seen from Fig.1a the cockscomb-like surface of FZ with uniform size of 400-500nm was successfully synthesized. After addition of TiO_2 the morphology of FZ was preserved (Fig 1b-d). However, denser dendrimeric fibre was observed after 5 wt% of TiO_2 doped on FZ.

Figure 1. FESEM images of (a) FZ (b) 1 wt% TiO_2/FZ (c) 3 wt% TiO_2/FZ (d) 5 wt% TiO_2/ FZ.

The FTIR spectra of the catalysts were conducted in the region of 4000 to 400 cm^{-1} in Fig 2A and the intensities of the bands were presented in Fig 2B. The band at 3460 cm^{-1} represents the hydroxyl group caused by incorporated humidity of KBr during the preparation of samples [17]. The peaks at 1090, 966, 800 and 465 cm^{-1} can be assigned to Si-O-T (T=Si or Al) assymetric vibration, external Si-OH group, symmetric vibration and Si-O-T bending, respectively [21]. Comparing with FZ spectra, it can be seen that all the peaks for TiO_2/FZ become shorter in intensity with increasing TiO_2 amount. This suggested that the interaction between FZ framework and TiO_2 due to dealumination, desilication and titanation during the preparation of sample by electrolysis. The weak peak of five-membered ring vibration of ZSM-5 can be seen at 550 cm^{-1} [22]. It was also observed that the peak of FZ at 966 cm^{-1} due to Si-O-H stretching decreased by the introduction of TiO_2, ascribed that relatively rearranged of the framework during the electrolysis. Similar trend was stated previously when TiO_2 was doped onto mesoporous silica nanoparticles (MSN) [23]. This band seems decreased with the increasing TiO_2, proving the probable restructuring of the hydroxyl groups and forming Si-O-Ti bonds [24]. The metal ion could be simply substituted with the H atom from the silanol groups of FZ during the electrolysis. This was approved by the presence of a weak band shoulder which is characteristic peak of Si-O-Ti vibration at 960 cm^{-1}, and it was increased with the increasing TiO_2 loading up to 3wt % [25]. The band then decreased due to the some Ti ions might not inserted into the system and it is proved by the agglomeration that happened in 5 wt% TiO_2/FZ [26]

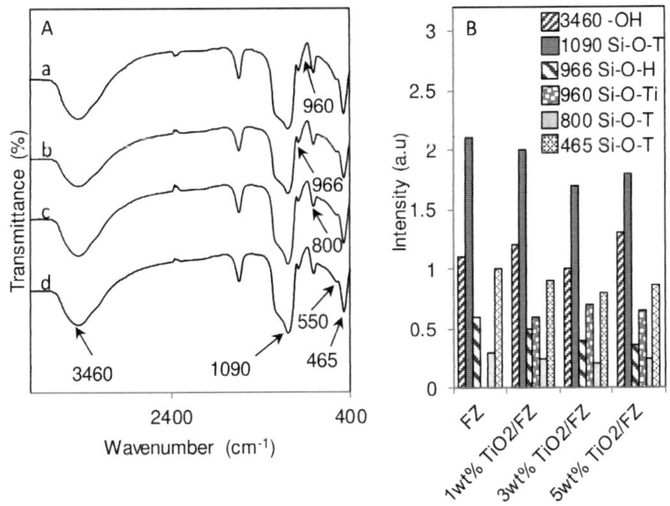

Figure 2. (A) FTIR spectra of (a) FZ (b) 1wt% TiO_2/FZ (c) 3wt% TiO_2/FZ (d) 5wt% TiO_2/FZ (B) Intensity.

The UV-Vis DRS analysis for FZ and TiO_2/FZ are shown in Fig. 3A. It shows that the absorption edge of all sample in the range between 300 to 400 nm. Compared with FZ, the absorption bands of all TiO_2/FZ catalysts acquired higher absorption band area assisted to the large number of photons being absorbed by the catalysts to produce higher amounts of electron-hole pairs to enhance the photocatalytic activity [27]. 3wt% TiO_2/FZ shows the highest absorption, which responsible in photocatalytic activity of catalyst. In addition, the TiO_2/FZ spectra shows that a blue shift in comparing with FZ, referring to the lower band gap energy. The red shift happened when increasing TiO_2 doping from 3 to 5% as higher loading TiO_2 may agglomerated on the surface of FZ and cover the active site which in turn causing to low penetration of light [28]. The band gap energies of all the catalysts were obtained from Kubelka-Munk theory (Fig. 3B) the calculated band-gap energies of FZ, 1 wt% TiO_2/FZ, 3 wt% TiO_2/FZ and 5 wt% TiO_2/FZ are 3.4, 3.1, 2.7 and 2.9 eV, respectively. This improvement offers great potential for photocatalytic reactions to be conducted under visible light irradiation [29].

IOP Conf. Series: Materials Science and Engineering **808** (2020) 012017 doi:10.1088/1757-899X/808/1/012017

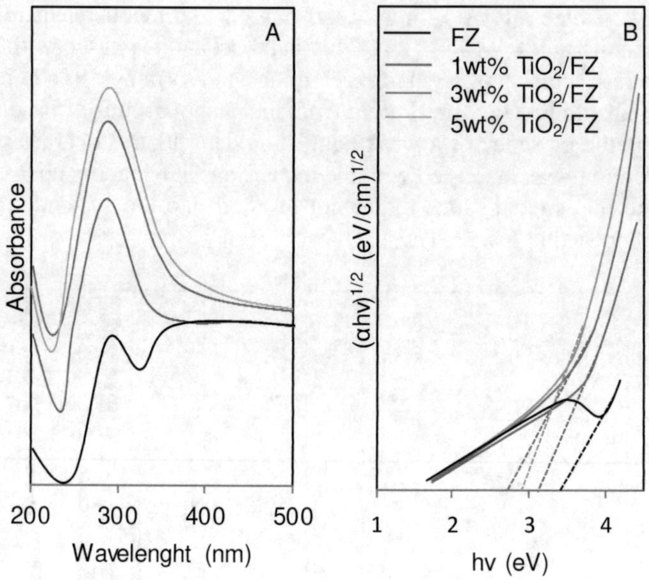

Figure 3. (A) Absorbance spectra (B) Kubelka-Munk spectra.

3.2 *Photocatalytic activity*

The pH shows a major role in the photodegradation of paracetamol since the charge of the photocatalyst surface influence the interaction with paracetamol. It can be seen that the photocatalytic performance enhanced along with increased in pH to achieve the maximum of 90% at pH 5 and then decreased for pH values in between 7 to 11 (Fig. 4A). The catalytic performances are influenced by the electrostatic interaction between the negatively charge paracetamol with the positively charge catalyst surface [30]. Since the zero point charge of the photocatalyst was at pH 6.1 (Fig. 4B), 3 wt% TiO_2/FZ carried positive charge and attracted to the negative charge of paracetamol. The large number of hydroxyl group in high alkaline solutions competed with the paracetamol ions and thus decreasing the photocatalytic efficiency.

Fig. 4C illustrated the phocatalytic trend of FZ, TiO_2, 1 wt% TiO_2/FZ, 3 wt% TiO_2/FZ and 5 wt% TiO_2/FZ on the photodegradation of paracetamol. The degradation of paracetamol for FZ is less than 10%, indicating that it had a small degradation effect of paracetamol. The degradation of paracetamol was lower for 1 and 5 wt% of TiO_2 doped due to the limitation of available reactive site and agglomeration of TiO_2 on the catalyst surface which reduce the light penetration for an efficient photocatalytic activity, respectively [31]. The result clearly showed that the photodegradation of paracetamol under visible light using 3 wt% TiO_2/FZ performed better, which may be due to its smaller band gap and the existence of the highest Si-O-Ti bonds. The 3 wt% TiO_2/FZ achieved the maximum degradation of paracetamol (90%), followed by 5 wt% TiO_2/FZ (71%), 1 wt% TiO_2/FZ (65%), TiO_2 (63%) and FZ (9%). This result suggested that FZ and Ti acted as a significant part in the photocatalytic degradation of paracetamol.

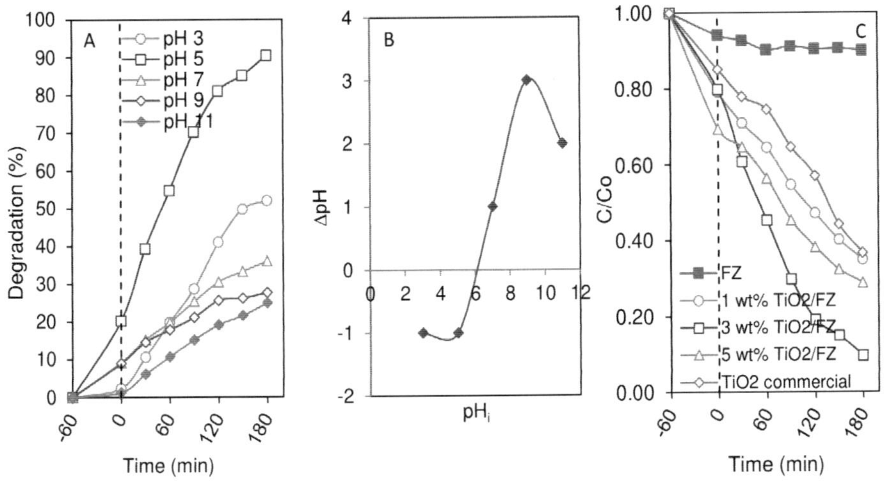

Figure 4. (A) Effect of pH for photodegradation of paracetamol under visible light irradiation [$C_{paracetamol}$= 10mg L^{-1}; W = 0.375g L^{-1}; 3wt% TiO_2/FZ] (B) pH_{pzc} of 3wt% TiO_2/FZ (C) Effect of TiO_2 percentage.

The kinetics of the degradation of paracetamol over the 3 wt% TiO_2/FZ was also studied. A series of reactions, with initial concentration of paracetamol in the scale of 10 and 100 mg L^{-1} were presented (Fig. 5A) to get the photodegradation rate for paracetamol (Fig. 5B). Generally, the pseudo-first order reaction kinetics be represented as equation (2):

$$\ln\left(\frac{C_o}{C}\right) = k_{app}t \tag{2}$$

where t is the time of the reaction, k_{app} is the reaction rate constant, C_0 is the initial concentration of paracetamol and C is the final concentration of paracetamol, respectively. The degradation rate constants were calculated as 0.0103, 0.0061, 0.0046, 0.0026 and 0.0011 min^{-1}, for solutions contained 10, 30, 50, 70 and 100 mg L^{-1}. The system was favourable at low concentration due to low rate constants for higher initial concentration [32,33]. In addition, large amounts of active sites are needed to occupied the larger amount of paracetamol compounds to improve photodegradation. Next, the Langmuir-Hinshelwood kinetics model was constructed based on the Eq. 3:

$$\frac{1}{k_{app}} = \left(\frac{1}{k_r K_{LH}}\right) + \frac{C_0}{k_r} \tag{3}$$

where C_0 is assigned as its concentration before the reaction (mg L^{-1}), the adsorption coefficient of the paracetamol presented as K_{LH} (L mg^{-1}) and the reaction rate constant is presented as k_r (mg L^{-1} min^{-1}). The values of k_r and K_{LH} were 0.115 mg L^{-1} min^{-1} and 0.089 L mg^{-1}, respectively. It is proved that the reaction would take place at the surface of the 3 wt% TiO_2/FZ catalyst due to the k_r value was higher than K_{LH}. Comparable trend has also been stated previously for photodegradation of ibuprofen [34].

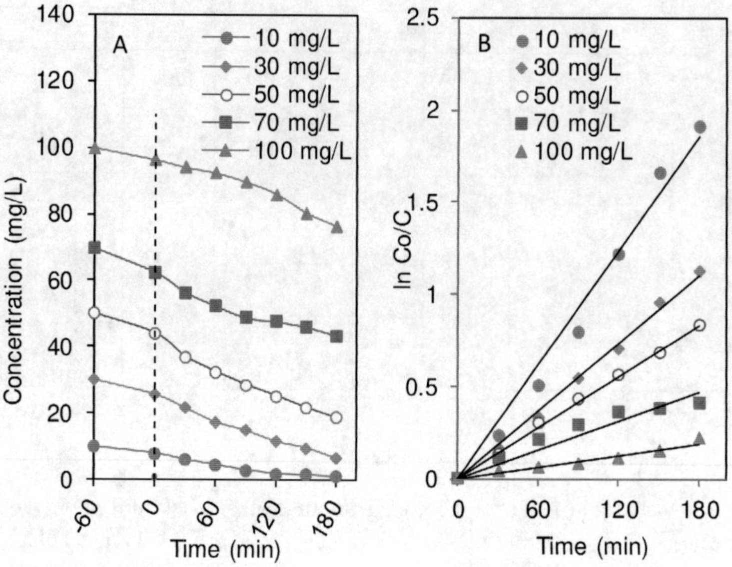

Figure 5. (A) Effect of initial concentration for photodegradation of paracetamol under visible light irradiation [pH= 5; W = 0.375g L^{-1}; 3wt% TiO_2/FZ] (B) kinetics degradation of paracetamol

4. Conclusions

In conclusion, a microemulsion technique and a simple electrolysis method were used to prepare TiO_2/FZ with different amounts of Ti to produce 1 wt% TiO_2/FZ, 3 wt% TiO_2/FZ and 5 wt% TiO_2/FZ catalysts. The results show, the wt% of TiO_2 gave a significant role in controlling the creation of Si-O-Ti bonds that confirmed by FTIR analysis. The higher the number of Si-O-Ti introduced in the framework, the smaller the band gap leading to great photodegradation activity of paracetamol. Thus, 3 wt% TiO_2/FZ showed the most efficient catalyst with the highest Si-O-Ti bonds and the lowest band gap energy. Furthermore, the kinetic study for photodegradation of paracetamol obeyed first-order kinetics that determine by Langmuir-Hinshelwood model. Finally, it could be concluded TiO_2/FZ synthesized using electrolysis method could be employed as promising way to treat other pollutants in waste water treatment.

Acknowledgments

The authors are grateful for the financial support by the Fundamental Research Grant Scheme (FRGS/1/2017/STG07/UTM/01/1).

References

[1] Le T X H, Charmette C, Bechelany M and Cretin M 2016 *Electrochim. Acta* **188** 7378
[2] Gusseme B D, Vanhaecke L, Verstraete W and Boon N 2011 *Water. Res* **45** 1829
[3] Zur J, Wojcieszyńska D, Hupert-Kocurek K, Marchlewicz A and Guzik U 2018 *Chemosphere* **206** 192
[4] Khusnun N F, Jalil A A, Triwahyono S, Hitam C N C, Hassan N S, Jamian N S, Nabgan W, Abdullah T A T, Kamarudin J and Hartanto D 2018 *Powder. Tecnol* **327** 170
[5] Mustapha F H, Jalil A A, Mohamed M, Triwahyono S, Hassan N S, Khusnun N F, Hitam C N C, Rahman A F A, Firmansyah L and Zolkifli A S 2017 *J. of Clean. Product* **168** 1150
[6] Ariyanti D, Dong J, Dong J and Gao W 2016 *Bull. of Chem. React. Eng & Cat* **11** 40
[7] Hosseini S A and Akbari M 2016 *Bull. of Chem. React. Eng & Cat* **11** 299

[8] Makama A B, Salmiaton A, Saion E B, Choong T S Y and Abdullah N 2017 *Bull. of Chem. React. Eng & Cat* **12** 62
[9] Hitam C N C, Jalil A A, Triwahyono S, Ahmad A, Jaafar N F, Salamun N, Fatah N A A, Teh L P, Khusnun N F and Ghazali Z 2016 *RSC Adv* **6** 76259.
[10] Araújo M M, Silva L K R, Sczancoski J C, Orlandi M O, Longo E, Santos A G D, Sá J L S, Santos R S, Luz Jr G E and Cavalcante L S 2016 *Appl. Surf. Sci.* **389** 1137
[11] Gomez S, Marchena C L, Pizzio L and Pierella L 2013 *J. Hazard. Mater* **258** 19
[12] Kanakaraju D, Kockler J, Motti C A, Glass B D and Oelgemöller M 2015 *Appl. Cat. B: Env* **166** 45
[13] Aziz F F A, Jalil A A, Triwahyono S and Mohamed M 2018 *Appl. Surf. Sc* **455** 84-95.
[14] Firmansyah M L, Jalil A A, Triwahyono S, Hamdan H, Salleh M M, Ahmad W F W and Kadja G T M 2016 *Catal. Sci. Technol* **6** 5178
[15] Jalil A A, Zolkifli A S, Triwahyono S, Rahman A F A, Ghani N N M, Hamid M Y S, Mustapha F H, Izan S M, Nabgan B and Ripin A 2018 *Ind. Eng. Chem. Res* **58** 553
[16] Sidik S M, Triwahyono S, Jalil A A, Aziz M A A, Fatah N A A and Teh L P 2016 *J. of CO_2 Utilization* **13** 71
[17] Rahman A F A, Jalil A A, Triwahyono S, Ripin A, Aziz F F A, Fatah N A A, Jaafar N F, Hitam C N C, Salleh N F M and Hassan N S 2017 J. of Clean. Product.143 948
[18] Jusoh N W C, Jalil AA, Triwahyono S and Mamat C R 2015 *Appl. Cat. A: Gen* **492** 169
[19] Teh L P, Triwahyono S, Jalil A A, Firmansyah M L, Mamat C R and Majid Z A 2016 *Appl. Cat. A: Gen* **523** 200
[20] Jaafar N F, Jalil A A, Triwahyono S, Efend J, Mukti, R R, Jusoh R, Jusoh N W C, Karim A H, Salleh N F M and Suendo V 2015 *Appl. Surf. Sc* **338** 75
[21] Zainudin N F, Abdullah A Z and Mohamed A R 2010 *J. Hazard. Mater* **174** 299
[22] Jalil A A, Zolkifli A S, Triwahyono S, Rahman A F A, Ghani N N M, Hamid M Y S, Mustapha F H, Izan S M, Nabgan B and Ripin A 2018 *Ind. Eng. Chem. Res* **58** 553
[23] Teh L P, Triwahyono S, Jalil A A, Firmansyah M L, Mamat C R and Majid Z A 2016 *Appl. Cat. A: Gen* **523** 200
[24] Liu X, Liu Y, Lu S, Guo W and Xi B 2018 *Chem. Eng. J* **350** 131
[25] Setthaya N, Chindaprasirt P, Yin S and Pimraksa K 2017 *Powder. Tecnol* **313** 417
[26] Landi S, Carneiro J, Ferdov S, Fonseca A M, Neves I C, Ferreira M, Parpo P, Soares O S G P, Pereira M F R, 2017 *J. Photochem. Photobio. A: Chem* **346** 60
[27] Wang Y, Chen J, Lei X and Ren Y 2017 *Micro. and Meso. Material* **250** 9
[28] Jiang C, Lee K Y, Parlett C M A, Bayazit M K, Lau C C, Ruan Q, Moniz S J A, Lee A F and Tang J 2016 *Appl. Cat. A: Gen* **521** 133
[29] Liu X, Liu Y, Lu S, Guo and Xi B 2018 *Chem. Eng. J* **350** 131-147
[30] Hassan N S, Jalil A A, Triwahyono S, Hitam C N C, Rahman A F A, Khusnun N F, Mamat C R, Asmadi M, Mohamed M, Ali M W and D. Prasetyoko *J. Taiwan. Ins. Chem. Eng* **82** 322
[31] Jusoh N W C, Jalil A A, Triwahyono S, Setiabudi H D, Sapawe N, Satar M A H, Karim A H, Kamarudin N N, Jusoh R, Jaafar N F, Salamun N and Efendi J 2013 *Appl. Cat. A: Gen* **468** 276
[32] Sharma A and Lee B -K 2016 *J. Env. Manage* **165** 1
[33] Jusoh N C W, Jalil A A, Triwahyono S, Setiabudi H D, Sapawe N, Satar M A H, Karim A H, Kamarudin N H N, Jusoh R, Jaafar N F, Salamun N and Efendi J 2013 *App. Cat. A: Gen* **468** 276
[34] Fauzi A A, Jalil A A, Mohamed M, Triwahyono S, Jusoh N W C, Rahman A F A, Aziz F F A, Hassan N S, Khusnun N F and Tanaka H 2018 *J of Env. Management* **227** 34

IOP Conf. Series: Materials Science and Engineering **808** (2020) 012018 doi:10.1088/1757-899X/808/1/012018

Beneficial interaction of copper oxide and fibrous silica for enhanced photocatalytic desulphurization

C N C Hitam[1], A A Jalil[1,2*], N N S M Shobri[1], A A Fauzi[1]

[1] School of Chemical and Energy Engineering, Faculty of Engineering, Universiti Teknologi Malaysia, Malaysia.

[2] Centre of Hydrogen Energy, Institute of Future Energy, Universiti Teknologi Malaysia, Malaysia.

*aishahaj@utm.my

Abstract. Photocatalytic desulphurization (PDS) is considered as a promising technique to remove hazardous sulphur compounds from fuel oil due to its high catalytic activity, safety, low energy consumption, and low cost. In this work, fibrous silica KCC-1 was synthesized by hydrothermal method followed by incorporation of CuO by electrolysis. The catalyst was used for PDS of dibenzothiophene (DBT). The presence of bicontinuous concentric lamellar morphology with uniform spherical shape was confirmed by FESEM analysis. The phase and crystallinity of the catalysts was investigated by XRD, while the band gap energy was determined by UV-Vis DRS. The PDS performance of CuO/KCC-1 was compared with KCC-1, CuO/SiO_2 and SiO_2. It was observed that the photoactivity of CuO/KCC-1 was higher than other catalysts (62.6%), which is mainly due to its fibrous morphology and appropriate band gap energy, as well as synergistic role of both CuO and KCC-1. This fibrous type catalyst could be a promising candidate for environmental purification specifically in the sulphur removal.

1. Introduction

Desulphurization of fuels is crucial for reducing sulphur content in order to solve the worldwide environmental pollution issue [1]. During combustion of fuel, sulphur emits as sulphur dioxide (SO_2) or sulphate which is one of the main sources of air pollution and acid rain [2]. The conventional hydrodesulphurization (HDS) method suffers the harsh reaction condition, expensive catalysts, and high purity hydrogen requirement. In recent time, photocatalytic desulphurization has become the main interest among researchers due to its greener approach for conversion of organic contaminants into harmless compounds [3]. The DBT and its derivatives could be converted into the sulfoxide (DBTO) and sulfone ($DBTO_2$) using this method [4].The widely used titanium dioxide (TiO_2) photocatalyst suffers several drawbacks including low specific surface area, poor charge carrier separation, and narrow utilization of solar light [5].

For those reason, various metal oxide photocatalysts with narrower band gap has been explored including AgO, CuO, Fe_2O_3, etc. Among them, CuO is considered as a promising candidate due to its appropriate band gap, economical, less toxicity, and abundant source [6]. Numerous studies has reported on the use of CuO for different photocatalytic applications such as decolourization of methyl orange (MO) [7], degradation of *p*- chloroaniline (PCA) [8], and removal of bisphenol A (BPA) [9].

However, this CuO require support materials for improved dispersion and surface area, as well as to prevent agglomeration.

In order to improve the dispersion of metal oxides for abundant active sites, several support materials has been utilized including mesostructured silica nanoparticles (MSN) [10], mesoporous titania nanoparticles (MTN) [11], carbon nanotubes (CNT) [12], etc. Fibrous silica (KCC-1) is considered as a new potential support candidate to design efficient photocatalysts for degradation of organic pollutants [13]. It was first discovered by Polshettiwar and co-workers, which then experiencing various modifications for enhanced performance in different applications [14-17]. Due to its several drawbacks, more modification is needed to improve its usage under visible light.

Herein we report the synthesis of KCC-1 and CuO supported on KCC-1 (CuO/KCC-1) via hydrothermal and electrochemical methods, respectively. The catalyst was used for visible-light photocatalytic removal of DBT. The physicochemical properties of the catalysts were well characterized by FESEM, XRD, and UV–Vis DRS. Thus, this present work is hoped to deliver initiative for future design approaches on photocatalytic removal of various organic contaminants as well as other application

2. Experimental

2.1 Materials

Cethyltrimethylammonium bromide (CTAB), urea, tetraethylorthosilicate (TEOS) and dibenzothiophene (DBT) were purchased from Merck. Toluene, butanol, acetonitrile and iso-octane were obtained from QRec. Meanwhile, tetraethylammonium perchlorate (TEAP) was synthesized by previous reported method [6].

2.2 Synthesis of catalyst

Fibrous silica KCC- 1 was fabricated by hydrothermal method as reported by previous study [15], while SiO_2 was obtained by calcining the commercial SiO_2. For KCC- 1, CTAB and urea were dissolved in distilled water and mixed with another solution consists of TEOS, toluene and butanol. Then, the mixture was vigorously stirred at 403 K, before cooled and centrifuged. After the precipitate was oven-dried for 12 h, the sample was calcined at 823 K for 6 h. Meanwhile, the 5 wt% CuO/KCC-1 and CuO/SiO_2 catalysts were synthesized by electrochemical method using KCC-1 and SiO_2 powders as support, respectively, according to earlier study [6]. An electrolysis cell was provided with a magnetic stirrer, a platinum plate (2 x 2 cm^2) as cathode and a copper plate (2 x 2 cm^2) as anode. Firstly, distilled water was used to dissolve TEAP supporting electrolyte, then KCC-1 or SiO_2 powder was added and stirred for 5 min. The electrosynthesis was conducted at 120 mA cm^{-2} and 273 K under atmospheric pressure. After the required CuO weight was reduced, the sample was impregnated and oven-dried, and calcined at 550°C for 3 hours. For pure CuO, the electrolysis was conducted as above without addition of any support material.

2.3 Characterization of the catalyst

The morphological properties of the catalysts were examined using FESEM (JEOL JSM-6701F). X-ray diffraction (XRD) analysis on powder diffractometer (Bruker Advance D8, USA) were used to detect the phase and crystallinity of the catalysts. UV–Vis diffuse reflectance spectra acquired by a PIKE Technologies DiffusIR was employed to determine the absorbance spectrum of the catalysts for band gap measurement.

2.4. Photocatalytic desulphurization reaction

The photocatalytic desulphurization of DBT was conducted as described in the previous report [18]. 100 mg L^{-1} of DBT in hexane was mixed with acetonitrile as extractant, followed by addition of 0.375 gL^{-1} catalyst, and stirred in dark treatment for 30 min. After 30 min, the reaction solution was irradiated with visible light, and sampling for every 10 min until 2 hours. The sample was centrifuged before being analyzed by UV-Vis spectrophotometer (Thermo Scientificc Genesys10uv Scanning) to measure the initial and final DBT concentration. The desulphurization percentage was obtained using this equation:

$$D = (1 - C/C_0) \times 100\% \quad (1)$$

Where C_0 is the sulphur content of the initial solution at $t = 0$, and C is the sulphur content after last sampling.

3. Results and Discussion

3.1 Morphological studies

Figure 1 displays the FESEM images of KCC-1 and CuO/KCC-1. Figure 1A reveals the presence of uniform spherical shape with bicontinuous concentric lamellar structure of KCC-1, with particle size ranging from 200 to 400 nm. Meanwhile, Figure B shows the preservation of fibrous silica KCC-1 upon addition of CuO, which is consistent with previous reported study [15]. For CuO/SiO_2, the irregular shaped silica morphology is expected, a similar observation with those reported for Ni/SiO_2 in dry reforming of methane due to the similar synthesis procedure [16].

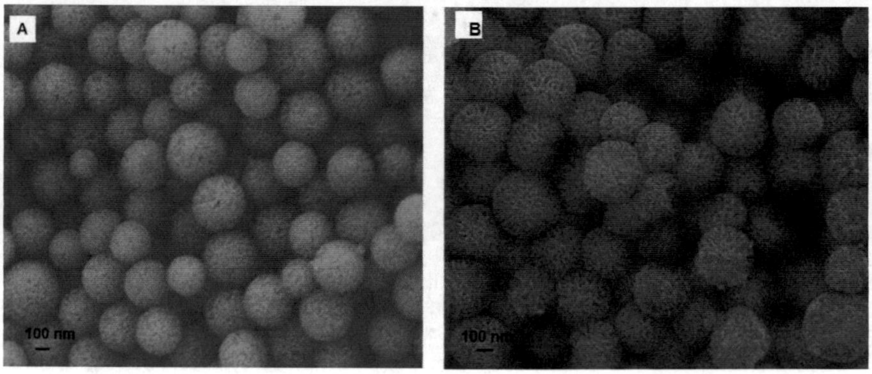

Figure 1. FESEM images of A) KCC-1 and B) CuO/KCC-1.

3.2 Structural properties

Figure 2 illustrates the XRD diffractogram of the bare and supported catalysts. The presence of amorphous structure in both KCC-1 and SiO_2 catalysts was signified by the existence of a broad peak between $2\theta = 20–30°$ [19].Figure 2 illustrates the XRD diffractogram of the bare and supported catalysts. The presence of amorphous structure in both KCC-1 and SiO_2 catalysts was signified by the existence of a broad peak between $2\theta = 20–30°$ [19].

The inset figure shows the XRD pattern of the electrosynthesized CuO with diffraction peaks located at $2\theta = 32.5$, 35.5, 38.7, 48.7, 53.4, 58.3, 61.5, 66.1 and 68.1°, which were attributed to reflection lines of monoclinic CuO nanoparticles [20].The decrease of the peak corresponding to amorphous silica of both KCC-1 and SiO_2 upon the introduction of CuO, suggesting partial blocking and distortion of the silica framework [21]. Besides, all distinct peaks of CuO were detected, which proving the successful loading of CuO dopant on both KCC-1 and SiO_2 catalysts, respectively.

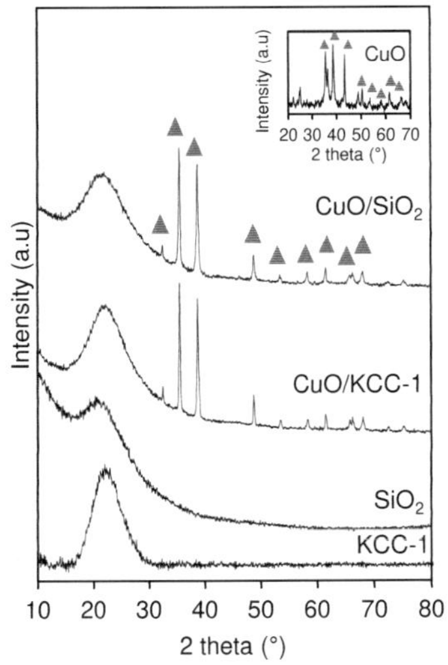

Figure 2. XRD pattern of the catalysts

3.3 Optical studies

Figure 3 displays the UV–vis DRS spectra of the catalysts. It was observed that KCC-1 had a significant absorption in the ultraviolet area at 320 nm and a red shift to a higher wavelength occurred after CuO was added.

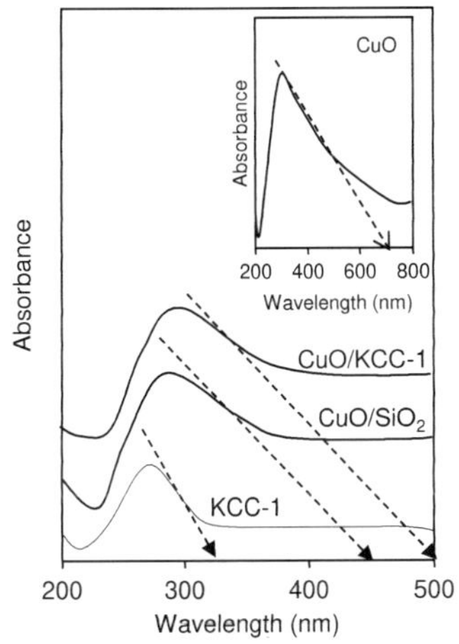

Figure 3. UV-Vis DRS spectra of the catalysts.

It was previously reported that the extension of the absorption spectrum is undoubtedly due to the interaction between the narrow band gap CuO (inset figure) and the support catalysts [22]. Based on this spectra, the band gap energies were calculated according to equation below;

$$E_g = \frac{1240}{\lambda} \tag{2}$$

The calculated band gap energy of the catalysts was tabulated in Table 1. The CuO/KCC-1 demonstrated a lower band gap energy than that of CuO/SiO_2, most probably due to the existence of more defect sites in KCC-1 as compared to SiO_2, as mentioned in earlier work [15]. It was also previously revealed that the amount of surface defect had significantly affected the band gap energy of the catalysts [23].

Table 1. Band gap energy of the catalysts

Catalysts	Band Gap Energy (E_g)
KCC-1	3.88
CuO/SiO_2	2.76
CuO/KCC-1	2.48
CuO	1.77

3.4 Photocatalytic desulphurization performance

Figure 4 displays the photocatalytic desulphurization performance of the catalysts under visible light irradiation. The CuO/KCC-1 demonstrated the highest photoactivity (62.6%) than that of CuO/SiO_2 (44.5%), KCC-1 (38.9%), CuO (31.4%) and SiO_2 (27%), mainly due to its fibrous morphology and narrowest band gap for the enhanced visible light absorption. Previous study has shown that the fibrous morphology contributed to the high surface area and high amount of surface defect [15]. These led to enhanced absorption of photon and adsorption of DBT molecules. Besides, it should be noted that the above mentioned criterion may also lead to an efficient charge carrier separation, which consequently resulted in an enhanced performance [24].

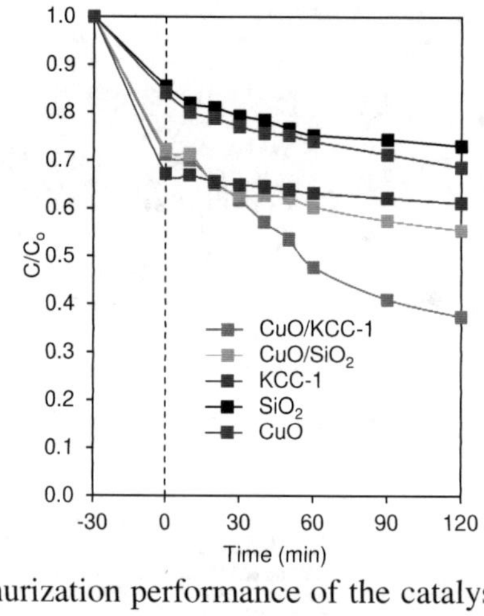

Figure 4. Photocatalytic desulphurization performance of the catalysts [0.375 g L^{-1} catalyst, 100 mg L^{-1} DBT, 120 min].

4.0 Conclusion

In this study, fibrous silica KCC-1 was fabricated via a hydrothermal method and loaded with CuO to form CuO/KCC-1 by electrochemical technique. The physicochemical properties of the catalysts were studied by FESEM, XRD and UV-Vis DRS analyses. The FESEM images reveals the existence of fibrous morphology KCC-1 in uniform spherical shape with bicontinuous concentric lamellar structure and particle size ranging from 200 to 400 nm. The XRD pattern shows that both KCC-1 and SiO_2 present as amorphous silica, and CuO was successfully loaded on both catalysts with the emerging of its characteristic peaks on the doped catalysts. Based on UV-Vis DRS, the absorption edge of KCC-1 shifted to higher wavelength by the addition of CuO, signifying a probable CuO/KCC-1 interaction and the role of CuO in altering the band energy of the catalysts. The photocatalytic desulphurization was conducted on 100 mgL^{-1} of DBT and as expected, the highest performance was achieved using CuO/KCC-1 (62.6%). This is due to the presence of fibrous morphology, incorporation of CuO, and the narrowest band gap. These criteria is synergistically crucial to achieve easier charge transfer and better charge carrier separation for enhanced photocatalytic performance.

Acknowledgments
This research study was sponsored by the Universiti Teknologi Malaysia through Professional Development Research University Grant (No. 04E33) and Collaborative Research Grant (Grant No. 07G62), and also by the Ministry of Higher Education Malaysia through Fundamental Research Grant No. FRGS/1/2017/STG07/UTM/01/1 (Grant No. 4F969).

References

[1] Abdi G, Ashokkumar M, Alizadeh A 2017 *Fuel.* **210**, 639.
[2] Abdullah G H and Xing Y 2018 *Energ. Fuels* **32**,8, 8254.
[3] Chen Y, Shen C, Wang J, Xiao G, and Luo G 2018 *ACS Sustainable Chem. Eng.* **6**, 10, 13276.
[4] Lu X, Li X, Chen F, Chen Z, Qian J, Zhang Q 2020 *J. Alloy. Compd* **815** 152326.
[5] Lu X, Li X, Qian J, Miao N, Yao C, Chen Z 2016 *J. Alloy. Compd.* **661**, 363.
[6] Hitam C N C, Jalil A A, Triwahyono S, Rahman A F A, Hassan N S, Khusnun N F, Jamian S F, Mamat C R, Nabgan W, Ahmad A 2018 *Fuel* **216**, 407.
[7] Jusoh N W C, Jalil A A, Triwahyono S, Mamat C R 2015 *Appl. Catal. A: Gen.* **492**, 169.
[8] Khusnun N F, Jalil A A, Triwahyono S, Jusoh N W C, Johari A and Kidam K 2016 *Phys. Chem. Chem. Phys.* **18**, 12323.
[9] Hassan N S, Jalil A A, Triwahyono S, Hitam C N C, Rahman A F A, Khusnun N F, Mamat C R, Asmadi M, Mohamed M, Ali, M W, Prasetyoko D 2018 *J. Taiwan Inst. Chem. E.* **82**, 322
[10] Rahman A F A, Jalil A A, Triwahyono S, Ripin A, Aziz F F A, Fatah N A A, Jaafar N F, Hitam C N C, Salleh N F M, Hassan N S 2017 *J. Clean Prod.* **143**, 948.
[11] Jaafar N F, Jalil A A, Triwahyono S 2017 *Appl. Surf. Sci.* **392**, 1068.
[12] Khusnun N F, Jalil A A, Triwahyono S, Hitam C N C, Hassan N S, Jamian F, Nabgan W, Abdullah T A T, Kamaruddin M J, Hartanto D 2018 *Powder Technol.* **327** 170.
[13] Singh R, Bapat R, Qin L, Feng H and Polshettiwar V 2016 *ACS Catal.* **6**,5, 2770.
[14] Polshettiwar V, Cha D, Zhang X and Basset J M 2010 *Angew. Chem., Int. Ed.* **49**, 9652.
[15] Hamid M Y S, Jalil A A, Rahman A F A, Abdullah T A T 2019 *React. Chem. Eng.* **4**, 1126.
[16] Abdulrasheed A A, Jalil A A, Hamid M Y S, Siang T J, Fatah N A A, Izan S M, Hassan N S 2019 *Int. J. Hydrogen Energ.* xx.
[17] Ibrahim M, Jalil A A, Khusnun N F, Fatah N A A, Hamid M Y S, Gambo Y, Abdulrasheed A A, Hassan N S *2019 Int. J. Hydrogen Energ.* xx.
[18] Hitam C N C, Jalil A A, Triwahyono S, Ahmad A, Jaafar N F, Salamun N, Fatah N A A, Teh L P, Khusnun N F and Ghazali Z 2016 *RSC Adv.* **16**, 76259.
[19] Fatah N A A, Triwahyono S, Jalil A A, Salamun N, Mamat C R, Majid Z A 2017 *Chem. Eng. J.* **314**, 650.
[20] Padil V V T and Černík M 2013 *Int. J. Nanomedicine* **8**, 889.
[21] Sidik S M, Triwahyono S, Jalil A A, Aziz M A A, Fatah N A A, Teh L P 2016 *J. CO_2 Util.* **13** 71
[22] Aazam E S 2014 *Ceram. Int,* **40** (5), 6705.
[23] Hitam C N C, Jalil A A, Abdulrasheed A A 2019 *J. Ind. Eng. Chem.* **74**, 172.
[24] Li X, Li F, Lu X, Zuo S, Li Z, Yao C, Ni C 2018 *Powder Technol.* **327**, 467.

IOP Conf. Series: Materials Science and Engineering **808** (2020) 012019 doi:10.1088/1757-899X/808/1/012019

Synthesis of fibrous silica tantalum (FSTa) for photooxidative desulphurization

M N S Rahim[1], **A A Jalil**[1,2*] **C N C Hitam**[1]

[1]School of Chemical and Energy Engineering, Faculty of Engineering, Universiti Teknologi Malaysia, Malaysia.
[2]Centre of Hydrogen Energy, Institute of Future Energy, Universiti Teknologi Malaysia, Malaysia.

*aishahaj@utm.my

Abstract. The photooxidative desulphurization (PODS) of dibenzothiophene (DBT) was examined under visible light using fibrous silica tantalum (FSTa), tantalum oxide doped fibrous silica (Ta/KCC-1) and commercial tantalum oxide (Ta_2O_5). FSTa was synthetized using hydrothermal method, while Ta/KCC-1 was obtained via a wet impregnation method. The catalysts were examined by field emission scanning electron microscopy (FESEM), X-ray Diffraction (XRD) and UV-Vis diffuse reflectance spectroscopy (UV-Vis DRS). It was shown that the FSTa possesses the highest photocatalytic performance (70.2%), due to its fibrous structure, well-dispersion of Ta, as well as its narrower band gap. These traits impact on the photocatalysis by promoting the deposition of the DBT on the catalyst, allowing the efficient transfer of charge carrier and preventing the electron-hole recombination.

1. Introduction

Sulphur oxides (SO_x) produced from the burning of fossil fuels has causes major impacts on the environment and human health. The emission of SO_x has led to atmospheric acidification, acidic deposition and aerosol production [1]. Thus, moving to low-sulfur fuels is an efficient approach to reduce SO_x emissions which has become a priority for many researchers and industries worldwide. Therefore, different desulphurization methods have been implemented to solve these issues throughout these years [2]. However, among all the treatment technologies, oxidative desulphurization (ODS) is considered as the most suitable technique to obtain an ultra-low sulphur containing in fuel oils due to its low cost, environmentally safe, low temperature and ambient pressure operating condition [3-7]. In order to improve the efficiency of this process, some assistance is often used like adsorptive-, microwave-, extractive-, electrochemical-, ultrasound- and photocatalytic-ODS (PODS).

In recent years, among the different heterogeneous catalysts, Ta_2O_5 is ranked as one of the most important transition metal oxides due to its suitable physical, chemical and photocatalytic properties [8]. However, its application under visible light is limited by high band gap energy (≥3.0 eV) which is more active toward UV light irradiation. With this in mind, visible light photocatalysts (λ= 400–800 nm) is crucial to be implemented for the practical use of PODS, hence, an optimum band-gap (< 3.0 eV) is required [9]. Previous studies have reported many techniques to improve photocatalyst to be used under visible light including the combination with other organic and/or inorganic compounds, addition of support material, incorporation of two or more suitable semiconductors, and the creation of defect structure for improved charge-carrier separation [2]. Besides, surface defect and fibrous morphology might also increase the visible light absorption by reducing the band gap energy [10]. Taking into consideration all the above-mentioned factors, this study aims to synthetize fibrous silica tantalum (FSTa) with narrow band gap and compare its photo activity over Ta/KCC-1 and commercial Ta_2O_5 for the PODS of DBT in model oil under visible light.

2. Experimental

2.1. Material

Cetyltrimethylammonium bromide (CTAB), urea, butanol, toluene, tetraethyl orthosilicate (TEOS), and dibenzothiophene were purchased from Merck Sdn.Bhd., Malaysia. Tantalum pentoxide (Ta_2O_5) were bought from Sigma Aldrich. Acetonitrile and hexane were obtained from QRec, Malaysia.

2.2. Catalysts synthesis

FSTa was prepared using the hydrothermal method as reported by previous work [11]. During synthesis of KCC-1, CTAB and urea were dissolved in distilled water. This solution was then mixed with other separated solution consisting TEOS, toluene and butanol, and stirred homogeneously under heating at 403 K. After cooled, the mixture was centrifuged and dried before being calcined at 823 K for 6 h. For FSTa, a similar procedure was used, except the Ta_2O_5 powder was added in the second solution. Ta/KCC-1 was synthesized using a wet impregnation method. Firstly, 1 g of KCC-1 was mixed with water in a beaker and stirred for 5 min, followed by addition of appropriate amount commercial Ta_2O_5. The mixture was continuously stirred and heated at 80°C until the fine powder obtained. The powder product was oven-dried for overnight and finally heat treated at 550°C for 3 hours. For Ta_2O_5 catalyst, commercial Ta_2O_5 was calcined at the same temperature and time as above.

2.3. Characterization

FESEM (JEOL JSM-6701F, Japan) was used to study the morphological properties of the catalysts. The X-ray diffraction (XRD) recorded on Bruker Advance D8, 40 kV, 40 mA powder diffractometer was utilized to investigate the structural properties of the catalysts. The PIKE Technologies DiffusIR was employed to obtain UV–Vis diffuse reflectance spectra (UV –Vis DRS) in the range of 200 to 800 nm for band gap energy measurement.

2.4. Photocatalytic oxidative desulphurization reaction

The photoactivities of the FSTa catalyst was investigated through the PODS of DBT. During reaction, a batch reactor containing 4 visible lamps and a chiller was used. 100 mg L^{-1}DBT was dissolved in hexane and used as model oil. Appropriate amount of catalyst was added to the mixed solution consisting model oil and acetonitrile as extractant (ratio 2:1), stirred constantly, and placed in the dark for 30 min to allow complete adsorption–desorption process. After that, the solution was visible light-illuminated and sampling for every 10 min. The sample was centrifuged and analyzed by UV-Vis spectrophotometer (Thermo Scientific Genesys10uv Scanning) at λ= 326 nm. Each set of experiments was repeated three times to ensure the accuracy. The same procedure was applied for other catalysts.

3. Results and discussion

3.1. Morphological study.

The morphology of FSTa was studied using FESEM as shown in Figure 1. It could be noticed from Figure 1A the presence of uniform cockscomb-like spherical shape of FSTa with dendrimeric fibers at its outer surface, with the size varies between 400-650nm (Figure 1B). This structure could offer a high adsorption of DBT molecule towards the catalyst and facilitate the absorption of light for enhanced photoactivity [12].

Figure 1. FESEM images of FSTa

3.2. Crystallinity. phase and structural studies.

XRD patterns of FSTa, Ta/KCC-1 and Ta_2O_5 catalysts are displayed in Figure 2. The commercial Ta_2O_5 (inset figure) presents a series of different major characteristic peaks at 2θ=22.9, 28.3, 36.7, 46.8, 49.8 and 55.5°. In fact, the FSTa exhibits the main peaks of Ta_2O_5 with a lower-intensity and noisy pattern, suggesting that Ta_2O_5 is barely losing its structural integrity [13]. This might be due to the presence of dendrimeric silica fibers surrounding the Ta_2O_5 seed. The highest intensity observed for the FSTa catalyst, implying the presence of larger Ta_2O_5 crystallites in the catalyst [14].

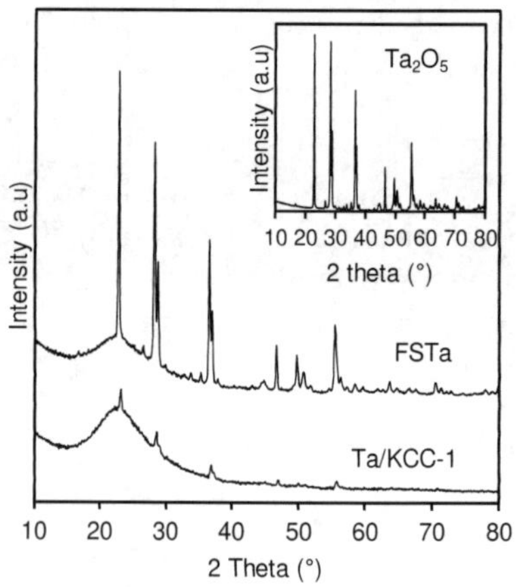

Figure 2. XRD patterns of FSTa, Ta/KCC-1 and Ta_2O_5 catalysts

3.3. Optical studies.

The UV –Vis DRS spectra and calculated band gap of the catalysts are displayed respectively in Figure 3 and Table 1. It could be observed that both Ta/KCC-1 and FSTa demonstrated narrowed band gap which are 2.88 and 2.25 eV, respectively. This is one of the factor to consider for a PODS under visible light. As expected, commercial Ta_2O_5 exhibits a wide band gap (3.54 eV) which is much higher than the maximum band gap required for visible light reaction. It is believed that the improved absorption edge may lead to a better charge carrier separation and high harvesting rate of solar light, which subsequently resulted in an enhanced performance.

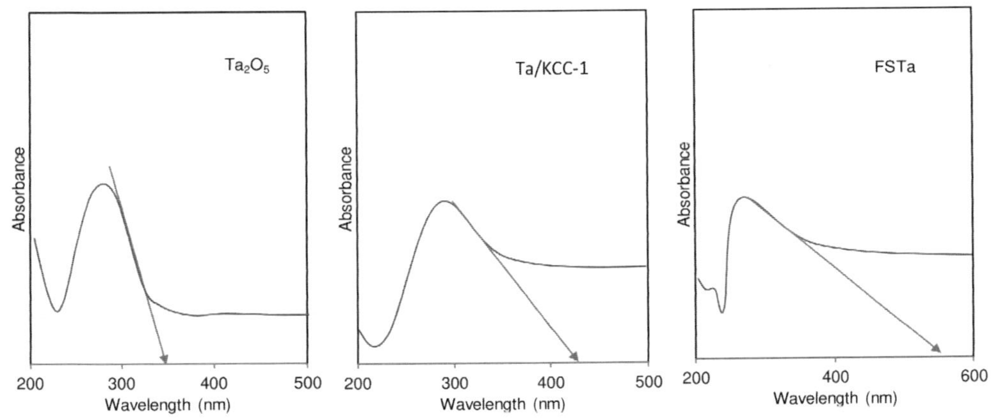

Figure 3. UV-Vis spectra of the composites

Table 1. Band gap energy of the catalysts

Catalyst	**E_g (eV)**
Ta_2O_5	3.54
Ta/KCC-1	2.88
FSTa	2.25

3.4. Photocatalytic oxidative desulfurization performance

The photocatalytic performance of the different catalysts was examined in the oxidation of DBT in model oil and their performance is shown in Figure 4.

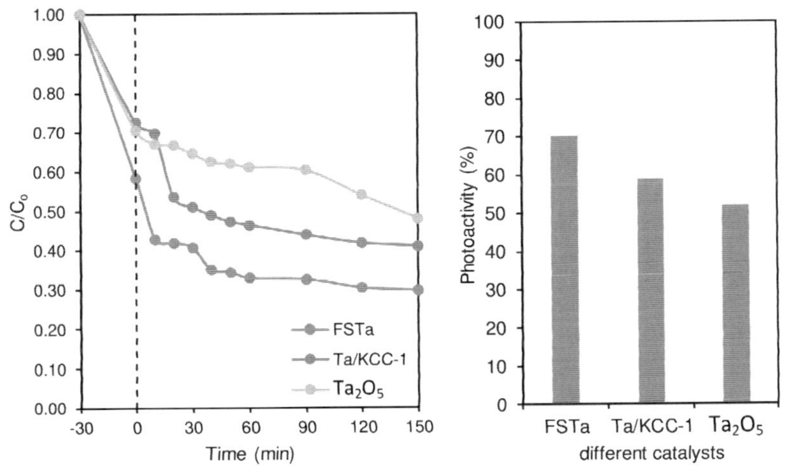

Figure 4. Photocatalytic oxidative desulphurization performance of the different catalysts [0.5 g L^{-1} catalyst, 100 mg L^{-1} DBT, 150 min].

It is clearly observed that the FSTa exhibited a better performance compared to Ta/KCC-1 and commercial Ta_2O_5 under visible light, and the PODS were in the following order: FSTa (70.2%) > Ta/KCC-1 (58.9%) > commercial Ta_2O_5 (51.9%). As predicted earlier, the superior performance demonstrated by FSTa was definitely due to its fibrous morphology for enhanced adsorption of DBT and absorption of visible light, structural completeness with higher crystallinity, well-dispersion of Ta, and narrowest band gap. These criteria might symbiotically play role in preventing the charge recombination for enhanced performance.

4. Conclusion

In summary, FSTa was successfully synthesized using a hydrothermal method, and has been successfully used for PODS of DBT under visible light irradiation. A cockscomb-like spherical shape with dendrimeric fibers and size ranging 400-650 nm of FSTa was detected via FESEM analysis. The XRD pattern proves the successfully loading of Ta_2O_5 on the surface of the FSTa and Ta/KCC-1 catalyst, in which a better dispersion was observed in FSTa. UV Vis DRS reveals the extended light absorption of both FSTa and Ta/KCC-1 towards visible region, which resulted in narrower band gap as compared to commercial Ta_2O_5. The fibrous morphology, well-dispersion of Ta_2O_5, as well as the narrowest band gap has allowed the best performance of FSTa than that of other catalysts. These criteria are crucial to improve DBT adsorption and light absorption, prevent the electron-hole recombination, and enhanced photocatalytic performance under visible light irradiation.

Acknowledgements

This research study was sponsored by the Universiti Teknologi Malaysia through Collaborative Research Grant (Grant No. 07G62).

References

[1] Hitam C N C, Jalil A A, Triwahyono S, Ahmad A, Jaafar N F, Salamun N, Fatah N A A, Teh L P, Khusnun N F and Ghazali Z 2016 *RSC Adv*, **6**, 76259
[2] Hitam C N C, Jalil A A, Abdulrasheed A A 2019 *J. Ind. Eng. Chem.* **74** pp 172
[3] Wang L, Cai H, Li S, Mominou N 2013 *Fuel*, **105** pp 752
[4] Lin F, Jiang Z, Tang N, Zhang C, Chen Z, Liu T, Dong B 2016 *Appl. Catal. B: Env.*, **188** pp 253
[5] Zhao D, Zhang J, Wang J, Liang W, Li H 2009 *Petrol. Sci. Technol.*, **27** pp 1
[6] Liu R, Zhang J, Xu Z, Zhao D, Sun S 2017, *J. Mater. Sci*, **53**(7), 4927
[7] Zhu W, Xu Y, Li H, Dai B, Xu H, Wang C, Liu H 2014 *Korean J Chem. Eng.*, **31**(2) pp 211
[8] Seisenbaeva G A, Cojocaru B, Jurca B, Tiseanu C, Nedelec J M, Kessler V G, Parvulescu V 2017 *Chemistry Select*. **2(1)** p 421
[9] Ahmed M, Xinxin G 2016 *Inorg. Chem. Front.* **3(5)** pp 578
[10] Hitam C N C, Jalil A A, Triwahyono S, Rahman A F A, Hassan N S, Khusnun N F, Jamian S F, Mamat C R, Nabgan W, Ahmad A 2018 *Fuel* **216** pp 407
[11] Hamid M Y S, Jalil A A, Rahman A F A, Abdullah T A T 2019 *React. Chem. Eng.*, **4** pp 1126
[12] Fauzi A A, Jalil A A, Mohamed M, Triwahyono S, Jusoh N W C, Rahman A F A, Aziz F F A, Hassan N S, Khusnun N F, Tanaka H 2018 *J. Environ. Manage*. **227** pp 34
[13] Aziz F F A, Jalil A A, Triwahyono S, Mohamed M 2018 *Appl. Surf. Sci.*, **455** pp 84
[14] Sidik S M, Triwahyono S, Jalil A A, Aziz M A A, Fatah N A A, Teh L P 2016 *J. CO_2 Util.* **13** p 71

IOP Conf. Series: Materials Science and Engineering **808** (2020) 012020 doi:10.1088/1757-899X/808/1/012020

Boosted Ag/g-C_3N_4 prepared under microwave irradiation for photocatalytic activity of RhB under visible light

M S Azami[1], A A Jalil[2,3*], C R Mamat[1], C N C Hitam[2], N S Hassan[2], I Hussain[1], A A Fauzi[2] and R H Adnan[1]

[1]Department of Chemistry, Faculty of Science, Universiti Teknologi Malaysia, 81310 UTM Johor Bahru, Johor, Malaysia.
[2]School of Chemical and Energy Engineering· Faculty of Engineering, Universiti Teknologi Malaysia, 81310 UTM Johor Bahru, Johor, Malaysia.
[3]Centre of Hydrogen Energy, Institute of Future Energy, Universiti Teknologi Malaysia, 81310 UTM Johor Bahru, Johor, Malaysia

*aishahaj@utm.my

Abstract. The effluent of Rhodamine B (RhB) from industries causes a high impact toward contamination of the environment. Photocatalytic degradation has become a promising destructive technique to remove those dye from industrial. In this study, Silver/graphitic carbon nitride (Ag/g-C_3N_4) prepared using simplistic preparation using microwave irradiation in its place of normal calcination under furnace as heating media. The prepared photocatalyst was characterized by using Fourier transform infrared (FTIR), N_2 adsorption-desorption, ultraviolet-visible diffuse reflectance spectra (UV-VIS/DRS). The deposited of Ag on g-C_3N_4 surface photocatalyst enhanced capability to extend the adsorption of visible light region. The texture of g-C_3N_4 affected after deposited with Ag and the bandgap energy of Ag/g-C_3N_4 narrowed from 2.7 eV to 2.31 eV significantly boosting and improved the efficiency of degradation RhB under visible light irradiation. The additional Ag on g-C_3N_4 revealed highest performance (98%) using RhB solution after exposed under visible light for 180 min. As a result, this works can give benefit to purify the effluent of toxic dyes as wastewater treatment.

1. Introduction

In recent years, worker have revealed a great interest in removal of dye pollutants from wastewater due to industrial development. The effluents discharged from industries have resulted in increasing wastewater to the environment. This wastewater needs to be treated as it contains highly carcinogenic aromatic amine that can cause a serious environmental pollution and bring negative impacts to all living creatures [1]. The discharged of dye from industries was estimated about over 200,000 tons per year cause by huge scale-production [2]. Rhodamine B (RhB) is one of the recognized dyes as colorant in textile due to its high stability [3]. However, this dye is very harmful to human being which can cause irritation of eyes and respiratory tract [4]. Various effort has been developed to remove this toxic dye including chemical treatment, biological treatment, physical adsorption, ozonation, and membrane filtration [5, 6]. However, these wastewater treatments have becoming less efficient to give complete decolorized of pollutant. Some of these treatment have potential to generate secondary product [7].

Nowadays, Advanced Oxidation process (AOPs) was considered as unique and effective technique in treating chemical dye pollutants. Heterogeneous photocatalysis was employed in AOPs give a promising destructive method to degrade and mineralize dye pollutant [8]. This method provided a powerful oxidizing agent which can oxidize any organic pollutants into harmless product such as carbon dioxide (CO_2) and water [9]. Generally, the most favourable photocatalysts are semiconductor materials (such as TiO_2, ZnO, SnO_2, g-C_3N_4) because of their band gap energy (E_g) [10].

Particularly, g-C_3N_4 has intensively becoming one of the most excellent materials which has been utilized by many researchers due to its strong ability to absorb visible light irradiation [11]. Besides, the g-C_3N_4 has biocompatible, resistant to oxidation and high thermal stability [12]. The constricted band gap of the g-C_3N_4 around (~2.70 eV) allow this material to penetrate under visible light more than 460 nm. Unfortunately, this carbonaceous also has their own drawbacks. This carbonaceous has great recombination of photogenerated electron-hole. This recombination can limit the productivity of the photocatalytic activity [13,14]. Various approach to modify this carbonaceous have been demonstrated to rise the potential of g-C_3N_4. One of promising technique by depositing g-C_3N_4 with noble metal such as Ag which can improve catalytic efficiency [15]. Various procedure have been utilized to synthesis Ag/g-C_3N_4 such as photodeposition, photo-assisted reduction method and microwave-assisted polyol process which undergoes wet preparation using ethylene glycol [16-18]. Most of these method required a specific equipment and involving a lots of chemical solvent. Thus, to best of our knowledge, there is no works has been done in synthesis Ag/g-C_3N_4 using dry preparation under microwave irradiation without using any solvent. Subsequently, the modified Ag with g-C_3N_4 was undergoes visible light irradiation using RhB as model pollutant.

2. Experimental

2.1. Catalyst preparation

Silver loaded on graphitic carbon nitride was prepared using microwave method with varying Ag loadings. 0.3 g of g-C_3N_4 prepared using pyrolysis was combined with $AgNO_3$ and the mixture was sonicated for 10 minutes [2]. Next, the mixture in the beaker was covered using aluminium foil and heated using microwave at 100 W for 20 minutes. Then, dried using oven for 2 hours.

2.2. Characterizations

Fourier Transform Infrared (FTIR) spectroscopy was used to investigate the catalyst functional group. The surface area of photocatalyst was analyzed via Nitrogen physisorption. The band gap of catalyst was determined using Ultraviolet-Visible Diffuse Reflectance (UV-Vis/DRS).

2.3. Photodegradation of RhB

The photocatalytic performance of RhB was conducted via a photocatalytic batch reactor with a cooling system. 400 W metal halide lamp was used for source of visible light. The reaction was undergoing adsorption for 30 min to achieve the equilibrium of adsorption and then continue the reaction under visible light for 3 hours. Then, the UV-Vis spectrophotometer was utilized to investigate the decolourization of RhB at the absorption band of 567 nm.

3. Result and Discussion

FTIR analysis was used to investigate the chemical structure of the modified catalyst. Figure 1 displays the FTIR spectra of prepared photocatalyst in the region amid 400 and 4000 cm^{-1}. For g-C_3N_4, there is several main characteristic absorption band that usually can be detected in the region of 4000-400 cm^{-1}. The signal appear at 3000-3400 cm^{-1} was allocated to the stretching vibration of nitrogen bonded to hydrogen (N-H) and surface adsorbed H_2O molecules [19]. Characteristic vibration of aromatic rings was assigned in the absorption signal between 1200 and 1465 cm^{-1} [20]. The signal around 1636 cm^{-1} is recognized to the occurrence of imine bonds [21]. Additionally, the peak at region 808 cm^{-1} is associated to the triazine units with breathing mode [19]. The depositing Ag on g-C_3N_4 might not influence the g-C_3N_4 structure but intensity Ag/g-C_3N_4 was decreased with additional of Ag. This probably the surface of g-C_3N_4 is directly deposited with Ag during synthesis. Therefore, the result demonstrated the preparation of catalyst using microwave was successful. This reduction intensity show similar behaviour by previous study on Ag loaded on g-C_3N_4 as indicated that Ag^+ could anchored on the amino group of g-C_3N_4 [22].

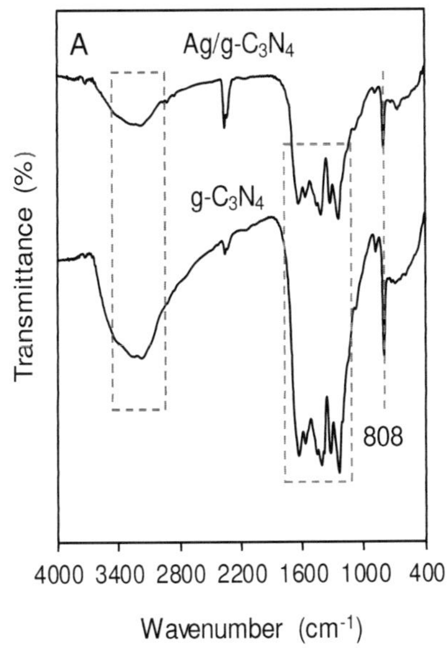

Figure 1. FTIR spectra of the catalysts

The UV-VIS/DRS spectroscopy was utilized measure the optical properties of prepared photocatalyst as depicted in Figure 2. The optical absorption of photocatalyst was converted into function of Kubelka-Munk in order to determine the bandgap of photocatalyst. Based on Figure 2 shows the bandgap of unmodified catalyst (g-C_3N_4) was measured exactly at 2.70 eV reliable with the preceding study [23]. The broad absorption band of Ag/g-C_3N_4 showed a drastically shift to lower energies from 2.70 eV to 2.31 eV after additional of Ag. This result suggested that the reduction in bandgap with introduce Ag in g-C_3N_4 led to increasing the utilization rate in visible light with efficient generation of electron-hole pairs which improve the ability of photocatalyst to harvest which favors the higher photocatalytic performance.

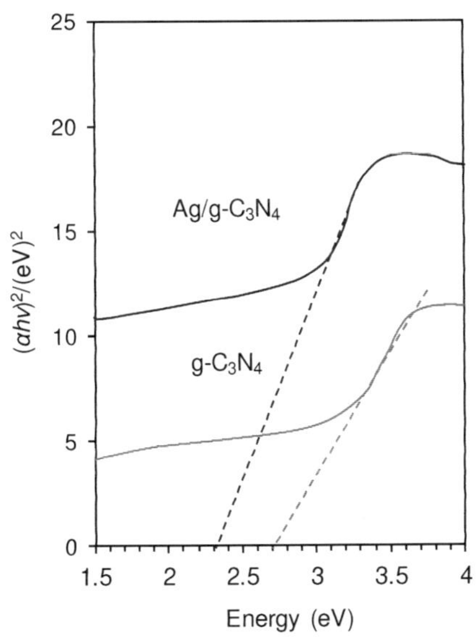

Figure 2. Kubelka-Munk function of the catalysts

The N_2 adsorption-desorption isotherm curve of prepared photocatalyst are depicted in Figure 3. Based on IUPAC classification, both catalyst demonstrated a typical type IV isotherm with hysteresis loop H3 which corresponds to the slit-like mesopores [16].

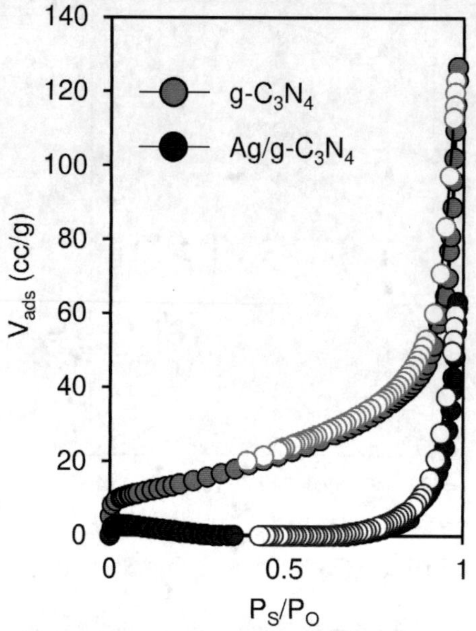

Figure 3. N_2 physisorption isotherm of the catalysts

Table 1 shows the information related with the surface area and pore catalyst. The surface area of Ag/g-C_3N_4 possesses a lower surface when comparing with g-C_3N_4. This phenomenon was corresponded to the g-C_3N_4 has been deposited by Ag. This deposition Ag might obstruction some of the mesopores in g-C_3N_4 [24]. Thus, the result significant to the high interface contact between both materials in the composite and subsequently improve the photocatalytic performance. The result also has similar behavior with previous studies observed when introduced Ag on carbonaceous [25].

Table 1. N_2 physisorption data of the prepared photocatalyst

sample	S_{BET} (m^2/g)	V_{pore} (cm^3/g)	Micropore volume (cm^3/g)	Mesopore volume (cm^3/g)
g-C_3N_4	49.4	0.185	0	0.185
Ag/g-C_3N_4	3.2	0.088	0.012	0.076

The photodegradation of the prepared photocatalyst was determined using RhB solution under visible light irradiation for 180 min. Figure 4 shows irradiation time versus of degradation rate of RhB using prepared photocatalyst. The adsorption behaviour of photocatalyst was investigated for 1 hr before visible light irradiation. After undergoes adsorption under dark for 60 min, g-C_3N_4 shows a high adsorption capacity (~21%) comparing with Ag/g-C_3N_4 (7%). The adsorption capacities of Ag/g-C_3N_4 was reduced comparing with g-C_3N_4. This might be owing to the smaller specific surface area [16]. In contrast, the degradation of g-C_3N_4 was exhibited with degradation rate 66.8% for 90 min and 84.8% for 180 min. The degradation of g-C_3N_4 might be accredited to its own strong absorption properties of visible-light. The photodegradation of RhB drastically improved after additional of Ag on g-C_3N_4 surface with degradation rate 97.1% for 90 min and 98.3% for 180 min. The Ag play a main factor for higher performance reaction of RhB. Thus, $Ag/g\text{-}C_3N_4$ demonstrated outstanding performance due to the additional Ag in $g\text{-}C_3N_4$ extend the absorption toward visible region as can be seen in Figure 2.

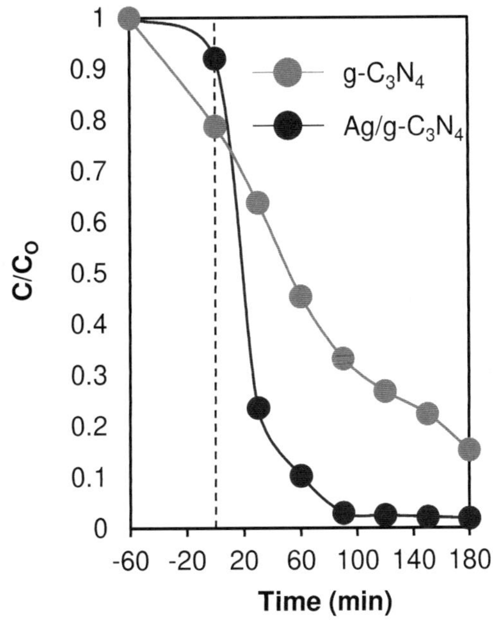

Figure 4. Photocatalytic performance on degradation of RhB using prepared photocatalyst using visible light irradiation

4. Conclusion

As conclusion, the composite of $Ag/g\text{-}C_3N_4$ was successfully prepared under facile preparation using microwave irradiation and characterized by FTIR, UV-VIS/DRS and N_2 physisorption. The FTIR and N_2 physisorption results proved the presence of Ag in $g\text{-}C_3N_4$. The bandgap of modified catalyst reduced from 2.70 eV to 2.31 eV significantly improved the performance of degradation RhB (98%) after exposed under visible light for 180 min. Thus, the synthesis of modified photocatalyst provided an excellent potential to be a promising photocatalyst for pollutants purifications.

Acknowledgments

This research work was sponsored and supported by the Universiti Teknologi Malaysia through the Ministry of Higher Education Malaysia through Fundamental Research Grant No. FRGS/1/2017/STG07/UTM/01/1 (Grant no: 4F969).

References

[1] Nawawi WI, Azami MS, Ang LS, Ishak MAM, and Ismail K 2017 *Water Qual. Res. J. Can* **52** 51-63
[2] Azami MS, Nawawi WI, Jawad H Ali, Ishak MAM, and Ismail K 2017 *Sains Malaysiana* **46** 1309-1316
[3] Hassan NS, Roslani NJ, Jalil AA, Triwahyono S, Salleh NF, and Jaafar NF 2015 *Mal. J. Appl. Sci* **11** 148-151
[4] Chandra MR, Rao TS, Pammi SVN, and Sreedhar B 2012 *Mater. Sci. Semicond. Process* **30** 672
[5] Taghdiri M 2017 *Int. J. Photoenergy* **2017** 1-15
[6] Elham OSJ, Muda SA, Hassan HS, and Abdullah STS 2018 *JKUKM* **2** 1-7
[7] Jaafar NF, Jalil AA, and Triwahyono S 2017 *Appl. Surf. Sci* **392** 1068-1077
[8] Azami MS, Ain SK, Zaharudin R, Bakar F, and Nawawi WI 2016 *Applied Mechanics and Materials* **835** 375-377

[9] Hitam CNC, Jalil AA, Triwahyono S, Rahman AFA, Hassan NS, Khusnun NF, Jamian SF, Mamat CR, Nabgan W, and Ahmed A 2018 *Fuel* **216** 407-417
[10] Azami MS, Nawawi WI, and Shukri DSM 2017 *Deswater* **92** 172-180
[11] Valentin CD, and Pacchioni G 2013 *Catal. Today* **206** 12-18
[12] Wang Y, Yang W, Chen X, Wang J, and Zhu Y 2018 *Appl. Catal. B* **220** 337-347
[13] Yan X, Gao Q, Qin J, Hui X, Ye Z, Li J, and Ma Z 2018 *Maters. Lett.* **217** 1-4
[14] Azami MS, Jalil AA, Hitam CNC, Mamat CR, Siang TJ, Hussain I, and Hambali HU 2019 *JEST* **2** 21-25
[15] Nagajyothi PC, Pandurangan M, Vattikuti SVP, Tettey CO, Sreekanth TVM, and Shim J 2017 *Sep. Purif. Technol* **188** 228-237
[16] Ye M, Wang R, Shao Y, Tian C, Zheng Z, Gu X, Wei W and Wei A 2018 *J. Photochem. Photobiol* **351** 145-153
[17] Sun T, Jiang HY, Ma CC, Mao F, and Xue B 2016 *Catal. Commun.* **76** 45-48
[18] Faisal M, Ismail AA, Harraz AF, Al-Sayari SA, El-Toni, AM, and Al-Assiri MS 2016 *Mater. Design* **98** 223-230
[19] Zhang JS, Zhang MW, Lin LH, and Wang XC 2015 *Angew. Chem. Int. Ed* **54** 6297-6301
[20] Kharlamov A, Marina B, Ganna K, and Nadezhda G 2016 *Diam. Relat. Mater.* **66** 16-22
[21] Samanta S, Martha S, and Parida K 2014 *Chem. Cat. Chem* **6** 1453-1462
[22] Yang Y, Guo Y, Liu F, Yuan X, Guo Y, Zhang S, Guo W, and Huo M 2013 *Appl. Catal. B* 142-**143** 828-837
[23] Zhang W, Li Z, and Deng H 2016 *J. Mol.Catal* **423** 270-276
[24] Zhu B, Xia P, Li Y, Ho W, and Yu J 2017 *Appl. Surf. Sci.* **391** 175-183
[25] Vignesh K, and Kang M 2015 *Mater. Sci. Eng. B* **199** 30-36

IOP Conf. Series: Materials Science and Engineering **808** (2020) 012021 doi:10.1088/1757-899X/808/1/012021

A Community-Based Emergency Response Management in Pasir Gudang, Johor: Theoretical Framework

Noor Afzan Ahmad[1,2], **Kamarizan Kidam**[1,2,3] **and Rahmat Mohsin**[1,2,3]

[1]School of Chemical and Energy Engineering, Universiti Teknologi Malaysia, 81310 Johor Bahru, Johor, Malaysia
[2]Faculty of Engineering Universiti Teknologi Malaysia, 81310 Johor Bahru, Johor, Malaysia
[3]UTM-MPRC Institute for Oil & Gas, Universiti Teknologi Malaysia, 81310 Johor Bahru, Johor, Malaysia

Abstract. Community-based emergency response management refers to a collaborative planning and engagement approach, which is designed to support communities and organizations in developing a safer, more resilient and sustainable future. That being mentioned, this study looked into the theoretical bases and practices in light of emergency response, apart from determining the functions of the community in line with emergency response. The primary goal of this study is to build a theoretical framework that enhances the functions of the community in times of emergency response. The mixed-method approach was adopted for data collection. The qualitative approach via interviews had been performed with organization and industry experts in Pasir Gudang, while the quantitative method was employed for the residents in Pasir Gudang. The theoretical framework embeds six elements of emergency response for the community to improve its emergency management. The identified community-based functions seemed to lack in activities, action, function, technique, knowledge, and skill. The proposed theoretical framework is bound to aid the management to comprehend the essence of community-based emergency response and to implement viable strategies in times of emergency.

1. Introduction

Pasir Gudang, being one of the biggest industrial cities, is located at the end east of the Johor Bahru District, Johor, Malaysia. It is known as a manufacturing center in Malaysia, a municipality region under the flagship zone of Iskandar Malaysia, and the third-largest province that has been envisaged to be a sustainable capital with worldwide stature in 2025 [1]. Due to rapid industrial development, the population at Pasir Gudang has been expected to escalate to 375,700, along with household population of 92,906 in year 2025 [2]. At present, Pasir Gudang is composed of 20 residential areas with the total households of 533,868, as tabulated in Table 1.

IOP Conf. Series: Materials Science and Engineering **808** (2020) 012021 doi:10.1088/1757-899X/808/1/012021

Table 1. Total households in Pasir Gudang (2013-2019).

Year	Malay	Chinese	Indian	Others	Non-citizen	Total
2013	123,634	69,716	14,667	1,053	18,219	227,290
2014	124,826	69,826	14,814	1,120	21,160	231,700
2015	133,288	73,895	15,809	1,233	23,070	247,295
2016	192,645	106,248	22,759	1,801	34,007	357,470
2017	198,931	108,832	23,463	1,852	35,106	368,185
2018	268,005	145,290	31,333	2,510	47,184	494,322
2019	288,289	154,822	32,032	2,723	53,387	533,868

Pasir Gudang visions to be a "Renowned Prosperous City" [3]. The key industries are petrochemical, oleo-chemical, transportation and logistics, shipbuilding, electrical and electronics, food, metal, and oil palm storage. Most of these industries involve dangerous and hazardous substances. The increasing number in population, together with over 45 high-risk industries, seems to draw attention to a number of safety issues towards the surrounding communities. A scenario of emergency occurs due to major incidents involving hazardous materials, explosions, transportation accidents, uncontrollable fires, chemical spillage, or combinations thereof, resulting in injuries, evacuations, and fatalities. According to Perry and Lindell [4], without an effective emergency response management, the community is exposed to high risks with hundreds and thousands of lives at stake if an accident happens. Prior to emergency, the administrative and local communities must be equipped with community-based emergency-preparedness measures, as reflected in the management plans.

2. Community engagement for emergency response

For better engagement during emergency response, it is essential to educate the communities about their involvement in emergency management practice [5]. An effective emergency response practice reflects the willingness displayed by the community to get involved in emergency management. Such involvement points out a top-down transmission model, instead of a single-layer conversation line amidst all stakeholders [6]. [7] asserted that the partnership shared between communities and emergency management practitioners is training-obligated and regulation-driven emergency management processes.

Recent years have witnessed the adverse effects of fatal accidents, thus gearing up communities to organize effective emergency service activities [8]. As part of effective emergency management, the public is provided vital information in the form of advice, instructions, and warning [9]. Upon evaluation, the communities seem to display effective multi-agency coordination and response in times of emergency [10-12]. Recent studies have highlighted the exceptional leadership and operational roles portrayed by the engaging communities as part of emergency management practice in times of distress [13]. [14] emphasized on the significance of the existing activities led by community that gave a hand during calamities. It is certain that knowledge, values, practices, and response arrangements of the community may improve emergency response [15].

3. Theories relevance to emergency response

The nature of emergency response is one that is heavily practical. From the stance of practice, emergency response, as defined by [16], refers to actions taken in readiness for, during, and immediately after an emergency to ascertain that the impacts of emergencies are minimized. In light of academic, [17] explained response as the involvement of several agencies in the attempt of reducing the adverse impacts caused by unanticipated perturbation in a society that potentially threatens lives, assets, social values, and the environment.

Table 2 showcases that emergency response is composed of several activities that include a range of resources, various stakeholders at differing levels, and the affected communities. [18] and [19] had warned against making isolated empirical deductions or providing references that limit association with operations of emergency management, such as managing people, comprehending procedures, and mobilizing resources. Hence, the theories selected for this study had been based on their correlation and relevance to both functions and activities that enhance response.

Table 2. Activities, actions, and functions for emergency response.

Emergency Management Phase	Activities	Actions	Functions
Response	Plan Implementation Public information	Emergency declaration Inform Higher Authorities	Warning Messages Activate coordination centers
	Registration and tracing Damage assessment	Evacuation Search and rescue	Mobilize resources Institute public health measures
	Provide medical support	Network Communication	Provide immediate relief

[20] claimed that several theories, namely broad perspectives, embryonic theories, micro theories, and normative theories, can be applied to identify emergency response in light of practitioners, along with their responsibilities during the response. [21], who had assessed a number of emergency response theories, namely decision, system, chaos, and management theories, depicted both the management of coordination of people, functions, and resources. The stated eight theories were employed to analyze emergency response and to identify gaps that exist in community-based emergency response. Table 3 describes emergency management and emergency response models.

Table 3. Emergency management and emergency response models.

Classification of theory	Theory Title	Description
Normative theory (F.S. Siebert, T. Peterson & W.Schramm, 1956)	Emergency management	This theory emphasizes on the application of preparedness concept for a response that involves the efforts of emergency managers and the development of operations plan of action during emergencies.
Perspectives theory (G.H.Mead, 1931)	Emergency management	This theory explains the actual problem that arises during the response phase and in emergency management.
Micro theory (R.Firisch 1933)	Emergency management	This theory explains prediction of human behavior on the effects of disasters.
Embryonic theory (E.Serres, 1820)	Emergency management	This theory initiates a framework conceptualized from elements derived from a real emergency response.
System theory	Emergency	This theory explains the integrative levels of elements

(L.Von Bertalanffy, 1940)	response	within a system, the interplay between systems, and their elements in identifying their respective functions.
Management theory (F.W. Taylor, 1911)	Emergency response	This theory emphasizes on the measurement and specification of all organizational tasks. These tasks are standardized and organized based on hierarchy divisions to define the lines of authority and control.
Chaos theory (Henri Pioncare, 1880)	Emergency response	This theory refers to a mathematical approach to study and predict the future behaviors of dynamic systems sensitive to initial settings. It has been adopted in philosophy, sociology, economics, and biology disciplines to depict issues stemming from the existing correlations in numerous social, organizational, and natural systems.
Decision theory (E.L.Lehmann 1950)	Emergency response	This theory embeds rational concepts regarding decision making in adherence to design, intelligence, review, and choice.

3.1. System and management theory
The system theory elaborates the integrative levels of elements found in a system, the interplay found between the systems, and the elements based on their functions [22]. This theory is vital to comprehend organizational management based on parts that are unified to attain a certain objective [23]. The management theory reflects careful measurement and specification of organizational tasks that are organized by adhering to hierarchy, as well as by determining both control and authority [24]. Both system and management theories are essential in outlining emergency response procedures that demand exceptional correlation between activities and elements, aside from the implementation of the activities based on hierarchy of authority. The standard operating procedure (SOP), for example, is a set of official documentation that depicts decisions, policies, and actions for execution by emergency responders [25].

3.2. Chaos theory
The chaos theory, referring to a mathematical approach, predicts and studies the future attributes of systems sensitive to their initial settings [26]. Despite solely applicable to deterministic system [27], the chaos theory is used in philosophy, biology, and sociology domains to elaborate issues that stem from existing correlations in organizational, natural, and social systems [28]. Calamities are intricate events that adversely affect many aspects, such as security, public safety, legal and public affairs, communication, and coordination of organizations. The management theory is vital to attain the demanded coordination level, so as to manage effectively any emergency. Nevertheless, it is common for emergency management (at response stage) to incorporate various parties and organizations to interact in an interrelated manner [29].

3.3. Decision theory
The decision theory is composed of several rational concepts related to decision making in adherence to review, design, choice, and intelligence [30]. Stemming from time constriction and stressful condition during disasters, the rational decision models could become ineffective at the response stage in emergency management [31]. Both demands and obstacles noted at this stage have led to the application of various decision-making models to coordinate resources and implement plans to gain

effective response [32]. The Markov Decision Process (MDP) models can be applied within the context of management due to their attributes that emphasize on the response objective [33].

The MDP model, which offers mathematical framework, is impacted by both outcomes and conditions that are either controlled or random to deliver decision in times of disaster [34]. The process of making decision is essential to study numerous optimization issues [35]. The attributes of MDP, at the stage of emergency response, adhere to various related and real-time decision making to meet the response objective, as well as to manage the event evolving state and the actions of other agencies involved [36]. The primary objective of response is to hinder the adverse effects of any disaster on assets, people, nature, and other essentialities [37].

4. Development of a theoretical framework for community-based emergency response

A theoretical framework is the foundation of a study, mainly because a researcher can uncover all factors related to a problem from observations and readings. According to [38], a theory refers to a systematic and coherent formulation of concepts, models, and ideas used to construct meaning to explain, interpret, and model practice. Through a theoretical framework, researchers can integrate these factors logically to provide a scientific basis for research. [39] asserted that a theoretical framework refers to a conceptual model of how one makes logical sense of the correlation between certain factors related to the issue at hand.

[40] assessed the bases of emergency response by implementing several theories, for instance, management, system, chaos, and decision theories, in order to describe both management and coordination of functions, resources, and people. Such enhanced comprehension contributes to the domain of community-based emergency response. Based on the above theories, a theoretical framework was developed for community-based emergency response to display the relevance of the selected theories, as illustrated in Figure 1.

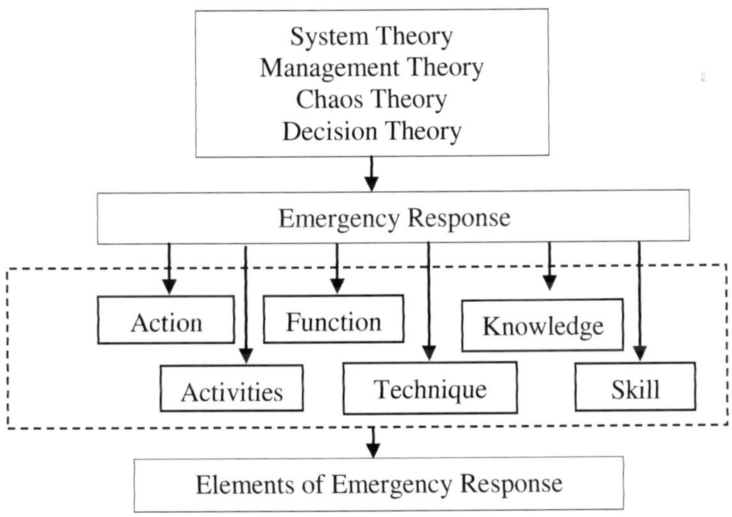

Figure 1. The theoretical framework of community-based emergency response.

5. Conclusion

Emergency response is indeed a critical phase in times of emergency. This paper, thus, assessed several emergency response theories to develop a theoretical framework for community-based emergency response in Pasir Gudang. Four theories were selected to evaluate the essence of emergency response, while six elements of emergency response were identified for the management to

IOP Conf. Series: Materials Science and Engineering **808** (2020) 012021 doi:10.1088/1757-899X/808/1/012021

implement viable and effective idea, so as to enhance the safety of the community during emergency events. It is better to be well-prepared and take appropriate measures in order to overcome any address impact on the community.

References

[1] Khazanah Nasional 2006 *Progress review of the GLC transformation programme*

[2] C S Ho, L W Chau, B T The, Y Matsuoka, Gomi K S, Nur Syazwani J, Nadzirah H, Muhammad Akmal H, Y Lv and A Rohayu (eds.) 2015 *Low carbon society Action Plan for Pasir Gudang 2025: Green and Clean Industrial City* (Johor Bahru: UTM-Low Carbon Asia Research Centre)

[3] Majlis Perbandaran Pasir Gudang 2016 *Jabatan Perancang dan Landskap*

[4] R W Perry and M . Lindell 2003 Preparedness for emergency response: Guidelines for the emergency planning process disasters (Blackwell Publishing) **27** 336

[5] Janssen M, Lee J, Bharosa N and Cresswell A 2010 Advances in multi-agency disaster management: Key elements in disaster research. Information Systems Frontiers **12**(1) 1-7

[6] Ozanne L and Ozanne J 2013 *Developing local partners in emergency planning and management: Lyttelton Time Bank as a builder and mobiliser of resources during the Canterbury Earthquakes*. Lyttelton Report

[7] Betts R 2007 *Community Engagement in Emergency Management.* 5th Flood Management Conference Warnanbool, 9-12 October Office of the Emergency Services Commissioner

[8] Mendonca D and Wallace W A 2004 Studying organizationally-situated improvisation in response to extreme events. Int. J of Mass Emergencies and Disasters **22**(2) 5-29

[9] Alexander D 2002 *Harpenden Principles of emergency planning and management*. Terra Publishing

[10] Fakuade D 2014 *New Zealand. Mass population response to major incidents and critical national infrastructure failure* Excerpt report from PhD thesis, University of Canterbury

[11] Betts R 2007 *Community Engagement in Emergency Management*. 5th Flood Management Conference Warnanbool, 9 – 12 October. Office of the Emergency Services Commissioner

[12] Crandall B, Klein G, Hoffman R, 2006 *Working Minds: A Practitioner's Guide to Cognitive Task Analysis* (Cambridge, MA: The MIT Press)

[13] Betts, R. 2007 *Community Engagement in Emergency Management*. 5th Flood Management Conference Warnanbool, 9 – 12 October Office of the Emergency Services Commissioner

[14] Cretney R M 2016 *Local responses to disaster: The value of community led post disaster response action in a resilience framework* Disaster Prevention and Management **25**(1) 27-40

[15] Kenney C M and Phibbs S 2015 *A Māori love story: Community-led disaster management in response to the Ōtautahi (Christchurch) earthquakes as a framework for action.*International Journal of Disaster Risk Reduction **14** 46-55

[16] EMA 2004 *Emergency Management in Australia: Concepts and Principle*s Australian Government

[17] Uhr C Johansson H and Fredholm L 2008 *Analysing Emergency Response Systems* Published in Journal of Contingencies and Crisis Management **16**(2) pp 80-90

[18] McEntire D.A. and Myers A 2004 *Preparing communities for disasters: issues and processes for government readiness* Disaster Prevention and Management **13**(2) 140-152

[19] Drabek T E 2004 *Emmitsburg, Maryland. Social Dimensions of Disaster* 2nd Edition Emergency Management Institute, Federal Emergency Management Agency

[20] Drabek T E 2004 *Emmitsburg, Maryland. Social Dimensions of Disaster* 2nd Edition Emergency Management Institute, Federal Emergency Management Agency

[21] McEntire D 2004 *Emmitsburg The status of Emergency Management Theory: issues, barriers, and recommendations for improved scholarship.* Paper presented at the FEMA Higher Education Conference
[22] Leydesdorff L. 2001 *A sociological theory of communication.* The self-organization of the knowledge-based society (USA: Universal Publishers)
[23] Osborne L. N. 2000 *Systems analysis for librarians and information professionals* 2nd ed Englewood (Colorado: Libraries Unlimited, Inc)
[24] Mele C., Pels J. and Polese F. 2010 *A Brief Review of Systems Theories and Their Managerial Applications.* Service Science **2**(1-2) 126-135
[25] Waugh, W. 2000 *Living with Hazards, Dealing with Disasters* (New York: M.E. Sharpe)
[26] Basener W.F *Topology and its applications (*John Willey & Sons, Inc)
[27] Kellert S. R. 1993 *Attitudes, Knowledge, and Behavior Toward Wildlife Among the Industrial* (Superpowers: United States, Japan, and Germany) Journal of Social Issues **49**(1) 53–69
[28] Werndl C 2009 *What are the new implications of chaos for unpredictability?* The British Journal for the Philosophy of Science **60** 195–220
[29] Pine J 2007 The contributions of management theory and practice to emergency management In
[30] Turpin S.M. and Marais M. A. 2004 Decision-making: theory and practice **20**(2) pp 143 - 160
[31] Boin A Hart P Stern E and Sundelius B 2005 *New York The politics of crisis management: Public leadership under pressur.* Cambridge University Press
[32] Klein G 2008 *Naturalistic decision making. Human Factor.* The Journal of the Human Factors and Ergonomics Society **50**(3) 456-460
[33] Yu Z Han C and Ma Y 2014 *USA Emergency decision making: a dynamic approach.* Proceedings of the 11th International ISCRAM Conference – University Park
[34] Meyn S P 2007 *Control Techniques for Complex Networks.* Cambridge University Press
[35] Guo X and Hernández-Lerma O 2009 *Continuous-Time Markov Decision Processes.* Springer
[36] Yu Z Han C and Ma Y 2014 *USA. Emergency decision making: a dynamic approach* Proceedings of the 11th International ISCRAM Conference – University Park
[37] Boin A Hart P Stern E and Sundelius B 2005 *The politics of crisis management: Public leadership under pressure (New York*: Cambridge University Press)
[38] Garrison R. 2000 *Theoretical challenges for distance education in the 21st century: A shift from structural to transactional issues* The International Review of Research in Open and distributed Learning **1**(1) pp 1-9
[39] Sakaran, U. (2000). *Research for Business: A Skill-Building Approach.* John Wiley.
[40] McEntire D A and Myers A 2004 *Preparing communities for disasters: issues and processes for government readiness* Disaster Prevention and Management **13**(2) 140-152

IOP Conf. Series: Materials Science and Engineering **808** (2020) 012022 doi:10.1088/1757-899X/808/1/012022

The Verification result of Permit to Work Assessment in Occupational Accident using Fault Tree Analysis

Nizar Jusoh[1], Mohamad Wijayanuddin Ali[1,2], Tuan Amran Tuan Abdullah[1,2], Alias Husain[3],

[1]School of Chemical and Energy Engineering, Faculty of Engineering, Universiti Teknologi Malaysia, 81310, Johor Bahru, Johor, Malaysia.
[2]Center of Hydrogen Energy, Institute of Future Energy, Universiti Teknologi Malaysia, 81310, Johor Bahru, Johor, Malaysia.
[3]Institut Pendidikan Guru Kampus Kota Bharu, Pengkalan Chepa, 16109 Kota Bharu, Kelantan, Malaysia.

Abstract. The objective of this paper is to elaborate and verified the results of the Permit To Work assessment that has been conducted using the Structural Equation Model (SEM) with Fault Tree Analysis (FTA) technique. The author used the Fault Tree Analysis to verify the result of five PTW elements i.e. Hazardous Activity, Worksite Inspection, Supporting Document, Work Description and Close Out. As it is known, the FTA is used to find the probability and cause of an accident or "Top Event". FTA is a relatively well-known technique used to find the probability of system failure. After the analysis using FTA technique, it was found that all the results of SEM and FTA gives the similar rank to Occupational Accident which is indicated that the result is correct and acceptable for this study. This is Part II of two paper, focusing on PTW Assessment result validation using Fault Tree Analysis.

1. Introduction

Permit To Work (PTW) is known as a special official document designed in such way, that being used intensively for control all the work activities in the Plant. With the increasing and importantly used of PTW in controlling Occupational Accidents at the Petrochemical Plant in routine and non-routine of daily activity, which in the earlier assessment of their element shows the impact of their element failures caused the Occupational accident. The study of the PTW on the industry showed that one the causes of occupational accidents is due to the weaknesses in PTW management in the industry in preventing the Occupational accidents. In the Part 1, the author analysed the PTW element failure using the Structural Equation Model (SEM) with AMOS graphic technique which gave a positive and definitive conclusion on the cause of occupational accidents at Petrochemical plants. The result of PTW element assessment has been achieved by using the SEM Amos results indicated that all the Permit To Work starting from the element of Hazardous Activity, Worksite Inspection, Supporting Document, Work Description and Close Out will be validated or evaluated using Fault Tree Analysis (FTA). The result from the Structural Equation Model) SEM) need to validate and verify that it will ensure the SEM result is correct and acceptable with the used data is accurate, quality, sensible and reasonable. In this Part 11, The Fault Tree Analysis (FTA) as a credible, quickness, accuracy and precision method used to validate the SEM Amos result that being performed earlier in Part 1. The FTA has been used by many researcher for describing in-depth analysis of cause and effect analysis of specific especially in chemical process, aerospace and others industry [1].

IOP Conf. Series: Materials Science and Engineering **808** (2020) 012022 doi:10.1088/1757-899X/808/1/012022

2. Methodology

2.1 Fault Tree Analysis (FTA)

Fault Tree are the deductive method extensively used in the risk of field analysis of process system [2]. It is the backward method to define top event or accident from various scenarios that can cause an accident. It also consist of basic event and intermediate event of an accident. The basic event cannot be defined anymore but the immediate event can be defined furthermore. The fault tree has two logic, the AND or OR gate. The AND gate is used for the process that occurs in parallel but the OR gate is used for either available options in the process can that may occur. The Fault Tree Objective is to provide of an integrated picture of some aspect of system operation beside it can be used to understand the level of protection for the design concept against failure. The use of the Fault Tree Analysis is to investigate the effect of safety barrier and the failure probability [3]. The various set of event that can lead to top event to occur are known as minimal cut set [4]. In Fault Tree Analysis the top event denotes or indicate the causal relationship leading to failure of safety barrier of the element and linked to failure of sub safety barrier or sub element. The FTA shows the sequence of logical relationships between elements of each stages in failure process [5]. The regression result of SEM Analysis can be analysed and verify using Fault Tree Analysis (FTA) [6]. The FTA is used to identify possible failure in a system and most commonly used for causal analysis in risk, safety and reliability. It provide a mean to identify the root cause of failure and to support the SEM analysis results. [7]. As per Process Accident Model which was developed with the series of safety element prevention failure and was placed in sequential order to depict the accident of hydrocarbon release accident [8]. The series of PTW element as barrier failure from the failure of the work description, hazardous activity, worksite inspection, supporting document and lastly the failure in closing out the PTW of a particular job for that particular time. The sequential of element failure is presented by using Fault Tree Analysis (FTA) and event tree diagram. The Top Event is the Occupational Accident, denote with the failure of all the safety barrier. [9]. In the FTA diagram, the second layer is associated with all the PTW element diagram on each of the PTW element also shown their escalating barrier failure.

3. Result

The item description is the item of each of PTW element construct in which the PTW element having five construct used in this study. The value of each items is being calculated using Fault Tree Analysis (FTA) formula of AND gate $P(A)= P(A_1)P(A_2)P(A_3)P(A_4)P(A_5)$. Formula OR gate $P(A)=1-[1-P(A_1)][1-P(A_2)][1-P(A_3)][1-P(A_4)][1-P(A_5)]$. Where the $P(A)$ is the Top Event, $P(A_1)$ to $P(A_5)$ is a immediate event.

The calculating of error probability of each construct items can be calculated from the logical relationship between immediate event to top event of every layer of the Fault Tree Analysis and the result on each item as below:

Table 1. Results Fault Tree by items/construct

No. Items	Items Description	Assigned Probability (OR Gate)
	Work Description Construct	
1	Employee fill up necessary info in PTW	0.1947
2	No valid PTW for work in the Plant	0.1959
3	Job carried out with right PTW	0.1379
4	Employee attend PTW training	0.1374
5	PTW does not conflict with other job	0.2222
	Hazardous Activity	
6	Improper chemical handling without PPE	0.84727
7	Lifting work with high load	0.8812

8	Gas testing not done at excavation area more than 1.5 m	0.1149
9	Electrical /mechanical t/shoot on pump	0.1527
10	Corroded piping replacement	0.1877
	Worksite Inspection	
11	Worksite not being check before approve the PTW	0.6946
12	Scaffolding pipe not check	0.8668
13	No worksite check on from H2s and Mercury	0.0996
14	No lock out and tag out on equipment	0.1679
15	Blinding and spading without procedure	0.1794
16	De pressure and draining without procedure	0.2417
17	No N2 purging done before vessel entry for cleaning	0.2759
18	Temporary barricade not install	0.1870
19	No regular gas testing in confine space	0.1992
20	Regular area check not done during work in progress	0.1571
	Supporting Document	
21	JHA not attach when submitting the PTW	0.1959
22	No risk assessment done on before submitting the PTW	0.1954
23	Wrong physical isolation on equipment	0.1374
24	Switch not bypass and no record in logbook	0.2222
25	No gas test result attached with PTW	0.2223
26	Inspection on crane not done before entry the plant	0.1947
	Close Out	
27	Housekeeping not done after job completed	0.1948
28	No proper shift handover	0.1571
29	Operator does not check worksite after job completed	0.1572
30	No sign off of PTW after job completed	0.1034

For further detail the description of all the barrier and their sub barrier as per below:

3.1 Work Description Element Failure

This is the first element in Permit To Work. Under this element, the work leader need to fill-up all the necessary and right information in this section. Wrong info will cause the wrong interpretation about the work that need to be performed. In most cases some work leader has put the wrong information. Example of the wrong info such as the plant area of work and description of work that they want to do in plant. From the Fault Tree Analysis each, the error probability of each sub-element has been calculated using the OR gate. The probability value for wrong work info failure was 0.1974. The another four sub elements was the no valid work permit (0.1999), the wrong selection of PTW (0.13794), workers not attending PTW training (0.1374) and work conflict between work group (0.2222). The total probability failure is 3.12966 E-05.

IOP Conf. Series: Materials Science and Engineering **808** (2020) 012022 doi:10.1088/1757-899X/808/1/012022

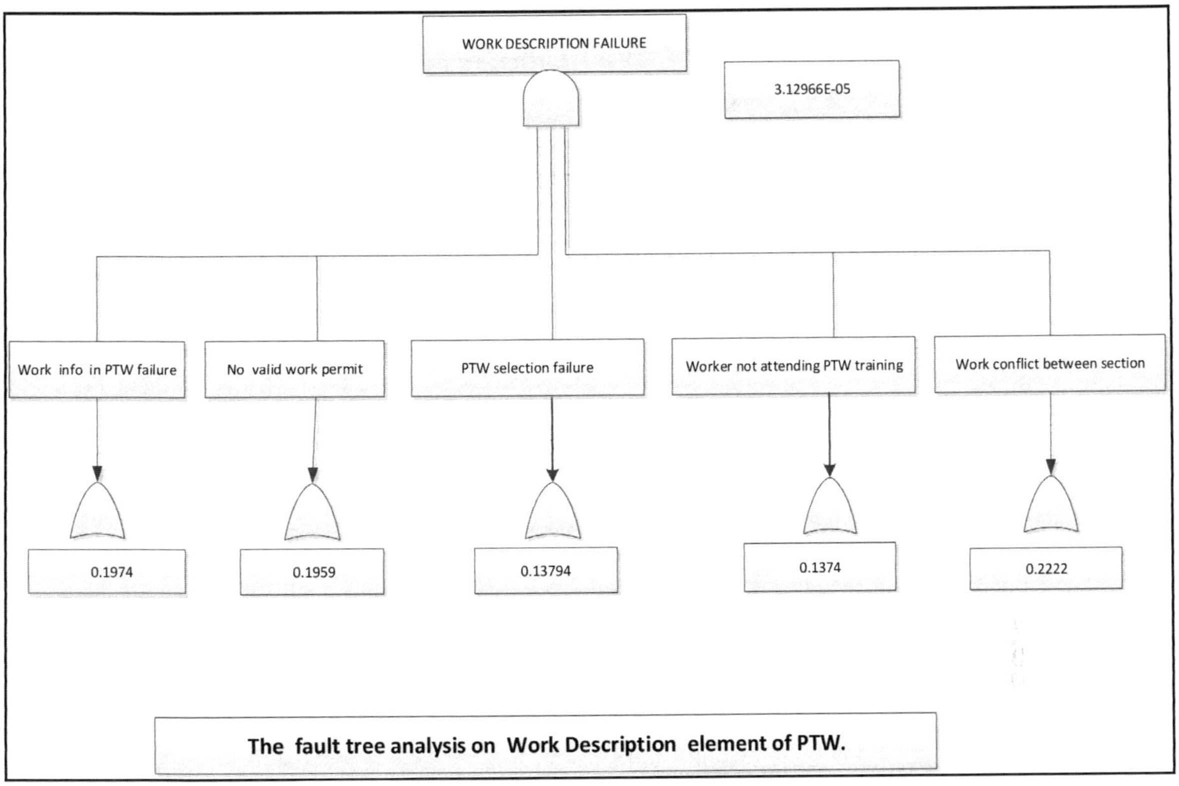

Figure 1. FTA for Work Description element

3.2 Hazardous Activity Element Failure

This is the second element of the PTW. The function of this element to ensure all the hazardous activities has been done the proper way. Example the checking and monitoring during hazardous activities in progress The example of hazardous activity in the Plant are working in confine space, lifting work, construction work during turn around, excavation, pig retrieval etc. Every work Leader and supervisor concern must be very careful when deal with hazardous activity work, because simple or silly mistake done will result in high consequences. There are six items under the Hazardous activity section. Those are the handling hazardous chemical with the error probability is 0.8472, lifting work activity is 0.8812, gas testing failure is 0.1149, electrical or mechanical work activity is 0.1527, replacement of structural or piping is 0.1877. The total probability failure is 2.4610E-03.

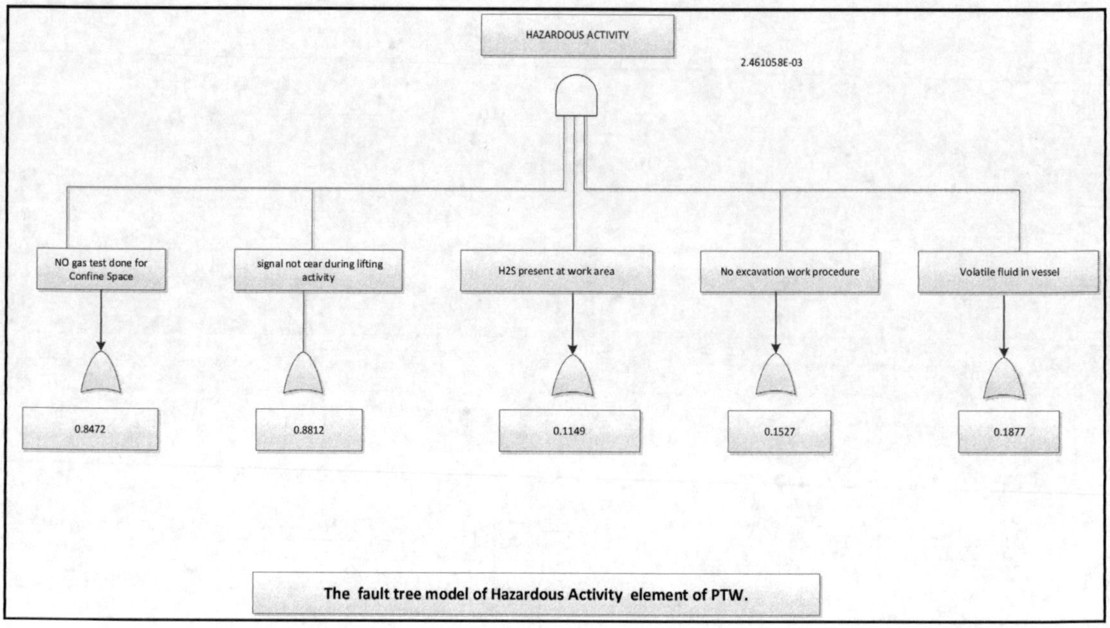

Figure 2. FTA for Hazardous Activity element

3.3 Worksite Inspection Element Failure

This is the third element of Permit To Work and has the 10 questionnaires items under this construct element. The Work leader and the area owner must be very diligent and carefully when fill-up this section. They are must performed thoroughly the area check before the particular PTW can be approved .The main item that must to check , but not limited to check such as scaffolding work, area free from combustible material, must do proper equipment isolation i.e. Lock Out and Tag Out, blinding or spading the equipment or vessel. If this barrier failed to check properly, the accident will happened. The used the FTA to calculate the error probability on each of items of each sub element. Error probability of failure to checking worksite is 0.6946, scaffolding error is 0.8668, failure to check area from H2S or mercury is 0.6996, failure to do Lock Out Tag Out is 0.1679, Blinding and spading failure is 0.1793, Failure to do N_2 purging is 0.27596, failure to install temporary sign and barricade is 0.18707, gas testing failure is 0.19924 and failure to do housekeeping is 0.15710. The total probability failure is 1.80821E-03.

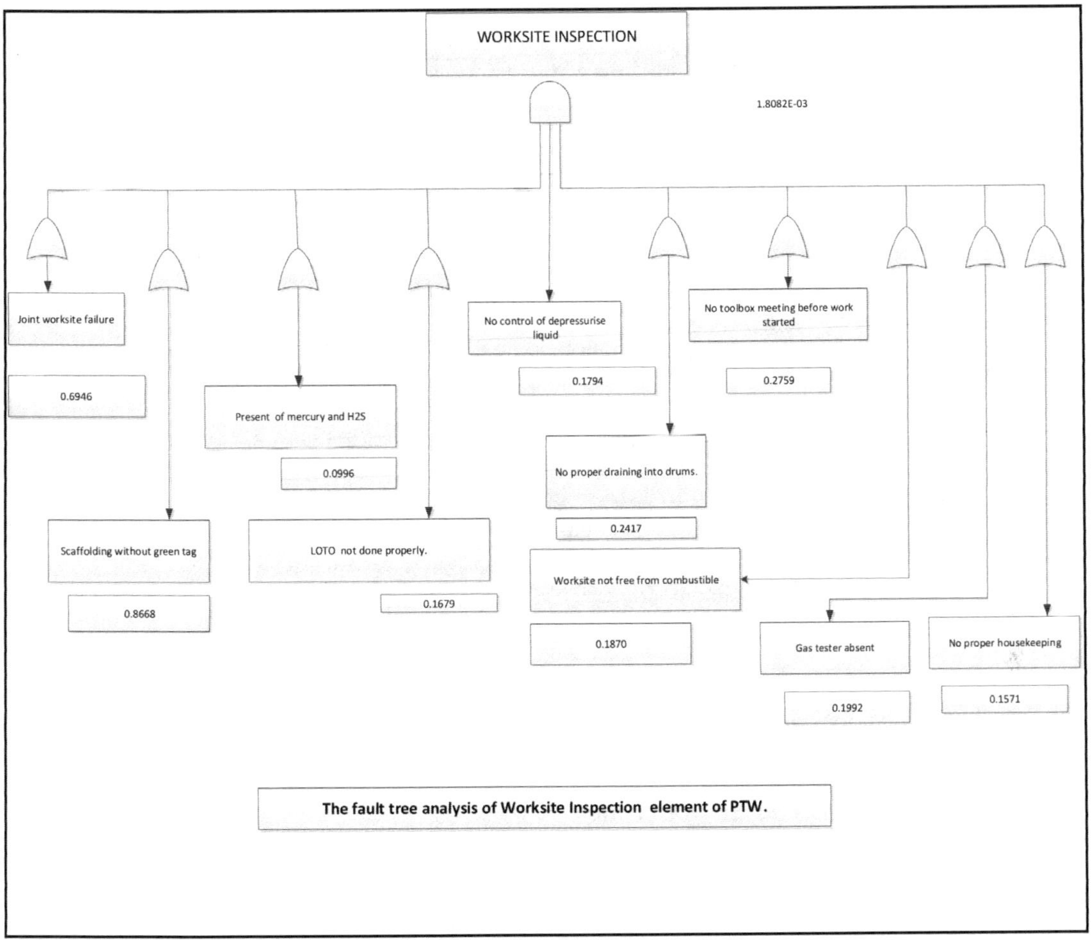

Figure 3. FTA for Worksite Inspection Element

3.4 Supporting Document Element Failure

Supporting document is the fourth element of PTW Accident model. This element is very important in supporting the PTW approval. All the document such as Job Hazard Analysis (JHA), Isolation certificate(Lock Out Tag Out) (LoTo), By pass certificate, Inspection certificate such s crane or vehicle inspection are attached together with PTW submission which are verifying that all the necessary equipment are fully isolated and secured to people to work on. For example the isolation job and Lock Out Tag Out (LoTo) must be done prior to approval of PTW and the isolation are done on equipment on site. Therefore in order to prevent any accident before it happen the focuses on all possible sources that is expecting to be existed in the plant area. The failure to comply with the supporting document will cause an accident with error probability for not attaching JHA is 0.1959, no vehicle inspection is 0.1954, failure to do risk assessment is 0.1374, physical isolation failure is 0.2222, failure to do switching bypass before do the job is 0.2233 and failure to performed gas testing is 0.1974. The total probability failure is 1.1695E-03.

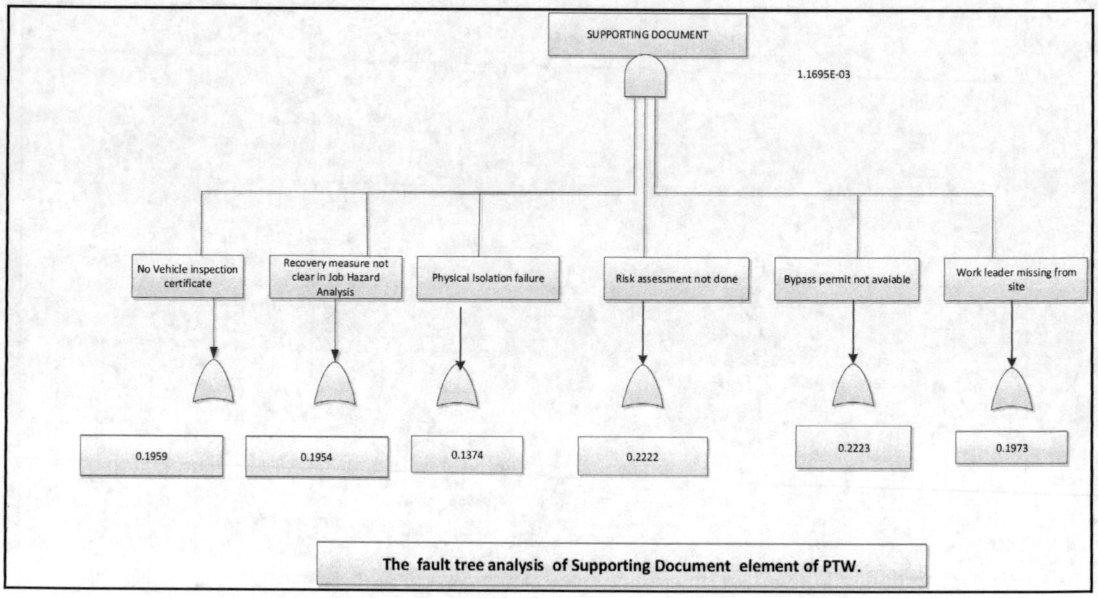

Figure 4. FTA for Supporting Document Element

3.5 Close Out Element Failure

This is the last element and last layer of protection in PTW. This element is to ensure all the items being check must to ensure permit user or work leader do housekeeping, proper handover if working on shift, ensure the area owner checking worksite after job completed before the permit being close out and sign off by Approval Party. Failure to do so will create error probability for housekeeping is 0.18143, handover failure is 0.15710, failure to do site checking is 0.15711 and failure to get job completion signature is 0.1036. At the end the total error probability is 5.2515E-08.

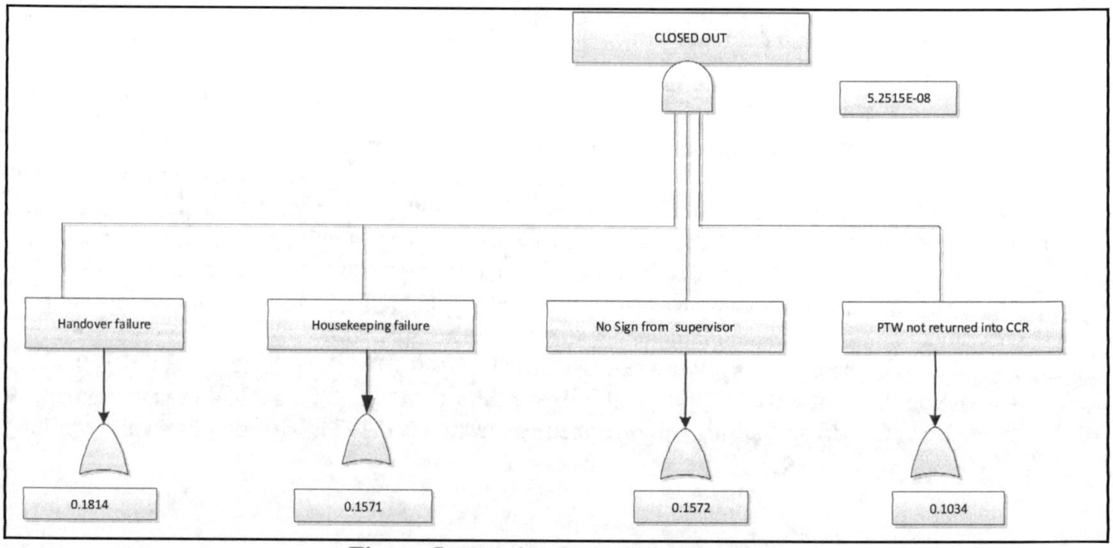

Figure 5. FTA for Close Out Element

4. Discussion

The whole Fault Tree Analysis for all PTW elements is shown in Figure 6 and results for comparison of SEM and FTA result is shown in Table 2 below:

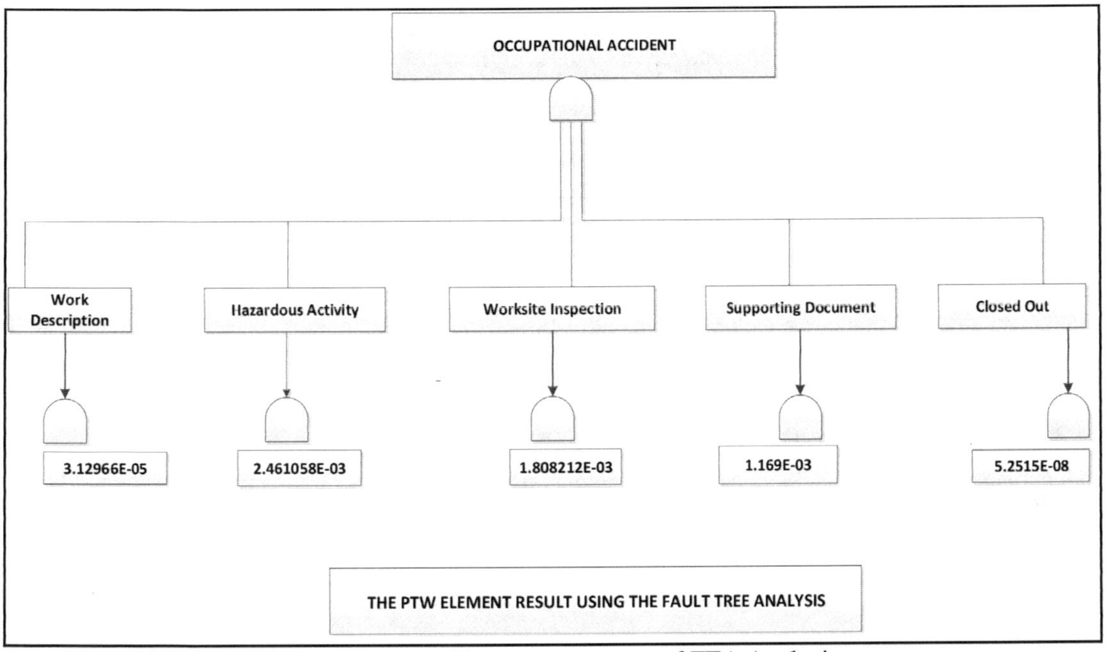

Figure 6. PTW Elements construct of FTA Analysis

The Permit To Work element results that has been performed by using Structural Equation Model (SEM) earlier that can be compared and found that from both results, its produced the same ranking with the Hazardous Activity is the main cause of the occupational accident. Even though the researcher used two different approaches either SEM or Fault Tree Analysis (FTA), both method still produced the same result and confirmed that the main cause of occupational accident occurring at site is Hazardous Activity and followed by other PTW element such as Worksite Inspection, Supporting Document, Work Description and Close Out.(refer to Table 4.1 below).

Table 2. Result SEM and FTA

No.	PTW Element	Result SEM	FTA Result	Rank
1	Hazardous Activity	0.56	2.4610 E-03	1
2	Worksite Inspection	0.32	1.8082 E-03	2
3	Supporting Document	0.22	1.1695 E-03	3
4	Work Description	0.16	3.1296 E-05	4
5	Close Out	0.05	5.2515 E-08	5

5. Conclusion

The depth study of an assessment Permit to Work element failure has been done earlier in Paper 1 using Structural Equation Model (SEM) analysis and the verification result was performed using the Fault Tree Analysis (FTA). The FTA results show the same rank with the SEM result i.e. Hazardous Activity is the main cause of occupational accident. From the result it was confirmed that the SEM results is correct and being proved and verified with FTA result and it meet the objective of the study.

Reference

[1] Faisal I. Khan, Iqbal, Ramesh, and Abbasi 2001 A new methodology for safety management based on feedback from credible accident-probabilistic fault tree analysis system *Process Safety and Environmental Protection* **92** 616-624

[2] Khakzad K and Amyotte 2011 Safety analysis in process facilities: Comparison of fault tree and Bayesian network approaches *Reliability Engineering and System Safety* **96** 925-932

[3] Al-Shanini, Ahmad, and Khan 2014 Accident modelling and safety measure design of a hydrogen station *International Journal of Hydrogen Energy* **39** 20362-20370

[4] Demichela and Camuncoli 2014 Risk based decision making. *Discussion on two methodological milestones Journal of Loss Prevention in the Process Industries* **28** 101-108

[5] Rathnayaka et.al 2011. SHIPP methodology *Process Safety and Environmental Protection* **89** 151-164

[6] Awang Z 2012 *Structural equation modeling using AMOS graphic*. Penerbit Universiti Teknologi MARA.

[7] Chua Yan Piaw 2014*Kaedah Dan Statistik Penyelidikan Buku 5: Ujian Regresi, Analisis Faktor dan Analisis SEM*. Mc Graw Hill Education.

[8] Kujath, Amyotte, and Khan 2010 A conceptual offshore oil and gas process accident model. *Journal of Loss Prevention in the Process Industries* **23** 323-330

[9] F I Khan, Sadiq and Husain 2002 Risk-based process safety assessment and control measures design for offshore process facilities *J. Hazard Meter.* **94** 1-36

[10] Fornell C and Larcker D F 1981 Structural equation models with unobservable variables and measurement error: Algebra and statistics *Journal of marketing research* 323-338

Facile Electrosynthesis of Fe_3O_4 Nanoparticles Mediated with Sodium Alginate for Paracetamol Degradation

A Masudi [1], N. W. C Jusoh [1,2], P Y Liew [1]

[1] Department of Chemical Process Engineering, Malaysia Japan International Institute ofTechnology, (MJIIT), Universiti Teknologi Malaysia Kuala Lumpur, Jalan Sultan Yahya Petra, 54100 Kuala Lumpur, Malaysia
[2] Center of Hydrogen Energy, Institute of Future Energy, Universiti Teknologi Malaysia Kuala Lumpur, Jalan Sultan Yahya Petra, 54100 Kuala Lumpur, Malaysia

*nurfatehah@utm.my

Abstract. High purity of Fe_3O_4 was prepared by electrochemical method in different concentration of sodium alginate (SA) as natural capping agent. The content of SA influenced the particle size and surface interaction to Fe_3O_4 as confirmed by X-ray diffraction (XRD) and Fourier Transform Infrared (FTIR) spectroscopy. The highest photocatalytic activity was obtained for Fe_3O_4 synthesized with 0.05% SA with initial concentration of 15 mg/L Paracetamol. This result contributed to electrochemical advancement to produce Fe_3O_4 with simple method.

1. Introduction

The occurrence of some organic contaminants in water became one of primary concern with increasing priority to environment sustainability. The contaminants include pharmaceutical, industrial chemical, pesticides and emerging compounds. The pharmaceutical industry increases almost 10% yearly in Malaysia. This continuous elevating demand, generated pharmaceutical discharge to environment. Paracetamol (PCT) is the largest analgesic and antipyretic drug without prescription worldwide which could be utilized as model of pharmaceutical waste. PCT was detected in the range of 0.22-6.8 μg/L in Spain [1], 0.033-0.071 μg/L in South Korean's surface water [2] and 0.01-0.07 μg/L in surface and sewage water in Malaysia [3].

Water quality could be improved by several techniques such as adsorption, biological and membrane. The former method actually only transfer pollutant to other materials and require additional post-treatment [4]. Biological route was ineffective for antibacterial drug [5], while membrane was limited by membrane fouling and high cost [6]. A promising method to degrade pollutants was based on advanced oxidation process (AOPs) which produced many reactive compounds. As part of AOPs, photocatalyst attract many researchers due to its simplicity to degrade pollutant with only assistance of light and a catalyst. However, the commercial catalysts were difficult to separate which increase the operational cost [7]. Fe_3O_4 is one of promising catalyst with easy recovery using external magnet. Zaidi et al [8] verified that the magnetic separation is more affordable than conventional filtration method.

There are many methods to synthesis nanosizes of Fe_3O_4 such as sol-gel [9], hydrothermal [10] and co-precipitation [11]. However, these routes require complex experimental set up and high temperature. Electrochemical method is one alternative to produce metal oxides and composites at low temperature. However, the existing electrochemical route require an ionic liquid such as 1-butyl-3-methyl-imidazolium chloride ([BMIM]Cl) [12]. Thus, its crucial to obtain low cost material to substitute the ionic liquid. The general role of ionic liquid is as a capping agent to produce nanoscale material. Some studies have been reported that urea and poly ethylene glycol (PEG) [13], seaweed [14] and sodium alginate [15] as promising capping agents. Sodium alginate (SA) not only reported as capping agent in single metal oxide or composite, but also assist adsorption to heavy metals [16]. To the best of our knowledge, a detail study on interaction between capping agent and catalyst still scarce in literatures. The aims of this study is to provide an insight to reagents interaction during the synthesis and photocatalytic activity to degrade PCT.

2. Experimental

2.1. Synthesis Method

The Fe_3O_4 synthesis method was conducted in one compartment and open air at 0 °C contain 1 M TEAP with a series concentration of SA (0, 0.01, 0.05 and 0.1 % w/v). The same size of iron and platinum (2×2 cm^2) as anode and cathode were placed in parallel position. Prior to electrolysis, pH was fixed at 9 in constant current density of 120 mA/cm^2 with continuous stir at 250 rpm. The obtained precipitate was washed with water and ethanol three times, respectively. The black powder was then dried at 60 °C overnight to remove the impurities. The sample was denoted as x% SA. Fe_3O_4, where x indicated the concentration of SA.

2.2. Characterization

The crystallinity and phase purity of the catalysts was measured by X-ray diffraction (XRD) D8 Advance Bruker in the range of 20-90 °. Joint committee on powder diffraction (JCPDS) was used to ensure the phases of the samples. Vibrational spectroscopy of the catalyst was recognized with Fourier transform infrared (FTIR) Shimadzu IRT Tracer-100 at wavelength scan of 400-4000 cm^{-1} using KBr pellet. The, the morphologies of the catalysts were visualized with FESEM JEOL JFC-1600.

2.3. Photodegradation Studies to Paracetamol

The photoactivity of the catalyst was evaluated to degrade paracetamol (PCT). The catalyst was placed in a dark condition for 1 h to obtain equilibrium of adsorption-desorption between catalyst and solution. Then, this suspension was exposed by simulated light for 2 h. The initial concentration of PCT was fixed at 15 mg/L. The concentration of PCT during reaction was then monitored by UV-vis Shimadzu at 244 nm. Each experiment was conducted three times to ensure the accuracy of the data. This procedure was replicated for all the studied parameter to elucidate the effect of pH, SA concentration and catalyst dosage.

3. Result and Discussion

3.1. Crystallinity and Phase Purity

Figure 1 shows the XRD pattern of Fe_3O_4 with a series concentration of SA. The peaks appeared at 30.2, 35.6, 43.2, 54.92, 57.4 and 62.9 ° assigned to (220), (311), (400), (422), (511) and (440) planes. These peaks match with JCPDS. card 19-062 attributed to Fe_3O_4 with inverse spinel structure [17]. As presented in Table 1, the crystallite size of Fe_3O_4 using Scherrer equation are in nanoscale. The use of low temperature during drying may cause larger crystallite size of Fe_3O_4 with SA than without SA (Fe_3O_4), due to the remaining of SA on Fe_3O_4 surface. The product was not calcined as the Fe_3O_4 may transform to other iron oxide phases. As displayed in Figure 1 (a), Fe_3O_4 without SA has very low crystallinity. Then, the crystallinity increased significantly after addition of low content of SA (0.01%). By increasing the content of SA to 0.05%, the crystal size of Fe_3O_4 decreased indicating that SA may cover entirely the Fe_3O_4 nuclei. However, at higher concentration of SA (0.1%), the particle size increased again as a result of attraction between excessive alginate chains to sodium ions which further agglomerated and became larger size [18].

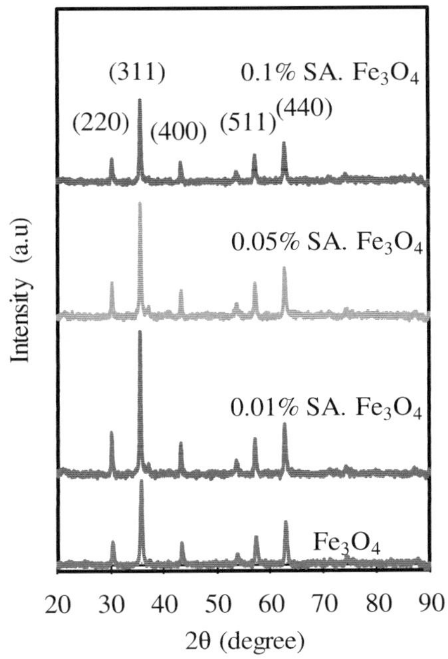

Figure 1. XRD pattern of Fe_3O_4 with various SA concentration

Table 1. Crystallite particle size of Fe_3O_4 in various SA concentration

Catalyst type	Concentration of sodium alginate (%w/v)	Size (nm)
Fe_3O_4	0	28.41
0.01% SA.Fe_3O_4	0.01	37.82
0.05% SA.Fe_3O_4	0.05	30.96
0.1% SA.Fe_3O_4	0.1	34.09

3.2. Vibrational Spectroscopy

Figure 2 displays FTIR spectra of the catalyst to obtain the chemical interaction between SA and Fe_3O_4. The spectra of SA comprised of 1093 and 1030 cm^{-1} assigned to C-O-C of saccharide structure. The band at 1417 and 1620 cm^{-1} are asymmetric and symmetric stretching vibration of carboxyl groups. Then, peaks at 2930 and 3340 cm^{-1} associated to stretching vibration of O-H and C-H [19]. The characteristic of Fe-O bond vibration appeared at 624 cm^{-1} in all catalysts. Then, new peaks were observed in 798 and 1120 cm^{-1} with low intensity in 0.01% catalyst which may be assigned to interaction between SA and Fe-O. These peaks became more intense at 0.05% catalyst which may be indicated as more SA bonding to Fe-O. The intensities of these new peaks were then decreased in 0.1% catalyst. This may be a sign that bonding between Fe_3O_4 and SA not as much as 0.05% catalyst [20].

IOP Conf. Series: Materials Science and Engineering **808** (2020) 012023 doi:10.1088/1757-899X/808/1/012023

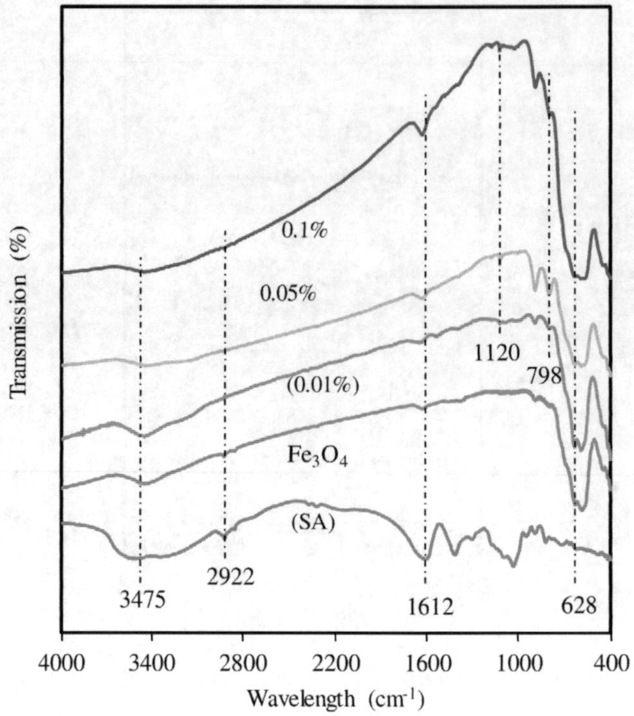

Figure 2. FTIR spectra of Fe_3O_4 with various SA concentration

3.3. Morphological Studies

The morphologies of 0.05% SA. Fe_3O_4 and Fe_3O_4 were investigated with field emission scanning electron microscopy (FESEM). The observation confirmed that the catalysts are spherical in shape. This result is similar with Esmat et al. From these figures, it was observed that Fe_3O_4 in presence of SA has lower agglomeration than Fe_3O_4.

Figure 3. FESEM images of (a) 0.05% SA. Fe_3O_4 and (b) Fe_3O_4

3.4. Photocatalytic Degradation

The photocatalytic activities were examined under several parameters such as effect of pH, SA concentration and catalyst dosage. These parameters were useful to obtain optimum condition for PCT degradation.

3.4.1. Effect of pH

pH of aqueous solution influenced the pollutant interaction to the catalyst surface. A typical semiconductor catalyst behaved as amphoteric depend on the environmental pH. In this study, the effect of pH was examined in the range of 5-9 as presented in Figure 3. The highest degradation rate was obtained at pH 9 where the Fe_3O_4 and PCT were negative and positively charged, respectively [21]. Meanwhile, both of Fe_3O_4 and PCT were in the same charge in other pH which further repel each other, then minimize the PCT degradation.

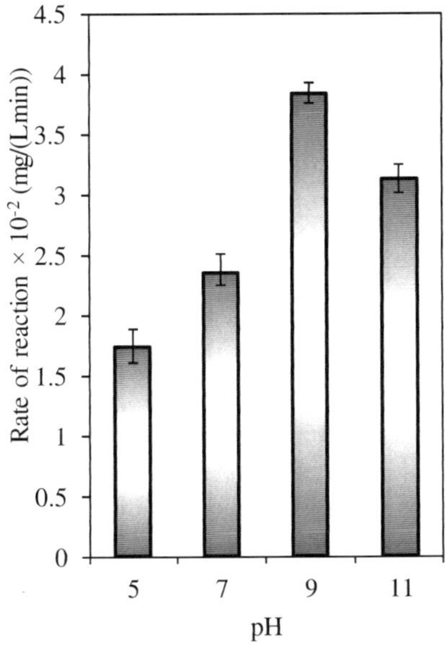

Figure 4. Effect of pH in PCT degradation

3.4.2. Effect of SA concentration

SA content played a crucial role for Fe_3O_4 formation. The different concentration of SA produced different size and interaction between catalyst and SA which further affected the photocatalytic activity as presented in Figure 5. The degradation rate of Fe_3O_4 without SA was higher than in presence of 0.01% SA may be attributed to its smaller particle size. Then, the degradation rate increased drastically as a result of small particle size when using 0.05% SA. Last, the degradation rate decreased when added more SA (0.1% w/v) due to their larger particle size. This assumption was strengthened by Aziz et al [22], where small particle size will result higher photodegradation performances. The presence of SA in Fe_3O_4 surface was crucial to enhance the adsorption of malachite green such as reported by [18], azo dye in gold-alginate [23] and selective removal of heavy metals (Pb^{2+}, Cu^{2+}, Cd^{2+}) [24]. Additionally, the formation of new bonds at 1120 cm^{-1} and 798 cm^{-1} indicated that SA was might be covalently bonded to Fe_3O_4 surface. These new bonds were considered as key role for enhanced paracetamol degradation. This result was aligned with research by Soares et al [25] where carrageenan was successfully grafted to siliceous Fe_3O_4 and produce robust, stable and reusable for methylene blue (MB) adsorption. Carrageenan is similar with SA in term of chemical structure and also family of polysaccharides. In this study, SA was attached to carboxyl and ether functional groups as hypothesized with FTIR. At low content of SA (0.01% w/v), new bonds were started to appear with small intensity, then become more intense using 0.05% of SA. However, at higher concentration (0.1%), the intensity reduced and followed

IOP Conf. Series: Materials Science and Engineering **808** (2020) 012023 doi:10.1088/1757-899X/808/1/012023

with sharper intensity in carboxyl group. This phenomena indicated that new bond formation with SA as active site for PCT degradation which may decrease at high concentration of SA due to accumulation of alginate chain attach to sodium ions as reported in the literature [18].

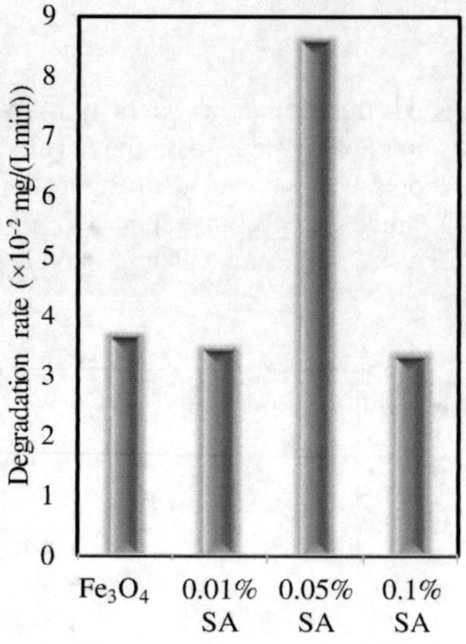

Figure 5. Effect of SA concentration in PCT degradation

3.4.3. Effect of catalyst dosage

The effect of catalyst dosage is presented in Figure 6. Catalyst dosage play vital role in pollutant degradation, especially in industrial scale. The presence of more Fe_3O_4 enhanced the photon absorption over catalyst surface and reach optimum degradation at 0.45 g/L. The degradation rate decreased at higher catalyst dosage due to excessive black opaque of Fe_3O_4 which hindered light penetration to Fe_3O_4 surface [26].

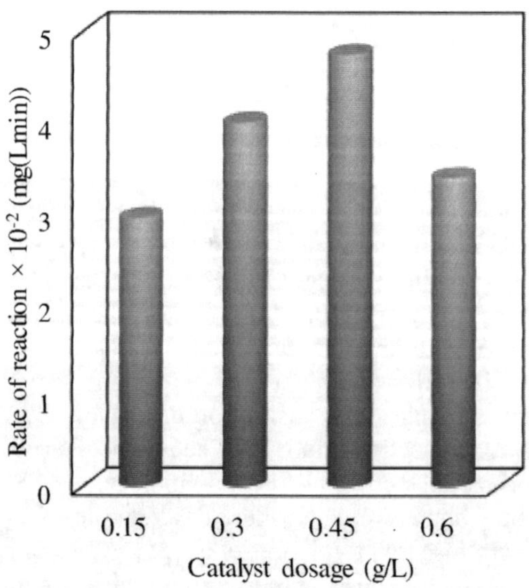

Figure 6. Effect of catalyst dosage in PCT degradation

4. Conclusion

A series of Fe_3O_4 was successfully prepared with electrochemical method with various concentration of SA. SA could control the particle size of Fe_3O_4 of which the interaction affected by the content of SA. The optimum degradation rate towards Paracetamol was obtained for 0.05% SA. Fe_3O_4 with 0.45 g/L of the catalyst at pH 9.

Acknowledgment

This research was supported by Ministry of Education (MOE) through Fundamental Research Grant Scheme (FRGS/1/2018/STG07/UTM/02/11). and MJIIT Incentive Scheme (Ahmad Masudi) from MJIIT UTM.

References

[1] Gómez M J, Martínez Bueno M J, Lacorte S, Fernández-Alba A R and Agüera A 2007 Pilot survey monitoring pharmaceuticals and related compounds in a sewage treatment plant located on the Mediterranean coast *Chemosphere* **66** 993-1002

[2] Kim S D, Cho J, Kim I S, Vanderford B J and Snyder S A 2007 Occurrence and removal of pharmaceuticals and endocrine disruptors in South Korean surface, drinking, and waste waters *Water Research* **41** 1013-21

[3] Al-Odaini N A, Zakaria M P, Yaziz M I and Surif S 2010 Multi-residue analytical method for human pharmaceuticals and synthetic hormones in river water and sewage effluents by solid-phase extraction and liquid chromatography–tandem mass spectrometry *Journal of Chromatography A* **1217** 6791-806

[4] Zou W, Gao B, Ok Y S and Dong L 2019 Integrated adsorption and photocatalytic degradation of volatile organic compounds (VOCs) using carbon-based nanocomposites: A critical review. Elsevier Ltd) pp 845-59

[5] Baêta B E L, Lima D R S, Silva S Q and Aquino S F 2015 Evaluation of soluble microbial products and aromatic amines accumulation during a combined anaerobic/aerobic treatment of a model azo dye *Chemical Engineering Journal* **259** 936-44

[6] Dharupaneedi S P, Nataraj S K, Nadagouda M, Reddy K R, Shukla S S and Aminabhavi T M 2019 Membrane-based separation of potential emerging pollutants. Elsevier) pp 850-66

[7] Sidik D A B, Hairom N H H, Zainuri N Z, Desa A L, Misdan N, Yusof N, Ong C B, Mohammad A W and Aripen N S M 2018 Photocatalytic degradation of industrial dye wastewater using zinc oxide-polyvinylpyrrolidone nanoparticles *Malaysian Journal of Analytical Sciences* **22** 693-701

[8] Zaidi N S, Sohaili J, Muda K and Sillanpää M 2014 Magnetic field application and its potential in water and wastewater treatment systems. pp 206-40

[9] Omri K, Sajieddine M, Bououdina M, El Mir L, Zhang B, Lemine O M and Alyamani A 2012 Sol–gel synthesis of 8nm magnetite (Fe3O4) nanoparticles and their magnetic properties *Superlattices and Microstructures* **52** 793-9

[10] Zhang H and Zhu G 2012 One-step hydrothermal synthesis of magnetic Fe 3 O 4 nanoparticles immobilized on polyamide fabric *Applied Surface Science* **258** 4952-9

[11] Aphesteguy J C, Kurlyandskaya G V, De Celis J P, Safronov A P and Schegoleva N N 2015 Magnetite nanoparticles prepared by co-precipitation method in different conditions *Materials Chemistry and Physics* **161** 243-9

[12] Li Y, Qiang Q, Zheng X and Wang Z 2015 Controllable electrochemical synthesis of Ag nanoparticles in ionic liquid microemulsions *Electrochem. Commun.* **58** 41-5

[13] Ibrahim N A, Nada A A, Hassabo A G, Eid B M, Noor El-Deen A M and Abou-Zeid N Y 2017 Effect of different capping agents on physicochemical and antimicrobial properties of ZnO nanoparticles *Chem. Pap.* **71** 1365-75

[14] Kannan R R R, Stirk W A and Van Staden J 2013 Synthesis of silver nanoparticles using the seaweed Codium capitatum P.C. Silva (Chlorophyceae) *South African Journal of Botany* **86** 1-4

[15] Chunfa D, Xianglin Z, Hao C and Chuanliang C 2016 Sodium Alginate Mediated Route for the Synthesis of Monodisperse Silver Nanoparticles Using Glucose as Reducing Agents *Rare Metal Materials and Engineering* **45** 261-6

[16] Esmat M, Farghali A A, Khedr M H and El-Sherbiny I M 2017 Alginate-based nanocomposites for efficient removal of heavy metal ions *International Journal of Biological Macromolecules* **102** 272-83

[17] Areerob Y, Cho J Y, Jang W K and Oh W-C 2018 Enhanced sonocatalytic degradation of organic dyes from aqueous solutions by novel synthesis of mesoporous Fe3O4-graphene/ZnO@SiO2 nanocomposites *Ultrasonics Sonochemistry* **41** 267-78

[18] Daemi H and Barikani M 2012 Synthesis and characterization of calcium alginate nanoparticles, sodium homopolymannuronate salt and its calcium nanoparticles *Scientia Iranica* **19** 2023-8

[19] Qiao X, Niu L, Zhang H, Wen X, Cao Y and Bai G 2017 Controllable fabrication of a novel porous Ni-alginate hybrid material for hydrogenation *Applied Catalysis B: Environmental* **218** 721-30

[20] Hussein-Al-Ali S H, El Zowalaty M E, Hussein M Z, Ismail M, Dorniani D and Webster T J 2014 Novel kojic acid-polymer-based magnetic nanocomposites for medical applications *International journal of nanomedicine* **9** 351-62

[21] Tan C, Gao N, Deng Y, Deng J, Zhou S, Li J and Xin X 2014 Radical induced degradation of acetaminophen with Fe3O4 magnetic nanoparticles as heterogeneous activator of peroxymonosulfate *Journal of Hazardous Materials* **276** 452-60

[22] Aziz F F A, Jalil A A, Triwahyono S and Mohamed M 2018 Controllable structure of fibrous SiO2–ZSM-5 support decorated with TiO2 catalysts for enhanced photodegradation of paracetamol *Applied Surface Science* **455** 84-95

[23] Li Y, Li G, Li W, Yang F and Liu H 2015 Greenly Synthesized Gold–Alginate Nanocomposites Catalyst for Reducing Decoloration of Azo-Dyes *Nano* **10** 1550108

[24] Wang F, Lu X and Li X-y 2016 Selective removals of heavy metals (Pb2+, Cu2+, and Cd2+) from wastewater by gelation with alginate for effective metal recovery *Journal of Hazardous Materials* **308** 75-83

[25] Soares S F, Simões T R, Trindade T and Daniel-da-Silva A L 2017 Highly Efficient Removal of Dye from Water Using Magnetic Carrageenan/Silica Hybrid Nano-adsorbents *Water, Air, & Soil Pollution* **228** 87

[26] Hassani A, Karaca C, Karaca S, Khataee A, Açışlı Ö and Yılmaz B 2018 Enhanced removal of basic violet 10 by heterogeneous sono-Fenton process using magnetite nanoparticles *Ultrasonics Sonochemistry* **42** 390-402

Enhanced Proton Conductivity of Porous UHMWPE membrane with Graphene-based material for Vanadium Redox Flow Battery

S S Sha'rani[1,2*], N. W. C. Jusoh[1,2], E Abouzari-Lotf[2,3], A Ahmad[2,3], and R R Ali[1,2]

[1]Malaysia-Japan International Institute of Technology, International Campus, Universiti Teknologi Malaysia, Jalan Semarak 54100, Kuala Lumpur, Malaysia.
[2]Advanced Materials Research Group, Center of Hydrogen Energy, Universiti Teknologi Malaysia, Jalan Semarak 54100, Kuala Lumpur, Malaysia.
[3]Department of Chemical Engineering, Universiti Teknologi Malaysia, 81310 Johor Bahru, Malaysia.

*ssophia2@graduate.utm.my

Abstract. Porous membrane with enhanced proton conductivity is designed and prepared for vanadium redox flow battery (VRFB). The membrane is composed of ultra height molecular weight polyethylene (UHMWPE) grafted with poly(vinylbenzyl chloride) (PVBC) at different degrees of grafting (DOG). Subsequently, the membrane is functionalized with graphene oxide (UHMWPE-g-PVB*GO*) under ultrasonication. The functionalization of the GO on the porous substrates is confirmed by Fourier transform infrared (FTIR) and Raman analyses. UHMWPE-g-PVB*GO* membranes exhibited ~37% higher proton conductivity compared to pristine UHMWPE upon modification at 30 °C and 100% relative humidity (RH). In relation to the degree of grafting, a higher degree of grafting possessed higher proton conductivity value for the porous membrane.

1. Introduction

The rapid progress of vanadium redox flow battery (VRFB) as one of the most promising electrochemical energy storage has influenced vigorous research on technology improvement. Comparing VRFB to other flow batteries, VRFB offers longer life cycle, flexible design, deep discharge capacity and low environmental risks [1]. In VRFB, the membrane plays a vital role in determining the potential commercialization and performance of the system. A membrane separates the positive and negative electrolytes while providing the diffusion of balancing ions such as protons (H^+) to complete the circuit [2]. Although VRFB has been successfully applied in various application fields, the expensive cost of the membrane hinders the commercial development of the battery system. In addition to this, rigorous research is also currently underway to further improve the performance system. Practically, an ideal membrane for VRFB should have high proton conductivity to avoid the ohmic losses, low vanadium permeability to prevent high self-discharge, excellent chemical and mechanical stability in the acidic and oxidative environment of the electrolytes and low cost for easy commercialization [3, 4].

Porous polyethylene (PE) is a porous membrane that has been used in many battery applications especially for lithium-ion batteries (LIB) [5, 6] and fuel cell [7, 8]. The use of PE in such applications can be regarded as versatile and promising due to its ability to withstand high temperatures and possessed superior chemical and mechanical stability. Additionally, the cost of these materials can be obtained as low as \$5–50 m^{-2} [9, 10]. The potential use of PE for VRFB has been discovered in many studies. A study conducted by Jung *et. al.* (2018) proved that the stability of the PE membrane is 27 times higher than Nafion with comparable performance in VRFB [11]. The excellent stability of such membrane against the vanadium electrolytes is due to its poor wettability [12]. Although this membrane is chemically stable, the high hydrophobicity of such membrane leads to a severe reduction in the proton conduction and simultaneously reduce the battery efficiency.

Generally, most porous membranes are hydrophobic; hence the proton conductivity of such membrane material is low (~10^{-12} S cm^{-1}) [13]. This is because, in the porous membrane, the hydrophilic functionalities like sulfonic acid, carboxylic acids, and quaternary ammonium pendants do not exist in the membrane structure. Hence, hydrophilic materials are required to make the membrane structure to be compatible with the water and to achieve excellent electrochemical properties in the VRFB. There have been extensive studies in improving the properties of the porous membranes. One of the most effective ways to enhancing the conductivity is by modifying the membranes with hydrophilic materials such as silica [14, 15], TiO_2 [16] and zeolite [17].

Graphene oxide (GO) is a carbon-based material that has plenty of functional groups, making this material to be readily surface modified with other materials. In an attempt to improve the proton conductivity of porous membranes for specific applications, GO served as one of the best choices. A study conducted by Karim *et al.* showed that GO (10^{-2} $S.cm^{-1}$) has higher conductivity compared to graphite oxide (~10^{-4} $S.m^{-1}$) and graphene oxide/proton hybrid (~10^{-5} $S.cm^{-1}$) [18]. The existence of many oxygenated-functional groups on the GO surface makes GO to exist as negatively charged, hence promoting the transportation of the positively charged H^+ through the hydrogen bonding network. Hence, combining this feature with the porous membrane would help to alleviate the low proton conductivity in the membrane.

In this paper, a commercial Ultra-Height Molecular Weight Polyethylene (UHMWPE) was used to fabricate high proton conductivity membranes. The UHMWPE was activated with polyvinyl benzene (PVBC) to provide an active site for the interaction with GO. The establishment of the modification technique was verified by using the chemical and morphological analyses to identify the presence of the GO on the membrane surface. Furthermore, the proton conductivity value of the pristine and modified membranes was measured and compared through electrical impedance analysis. The results demonstrated enhanced proton conductivity with lower area resistance for the developed membranes.

2. Methodology

2.1. GO Functionalisation

Few different surface-modified composite membranes were prepared in this research. To introduce the functional site for the UHMWPE, the porous substrates were first activated by poly(vinyl benzyl chloride) (PVBC) of different degree of grafting (DOG) by using the following procedure [19, 20]. Electron beam accelerator (EPS 3000) was used to irradiate the samples. The degree of grafting was calculated using the following equation:

$$DOG\ (\%) = \frac{W_g - W_s}{W_s} \text{ x } 100 \qquad (1)$$

Briefly, the activated UHMWPE is immersed in 3.0 wt% of GO under sonication for 24 hours. The samples were then washed for 5 times with deionised water to removes the weakly bonded GO. Subsequently, the prepared samples were vacuum-dried overnight at 60°C. The membranes were denoted as UHMWPE-*g*-PVB*GO* (*x*%) where *x* represents the DOG. Figure 1 shows the schematic representation of the membrane preparation steps.

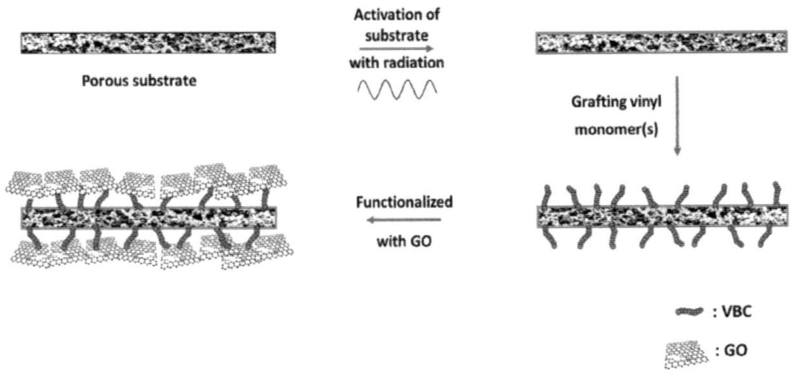

Figure 1. Schematic representation of the membrane preparations steps

2.2. Characterisation and Analysis

The analyses of the chemical and morphological surface of the prepared membranes were investigated. The infrared (IR) absorption was recorded in the region of 6000 – 400 cm^{-1} by using FTIR spectrophotometer (Perkin Elmer, USA). Raman spectroscopy was also carried out to complement the IR analysis on the presence of the GO on the membrane surface.

2.3. Proton Conductivity Measurement

The conductivity of the membranes was measured before and after modification. Electrochemical impedance spectroscopy (EIS) was used to record the conductivity and the area resistance of the membrane from ZPlot software. Before the measurement, the membranes were equilibrated with the acid solution overnight. The conductivity value obtained for all samples were measured at 100% relative humidity at 30°C. Equation 3 shows the formula used in calculating the conductivity and area resistance of the membranes:

$$\sigma\ (S\ cm^{-1}) = \frac{L(cm)}{R(\Omega)\ x\ S(cm^2)} \tag{2}$$

Where L is the membrane thickness, R is the resistance and S is the membrane active surface area. The denominator of this equation can also serve as the area resistance of the membrane.

3. Results and Discussion

Few types of porous membranes with different DOG were obtained by surface modification with GO layer. Figure 2 shows the schematic representation of the chemical interaction of the activated porous substrates with GO at the surface. Three prominent peaks are obtained from the FTIR analysis which corresponds to $-CH_2$ (symmetric and asymmetric stretching at 2849 cm^{-1} and 2916 cm^{-1} respectively) and $-CH_2$ deformation (1463 cm^{-1}). The traces of GO on the membrane surface can be seen on the additional peaks of oxygen-containing groups shown by the modified membranes. The vibration of C-O stretching of –COOH appeared at around 1250 cm^{-1}. The bands that can be observed at 1050 cm^{-1} corresponds to the stretching vibration of epoxy group (C-O-C). The peak at 1727 cm^{-1} represents the C=O stretching [21].

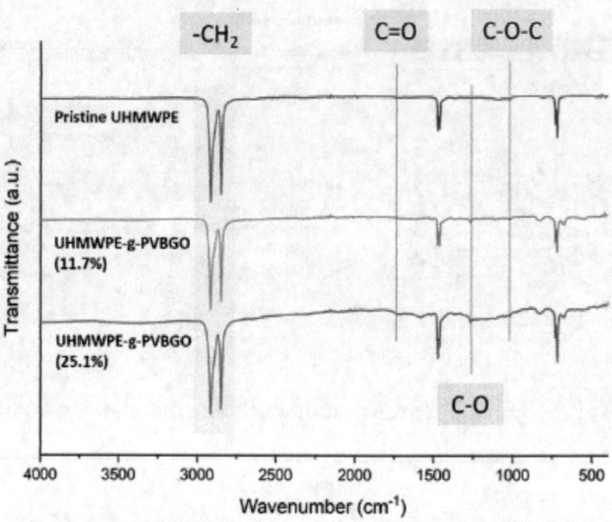

Figure 2. FTIR of the pristine and modified membrane

To further confirm the GO functionalization procedure, Raman analysis was carried out. Figure 3 shows the Raman analysis of the pristine and modified membranes. Comparing the pristine and modified membrane, the modified membrane shows fundamental peaks in 1250 to 1750 cm^{-1} region. The peak at 1353 cm^{-1} can be associated with the D vibration band caused by the disordered structure of GO. Meanwhile, the G vibration band at 1587 cm^{-1} correspond to sp^2 carbon atoms plane stretching motion. In addition, the peak intensity of the modified membrane at around 2500 – 3000 cm^{-1} regions become widen as the result of the interaction between GO and the grafted substrates [22].

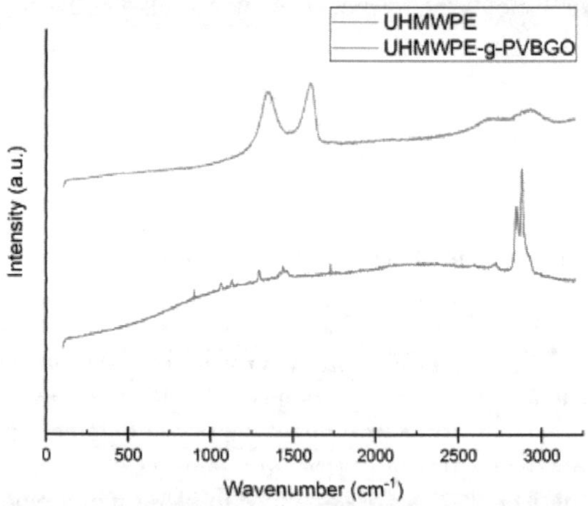

Figure 3. Raman spectra of pristine and modified UHMWPE membrane

Proton conductivity is important parameter that directly affect the voltage efficiency (VE) of a VRFB system. For porous substrates, usually the proton conductivity is very small (~10^{-12} mS cm^{-2}) since there is no ion exchange that can serve as the ion exchange site to carry the H^+ ions. Hence, the conductivity value of the porous membranes depends on the size and pore exclusion effect of the porous membrane. In addition to the low proton conductivity, porous membrane also highly hydrophobic in nature with low wettability to carry the water transport through the membranes.

Table 1 shows the proton conductivity and area resistance of the membranes. Upon modification with GO under sonication, the membrane surface is covered with layers of GO. The proton conductivity of the modified membranes was found to be higher than the pristine membrane. At higher DOG, the proton conductivity increase. This DOG represents the amount of available site for OH from the GO to be interconnected with CH_2-Cl bond through the normal covalent bond [18]. As the results, the negatively charged GO form hydrogen bond with the water molecule to facilitate faster proton transportation through the porous structure of the membrane. Area resistance of the modified membranes showed a reverse trend since more protons can move freely through the membrane. Table 1 shows the proton transfer through the modified membranes.

Table 1. proton conductivity and area resistance of the membranes

Membranes	Thickness (mm)	Conductivity (mS cm^{-1})	Area Resistance (Ω cm^2)
Pristine UHMPWE	0.07	27.40	0.255
UHMWPPE-*g*-PVB*GO* (10.5%)	0.07	30.44	0.230
UHMWPPE-*g*-PVB*GO* (11.7%)	0.07	30.61	0.229
UHMWPPE-*g*-PVB*GO* (25.1%)	0.08	37.71	0.212

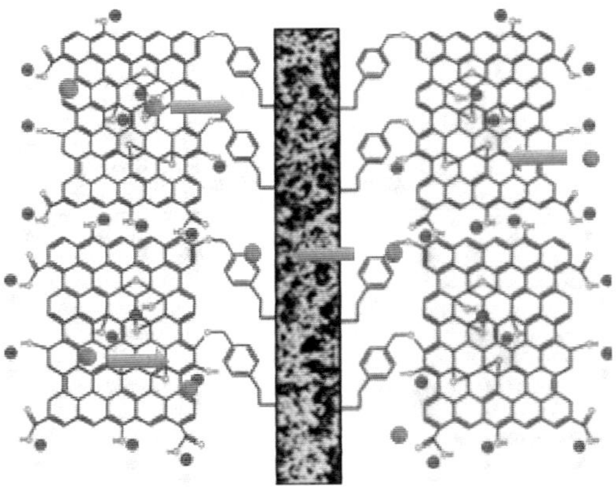

Figure 4. the proton transfer through the modified membranes

4. Conclusion

The modification of UHMWPE membrane was successfully carried out using GO to enhance the proton conductivity of the membrane. FTIR and Raman analyses was carried out for both pristine and modified membrane to confirm the presence of GO on the membrane surface. The proton conductivity value of the membrane increased with the increased in the DOG. The GO on the membrane surface increases the ionic functionalities of the porous membrane. Hence, resulting in an increase in the proton conductivity value.

Acknowledgement

The research was primarily supported by Universiti Teknologi Malaysia and ministry of higher education under Fundamental Research Grants (vote no. #4C116) and partnership research program of Innovative Research Universities of Australia and Malaysia Research University Network (IRU-MRUN) (vote no. #00M84).

References

[1] Li X, Zhang H, Mai Z, Zhang H and Vankelecom I 2011 Ion exchange membranes for vanadium redox flow battery (VRB) applications *Energy & Environmental Science* **4** 1147

[2] Prifti H, Parasuraman A, Winardi S, Lim T M and Skyllas-Kazacos M 2012 Membranes for redox flow battery applications *Membranes (Basel)* **2** 275-306
[3] Shi Y, Eze C, Xiong B, He W, Zhang H, Lim T M, Ukil A and Zhao J 2019 Recent development of membrane for vanadium redox flow battery applications: A review *Appl. Energy* **238** 202-24
[4] Parasuraman A, Lim T M, Menictas C and Skyllas-Kazacos M 2013 Review of material research and development for vanadium redox flow battery applications *Electrochim. Acta* **101** 27-40
[5] Yang C, Tong H, Luo C, Yuan S, Chen G and Yang Y 2017 Boehmite particle coating modified microporous polyethylene membrane: A promising separator for lithium ion batteries *J. Power Sources* **348** 80-6
[6] Li R and Gao P 2017 Nanoporous UHMWPE Membrane Separators for Safer and High-Power-Density Rechargeable Batteries *Global Challenges* **1** 1700020
[7] Espiritu R, Mamlouk M and Scott K 2016 Study on the effect of the degree of grafting on the performance of polyethylene-based anion exchange membrane for fuel cell application *Int. J. Hydrogen Energy* **41** 1120-33
[8] Yildirim M H, Stamatialis D and Wessling M 2008 Dimensionally stable Nafion–polyethylene composite membranes for direct methanol fuel cell applications *J. Membr. Sci.* **321** 364-72
[9] Minke C, Kunz U and Turek T 2017 Techno-economic assessment of novel vanadium redox flow batteries with large-area cells *J. Power Sources* **361** 105-14
[10] Mu D, Yu L, Yu L and Xi J 2018 Toward Cheaper Vanadium Flow Batteries: Porous Polyethylene Reinforced Membrane with Superior Durability *ACS Applied Energy Materials* **1** 1641-8
[11] Jung H-Y, Moon G-O, Sadhasivam T, Jin C-S, Park W-S, Kim H-T and Roh S-H 2018 Ionic transportation and chemical stability of high-endurance porous polyethylene separator for vanadium redox flow batteries *Solid State Ionics* **327** 110-6
[12] Cao J, Yuan Z, Li X, Xu W and Zhang H 2015 Hydrophilic poly(vinylidene fluoride) porous membrane with well connected ion transport networks for vanadium flow battery *J. Power Sources* **298** 228-35
[13] Zhou X L, Zhao T S, An L, Wei L and Zhang C 2015 The use of polybenzimidazole membranes in vanadium redox flow batteries leading to increased coulombic efficiency and cycling performance *Electrochim. Acta* **153** 492-8
[14] Zhang H, Zhang H, Li X, Mai Z and Wei W 2012 Silica modified nanofiltration membranes with improved selectivity for redox flow battery application *Energy Environ. Sci.* **5** 6299-303
[15] Wei X, Nie Z, Luo Q, Li B, Chen B, Simmons K, Sprenkle V and Wang W 2013 Nanoporous Polytetrafluoroethylene/Silica Composite Separator as a High-Performance All-Vanadium Redox Flow Battery Membrane *Advanced Energy Materials* **3** 1215-20
[16] Bae T-H and Tak T-M 2005 Effect of TiO2 nanoparticles on fouling mitigation of ultrafiltration membranes for activated sludge filtration *J. Membr. Sci.* **249** 1-8
[17] Yuan Z, Zhu X, Li M, Lu W, Li X and Zhang H 2016 A Highly Ion-Selective Zeolite Flake Layer on Porous Membranes for Flow Battery Applications *Angew. Chem.* **128** 3110-4
[18] Karim M R, Hatakeyama K, Matsui T, Takehira H, Taniguchi T, Koinuma M, Matsumoto Y, Akutagawa T, Nakamura T, Noro S, Yamada T, Kitagawa H and Hayami S 2013 Graphene oxide nanosheet with high proton conductivity *J. Am. Chem. Soc.* **135** 8097-100
[19] Abdiani M, Abouzari-Lotf E, Ting T M, Moozarm Nia P, Sha'rani S S, Shockravi A and Ahmad A 2019 Novel polyolefin based alkaline polymer electrolyte membrane for vanadium redox flow batteries *J. Power Sources* **424** 245-53
[20] Mohamad N A, Abouzari Lotf E, Mohamed M N, Arshad A and Abdullah T A T 2019 A comparison of CO2 adsorption behaviour of mono- and diamine-functionalised adsorbents *E3S Web of Conferences* **90** 01010
[21] Monteserín C, Blanco M, Aranzabe E, Aranzabe A, Laza J M, Larrañaga-Varga A and Vilas J L 2017 Effects of Graphene Oxide and Chemically-Reduced Graphene Oxide on the Dynamic Mechanical Properties of Epoxy Amine Composites *Polymers* **9** 449

[22] Dresselhaus M S, Jorio A, Hofmann M, Dresselhaus G and Saito R 2010 Perspectives on carbon nanotubes and graphene Raman spectroscopy *Nano Lett.* **10** 751-8

IOP Conf. Series: Materials Science and Engineering **808** (2020) 012025 doi:10.1088/1757-899X/808/1/012025

Catalytic Cracking of High Density Polyethylene Pyrolysis Vapor over Zeolite ZSM-5 towards Production of Diesel

N F A Ghaffar[1], **A Johari***[1,2], **T A T Abdullah**[1,2] and **A Ripin**[1,2]

[1] School of Chemical and Energy Engineering, Faculty of Engineering, Universiti Teknologi Malaysia, 81310 Skudai, Johor, Malaysia
[2] Centre of Hydrogen Energy, Institute of Future Energy, Universiti Teknologi Malaysia, 81310 Skudai, Johor, Malaysia

*anwar@utm.my

Abstract. The upsurge in the amount of plastic pollution has drawn the attention of various researchers in finding the best solution for plastic waste treatment. Pyrolysis of high density polyethylene using ZSM-5 catalyst has been investigated using two-staged reactors consisting of a pyrolysis reactor and a catalytic fixed bed reactor. The effect of the heating rate (2 °C/min, 6 °C/min, 10 °C/min, 14 °C/min, 18 °C/min) of the pyrolysis reactor and reaction time (75 min, 90 min, 105 min) were studied on the yield of products and the quality of pyrolytic oil products. The quality of pyrolytic oil produced was evaluated based on the gross calorific values (GCV) and the paraffin, olefin, naphthenic and aromatic (PONA) compounds distribution, which were used to compare with conventional fuel. The highest yield of oil product was obtained at a reaction time of 75 minutes, where it was 61.05 wt.%, with solid product yield of 0.41wt.% and gas product yield of 38.54wt.%, at a heating rate of 2°C/min. The GCV of oil products were within the range of that of conventional diesel (44 – 45 MJ/kg). The pyrolytic oil products contained mainly C_{10} to C_{24} hydrocarbon compounds, which were diesel range hydrocarbon. In terms of the PONA distribution, the composition of paraffin was larger for most of the products, while the composition of naphthenes and aromatic were relatively small with respect to the range for conventional diesel.

1. Introduction

Plastics are one of the most commonly used materials for the production of consumer use objects due to its low cost, durability, malleability, resistance to most chemicals and most importantly, lightweight. However, the disadvantage of plastics is far more alarming. This is due to the inability of plastics to decompose in a short period of time. "Over 300 million metric tons of plastics are produced in the world annually, and about 50% of this volume is for disposal applications, product that are discarded within a year of their purchase" [1]. In fact, in Malaysia, the plastics manufacturing industry contributes to RM12.96 billion export sales in 2015 compared to 9.4 billion in 2010 [2]. Hence, the local production of plastics in Malaysia has been growing steadily and is expected to further expand due to its economic contribution to Malaysia's overall economy. As a consequence, there will be an upsurge in the number of plastic wastes, especially for single-use plastic products.

Currently, the conventional management of municipal solid waste (MSW) including plastic waste in Malaysia is through landfilling [3]. Unfortunately, this type of waste management, especially non-sanitary landfill, is not an efficient way for the treatment of plastic waste since it takes up too much space and does not fully contain the plastic from polluting the land and ocean. Besides that, mechanical recycling is also another type of treatment of plastic waste in Malaysia. It is a process whereby the plastic wastes are recycled to produce new plastic resin. Nevertheless, it can only handle a single type of plastic and thus, requires rigorous presorting steps before the process takes place. Also, a water treatment facility is required to treat the contaminated water from cleaning the feedstock, which will be an added cost for installation and operation [4].

Furthermore, plastic waste can also be treated thermally whether through incineration, gasification or pyrolysis [5]. Incineration of waste is a process used to generate electricity by using waste as fuel to produce steam. Typically, incineration of waste is performed on MSW which consists of non-recyclable plastics rather than all types of plastic waste. However, this method causes the emission of toxic and hazardous gases from the combustion of plastic, which will necessitate an air purification system in order to filter the pollutants from entering the atmosphere [6]. Besides that, gasification is another process of thermal treatment performed by heating waste at high temperatures to produce syngas. The feedstock can be either purely plastic waste or co-fed with biomass to manipulate the quality of syngas. Similar to incineration, gasification requires a series of syngas purification system due to the formation of CO_2, NOx, and SOx from the reaction with air [6]. Feed pretreatment steps are also required since the process occurs at high temperatures and short residence times. Consequently, additional cost is needed for the overall gasification plant operation [6].

Pyrolysis is a thermal treatment of plastic waste at high temperatures without the presence of oxygen, but with an inert environment. Unlike other types of thermal treatment, this option can handle highly contaminated and heterogeneous plastic feedstock and requires few pretreatment steps making it more economically attractive among the others. Flue gas clean up system will not be a main concern since the gas product will be contained and treated prior to its utilization as fuel [5]. Pyrolysis of plastic also produces high calorific value pyrolytic oil and gas which matches conventional fuels that could be easily marketed and used in gas engines to produce electricity and heat [5]. Catalytic pyrolysis seems to be more efficient in producing high-quality oil due to the ability of the catalyst to reduce the amount of impurity, dictate the chemical composition in the product and manipulate the amount of yield of certain products [7]. The results of product yields also vary depending on the type of catalyst used, whereby highly acidic catalyst typically yield higher gas product [8] while catalyst with lower acidity yield higher liquid product [9-10]. Thus, in order to increase the yield of liquid product, a lower acidity catalyst is preferred, which resulted in the use of zeolite ZSM-5 catalyst of SiO_2:Al_2O_3 ratio of 1000 since a larger ratio indicates lower acidity as studied by Shirazi et al., [11].

Most of the experiments performed on catalytic pyrolysis using zeolites are unable to emulate the chemical composition of real diesel or fuel with respect to the paraffin, olefin, naphthenic and aromatic (PONA) compounds distribution since some of them either have an unbalanced amount of paraffin, olefin, naphthene, and aromatics or contain mostly aromatics or olefin compounds [12-16]. Therefore, it is vital to analyze the PONA distribution of pyrolytic oil product and compare with that of conventional diesel. This will help to determine the necessary compounds required to be manipulated to best replicate the PONA distribution of conventional diesel to imitate its physical and chemical properties [17]. This study will explore the possibility of producing diesel range pyrolytic oil from the catalytic cracking of high density polyethylene (HDPE) pyrolysis vapor using ZSM-5 catalyst and performing analysis on the pyrolytic oil products with respect to the PONA compounds distribution.

2. Experimental

2.1. Raw material and catalyst

HDPE pellets of size 3 x 4 x 1 mm (W x L x H) were obtained from Polyethylene Malaysia Sdn. Bhd. Zeolite ZSM-5 catalyst with Si/Al ratio of 1000 in powder form, was procured from Acros Organics. The catalyst was pressed and sieved to a particle size of 1.0 – 1.4 mm and was heated for 4 hours at 500°C to remove the adsorbed gas or other impurities. The HDPE was characterized on the proximate analysis according to a method described by Wong [18] and thermal analysis by the thermogravimetric

IOP Conf. Series: Materials Science and Engineering **808** (2020) 012025 doi:10.1088/1757-899X/808/1/012025

analyzer (TGA-50, Shimadzu). The ZSM-5 catalyst was characterized based on BET surface area analysis (SA3100 Beckman Coulter) and its acidity properties by temperature-programmed desorption of ammonia (TPD-NH_3) using chemisorption analyzer unit (Micromeritics Autochem II 2920).

2.2. Pyrolysis experiment
The experiment was carried out in a two-staged pyrolysis catalysis batch reactor as shown in Figure 1. Nitrogen gas, which was used as a carrier gas, was flown for 5 minutes through the reactor, where the pyrolysis took place at the horizontal reactor while the catalytic cracking occurred at the vertical reactor. The plastic sample, weighing 5g, was placed on two ceramic combustion boats in the horizontal reactor, where it was heated using an electrical heater at a heating rate of 2°C/ min until the desired temperature of 525 °C. The nitrogen flowrate was set at 60 cm^3/min. After the temperature in the pyrolysis reactor reached 525°C, the reaction time was set at 90 minutes (for the heating rate variable). 0.5 g ZSM-5 catalyst was placed on a stainless steel mesh in the vertical fixed bed reactor, held at the middle part of the reactor by a steel ring, where it was preheated using an electrical heater at the final temperature, 525°C, of the pyrolysis reaction. Once pyrolysis took place, the volatile components flowed out of the pyrolysis reactor and through the ZSM-5 catalyst placed inside the fixed bed reactor. The reacted condensable products from the fixed bed reactor were liquefied in three units of cold trap condensers. The first condenser contained liquid nitrogen as the cooling agent while the subsequent units contained an ice bath with added ethanol. The residual solid of HDPE in the pyrolysis reactor and the oil products in the condenser collected were weighed to determine the yield of the solid and oil products. The oil product was kept at 5°C in a refrigerator prior to analysis.

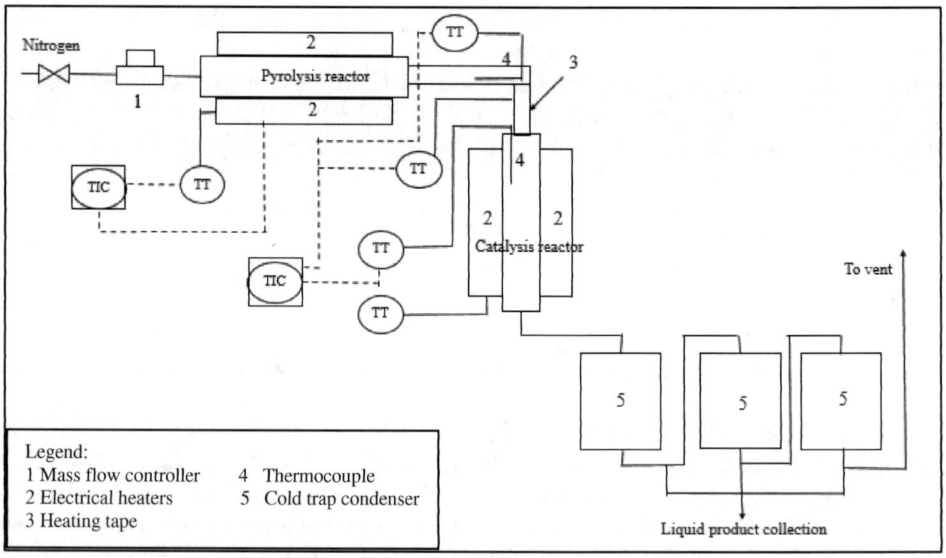

Figure 1. Experimental set-up diagram

The experiment was then repeated at heating rates of 6°C/min, 10°C/min, 14°C/min and 18°C/min. The experiment was also repeated based on the heating rate that gives the maximum yield of oil product where, it was performed at reaction times of 75 minutes and 105 minutes, to investigate the effect of reaction time on the pyrolysis experiment.

2.3. Analysis of Products

2.3.1. Yield of Products. The yield of solid, oil and gaseous products collected in each run was calculated using Equations (1), (2), and (3) as follows.

Solid yield (wt.%) = [Mass of solid residue in the ceramic boat (g)] [Mass of HDPE feedstock (g)] ×100% (1)

Oil yield (wt.%) = [Mass of condensed product in the cold trap (g)]/ [Mass of HDPE feedstock (g)] × 100% (2)

Gas yield (wt.%) = [Mass of HDPE feedstock (g) - Mass of oil product(g) - Mass of solid product(g)] / [Mass of HDPE feedstock (g)] × 100% (3)

2.3.2. Analysis of Oil Products. A bomb calorimeter manufactured by IKA C2000, was used to determine the Gross Calorific Values (GCV) of the oil products. The chemical composition of the oil products was determined using gas chromatography coupled with a mass selectivity detector (GCMSD). Prior to the analysis, the oil products were diluted with n-hexane solution (Analytical Reagent) at a volume ratio of 1 to 9. The GCMSD was equipped with a 30 m × 0.25 mm capillary column of 5% phenyl-methylpolysiloxane (HP-5). Chromatographic peaks were identified by means of NIST standard reference database. The compounds detected in the oil products were classified based on the carbon number distribution and the paraffin, olefin, naphthene and aromatic chemical groups (PONA) distribution. These results were compared with that of conventional diesel based on a study by Farrell et al., [17].

3. Results and Discussion

3.1. Proximate and thermal analysis of HDPE polymer

Table 1 summarized the proximate analysis of HDPE feedstock. The result showed that HDPE feedstock contained mostly volatile matter, which consisted of 99.52 wt.%. The high amount of volatile matter signified a large amount of volatile components from HDPE that can be condensed into liquid product [19]. A small amount of moisture content, was also observed, which might be due to the moisture from the air onto the sample during storage. The ash content was minimal, which was about 0.10 wt.%. Besides that, there was no amount of fixed carbon observed in the HDPE sample.

Table 1. Proximate analysis of HDPE

Type of Content	Weight (%)
Moisture	0.38
Volatile Matter	99.52
Ash	0.10
Fixed Carbon	0.00
Total	100.00

Figure 2 shows the thermogravimetric (TG) and derivative thermogravimetric (DTG) curves of the HDPE feedstock. The TG curve showed that the decomposition of HDPE occurred in a single step pattern. Based on the DTG curve, the peak decomposition temperature was 473°C. Therefore, the minimum temperature for the pyrolysis was set at 525°C, which is higher than the temperature of 473°C, to ensure the feedstock had been fully decomposed. The shoulder or smaller peak before the decomposition peak was due to the presence of impurities such as plasticizers or additives that are common in polymer plastics in addition to low heating rate usage [35].

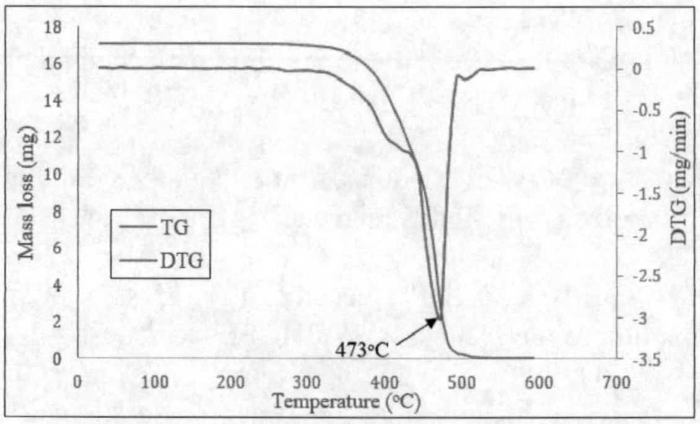

Figure 2. Thermograph of HDPE feedstock

3.2. Surface area and TPD-NH$_3$ analysis of ZSM-5 catalyst

The result for the surface area analysis was reported in Table 2. The result was compared with other ZSM-5 catalyst used in a similar study for the catalytic cracking of HDPE pyrolysis vapor. The BET surface area for the catalyst used in this study was 359.85m^2/g, with a total pore volume of 0.2261 cm^3/g and a micropore volume of 0.0901 cm^3/g. On the other hand, the BET surface area for the catalyst from the other study was lower, which was 266 m^2/g. Besides that, the total pore volume and micropore volume of the catalyst used were higher than the one in this study, which were 0.3500 cm^3/g and 0.2300 cm^3/g, respectively.

Table 2. Nitrogen adsorption-desorption analysis of ZSM-5 catalyst

Catalyst	Si/ Al ratio	BET Surface Area (m^2/g)	Micropore Surface Area (m^2/g)	Total Pore Volume (cm^3/g)	Micropore Volume (cm^3/g)	Reference
ZSM-5	1000	359.85	190.22	0.2261	0.0901	This study
ZSM-5	20	266.00	-	0.3500	0.2300	[37]

The total amount of ammonia desorbed by the ZSM-5 catalyst was indicated by the peak area of the curve in Figure 3. Based on the result obtained, the total ammonia desorbed was 0.051 mmol NH$_3$/g of catalyst. Since the ZSM-5 catalyst used had a Si/Al ratio of 1000, then a smaller value was anticipated, which indicated low acidity [11].

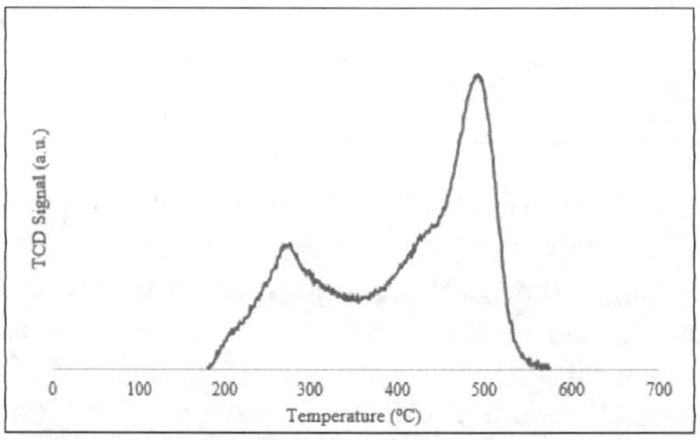

Figure 3. TPD-NH$_3$ profile from ZSM-5 catalyst

3.3. Product Yield

The result of product yields from different heating rates was summarized in Table 3. Note that the time taken to reach the final temperature in Table 3, corresponds to the time taken for the temperature to increase from room temperature of 25°C to the designated final temperature of 525°C in the horizontal reactor, based on varying heating rate. As the heating rate increased from 2°C/min to 18°C/min, the solid yield decreased from 0.32 to 0.02 wt.%. The decrease in solid yield was due to the rise in the rate of degradation of HDPE sample feedstock as the heating rate increased. Since at 18°C/min, the HDPE sample feedstock degraded faster, the amount of solid residue left was smaller compared to lower heating rates. This was in agreement with studies by Singh et al., [20] and Efika et al., [21] who conducted pyrolysis of mixed plastic waste and refuse-derived fuel, respectively.

Table 3. Effect of heating rate on product yields

Heating rate (°C/min)	Time taken to reach final temperature (min)	Solid yield (wt.%)	Oil yield (wt.%)	Gas yield (wt.%)
2	250	0.32	59.37	40.31
6	83.3	0.19	55.27	44.54
10	50.0	0.11	50.48	49.41
14	35.7	0.10	47.88	52.02
18	27.8	0.02	51.52	48.46

However, in terms of oil and gas yields, the oil product yield decreased while the gas product yield increased overall, as the heating rate increased. At a higher heating rate, the rapid degradation of polymers allowed further cracking of the hydrocarbons to form short-chain polymers with low carbon numbers, which consisted of mostly gases. As indicated in Table 3, it took only 27.8 minutes for the pyrolysis reactor to reach the final temperature of 525°C, at a heating rate of 18°C/min. As a result, at a higher heating rate, the decomposition rate of HDPE is higher, which resulted in a higher yield of gas products. This also resulted in a decrease in oil products since more HDPE is vaporized and flowed out through the gas outlet in the sweeping gas environment. Thus, higher heating rates increased the yield of gas products and decreased the yield of oil products.

Table 4 summarized the product yields based on different reaction times at a constant heating rate of 2°C/min. This heating rate was chosen due to the highest oil yield produced from the previous experimental run as per Table 3. The result showed that the solid yield declined from 0.41wt.% to 0.16wt.% as the reaction time increased from 75 minutes to 105 minutes. Since more heating time was allocated, then more HDPE polymer was degraded into volatiles in the pyrolysis reactor, leaving a small amount of the polymer as solid product. Thus, increasing the reaction time decreased the yield of solid product in the reactor.

Table 4. Effect of reaction time on product yields

Reaction time (minute)	Solid yield (wt.%)	Oil yield (wt.%)	Gas yield (wt.%)
75	0.41	61.05	38.54
90	0.32	59.37	40.31
105	0.16	55.72	44.12

In terms of oil yield, the highest oil yield, 61.05 wt.% was obtained at the lowest reaction time, which was 75 minutes. As the reaction time increased, the yield of oil products decreased. Reversibly, as the reaction time increased, the yield of gas products increased. This was due to the increase in heating time in the pyrolysis reactor causing more volatile products to flow through the catalysis reactor and form shorter chain hydrocarbon. This is because the heating time of the HDPE feedstock in the reactor is longer which leads to secondary reactions among the volatiles [23-24]. This includes further cracking of volatiles to non-condensable gases, which are compounds of carbon number C1 to C4, rather than condensable gases, which are compounds of carbon number C5 and above. As a result, longer reaction time leads to more formation of gas products which typically consists of compounds of carbon number C1 to C4 [25] and hydrogen. These findings, whereby the gas yields were higher, and the oil yields are lower at higher reaction time, were in accord with studies by Adnan et al., [26] and Miandad et al., [27].

3.4. Analysis of Oil Products

3.4.1. Gross Caloric Values (GCV) of Oil Products. Table 5 summarized the Gross Calorific Values (GCV) of the pyrolytic oil products for the parameters of heating rate and reaction time. The GCV of the oil products obtained were between 44 to 45 MJ/kg. All the GCV values were within the range of that of conventional diesel [28]. This is expected because HDPE had a high amount of volatile matter, which was 99.52 wt.%, as indicated in Table 1. As a result, the volatile matter contains hydrocarbon gases which have high energy content once the HDPE is combusted in a bomb calorimeter [16]. This lead to the high GCV values obtained for the pyrolytic oil products. Plastics, which originated from petroleum crude have high GCV values in the range of 44 – 46 MJ/kg, specifically for polyethylene derived pyrolysis products [22], [36].

Table 5. Summary of Gross Calorific Values (GCV) of the Oil Products

Parameters	**Gross Calorific Value (GCV)/ MJ/kg**	
Heating rate (°C/min)	**Pyrolytic oil products**	**Conventional diesel**
2	44.99	
6	44.81	
10	45.24	
14	44.77	
18	45.12	44 - 46
Reaction time (min)		
75	45.10	
90	44.99	
105	43.88	

3.4.2. Chemical Composition of Oil Products. The chemical compounds in the oil products were classified based on carbon number and the Paraffin, Olefin, Naphthenic, and Aromatic (PONA) compound distribution. These classifications were selected as a means for comparison with the current standards of conventional fuel.

Figure 4(a) represents the effect of heating rate on the carbon number distribution of oil products, while Figure 4(b) represents the effect of reaction time. Based on the oil product analysis using GCMSD, the compounds detected in the oil products were classified based on the carbon number. The y-axis indicated the peak area percentage of compounds detected in the GCMSD analysis. In general, the oil products contained a high amount of C_{10} to C_{24} hydrocarbon compounds, which are diesel range compounds, while a minimum amount of C_{24+} hydrocarbon compounds, which are heavy oil range compounds as observed by other researchers using zeolite based catalysts [16], [9]. This shows that the ZSM-5 catalyst was effective in increasing the decomposition of HDPE to short-chain hydrocarbon compounds. With reference to Figure 4(a) and 4(b), the parameter of heating rate and reaction time had little or no direct influence on the carbon number distribution in the oil products. A similar result was also obtained by Tamri et al., [34], whereby the heating rate varied had little or no direct influence on the carbon number distribution of the pyrolytic oil products.

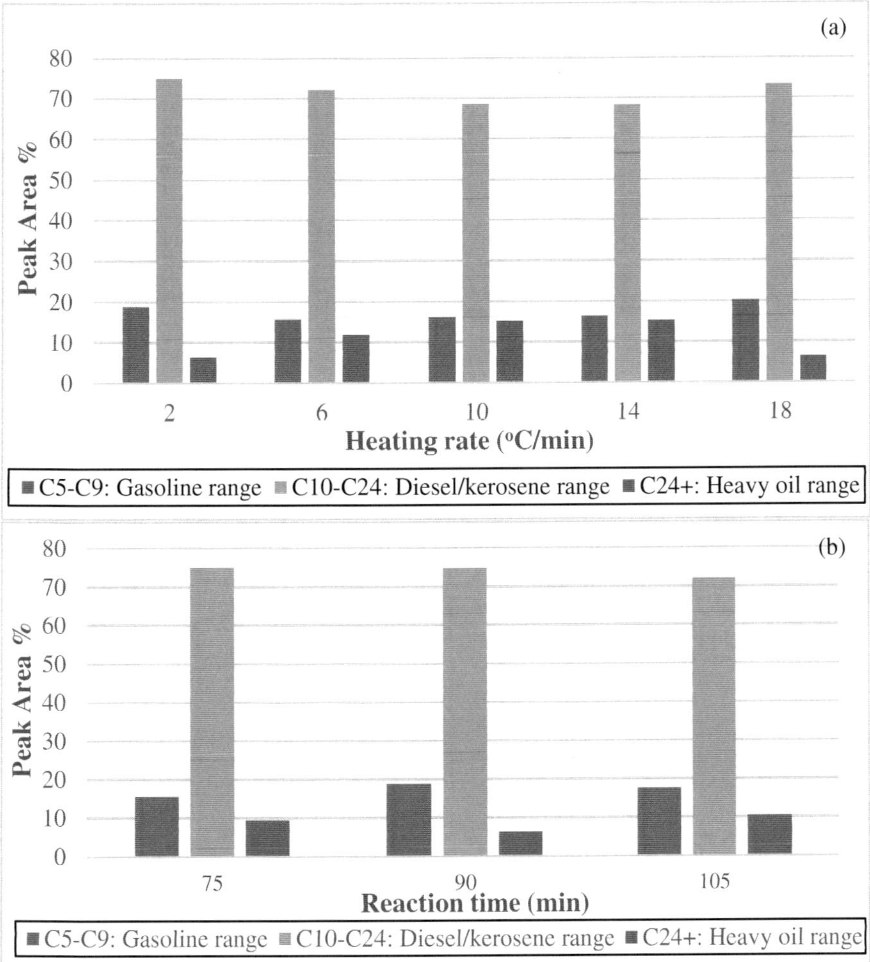

Figure 4. a) Effect of heating rate on the carbon number distribution, b) Effect of reaction time on the carbon number distribution

Tables 6 and 7 indicate the PONA distribution of the oil products with respect to conventional diesel. The oil products contained a high amount of paraffin compounds which ranged from 46 – 61 % in the composition. Meanwhile, olefin, naphthenic and aromatic compounds were between the range of 9 – 22% depending on the parameters varied. Comparing the PONA distribution of oil products with the specification for conventional diesel, only a few of the oil products were able to meet the specification, which was from Table 6, oil products for heating rate 10°C/min, 14°C/min and 18°C/min. These oil products were only able to meet the specification for the paraffin and the aromatic compound groups. For Table 7, the pyrolytic oil products were not able to meet the specification for diesel where it contained a large amount of paraffin and a small amount of aromatics.

Table 6. Effect of heating rate on the PONA distribution of the oil products

Heating rate (°C/min)	Compound composition (%)			
	Paraffin	**Olefin**	**Naphthene**	**Aromatic**
2	61.93	9.81	12.25	13.90
6	58.66	11.34	16.10	12.41
10	48.27	19.92	15.80	14.85
14	46.45	16.94	14.41	20.84
18	51.00	22.19	10.24	15.66
Conventional diesel	25 - 50	-	20 - 40	15 - 40

Table 7. Effect of reaction time on the PONA distribution of the oil products

Reaction time (min)	Compound composition (%)			
	Paraffin	**Olefin**	**Naphthene**	**Aromatic**
75	60.97	17.82	11.75	8.04
90	61.93	9.81	12.25	13.90
105	61.68	13.33	12.83	10.16
Conventional diesel	25 - 50	-	20 - 40	15 - 40

Since HDPE is a polyolefinic compound, the depolymerization of HDPE produces a large number of olefins; whereby these olefins can further react to form aromatics through a tertiary reaction [29]. Hydrogenation of olefin compounds can also occur to form paraffin compounds [12]. Since a large amount of paraffin compounds were observed, it is suggested that hydrogenation of olefins occurred rapidly from the process. This finding was also observed in the catalytic upgrading of pyrolysis wax oil from municipal plastic waste, where the composition of paraffin compounds accounted for approximately 43wt.% using 80% Zeolite Y and 20% Clay as the catalyst [27].

Besides the formation of paraffin, the olefin compounds also played a major role in the formation of aromatic compounds. This can be achieved through the cyclization of olefin compounds through the Diels Alder reaction before undergoing aromatization to form benzene and other aromatic compounds. However, as seen in Tables 6 and 7, the aromatic compounds was still minimum compared to the range of 15 – 40% in conventional diesel. Small amounts of cycloolefin compounds were detected, which indicated that cyclization occurred on the catalyst pores but not significantly. Typically, cyclization and aromatization of the volatile products occur at the acidic sites inside the catalyst pores, to form aromatic compounds [18]. Kim et al., [31] were able to obtain higher formation of aromatic compounds using desilicated hierarchical ZSM-5, which is a more acidic catalyst rather than the commercial ZSM-5, which is a lower acidic catalyst from the co-pyrolysis of wood with a plastic polymer.

In addition to catalyst acidity, catalyst mesoporosity also affected the formation of aromatics hydrocarbon in the pyrolytic products [31]. This is because conventional ZSM-5 zeolites, which are typically microporous in structure limits the mass transfer of molecule reactants to only minuscule sized molecules [32]. Therefore, by increasing the size of catalyst pores, a larger amount of molecules can enter the pores and thus react with the acidic sites to form aromatic compounds. In this case, the aromatic compounds was minimum due to the lack of acidic sites on the ZSM-5 zeolite catalyst as a result of high SiO_2: Al_2O_3 ratio within the zeolite framework. The high SiO_2: Al_2O_3 ratio indicates that the zeolite framework contains a minimal amount of Al atom, which contributes towards the acidity of the catalyst. Since the amount of Al atom is low, the acidity or the density of acid sites within the catalyst is low [11]. Besides that, the ZSM-5 catalyst used was microporous in structure as shown in Table 2 earlier. This indicates that a limited amount of polymer reactants were able to enter the catalyst pores for the catalytic reaction. Thus, the acidity and the porosity of the catalyst used needs to be improved to modify the selectivity of PONA compound distribution in the oil products.

4. Conclusion

For both parameters, as the heating rate and reaction time increased, the solid and oil product yield decreased while the gas product yield increased. The GCV of all the oil products were within the specification for conventional diesel (44 - 45MJ/kg). For the carbon number distribution, all of the oil products contained more than 68% in peak area, of C_{10} to C_{24} hydrocarbon compounds, which are diesel/ kerosene range hydrocarbon compounds. The PONA distribution of the oil products showed large amount of paraffin compounds while varying amount of olefin, naphthenic and aromatic compounds. Few of the oil products were able to meet the specified range for PONA distribution with respect to conventional diesel. It is suggested that more studies to be conducted in the future on the possibility of manipulating the PONA distribution of the pyrolytic oil products to match with that of conventional fuel for its potential application as alternative fuel.

Acknowledgments

The author would like to thank Universiti Teknologi Malaysia for the financial support provided through the Research University Grant 20H52 and 04G13.

References

[1] Singh P and Sharma V P 2016 Integrated Plastic Waste Management: Env. and Improv. Health Approaches *Proc. Env. Sciences* **35**, 692 – 700
[2] Malaysian Plastics Man. Assoc. (2016, April 21). Perf. of the M'sian Plastics Ind. Retrieved 15th Sept, 2018 from: http://www.mpma.org.my
[3] Johari A, Ahmad S I, Hashem H, Alkali H and Mat R 2012 Econ. and env. benefits of landfill gas from MSW in Malaysia *Renew. and Sust. Energy Reviews* **16**, 2907 -12
[4] Sharudin S D A, Abnisa F, Daud W M and Aroua M K 2016 A review on pyro. of plastic wastes *Energy Conv. and Mgmt* **115**, 308–326
[5] Al-Salem S M, Lettieri P and Baeyens J. 2009 Recyc. and rec. routes of plastic solid waste (PSW): A review *Waste Mgmt* **29**, 2625–43
[6] Al-Salem S M, Antelava A, Constantinou A, Manos G and Dutta A 2017 A review on thermal and cat. pyro. of PSW *J. Env. Mgmt* **197**, 177-198
[7] Rehan M, Miandad R, Barakat M A, Ismail I M I, Almeelbi T, Gardy J, Hassanpour A, Khan M Z, Demirbas A and Nizami A S 2016 Effect of zeolite cat. on pyro. liq. oil *Int. Biodeterioration & Biodegrad.* **119**, 162-175
[8] Marcilla A, Beltran M I and Navarro R 2009 Thermal and cat. pyro. of polyeth. over HZSM5 and HUSY zeolites in a batch reactor und. dynamic cond. *Appl. Catalysis B: Environmental* **86**, *78–86*
[9] Olazar M, Lopez G, Amutio M, Elordi G, Aguado R, Bilbao J 2009 Influence of FCC cat. steaming on HDPE pyro. prod. dist. *J. Anal. and Appl. Pyrolysis* **85**, *359–365*
[10] Wong S L, Ngadi N, Abdullah T A T and Inuwa I M 2016 Cat. Cracking of LDPE Dissolved in Benzene Using Nickel Impr. Zeolites *Ind. & Eng. Chem. Research* **55**, 2543-55
[11] Shirazi L, Jamshidi E and Ghasemi M R 2008 The effect of Si/Al ratio of ZSM-5 zeolite on its morphology, acidity and crystal size *Cryst. Res. Technol.* **43**, No. 12, 1300 – 06
[12] Lee K H, Noha N S, Shina D H and Seo Y 2002 Comparison of plastic types for cat. degrad. of waste plastics into liq. prod. with spent FCC cat. *Polymer Degrad. and Stab.* **78**, 539–544
[13] Lee K H, Shin D H and Seo Y H 2004 Liq.-phase cat. degrad. of mix. of waste HDPE and polystyrene over spent FCC cat. Effect of mix. prop. of reactants *Polymer Degrad. and Stab.* **84**, 123–7
[14] Abbas-Abadi M S, Haghighi M N and Yeganeh H 2013 Eval. of pyro. prod. of virgin HDPE degrad. using diff. process para. in a stirred reactor *Fuel Proc. Tech.* **109**, 90–95

[15] Seo Y H, Lee K H and Shin D H 2003 Inves. of cat. degrad. of HDPE by hydrocarbon group type analysis. *J. Anal. Appl. Pyrolysis* **70**, 383–398
[16] Syamsiro M, Saptoadi H, Norsujianto T, Noviasria P, Chenga S, Alimuddin Z and Yoshikawa K 2014 Fuel Oil Prod. from MPW in Seq. Pyro. and Cat. Reforming Reactors *Energy Procedia* **47**, 180–8
[17] Farrell J T, Cernansky N P, Dryer F L, Friend D G, Hergart C A, Law C K, McDavid R M, Mueller C J, Patel A K and Pitsch H 2007 Dev. of an Exp. Database and Kinetic Models for Sur. Diesel Fuels *SAE Technical Papers*, 1-27
[18] Wong S L 2016 Crack. of Low Density Polyethy. Diss. In Benzene To Liq. Fuels Using Zeolite-Based Cat. Ph.D. Thesis. Universiti Teknologi Malaysia
[19] Sharudin S D A, Abnisa F, Daud W M and Aroua M K 2017 Energy rec. from pyro. of plastic waste: Study on non-recycled plastics (NRP) data as the real measure of plastic waste *Energy Conv. and Mgmt* **148**, 925–934
[20] Singh R K, Ruj B, Sadhukhan A K and Gupta P 2019 Impact of fast and slow pyro. on the deg. of mixed plastic waste: Prod. yield anal. and their char. *J. of the Energy Institute* **92,** Issue 6, 1647-57
[21] Efika E C, Onwudili J A and Williams P T 2015 Prod. from the high temp. pyro. of RDF at slow and rapid heating rates *J. of Anal. and App. Pyro.* **112**, 14–22
[22] Pinto F, Costa P, Gulyurtlu I, Cabrita I 1999 Pyro. of plastic wastes. 1. Effect of plastic waste comp. on prod. yield *J. Anal. and Appl. Pyro.* **51**, 39–55
[23] Biswas B, Singh R, Krishna B B, Kumar J and Bhaskar T 2017 Pyro. of azolla, sargassum tenerrimum and water hyacinth for prod. of bio-oil *Bioresource Tech.* **242**, 139–145
[24] Singh R K, Mondala S, Ruj B, Sadhukhana A K and Gupta P 2019 Interaction of three categ. of tyre waste during co-pyro.: Effect on prod. yield and qual. *J. Anal. and Appl. Pyro.* **141**, 104618
[25] Singh R K and Ruj B 2016 Time and temp. depended fuel gas gen. from pyro. of real world MPW *Fuel* **174**, 164–171
[26] Adnan, Shah J and Jan M R 2014 Thermo-catalytic pyro. of polystyrene in the pres. of zinc bulk cat. *J. of the Taiwan Inst. of Chem. Eng.* **45**, 2494–2500
[27] Miandad R, Nizami A S, Rehan M, Barakat M A, Khan M I, Mustafa A, Ismail I M I and Murphy J D 2016 Influence of temp. and reaction time on the conv. of polystyrene waste to pyro. liq. oil *Waste Mgmt* **58,** 250–9
[28] Staffell I March 2011 The Energy and Fuel Datasheet. Claverton Energy. Retrieved on 5th July 2019 from:http://www.claverton-energy.com/wordpress/wp-content/uploads/2012/08/the_energy_and_fuel_data_sheet1.pdf
[29] Onwudili J A, Insura N and Williams P T 2009 Comp. of prod. from the pyro. of polyethylene and polystyrene in a closed batch reactor: Effects of temperature and residence time *J. Anal. and Appl. Pyrolysis* **86**, 293–303
[30] Lee K H 2012 Effects of the types of zeolites on cat. upgrading of pyro. wax oil *J. Anal. and Appl. Pyrolysis* **94**, 209–14
[31] Kim Y M, Jeong J, Ryu S, Lee H W, Jung J S, Siddiqui M Z, Jung S C, Jeone J K, Jae J and Park Y K 2019 Cat. pyro. of wood polymer composites over hierarchical mesoporous zeolites *Energy Conver. and Mgmt* **195**, 727–37
[32] Lee H, Kim Y M, Jung K B, Jae J, Jung S C and Jeon J K 2018 Catalytic hydrodeoxygenation of Geodae-Uksae pyrolysis oil over Ni/desilicated HZSM-5 *J. Clean Prod.* **174**, 763–70
[33] Abbas A S and Shubar S D S 2008 Pyrolysis of High-density Polyethylene for the Prod. of Fuel-like Liquid Hydrocarbon *Iraqi J. of Chem. and Petrol. Eng.* **9** No.1, 23-29
[34] Tamri Z, Yazdi A V, Haghighi M K, Abbas-Abadi M S and Heidarinasab A 2018 The effect of temp., heating rate, initial cross-linking and zeolitic cat. as key process and structural para. on the degrad. of natural rubber (NR) to produce the valuable HC *J. Anal. and Appl. Pyro.* **134**, 35–42

[35] Das P and Tiwari P 2018 The effect of slow pyro. on the conv. of packag. waste plastics (PE and PP) into fuel *Waste Mgmt* **79**, 615-624
[36] Kamal D M and Zainuri F 2015 Green Product of Liquid Fuel from Plastic Waste by Pyrolysis at 900 °C *J. Energy and Power Eng.* **9**, 40-44
[37] Ratnasari D K, Nahil M A and Williams P T 2017 Cat. pyro. of waste plastics using staged catalysis for prod. of gasoline range HC oils *J. Anal. and Appl. Pyro.* **124**, 631-37

IOP Conf. Series: Materials Science and Engineering **808** (2020) 012026 doi:10.1088/1757-899X/808/1/012026

Evaluation of Perfluorinated Sulfonic Acid Membranes for Vanadium Redox

S S Sha'rani[1,2*], N. W. C. Jusoh[1,2], E Abouzari-Lotf[2,3], A Ahmad[2,3], and R R Ali[1,2]

[1]Malaysia-Japan International Institute of Technology, International Campus, Universiti Teknologi Malaysia, Jalan Semarak 54100, Kuala Lumpur, Malaysia.[2]Advanced Materials Research Group, Center of Hydrogen Energy, Universiti Teknologi Malaysia, Jalan Semarak 54100, Kuala Lumpur, Malaysia. [3]Department of Chemical Engineering, Universiti Teknologi Malaysia, 81310 Johor Bahru, Malaysia.

*ssophia2@graduate.utm.my

Abstract. The performance and cost of the membranes have always been crucial for the utilization of electrochemical energy devices. This article presents the properties and the performance of two commercially available and low-cost perfluorinated sulfonic acid membranes of GN115 and GN212C for vanadium redox flow battery (VRFB) application in comparison with Nafion117 (N117) membrane. The vanadium (IV) permeability of GN115 membrane was found to be close to N117, unlike GN212C membrane which showed 4 times higher vanadium (IV) permeability than N117 under similar conditions. Both GN115 and GN212C membranes showed isotropic conductivity and higher values than N117. The battery test results indicated that the high coulombic efficiency (CE) and high voltage efficiency (VE) followed the vanadium (IV) permeability and proton conductivity trends. Both membranes revealed outstanding stability in long-term charge-discharge testing under various current densities. The energy efficiency was found to be higher than N117 with values of 76.6% and 76.8% for GN115 and GN212C, respectively. The overall results suggest that both lower-cost membranes have a strong potential for VRFB application as the single-cell performance results were close to N117. The high value of proton conductivity of GN212C membrane compensates its vanadium (IV) permeability leaving reasonable battery performance.

1. Introduction

Vanadium redox flow battery (VRFB) has attracted excessive attention for renewable energy storage due to its unique set of advantages including long life cycle, absence of electrolytes contamination, deep charge and discharge capacity, fast response time, low maintenance cost and relatively large energy and power ratings [1-4]. Compared to traditional rechargeable batteries such as zinc-bromide flow battery and iron-chromium flow battery, VRFB allows the battery's power and capacity to be sized independently. However, widespread commercialization of the battery being postponed due to the high capital cost. Such high cost is mainly attributed to the cost of components' materials including membranes [5]. Therefore, developing low-cost membranes is one of the major challenges to prompt the development of high-performance energy storage systems.

In VRFB, the membrane acts as a barrier to prohibit the undesired permeability of negative and positive electrolytes and transfer protons during the passage of current [1, 6]. Therefore, to achieve a high-performance battery, the membrane needs to possess low vanadium permeability without scarifying the desired proton conductivity and excellent mechanical and chemical stability in harsh acidic electrolytes in addition to having an affordable cost. Currently, perfluorinated sulfonic acid (PFSA) membranes representing by Nafion® are the most widely used material in VRFB, displaying outstanding mechanical and chemical stability and high conductivity [8-11]. However, despite its desired advantages, Nafion based membranes are challenged by high vanadium ion crossover in addition to high cost [7, 8]. Undesired crossover leads to a reduction in coulombic efficiency due to the self-discharge phenomenon [9]. This situation further emphasizes the need for the development of new alternative cost-effective membranes to prompt the mass deployment of VRFB technology in the market place.

The majority of newly researched membranes for VRFB are hydrocarbon-based materials [10-15]. Typically, sulfonated aromatic polymers based on poly(ether ether ketone) [16-18], poly(fluorenyl ether ketone)s [13, 14], polysulfones [12, 19], poly(arylene ether sulfone)s [12, 20, 21] and poly(phthalazinone ether ketone ketone)s [22, 23] were developed and tested in the battery. Although some improvements have been achieved, most of these new membranes did not meet all requirements for commercial applications. Particularly, the polar groups of the membranes are not stable enough in the highly oxidative environment of the battery [24-26]. Moreover, most of these membranes are prepared in a lengthy multistep procedure that adversely affects the cost-effectiveness [5, 14]. Therefore, it seems that the technically and industrially viable alternative to improve the battery performance is the use of commercially available and low-cost ion exchange materials with established technology.

In this work, two newly developed low-cost commercial PFSA membranes named as GN115 and GN212C were investigated for application in VRFB. Unlike Nafion117, these membranes were manufactured through the casting method instead of extrusion. The structures and properties of the membranes were initially investigated using Scanning Electron Microscopy (SEM), X-Ray Diffraction (XRD), Fourier Transform Infrared (FTIR) spectroscopy and Thermal Analysis (TGA). The efficiency of the membranes was evaluated through vanadium ion permeability and proton conductivity analysis. The preliminary behavior of the membranes in VRFB was also studied using a single cell test in comparison to Nafion 117.

2. Experimental section

2.1. Materials

The GN115 and GN212C membranes with thickness of 115 and 33 μm were provided in-kind by General Energy, Co., Ltd. GN212C is a composite membrane with special graphene formula. N117 (177 μm) was selected for comparison in this study due to the well-balanced properties of vanadium permeability and conductivity among other Nafion series. Vanadium (IV) oxide sulfate hydrate, $VOSO_4.nH_2O$ of 97% purity was purchased from Sigma Aldrich. The other chemicals used in this work were purchased from Mallinckrodt Baker. The as-received GN115 and GN212C membranes were pretreated according to the standard protocol used for N117 membrane. This involved immersion of the membranes in 3 wt. % H_2O_2 solution for 1 h at 100 °C followed by immersion in distilled water for 1 h at 100 °C and immersion in 2.0 M H_2SO_4 solution for 1 hour at 100 °C. Finally, the membranes were immersed in distilled water for 1 h at 100 °C and kept in the distilled water prior to use.

2.2 Membrane Characterization

The infrared spectra were recorded on Perkin Elmer FTIR spectrophotometer (Spectrum 100) at wave number range 600–4000 cm^{-1}. XRD of the membranes were performed on PANatytic XPert Pro (λ = 0.15406 nm) at 45 kV and 20 mA and diffraction angle range of $2\theta = 10–80^{\circ}$ with 2° min $^{-1}$. Thermal degradation of the membranes was recorded on a TGA Q50 (TA Instruments). Prior to the analysis, the membranes were vacuum-dried at 60 °C overnight and analysed with a 10 °C min^{-1} heating rate under N_2 atmosphere. Scanning Electron Microscope of GEMINI500 was used to evaluate the surface morphology of the membranes. Prior to imaging, the membranes were coated with Platinum by using Auto-fine coater (JEC-3000FC, JEOL, Japan) for 30 s at 20 mA.

The in-plane ($\boldsymbol{\sigma}_{\parallel}$)and through-plane ($\boldsymbol{\sigma}_{\perp}$) proton conductivity values of the membranes were recorded by BekkTech conductivity (BT-112, four-point probe) and home-made cells, respectively as shown in Figure 1. The home-made cell used stainless steel electrodes freshly polished with a diamond polishing machine. The membranes were tested by clamping in the cells and subsequently immersing in a beaker of 1.0 M H_2SO_4 at room temperature. A DC conductivity testing (Keithley Kickstart SMU 2400) was set to provide the voltage between the central probes and to measure the resulting current in the membrane. The resistance (R) obtained from the slope of voltage versus current plot and the proton conductivity (σ) were calculated by using equation 1:

$$\sigma(\mathrm{S/cm}) = \frac{\mathrm{L\ (cm)}}{\mathrm{R}(\Omega)\mathrm{xA}(\mathrm{cm}^2)} \quad (1)$$

where A is the area of the membranes and L is the distance between the two inner probes. The denominator of the equation is the area times resistance (AR). Anisotropic proton conductivity ratio was calculated from the values of conductivity obtained by in-plane and through-plane methods according to equation 2.

$$\boldsymbol{\sigma}_{\parallel/\perp} = \frac{\boldsymbol{\sigma}_{\parallel}}{\boldsymbol{\sigma}_{\perp}} \quad (2)$$

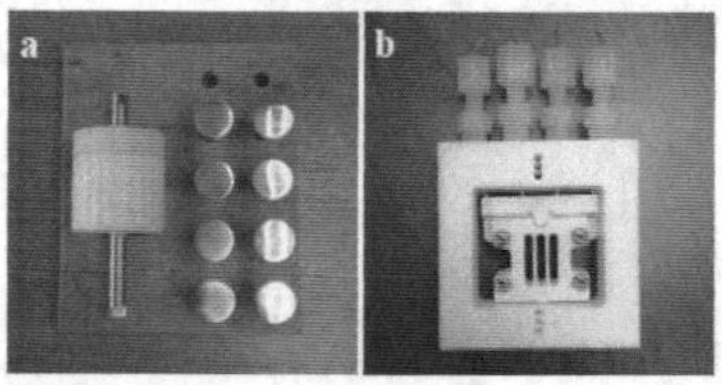

Figure 1. Conductivity cells for (a) trough-plane and (b) in-plane measurements

The permeability of vanadium (IV) across the membranes was measured with the method described elsewhere [11]. The membrane with an effective area of 1.72 cm^2 was clamped in a side-bi-side cell (Perme Gear, Inc.). The left side of the cell was filled with 1.0 M $VOSO_4$ in 2.5 M H_2SO_4 solution while the right side was filled with 1.0 Na_2SO_4 in 2.5 M H_2SO_4 solution. The solution in the right side was termed as blank solution. The solutions in both half-cells were continuously stirred at 220 RPM to avoid concentration polarization. UV-Vis spectrophotometer (UV-1800 Shimadzu) was used to monitor the concentration of permeated vanadium ion. The permeability of the vanadium ion (*P*) was obtained from the slope of the plot of the vanadium (IV) concentration ($C_{V(IV)}$) versus diffusion time (*t*) using equation 3.

$$\boldsymbol{C_{V(IV)}} = \frac{\boldsymbol{AP}}{\boldsymbol{V_R T}}\boldsymbol{C_L(t - t_0)} \quad (3)$$

where, t_0 and C_0 are the initial time and vanadium (IV) concentration in the right side of the cell, V_R is the total volume of solutions. *T* and *A* are the thickness and membrane effective area, respectively. The selectivity of ion transport in the membrane is defined as the ratio of proton conductivity and vanadium permeability.

2.3 VRFB single cell performance

The VRFB single cell was assembled by clamping the membrane with two carbon felts (5 cm x 5 cm), two PVC plates and two rubber sheets to provide the necessary pressure. The electrolyte of 1.5 $m^{V3.5+}$ ($m_{VO^{2+}} = m_{V^{3+}}$) in 3.0 m H_2SO_4 solution was used for positive and negative sides [27]. The volume of the electrolytes on each side was 22 mL and both electrolytes were cyclically pumped from the storage tank into respective half-cells with a flow rate of 10 mL min^{-1}. The temperature of the electrolytes and the cell was kept at 30 °C. The charge-discharged tests were performed under the same operating conditions and controlled by a battery testing system (MTI-BST8) with a cut-off voltage of 1.72 V and 0.8 V. Voltage efficiency (VE), coulombic efficiency (CE) and energy efficiency (EE) for a single charge-discharge cycle were calculated as follows [21, 23]:

$$CE = \frac{I_{constant}t_{discharge}}{I_{constant}t_{charge}} \text{ x } 100\% \quad (4)$$

$$VE = \frac{V_{discharge}}{V_{charge}} \text{ x } 100\% \quad (5)$$

$$EE = CE \text{ x } VE \quad (6)$$

3. Results and discussion

3.1 Characterization of membranes

The IR spectra for GN115 and GN212C membranes is shown in the Figure 2. The peaks between 1000 and 1360 cm^{-1} are attributed to the C-F absorption related bonds. Peaks at 1060 and 976 cm^{-1} represent the S–O and C–O–C groups of the PFSAs, respectively. The peak at 1637 cm^{-1} is assigned to the bending vibration of O–H due to absorbed water. It can be seen that the spectra of GN115 and GN212C membranes are identical to N117 representing perfluorinated sulfonic acid membranes and the band positions are in a good agreement with the literature values [28-30].

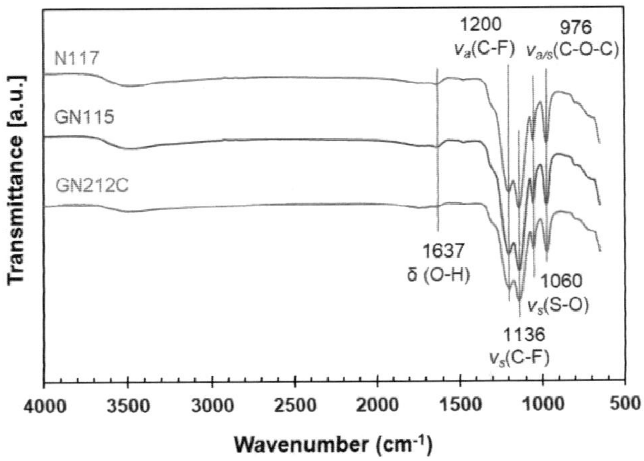

Figure 2. FT-IR spectra of GN115, GN212C and N117 membranes

The XRD curves of GN115, GN212C membranes are displayed in Figure 3 is compared with N117. The peak at 2 θ of 16–18° and the halo at 37–40°, shown in all membranes, are attributed to the PTFE-like backbone and amorphous structure of the PFSA type membranes, respectively [31]. Comparing the curves of N117, GN115 and GN212C reveals that the GN212C has slightly higher peak intensity at 22°. Such minor difference could be ascribed to the presence of graphene in the GN212C composite membrane [32, 33].

The TGA curves of GN115, GN212C and N117 membranes are shown in Figure 3a. The TGA curves of the GN115 seem to be similar to that of N117 and overlapping. Unlikely, the GN212C membrane shows a slightly different curve with the thermal degradation started at lower temperature compared to N117. This slightly lower thermal stability of GN212C composite membrane could be due to the introduction of graphene filler [34].

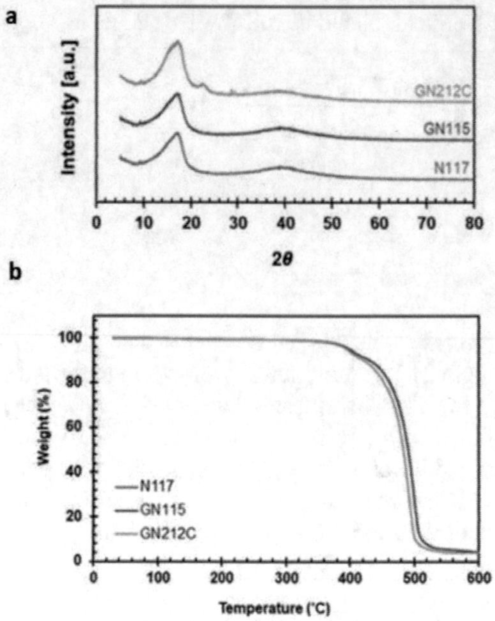

Figure 3. XRD (a) and TGA (b) curves of GN115 and GN212C membranes in comparing with N117

The surface morphology of the membranes was evaluated using SEM as shown in the Figure 4. Comparison of the SEM images of N117 with GN115 and GN212C membranes indicates that the casted membranes of GN212C and GN115 membranes have a rougher surface in compared to the N117.

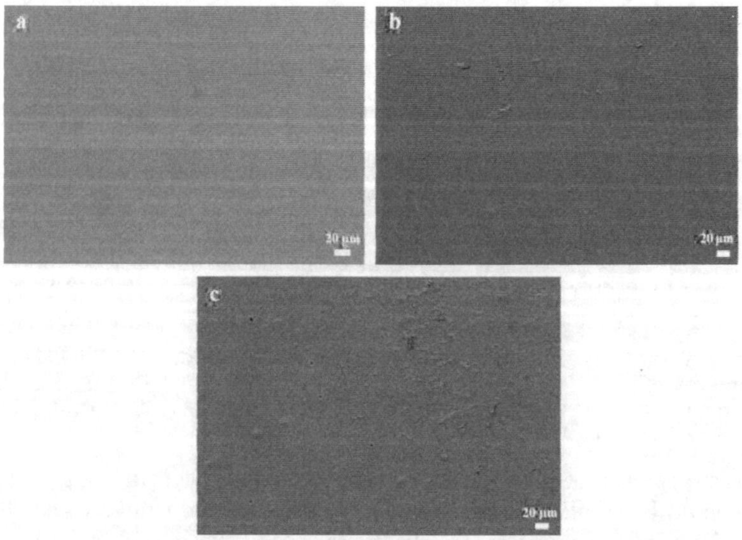

Figure 4. The surface SEM images of N117(a), GN115 (b) and GN212C (c) membranes.

3.2 Proton Conductivity, vanadium ion permeability and selectivity

To investigate the overall efficiency of the membranes in VRFB, proton conductivity, vanadium ion permeability and the single cell tests were evaluated and compared with N117. As shown in the Table 1, both of these low-cost PFSAs showing higher conductivity values and lower area resistance than N117. The composite GN212C is showing considerably higher conductivity values of as high as 108 mS cm^{-1} at room temperature. This remarkable high conductivity value is possibly due to the lower thickness and introducing of graphene-based filler. Area resistance is an imperative parameter that determine the voltage efficiency of the VRFB [1]. Therefore, the results give a good indication that the membranes could perform better than N117 in VRFB. Comparing the through-plane and in-plane conductivity values of these new PFSAs were performed to clarify the isotropic or anisotropic behaviour. The conductivity values for N117 showed anisotropic behaviour with comparable levels with the reported values of through-plane and in-plane [35, 36]. In contrast, both GN212C and GN115 show almost isotropic conductivity as only slightly differences in the through-plane and in-plane conductivity values were observed. As reported in the literature, the anisotropic proton conductivity of PFSAs is mainly depends on the membrane preparation and pre-treatment such as casting, stretching, extrusion and hot-pressing [37-40]. It could be concluded that the collection of isotropic grains is randomly oriented in the GN212C and GN115 membranes upon solution casting assisted film preparation. Whereas, the conducting domains are mainly in the plane or perpendicular to the plane of the film in Nafion117 [41].

Different vanadium ions of V^{2+}, V^{3+}, VO^{2+}, and VO_2^+ are present in a VRFB electrolyte. Permeability of VO^{2+} was evaluated to observe the trends in the new low-cost PFSAs in comparison to N117. The permeability was measured using side-bi-side cell at room temperature. Prior to the measurement, a standard curve of VO^{2+} was established using UV- Vis spectrophotometry technique and the peak of VO^{2+} was recorded at 765 nm. Figure 5 shows the plot for the variation of VO^{2+} concentration in the blank solution after crossing different membranes as a function of time. Among all membranes, GN212C showed the highest VO^{2+} permeability of $95.5 \times 10–7$ cm^2.min^{-1} which is almost 4 times higher than the values for GN115 and N117. On the other hand, the permeability of GN115 was close to N117. In fact, the composition, fabrication procedure and thickness of the membranes are controlling the vanadium permeability. In general, membranes prepared by extrusion are showing high degree of amorphous structure which results in larger fractional free volume and higher vanadium permeability than casted membranes. The extent of fabrication procedure on the permeability could be reduced due the pre-treatments of soaking in acid and water as well as exposure to electrolyte. The permeability increase trend of GN212C > GN115 > N117 that the differences in the membrane thickness and possibly microstructures are dominating the electrolyte permeability. Therefore, it is reasonable to state that the increase in the thickness cause a gradual decrease in the VO^{2+} permeability. This is due to the longer and more complicated transfer channels in thicker membranes.

The ion selectivity of the membranes, which is the ratio of proton conductivity over vanadium ion permeability, was investigated to allow an efficient determination of membrane appropriateness for VRFB that goes beyond the routine determination of the conductivity and electrolyte permeability. A higher value for ion selectivity implies enhanced performance of the membrane in VRFB. Table 1 summarizes the data for ion selectivity together with the proton conductivity, area resistance and vanadium permeability of the new PFSA membranes compared to N117. The area resistance of the membrane was calculated based on the through-plane value. The highest ion selectivity of 3.84×10^4 S min cm^{-3} was calculated for GN115 membrane which is substantially higher than the values of N117 and GN212C membranes. This confirms that the GN115 membrane has a good balance between the vanadium (VO^{2+}) permeability and the proton conductivity.

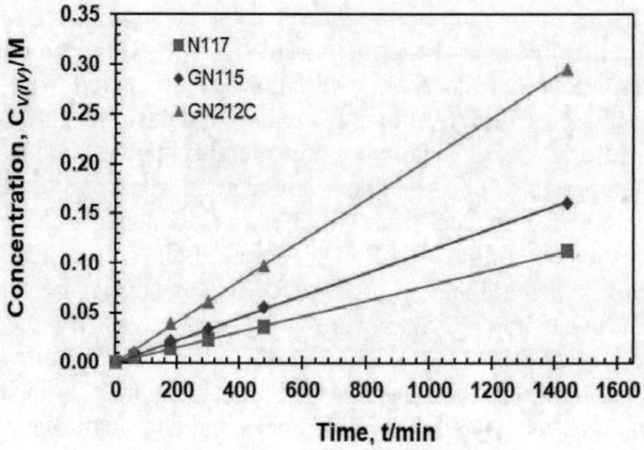

Figure 5. Time-dependent variation of VO^{2+} concentration in blank solution across the investigated membranes.

Table 1. Proton conductivity, area resistance, VO^{2+} permeability and ionic selectivity of the membranes.

Membrane	Thickness, (μm)	Proton conductivity, σ (mS/cm)		Area Resistance (Ω cm²)	Vanadium permeability ($x10^{-7}$ cm² min⁻¹)	Ion selectivity ($x10^4$ S min cm⁻³)
		Through-plane ($\sigma_{|}$)	In-plane ($\sigma_{||}$)			
GN115	115	96.1	96.0	0.12	25.0	3.84
GN212C	33	108.3	108.5	0.03	95.5	1.14
Nafion 117	177	88.0	82.6	0.20	24.0	3.52

3.3 VRFB single cell performance

The cycling performance of the battery assembled with both membranes was investigated using a charge-discharge test at current densities of 40, 60, 80 and 120 mA cm^{-2} for 10 cycles in compared with commercial N117. As shown in the typical charge-discharge cycles at 40 mA.cm^{-2} (Figure 6), both GN115 and GN212C membranes have good and stable performance in VRFB as of N117. Such cycling data were used to calculate the cell efficiency parameters of CE, VE and EE. Figure 7 shows typical efficiency parameters for the membranes at 40 mA cm^{-2}. At this current density, GN115 membrane shows the highest CE of 89.5% and highest VE of 88.2% was achieved for GN212C membrane. Substantially improvements in the CE and VE of GN115 and GN212C membranes could be attributed to the is mainly due to lower vanadium crossover in GN115 and area resistance in GN212C, respectively. In fact, very high vanadium permeability and very low area resistance of GN212C are direct consequences of membrane thickness. On the other hand, the EE of GN115 and GN212C was found to be close to N117 at 40 mA cm^{-2}, suggesting a high potential of these low-cost membranes due to the proper balance between the proton conductivity and the vanadium permeability.

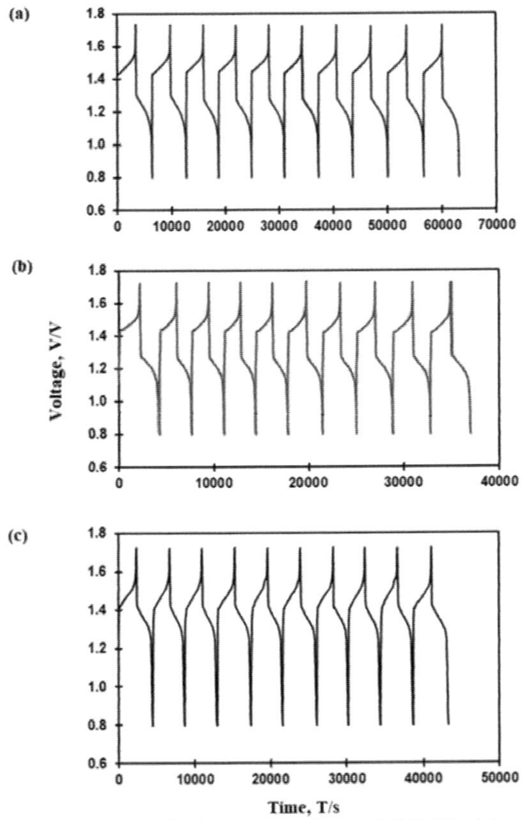

Figure 6. Comparison of the charge-discharge cycles of N117 (a), GN115 (b) and GN212C (c) membranes at 40 mA cm^{-2}.

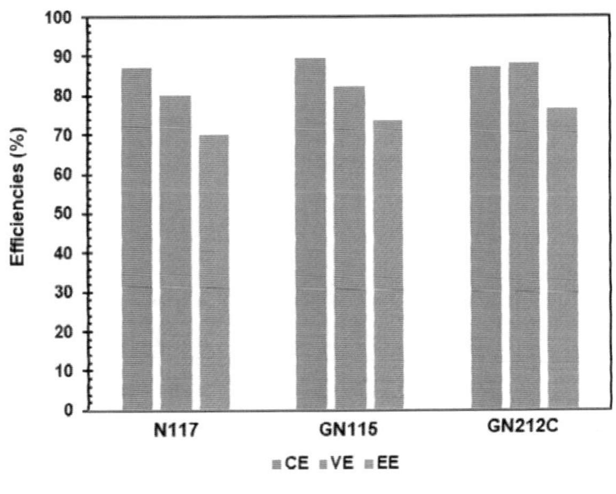

Figure 7. Efficiencies of the VRB employing low-cost PFSA membranes in compared to N117 at 40 mA cm^{-2}.

In order to evaluate the chemical and electrochemical stability of these two new membranes over long time and higher current densities, 10 cycles of charge-discharge test were conducted at each current density of 40, 60, 80 and 120 mA cm^{-2}. Stable CE, VE and EE values were obtained and the average values for the parameters were plotted in Figure 8.

In general, all membranes show increased in CE when the current density value increased. At lowest current density, the CE of GN115 is slightly higher than N117, however the CE value of N117 overtake the CE value of GN115 when the current density increased to 60 mA cm^{-2}. The CE of the membranes is greater at higher current density due to shorter time required to achieve an equal state of charge [42]. In comparison to the reported membranes in the literature, average CE was comparable. Typically, the CE of GN115 is higher than Nafion 115 [43] and porous polyvinylidene fluoride membranes [44] at 80 mA cm^{-2} and sulfonated poly(ether ether ketone) (SPEEK) at 40 mA cm^{-2} [45]. Unlike CE, the values of VE decreased rapidly when the current density is increased. This can be due to a higher ohmic losses when applying higher current. In addition, the values for EE of N117, GN115 and GN212C decreased steadily with higher current density due to the decreased in the VE. From the results, it can be concluded that the major factor for the EE of the membranes, comes from value of VE. However, the EE is still comparable with diverse cation exchange membranes reported in the literature e.g. N115, N112, N1135 [43], sulfonated polyimide [46] amphoteric ion exchange membrane [47] and SPEEK [45, 48]. Generally, the results show that the both low-cost PFSA membranes have excellent stability and good performance in the redox flow battery and can be employed in industrial applications.

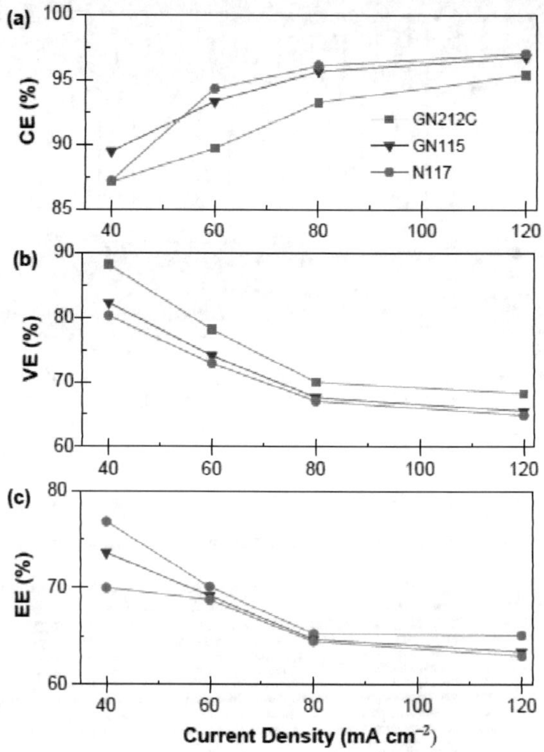

Figure 8. Cyclic performance of the VRB single cell assembled N117, GN115 and GN212C.

4. Conclusion

In this work, newly commercialized low-cost perfluorinated sulfonic acid membranes of GN115 and GN212C were investigated for potential VRFB application in comparison with well-established N117 membrane. The presence of graphene filler in GN212C membrane was led to an obvious enhancement in its conductivity. The results of SEM revealed that these new solution casted membranes have more roughed surface compared to the N 117. The GN212C membrane showed the highest proton conductivity compared to both N117 and GN115 membranes. The GN115 membrane showed a vanadium permeability of at par with N117 membrane. Although GN212C membrane has the highest vanadium permeability, the high proton conductivity and small area resistance can still be considered for VRFB application. The efficiencies of the GN115 and GN212C at 40 mA.cm^{-2} were also close to the N117 suggesting a strong potential for the membrane to be used in VRFB.

Acknowledgment
The authors wish to acknowledge the financial support from Fundamental Research Grant (FRGS) awarded by Ministry of higher education (Vot.#4F775). The authors also wish to thank General Energy Co., Ltd for providing the samples of GN212C and GN115 membranes.

References

[1] B. Schwenzer, J. Zhang, S. Kim, L. Li, J. Liu, Z. Yang 2011 Membrane development for vanadium redox flow batteries, *ChemSusChem* **4** 1388-1406
[2] A.Z. Weber, M.M. Mench, J.P. Meyers, P.N. Ross, J.T. Gostick, Q.H. Liu 2011 Redox flow batteries: a review *Journal of Applied Electrochemistry* **41** 1137-1164
[3] L.H. Yu, F. Lin, L. Xu, J.Y. Xi 2016 A recast Nafion/graphene oxide composite membrane for advanced vanadium redox flow batteries, *Rsc Advances* **6** 3756-3763
[4] R. Ye, D. Henkensmeier, S.J. Yoon, Z. Huang, D.K. Kim, Z. Chang, S. Kim, R. Chen 2017 Redox Flow Batteries for Energy Storage: A Technology Review *Journal of Electrochemical Energy Conversion and Storage* **15** 010801-010801-010821.
[5] R. Ye, D. Henkensmeier, S.J. Yoon, Z. Huang, D.K. Kim, Z. Chang, S. Kim, R. Chen 2017 Redox Flow Batteries for Energy Storage: A Technology Review *Journal of Electrochemical Energy Conversion and Storage* **15** 2017
[6] L. Cao, A. Kronander, A. Tang, D.W. Wang, M. Skyllas-Kazacos 2016 Membrane Permeability Rates of Vanadium Ions and Their Effects on Temperature Variation in Vanadium Redox Batteries *Energies* **9** 2016
[7] G.-J. Hwang, S.-W. Kim, D.-M. In, D.-Y. Lee, C.-H. Ryu 2018 Application of the commercial ion exchange membranes in the all-vanadium redox flow battery *Journal of Industrial and Engineering Chemistry* **60** 360-365
[8] J. Vrána, J. Charvát, P. Mazúr, P. Bělský, J. Dundálek, J. Pocedič, J. Kosek 2018 Commercial perfluorosulfonic acid membranes for vanadium redox flow battery: Effect of ion-exchange capacity and membrane internal structure *Journal of Membrane Science* **552** 202-212
[9] J. Dai, Y. Dong, P. Gao, J. Ren, C. Yu, H. Hu, Y. Zhu, X. Teng 2018 A sandwiched bipolar membrane for all vanadium redox flow battery with high coulombic efficiency *Polymer* **140** 233-239
[10] I. Strużyńska-Piron, M. Jung, A. Maljusch, O. Conradi, S. Kim, J.H. Jang, H.-J. Kim, Y. Kwon, S.W. Nam, D. Henkensmeier 2017 Imidazole based ionenes, their blends with PBI-OO and applicability as membrane in a vanadium Redox flow battery *European Polymer Journal* **96** 383-392
[11] S.S. Sha'rani, E.A. Lotf, A. Ahmad, W.A.W. Ibrahim, M.M.E.-s. Nasef, R.R. Ali 2016 Stability and performance evaluation of ion-exchange membranes for vanadium redox flow battery *Jurnal Teknologi* **78** 7-12
[12] S.-W. Choi, T.-H. Kim, S.-W. Jo, J.Y. Lee, S.-H. Cha, Y.T. Hong 2018 Hydrocarbon membranes with high selectivity and enhanced stability for vanadium redox flow battery applications: Comparative study with sulfonated poly(ether sulfone)s and sulfonated poly(thioether ether sulfone)s *Electrochimica Acta* **259** 427-439
[13] Y. Wang, S. Wang, M. Xiao, D. Han, M.A. Hickner, Y. Meng 2013 Layer-by-layer self- assembly of PDDA/PSS-SPFEK composite membrane with low vanadium permeability for vanadium redox flow battery *RSC Advances* **3** 15467
[14] W.M. Xie, R.X. Zheng, X. Song, F.C. Ding, X.H. Huang, Q.D. Ling 2014 Sulfonated Poly(fluorenyl ether ketone) Membranes with Suppressed Semi-Interpenetrating Crossover

and Enhanced Proton Selectivity for Vanadium Redox Flow Batteries *Advanced Materials Research* **1004-1005** 692-695.

[15] J. Liao, Y. Chu, Q. Zhang, K. Wu, J. Tang, M. Lu, J. Wang 2017 Fluoro-methyl sulfonated poly(arylene ether ketone- co -benzimidazole) amphoteric ion-exchange membranes for vanadium redox flow battery *Electrochimica Acta* **258** 360-370

[16] D.H. Hyeon, J.H. Chun, C.H. Lee, H.C. Jung, S.H. Kim 2015 Composite membranes based on sulfonated poly(ether ether ketone) and SiO2 for a vanadium redox flow battery *Korean Journal of Chemical Engineering* **32** 1554-1563

[17] H.Y. Jung, S. Jeong, Y. Kwon 2016 The Effects of Different Thick Sulfonated Poly (Ether Ether Ketone) Membranes on Performance of Vanadium Redox Flow Battery *Journal of the Electrochemical Society* **163** A5090-A5096

[18] F. Wang, G. Wang, J. Zhang, B. Li, J. Zhang, J. Deng, J. Chen, R. Wang 2017 Novel sulfonated poly(ether ether ketone)/oxidized g-C 3 N 4 composite membrane for vanadium redox flow battery applications *Journal of Electroanalytical Chemistry* **797** 107-112

[19] B.P. Gindt, Z.J. Tang, D.L. Watkins, D.G. Abebe, S. Seo, S. Tuli, H. Ghassemi, T.A. Zawodzinski, T. Fujiwara 2017 Effects of sulfonated side chains used in polysulfone based PEMs for VRFB separator *Journal of Membrane Science* **532** 58-67

[20] L. Semiz, N.D. Sankir, M. Sankir 2014 Directly Copolymerized Disulfonated Poly(arylene ether sulfone) Membranes for Vanadium Redox Flow Batteries *Int. J. Electrochem. Sc.* **9** 3060-3067

[21] D.Y. Chen, S.J. Wang, M. Xiao, Y.Z. Meng 2010 Synthesis and properties of novel sulfonated poly(arylene ether sulfone) ionomers for vanadium redox flow battery *Energ Convers Manage* **51** 2816-2824

[22] B.G. Zhang, S.H. Zhang, D.B. Xing, R.L. Han, C.X. Yin, X.G. Jian 2012 Quaternized poly(phthalazinone ether ketone ketone) anion exchange membrane with low permeability of vanadium ions for vanadium redox flow battery application *Journal of Power Sources* **217** 296-302.

[23] S.H. Zhang, B.G. Zhang, D.B. Xing, X.G. Jian 2013 Poly(phthalazinone ether ketone ketone) anion exchange membranes with pyridinium as ion exchange groups for vanadium redox flow battery applications *J. Mater. Chem. A* **1** 12246-12254.

[24] S. Kim, T.B. Tighe, B. Schwenzer, J. Yan, J. Zhang, J. Liu, Z. Yang, M.A. Hickner 2011 Chemical and mechanical degradation of sulfonated poly(sulfone) membranes in vanadium redox flow batteries *Journal of Applied Electrochemistry* **41** 1201-1213

[25] Z. Yuan, X. Li, J. Hu, W. Xu, J. Cao, H. Zhang 2014 Degradation mechanism of sulfonated poly(ether ether ketone) (SPEEK) ion exchange membranes under vanadium flow battery medium *Physical chemistry chemical physics : PCCP* **16** 19841-19847

[26] Z. Yuan, X. Li, Y. Zhao, H. Zhang 2015 Mechanism of Polysulfone-Based Anion Exchange Membranes Degradation in Vanadium Flow Battery *ACS applied materials & interfaces* **7** 19446-19454

[27] M. Etesami, E. Abouzari-Lotf, A. Ripin, M. Mahmoud Nasef, T.M. Ting, A. Saharkhiz, A. Ahmad 2018 Phosphonated graphene oxide with high electrocatalytic performance for vanadium redox flow battery *International Journal of Hydrogen Energy* **43** 189-197

[28] K. Feng, L. Hou, B. Tang, P. Wu 2015 Does thermal treatment merely make a H2O-saturated Nafion membrane lose its absorbed water at high temperature? *Physical chemistry chemical physics : PCCP* **17** 9106-9115

[29] J. Grosse Austing, C. Nunes Kirchner, L. Komsiyska, G. Wittstock 2016 Layer-by-layer modification of Nafion membranes for increased life-time and efficiency of vanadium/air redox flow batteries *Journal of Membrane Science* **510** 259-269

[30] Z. Liang, W. Chen, J. Liu, S. Wang, Z. Zhou, W. Li, G. Sun, Q. Xin 2004 FT-IR study of the microstructure of Nafion® membrane *Journal of Membrane Science* **233** 39-44

[31] T. Li, G.M. Zhong, R.Q. Fu, Y. Yang 2010 Synthesis and characterization of Nafion/cross- linked PVP semi-interpenetrating polymer network membrane for direct methanol fuel cell *Journal of Membrane Science* **354** 189-197

[32] C.F. Chang, Q.D. Truong, J.R. Chen 2013 Graphene as excellent support for rapid and efficient near infrared-assisted tryptic proteolysis *Colloids and surfaces. B, Biointerfaces* **104** 221-228

[33] J. Ding, W. Yan, W. Xie, S. Sun, J. Bao, C. Gao 2014 Highly efficient photocatalytic hydrogen evolution of graphene/YInO3 nanocomposites under visible light irradiation *Nanoscale* **6** 2299-2306

[34] R. Kumar, M. Mamlouk, K. Scott 2011 A Graphite Oxide Paper Polymer Electrolyte for Direct Methanol Fuel Cells *International Journal of Electrochemistry* **2011** 1-7.

[35] M.W. Verbrugge 1990 Analysis of Promising Perfluorosulfonic Acid Membranes for Fuel-Cell Electrolytes *Journal of The Electrochemical Society* **137** 3770

[36] N. Yoshida, T. Ishisaki, A. Watakabe, M. Yoshitake 1998 Characterization of Flemion® membranes for PEFC *Electrochimica Acta* **43** 3749-3754

[37] T. Soboleva, Z. Xie, Z. Shi, E. Tsang, T. Navessin, S. Holdcroft 2008 Investigation of the through-plane impedance technique for evaluation of anisotropy of proton conducting polymer membranes *Journal of Electroanalytical Chemistry* **622** 145-152

[38] K.A. Mauritz, R.B. Moore 2004 State of understanding of Nafion *Chemical reviews* **104** 4535-4586

[39] J. Hou, J. Li, K.G. Wilmsmeyer, Z. Zhang, L.A. Madsen 2011 Understanding Anisotropy, Transport, and Ion Associations Inside Ionic Polymers, in: NMR Spectroscopy of Polymers: Innovative Strategies for Complex Macromolecules *American Chemical Society* **2011** 251-263

[40] J.K. Park, J. Li, G.M. Divoux, L.A. Madsen, R.B. Moore 2011 Oriented Morphology and Anisotropic Transport in Uniaxially Stretched Perfluorosulfonate Ionomer Membranes *Macromolecules* **44** 5701-5710

[41] M.J. Park, N.P. Balsara 2010 Anisotropic Proton Conduction in Aligned Block Copolymer Electrolyte Membranes at Equilibrium with Humid Air *Macromolecules* **43** 292-298

[42] Y. Li, J. Sniekers, J.C. Malaquias, C. Van Goethem, K. Binnemans, J. Fransaer, I.F.J. Vankelecom 2018 Crosslinked anion exchange membranes prepared from poly(phenylene oxide) (PPO) for non-aqueous redox flow batteries *Journal of Power Sources* **378** 338-344

[43] B. Jiang, L. Wu, L. Yu, X. Qiu, J. Xi 2016 A comparative study of Nafion series membranes for vanadium redox flow batteries *Journal of Membrane Science* **510** 18-26

[44] W. Wei, H. Zhang, X. Li, H. Zhang, Y. Li, I. Vankelecom 2013 Hydrophobic asymmetric ultrafiltration PVDF membranes: an alternative separator for VFB with excellent stability *Physical chemistry chemical physics : PCCP* **15** 1766-1771

[45] S. Winardi, S.C. Raghu, M.O. Oo, Q. Yan, N. Wai, T.M. Lim, M. Skyllas-Kazacos 2014 Sulfonated poly (ether ether ketone)-based proton exchange membranes for vanadium redox battery applications *Journal of Membrane Science* **450** 313-322

[46] J. Li, X. Yuan, S. Liu, Z. He, Z. Zhou, A. Li 2017 A Low-Cost and High-Performance Sulfonated Polyimide Proton-Conductive Membrane for Vanadium Redox Flow/Static Batteries *ACS applied materials & interfaces* **9** 32643-32651

[47] J. Qiu, M. Zhai, J. Chen, Y. Wang, J. Peng, L. Xu, J. Li, G. Wei 2009 Performance of vanadium redox flow battery with a novel amphoteric ion exchange membrane synthesized by two-step grafting method *Journal of Membrane Science* **342** 215-220

[48] R. Gan, Y. Ma, S. Li, F. Zhang, G. He 2017 Facile fabrication of amphoteric semi-interpenetrating network membranes for vanadium flow battery applications *Journal of Energy Chemistry* **2017**

IOP Conf. Series: Materials Science and Engineering **808** (2020) 012027 doi:10.1088/1757-899X/808/1/012027

Characterization of ozone production from multi-cylinder reactor in non-thermal plasma device using multivariable power least squares method

L S Tan[1], M T Lim[2,*] and W Y Tey[1,3]

[1]Malaysia-Japan International Institute of Technology, Universiti Teknologi Malaysia, 54100, Kuala Lumpur, Malaysia.

[2] Biomass Plasma Technologies, Renewable Energy and Green Technology, TNB Research Sdn. Bhd., 43000 Kajang, Selangor, Malaysia.

[3]Faculty of Engineering, UCSI University, 56000 Kuala Lumpur, Malaysia.

* mook.tzeng@tnb.com.my

Abstract. Other than usage for environmental remediation, ozone is also increasingly studied for its potential in combustion enhancement. In this study, the characterization of the ozone production at varied voltage, duty cycle and cylinder configuration for reactor was conducted using multivariable power least squares method (MPLSM). This alternative correlation method in the form of power function was applied in this study to determine the dominant factor affecting the ozone production using the multi-cylinder reactor. The regressed equation using MPLSM method indicated voltage as the dominant factor in the production of ozone compared to the effect of duty cycle. The correlations generated from MPLSM for both reactor configurations were able to predict most of the ozone concentration results within 25% deviation from the actual experimental data. As such, MPLSM could be considered as an alternative method to be used for correlations of non-polynomial results.

1. Introduction

With the current trend of intensified, clean and safe process [1, 2], ozonation is gaining attention as alternative for environmental remediation [3-5]. Ozone was also found to effectively decompose volatile organic compounds (VOC) and removing pollutant gases as well as odour treatment [6-9].

Recently, ozone is also studied for potential in combustion enhancement [10-13]. Ombrello *et* al. [10] identified that development of new technologies to enhance combustion should also cover beyond high speed propulsion devices. Stationary power generation and ultra-lean, ultra-low emission internal combustion engines are some of the areas which can be further improved through new combustion technologies such as plasma-assisted combustion technique [10]. Plasma activation offers the benefit of decreased ignition times and lower ignition temperatures, enhanced flame stabilization, increased flame propagation and increased flammability limit in combustion systems [11, 12, 14]. The improved combustion could be achieved via the presence of free electrons, ions, active radicals and excited molecules as well as thermal enhancement generated from the corona discharge [10, 12]. As such, modification of traditional combustion techniques by incorporating non-thermal plasma in the system is a promising area to be explored. Non-thermal plasma was found to be effective in producing ozone through excitation of energetic electrons [7].

This work is a continuation from our previous study on characterization of ozone production from multi-cylinder reactor in a non-thermal plasma device [15]. In our previous study, ozone production experiment was conducted, and the data collected was characterized using response surface methodology (RSM). However, it was found that the configuration was biased towards one of configuration. Consequently, the predicted results for another configuration was out of range.

MPLSM was selected as an alternative method for the characterization in this study because the trend of ozone generation is not always in polynomial form as it depends on the parameter or combination of parameters varied [8, 17, 18]. However, RSM usually analyses responses based on polynomial trend only. MPLSM is able to solve multivariables problems using power function, in which the indices of the parameters are indicated as real numbers. It is able to approximate the indices of the variables using simple matrix solution. The magnitude of the indices could also be used to indicate the variable dominance ranking.

In this study, the same data which was collected in our previous study was characterized for the ozone production using multivariable power least squares method (MPLSM) [16]. The accuracy of the regressed equation from the method would be compared with the regression from RSM.

2. Methodology

2.1 Regression using MPLSM

The experimental parameters from our previous study [15], as shown in Table 1, was used in this study. The data was obtained from variation of numerical factors, i.e. the range of voltage between of 12 kV to 16 kV and duty cycle from 10 to 30. The similar set of variation was conducted in categorical factor of multi-cylinder reactor configuration, i.e. 3 x 40 mm and 10 x 10 mm.

The evaluation on the effect of voltage and duty cycle to the ozone concentration produced from the multi-cylinder reactor was conducted for each configuration of cylinders using MPLSM [16]. The method considered the independent numerical variables, x1, x2... xm, in which m is the total number of set in the independent variables, can be correlated with the dependent variable, y using the multiplication of respective power function of the independent variables as shown in Equation (1):

$$y \approx y^h = a x_1^{b_1} x_2^{b_2} \cdots x_m^{b_m} \tag{1}$$

Equation (1) was then expanded in logarithmic form as shown in Equation (2) before the deployment of least squares method

$$\text{In}\left(y^h\right) = \text{In}(a) + b_1 \text{In}(x_1) + b_2 \text{In}(x_2) + \cdots + b_m \text{In}(x_m) \tag{2}$$

By taking A = In(a) and $B_j = b_j$, with the subscript j as the number of set of variables, the square of least squares residual was formulated as:

$$R^2 = \sum_{i=1}^{n} \left(In(y_i) - In\left(y_i^h\right) \right)^2 = \sum_{i=1}^{n} \left(In(y_i) - \text{A} - \text{B}_1 In(x_{1,i}) - \text{B}_2 In(x_{2,i}) - \cdots - \text{B}_m In(x_{m,i}) \right)^2 \tag{3}$$

Upon expansion of Equation (3), solution for A was obtained by taking the minimization of R^2 with respect to A, in which:

$$\text{A} = \frac{1}{n} \left(\left(\sum_{i=1}^{n} \text{In}(y_i) \right) - \sum_{j=1}^{m} \left(\text{B}_j \sum_{i=1}^{n} \sum_{j=1}^{m} \text{In}\left(x_{j,i}\right) \right) \right) \tag{4}$$

IOP Conf. Series: Materials Science and Engineering **808** (2020) 012027 doi:10.1088/1757-899X/808/1/012027

where i is the ith elements within the set of variables while n is the total number of elements within the set of variables. The minimization of R^2 with respect to $B_1, B_2, \ldots, B_m$ would yield the set of equations which was solved using matrix.

Table 1. Parameters varied in this study [15].

Run	Voltage (kV)	Duty cycle	Configuration (mm)
1	16	20	10x10
2	16	10	3x40
3	14	20	10x10
4	12	10	3x40
5	12	30	3x40
6	16	30	10x10
7	14	30	10x10
8	12	30	10x10
9	14	20	3x40
10	14	20	3x40
11	12	10	10x10
12	16	20	3x40
13	14	20	10x10
14	12	20	10x10
15	16	10	10x10
16	14	10	3x40
17	16	30	3x40
18	14	20	3x40
19	14	30	3x40
20	14	10	10x10
21	14	20	10x10
22	14	20	10x10
23	14	20	10x10
24	14	20	3x40
25	14	20	3x40
26	12	20	3x40

$$\begin{pmatrix} \gamma_1 & -\lambda_{12} & \cdots & -\lambda_{1m} \\ -\lambda_{21} & \gamma_2 & \cdots & -\lambda_{2m} \\ \vdots & \vdots & \ddots & \vdots \\ -\lambda_{m1} & -\lambda_{m2} & \cdots & \gamma_m \end{pmatrix} \begin{pmatrix} B_1 \\ B_2 \\ \vdots \\ B_m \end{pmatrix} = \begin{pmatrix} \xi_1 \\ \xi_2 \\ \vdots \\ \xi_n \end{pmatrix} \quad (5)$$

$$\gamma_j = n\left\{\sum_{i=1}^{n}\sum_{j=1}^{m}\left[\mathrm{In}\left(x_{j,i}\right)\right]^2\right\} - \left\{\sum_{i=1}^{n}\sum_{j=1}^{m}\left[\mathrm{In}\left(x_{j,i}\right)\right]\right\}^2 \quad (6)$$

$$\xi_j = n\left[\sum_{i=1}^{n}\sum_{j=1}^{m}\text{In}\left(x_{j,i}\right)\text{In}\left(y_j\right)\right]-\left[\sum_{i=1}^{n}\sum_{j=1}^{m}\text{In}\left(x_{j,i}\right)\right]\left[\sum_{i=1}^{n}\text{In}\left(y_i\right)\right] \quad (7)$$

$$\lambda_{jk} = \left[\sum_{i=1}^{n}\sum_{j=1}^{m}\text{In}\left(x_{j,i}\right)\right]\left[\sum_{i=1}^{n}\sum_{j=1}^{m}\text{In}\left(x_{k,i}\right)\right]-n\left[\sum_{i=1}^{n}\sum_{k=1}^{m}\text{In}\left(x_{j,i}\right)\text{In}\left(x_{k,i}\right)\right] \quad (8)$$

The subscript k represents the value of second subscript of the term ρ. The matrix in Equation (5) was solved using Matlab R2013a software.

3. Results and discussion

3.1 Ozone production model based on MPLSM

Based on the method described in Section 2.1, the coefficients of the model with power based correlations generated from MPLSM for configuration of 3 x 40 mm and 10 x 10 mm are shown in Equation (9) and (10), respectively.

$$\text{Ozone concentration} = 0.01\text{Voltage}^{1.84}\text{Duty cycle}^{0.0947} \quad (9)$$

$$\text{Ozone concentration} = 0.008\text{Voltage}^{1.92}\text{Duty cycle}^{0.0451} \quad (10)$$

where unit for ozone concentration is ppm, unit for voltage is kV and duty cycle is unitless.

Based on the correlations, the predicted versus actual data for multi-cylinder reactor with configuration of 3 x 40 mm and 10 x 10 mm were plotted and presented in Figure 1 and Figure 2 respectively. The predicted data from our previous correlation based on RSM is also included in the graph for comparison [15].

It is observed that the correlations generated from MPLSM for both configurations was able to predict the response close within the 25% range of prediction target. Several points were slightly outside the targeted range. However, MPLSM managed to correct the range of the predicted ozone concentration for configuration of 3 x 40 mm closer to actual data.

IOP Conf. Series: Materials Science and Engineering **808** (2020) 012027 doi:10.1088/1757-899X/808/1/012027

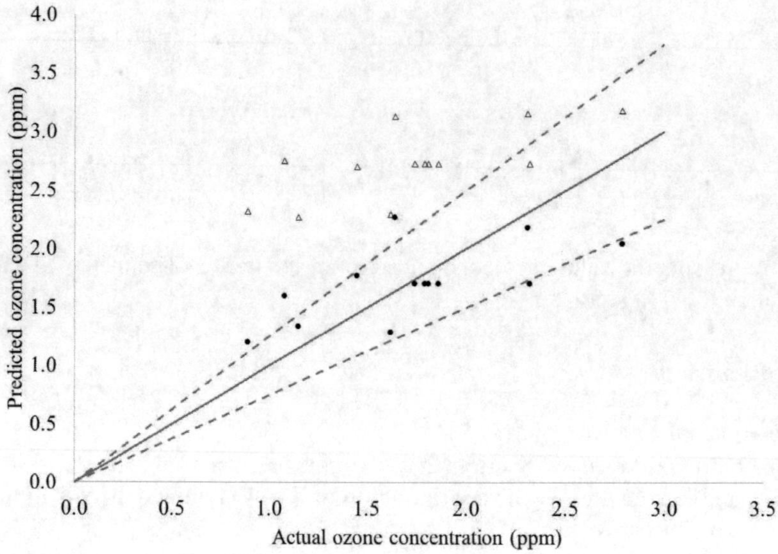

Figure 1. Comparison of predicted results using correlations based on MPLSM (•) and RSM (Δ) regression versus actual ozone concentration data generated from 3 x 40 mm electrode configuration. The solid line represents 100% accuracy while the dotted lines represent the 25% deviation from perfect accuracy.

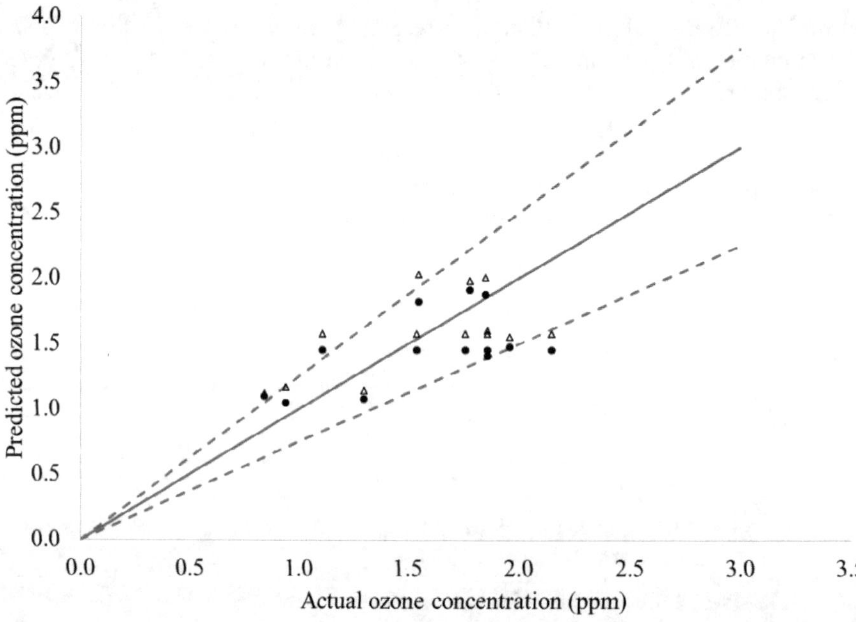

Figure 2. Comparison of predicted results using correlations based on MPLSM (•) and RSM (Δ) regression versus actual ozone concentration data generated from 10 x 10 mm electrode configuration. The solid line represents 100% accuracy while the dotted lines represent the 25% deviation from perfect accuracy.

From the indices of the variables in Equation (9) and (10), voltage was the dominant variable since it has the higher value of power compared to duty cycle. This is consistent with the ANOVA results from our previous RSM analysis [15].

Based on Figure 1 and Figure 2, it is also observed that cylinders configuration of 3 x 40 mm could produce higher ozone concentration compared to cylinders configuration of 10 x 10 mm. This was because stronger electric field was generated with the lower number of cylinders in the reactor as the same input power was divided across lesser cylinders. This led to increased electron number density which enhanced ionization and collisional excitation of plasma electrons [19].

3.2 Comparison of RSM and MPLSM for Characterization of Ozone Production

The advantage of RSM was that it could analyse the significance of both numerical and categorical factor at once. Hence, it would be beneficial to be used in analysis which required conformation of categorical factor based on statistical analysis. However, the resulting model from simultaneous numerical and categorical factor needs to be graphically verified based on set deviation target such as in Figure 1 and Figure 2. Otherwise, the biasness of model towards a certain categorical factor might be undetected.

As explained in Section 2.1, the trend of ozone production depends on the parameter or combination of parameters varied and it is not always in polynomial form. Under this circumstance, MPLSM is an alternative method to be used to correlate the variables with the response. The indices for each variable can also be used to determine the dominance of the factor in influencing the response. Based on the results obtained in this study, MPLSM appeared to be reliable in correlation of the data. However, MPLSM was unable to conduct numerical analysis for categorical and numerical factor simultaneously. As such, the significance of categorical factor would need to be analysed separately if MPLSM is employed.

4. Conclusion

The effect of voltage, duty cycle and different configuration on ozone production in multi-cylinders reactor in non-thermal plasma was characterized using MPLSM. Voltage was identified as dominant factor via analysis using MPLSM. The correlations generated from MPLSM for both configurations were able to predict the ozone concentration results close within the prediction target range. As such, MPLSM could be considered as an alternative method to be used for correlations of non-polynomial trend of results.

Acknowledgements

The authors would like to acknowledge the contribution of data from TNB Research Sdn. Bhd. and also funding from Tenaga Nasional Berhad in relation to this work (TNBR/SF195/2015). The authors are also grateful for the financial support provided by UTM (RUG Tier 2 PY/2019/00317).

References

[1] Kumar V and Deo P N K 2012 *Green Process. Synt.* **1** 79
[2] Hessel V 2015 *Green Process. Synt.* **4** 257
[3] Zeng Z, Zou H, Li X, Arowo M, Sun B, Chen J, Chu G and Shao L 2013 *Chem. Eng. J.* **229** 404
[4] Zhou S, Bu L, Shi Z, Bi C and Yi Q 2016 *Chem. Eng. J.* **306** 719
[5] Dudasova H, Derco J, Sumegova L, Dercova K and Laszlova K 2017 *J. Hazard. Mat.* **321** 54
[6] Chae J O, Moon S I, Sun H S, Kim K Y, Vassiliev V A and Mikholap E M 1999 *KSME Int. J.* **13** 647
[7] Moon J D and Jung J S 2007 *J. Electrostatics* **65** 660
[8] Takaki K, Hatanaka Y, Arima K, Mukaigawa S and Fujiwara T 2008 *Vacuum* **83** 128
[9] Jõgi I, Erme K, Raud J and Laan M 2016 *Fuel* **173** 45
[10] Ombrello T, Won S H, Ju Y and Williams S 2010 *Combust. Flame* **157** 1906

[11] Wang Z H, Yang L, Li B, Li Z S, Sun Z W, Aldén M, Cen K F and Konnov A A 2012 *Combust. Flame* **159** 120
[12] Liang X, Wang Z, Weng W, Zhou Z, Huang Z, Zhou J and Cen K 2013 *Int. J. Hydrogen Energy* **38** 1177
[13] Zhao F, Li S, Ren Y, Yao Q and Yuan Y 2016 *Fuel* **186** 518
[14] Zhang Y, Zhu M, Zhang Z, Shang R and Zhang D 2016 *Fuel* **186** 414
[15] Tan L S, Zulkifli A Z S and Lim M T 2018 *IOP Conf. Series: Mat. Sci. Eng.* **342** 012087
[16] Tey W Y, Lee K M, Asako Y, Tan L K and Arai N 2019 *Ain Shams Eng. J.* **In Press**
[17] Moon J D and Geum S T 1998 *IEEE Trans. Ind. App.* **34** 1206
[18] Petr L, Martin C, Vaclav B, Vaclav J and Pavel S 2005 *J. Phys. D: Applied Phys.* **38** 409
[19] Lim M, Jayapalan K K, Zulkifli A Z S, Hamdan H K L and Chin O 2017 *Int. J. Env. Sci. Dev.* **8** 50

IOP Conf. Series: Materials Science and Engineering **808** (2020) 012028 doi:10.1088/1757-899X/808/1/012028

Radiation induced graft copolymerization of amine-containing monomer onto polyethylene coated propylene for CO_2 adsorption

N A Zubair[1,2*], M M Nasef[3], T M Ting[4], E C Abdullah[1] and A Ahmad[2]

[1] Malaysia-Japan International Institute of Technology, Universiti Teknologi Malaysia, Jalan Sultan Yahya Petra, 54100 Kuala Lumpur, Malaysia
[2] Center of Hydrogen Energy, Institute of Future Energy, Universiti Teknologi Malaysia, Jalan Sultan Yahya Petra, 54100 Kuala Lumpur, Malaysia
[3] Department of Chemical Engineering, Faculty of Engineering, Universiti Teknologi Petronas, Bandar Seri Iskandar, 31750 Tronoh, Perak, Malaysia
[4] Radiation Processing Technology Division, Malaysian Nuclear Agency, 43000 Kajang, Selangor, Malaysia

* afifahzubair@gmail.com

Abstract. Modification of polyethylene coated polypropylene by radiation-induced graft copolymerization (RIGP) of N-vinylformamide film followed by hydrolysis was studied to produce polyvinylamine (PVAm) containing adsorbent for CO_2 adsorption. A number of grafting parameters such as irradiation dose, monomer concentration, and reaction temperature were investigated to obtain desired degree of grafting (DOG) prior to hydrolysis. The DOG was found to be a function of reaction parameters and achieved a desired value of 131% at 40 wt% NVF concentration in toluene, 300 kGy dose, 70°C temperature and 1h reaction time. Subsequently the hydrolysis of formamide groups in the grafted films into amine functionalities was carried out in basic medium. Fourier-transform infrared spectroscopy (FTIR) and scanning electron microscopy (SEM) confirmed the NVF monomer was successfully grafted on PE/PP and consequently the hydrolysis of amide into amine functionalities has occurred in the grafted substrates. The CO_2 adsorption capacity of PVAm modified adsorbent was brought to be a function of DOG, which is corresponding to the amine content after hydrolysis. The adsorbent showed the highest CO_2 adsorption capacity of 1.32 mmol/g when tested with gas composition of 40% CO_2 and 60% nitrogen at room temperature.

1. Introduction

CO_2 is the most serious greenhouse gas released into the atmosphere and become a critical issue over the past 50 years. This vast amount of CO_2 causes global warming that contribute to climate change include the sun's intensity, volcanic eruptions, and changes in naturally occurring greenhouse gas concentrations. Currently, liquid amine absorption, also known as amine scrubbing is the main commercial method used for CO_2 capturing. This method involves the use of monoethanolamine (EA), diethanolamine, or methyldiethanolamine solutions for CO_2 removal [1]. However, the process involves a number of serious challenges including high regeneration energy, large equipment size, amine degradation, and equipment corrosion [2].

Recently, solid adsorption has become the most preferable technique for CO_2 capturing. Among solid adsorbents, polymeric fibres containing amine functional groups have shown some advantages compared to other adsorbents [3, 4]. Modification of polymers by radiation induced grafting polymerization (RIGP) of functional monomers is a flexible method for imparting desired chemical and physical properties to various substrates and allowing tuned compositions. The grafted materials were found suitable for adsorption and separation processes and energy applications [5]. This method offers some advantages in preparation of functionalised polymeric materials such as the ability to control the graft copolymer composition by selecting the suitable grafting conditions [6], in addition to its procedural simplicity and environmental friendliness [7].

In RIGP, various substrates (sheets and fibres) have been frequently used as starting materials for development of variety of membranes and adsorbents. For example grafting of amide or amine-containing groups onto different substrates have been reported in literature and recently proposed for important materials such as adsorbents [8] and catalyst [9]. Particularly, an adsorbent for CO_2 capturing was obtained by RIG of allylamine onto polyacrylonitrile fibres and demonstrated good adsorption capacity [10]. Another CO_2 adsorbent was prepared by RIG of GMA onto PE/PP non-woven sheet followed by amination [11]. The selection of PE/PP as a substrate was motivated by its high chemical stability, mechanical strength, thermal resistance and reasonable cost and ability to form relatively stable radicals upon radiation at room temperature [12].

Recently Rojek *et al.* (2017) studied RIG of N-vinylformaide (NVF) monomer into porous ultrahigh molecular weight polyethylene (UMWPE) films and the obtained membrane was used as a CO_2 adsorbent. A noticeably high concentrations of monomer in the range of 80-100% were used [8]. Thus, it would be highly interesting to graft NVF at lower concentrations and modified with amine functionalities for CO_2 adsorbent to reduce the cost of the graft copolymer. Therefore, the objective of the present study is to investigate the behaviour of grafting reaction parameter of NVF onto fibrous PE/PP sheets and testing its hydrolysed copolymer for CO_2 adsorption. The evidence of formation of poly(NVF) in the grafted fibrous sheets is provided using Fourier-transform infrared spectroscopy (FTIR) and scanning electron microscopy (SEM). Moreover, the potential of using the obtained functionalised adsorbent for CO_2 adsorption was evaluated.

2. Materials and methodology

2.1. Materials

PE/PP fibrous sheets were obtained from Kurashiki Co., Japan. NVF with a purity of 98% was obtained from Sigma Aldrich and used without any further purification. Ethanol and toluene (reagent grade) were purchased from Merck. All chemicals were used as received without further treatment.

2.2 Irradiation of PE/PP substrates

PE/PP substrates were rinsed in ethanol for few hours under sonication, then removed and vacuum dried and placed in zipped plastic bags. The substrates were irradiated by an electron beam [4] accelerator (NHV-Nissin High Voltage, EPS3000) operated at an acceleration energy of 2 MeV, a beam current of 10 mA and a dose of 25kGy/pass.

2.3. Graft polymerization

The irradiated PE/PP substrates were transferred to an ampoule containing de-oxygenated NVF solutions diluted with toluene to various concentrations in the range of 10 wt% to 40 wt% and bubbled with N_2 for 30 minutes. Subsequently, the NVF solution was transferred to evacuated glass ampoules hosting the irradiated PE/PP substrates. The glass ampoule was sealed and kept in water bath at different temperatures (ranged from 40 to 70°C) to initiate the graft polymerization. After completion of the reaction, the grafted PE/PP substrates were removed, washed with methanol and rinsed in a methanol-containing vessel under sonication for 1h to remove the residual monomer and occluded homopolymer. The grafted substrates were dried under vacuum oven and the amount of NVF monomer grafted onto PE/PP substrates was evaluated by the DOG using equation (1).

$$Degree\ of\ grafting, \qquad DOG\ (\%) = \frac{W_f - W_i}{W_i} \times 100\% \tag{1}$$

where, W_i and W_f are the initial weights of substrates and their weight after grafting, respectively.

2.4. Hydrolysis of poly(NVF) grafted PEPP

Hydrolysis of formamide group in the grafted substrates into amine group was carried out in basic medium using 2M NaOH solution at 80°C for 4h under reflux as reported in literature [8]. After treatment, the substrates were removed, washed with deionized water repeatedly and immersed in ethanol overnight. The substrates were removed and vacuum dried at 80°C and eventually weighed.

2.5. Characterization of the grafted fibres

A phenom PRO scanning electron microscope (SEM) was used to obtain images of the substrates before and after grafting. A Perkin Elmer Spectrum One FTIR Spectrometer equipped with an attenuated total reflectance was used to analyse the substrates in the frequency range 500 – 4000 cm^{-1} at a resolution of 4 cm^{-1}. Prior to FTIR measurements, the substrates were dried in a vacuum at 40°C overnight.

2.6. CO_2 adsorption test

The adsorption measurement was performed using a Rubotherm gravimetric sorption analyser. Substrates weigh of 0.1-0.2 g of hydrolysed PE/PP-*g*-poly(VAM) were placed inside the substrate container. The substrates were pre-treated at 80°C for 2h under vacuum prior to adsorption test. The buoyancy mode was selected to obtain the weight and volume of the substrate. The buoyancy measurements were performed in N_2 gas with 500 ml/min flow rate and operating temperature of 30°C with pressure range of 0 to 30 bar. In the adsorption test, the mixture of CO_2 and N_2 gases with a total flow rate of 500 ml/min was used in 30°C temperature and pressure range of 0-30 bar.

3. Results and discussion

3.1. Effect of grafting parameters on degree of grafting

3.1.1. Effect of irradiation dose. Figure 1 shows the effect of absorbed dose on DOG of NVF onto PE/PP substrates. The grafting parameters including time, temperature and monomer concentration were set at 1h, 70°C and 20 wt% NVF. As can be seen, the DOG increases with the increase in the dose from 50 to 300kGy. This is due to the increase in the number of radicals generated in the system, and as a result, more radicals contribute to initiation reaction. This high number of radical will participate in the graft polymerisation reaction leading to higher DOG. Similar dose trends were observed for RIGP of acrylicacid onto nylon-6membrane [13] and grafting of acrylamide and acrylic acid onto nylon-6 fabric [14].

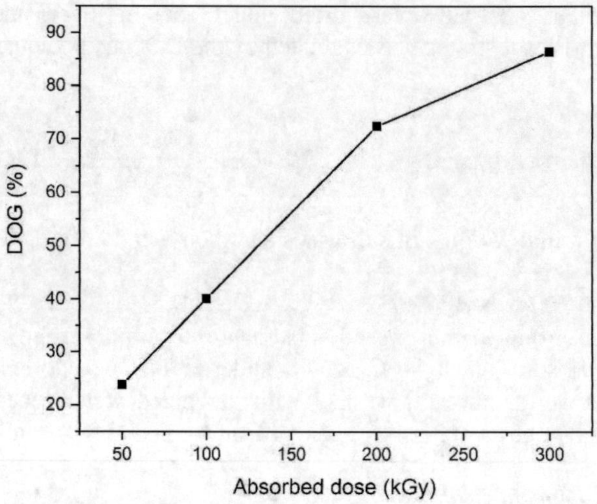

Figure 1. Variation of the degree of grafting with the absorbed dose for grafting of NVF onto PE/PP. Grafting parameters: 20 wt% GMA, 70°C and 1h

3.1.2. Effect of monomer concentration. Figure 2 shows the effect of variation of monomer concentration on DOG. It can be seen that DOG increases with the increase in monomer concentration from 10 wt% to 40 wt%.This can be attributed to the increase of the number of monomer molecules diffused to the grafting zone, leading to abundant monomers for grafting percentage. A further increase in the monomer concentration leads to arise in the viscosity that hinders the monomer diffusion to the radicals' sites and eventually reduces the rate of grafting reaction. The increasing viscosity of reaction medium enhanced the homopolymer formation as indicated by the increase in the viscosity of the grafting residues. Similar behaviour was reported for grafting of VBC onto poly(ether ketone) film [15] and butyl-spacer styrenic monomer onto ETFE [16].

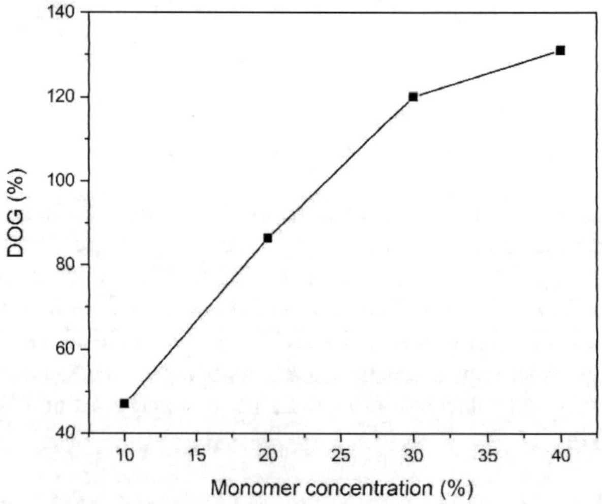

Figure 2. Variation of the degree of grafting with the monomer concentration for grafting of NVF onto PE/PP. Grafting parameters: 300kGy, 70°C and 1h.

3.1.3. Effect of reaction temperature. Figure 3 shows the effect of variation of reaction temperature on DOG of NVF onto PE/PP. The DOG increased with the increase in temperature from 40°C to 70°C. The results have shown that the rate of grafting increased sharply during the initial period of the reaction. This trend is very popular in various grafting systems and can be attributed to the acceleration in chain mobility leading to the enhancement in the monomer diffusion to the grafting zone, which is accompanied by an increase in the rate of initiation and propagation. This behaviour is going along with grafting of VBC onto poly(hexafluoropropylene- co-tetrafluoroethylene) films [17] and grafting of dimethylamino ethylmethacrylate onto PE/PP nonwoven fabric [18]. Further temperature increases up to 70°C resulting in a higher amount of homopolymer contamination on the PE/PP substrates.

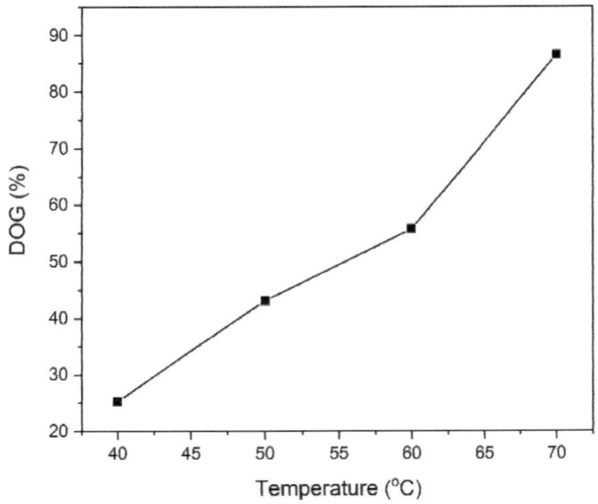

Figure 3. Variation of the degree of grafting with temperature for grafting of NVF onto PE/PP. Grafting parameters: 20 wt%, 300kGyand 1h.

3.2 Evidence of grafting

SEM images of pristine PE/PP and the corresponding poly(NVF) grafted PE/PP with two different DOG are shown in Figure 4. The average fibre diameter was found to increase by the incorporation of poly(NVF) grafts. For instance, the average diameter of fibre of the pristine PE/PP increased from 15 to 20 mm by incorporation of poly(NVF) and rise to 22 mm when DOG increased from 40 to 60. The hydrolysis resulted in no significant changes in fiber size. It can be resolved that the grafted monomer unit is assumed to remain the same before and after hydrolysis.

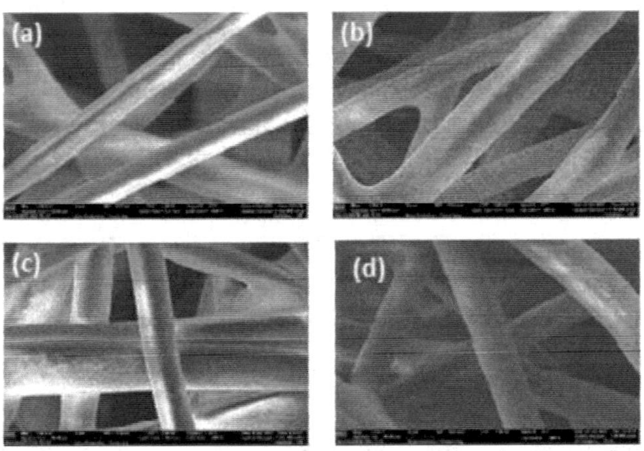

Figure 4. The SEM images of a) pristine PE/PP b) poly(NVF) grafted PE/PP with DOG of 40% and c) poly(NVF) grafted PE/PP with DOG of 60% (d) hydrolysed poly(VAm).

Figure 5 shows FTIR spectra of pristine PE/PP and grafted PE/PP substrates at different degrees of grafting. The incorporation of poly(NVF) grafts to PE/PP substrate was confirmed by the appearance of new characteristic peaks such as the absorption band at 1270 and 1354 cm^{-1} which is due to N–H stretching vibration of pendant chain of NVF molecule resulted from the interaction between N–H bending vibration and –CN stretching vibration, respectively. The peak at 1750 cm^{-1} which appeared in both grafted PE/PP with different DOG was assigned for to C=O stretching vibration of secondary amide present in the pendant chain attached to PE/PP. It is evident that NVF was successfully grafted on PE/PP fibrous substrates. The hydrolysis of amide to amine is confirmed by the disappearance of a sharp peak at 1750 cm^{-1} and the appearance of a broad peak at 3360 cm^{-1}.

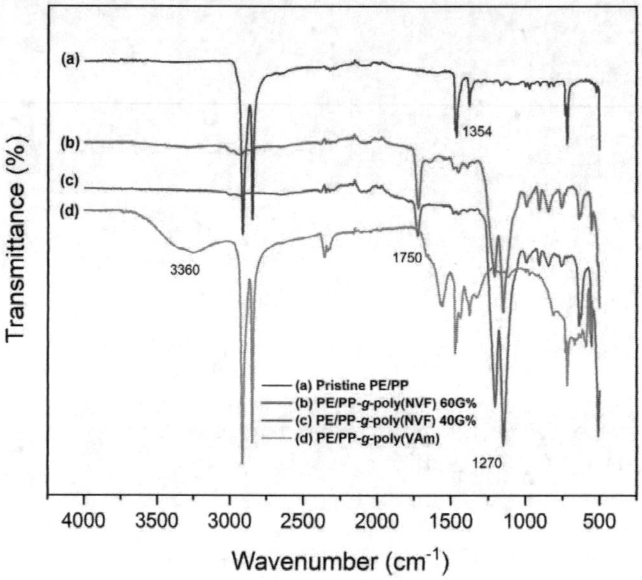

Figure 5.FTIR spectra of original PE/PP substrate, grafted PE/PP with different degree of grafting and PE/PP-*g*-poly(VAm).

3.3. CO_2 adsorption

Figure 6 shows CO_2 adsorption isotherms for pristine PE/PP substrate, PE/PP-*g*-poly(NVF) and PE/PP-*g*-poly(VAm) fibrous adsorbents (DOG = 131%) under different pressures at 30°C. As can be seen the amounts of CO_2 adsorbent increased with the increase in the pressure in all substrates. However, it can be obviously seen that CO_2 adsorption capacity of PE/PP-*g*-poly(VAm) fibrous adsorbent is far greater than that of PE/PP-g-poly(NVF) substrate which in turn has 3 times (0.25 $mmolg^{-1}$) higher CO_2 capacity than pristine PE/PP substrate (0.08 $mmolg^{-1}$) at all pressure values. The hydrolysis of amide to amine provides basic active sites on adsorbent surfaces that improves the affinity for CO_2 adsorbates.

Particularly, the isotherm of PE/PP-g-poly(NVF) showed a sharp increase until a pressure of 5 bar beyond which it continued to increase but at a slower rate and tended to level off with the pressure increase up to 30 bar. This isotherm seems to follow type I, which resembles Langmuir adsorption model in which monolayer chemisorption in microporous structure is taking place [19]. A maximum CO_2 adsorption capacity of 1.32 $mmolg^{-1}$ was achieved at 30 bar and at room temperature. It can be concluded that RIG of NVF onto PE/PP substrates and subsequent hydrolysis reaction is highly effective and shorter route for imparting CO_2 capturing group. Moreover, the CO_2 adsorption capacity of this fibrous adsorbent is a pressure dependent.

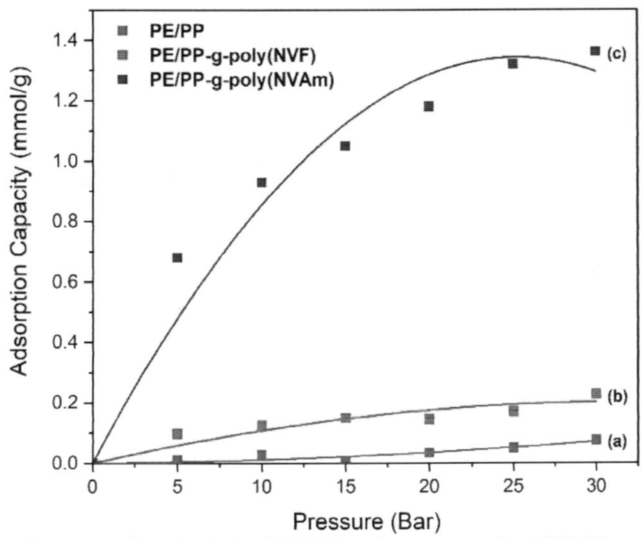

Figure 6. The adsorption capacityof original PE/PP substrate, grafted PE/PP and hydrolysed PE/PP

4. Conclusion

In this study, NVF was grafted onto pre-irradiated PE/PP by RIGP followed by hydrolysis to introduce amine functional group for CO_2 adsorption. The reaction parameter during the grafting was optimised to obtain desired DOG. The DOG was found to be the function of the reaction parameters suchas irradiation dose, monomer concentration andreaction temperature. The results of SEM and FT-IR revealed that NVF was successfully grafted onto the PE/PP nonwoven sheet, and the hydrolysed copolymer has high potential to be applied as CO_2 adsorbent.

Acknowledgement

This work was supported by Universiti Teknologi Malaysia (UTM) and Malaysia Thailand Joint Authority (MTJA) through the cost centre #R.K130000.7643.4C116.

References

[1] Singh A, Sharma Y, Wupardrasta Y and Desai K 2016 Selection of amine combination for CO2 capture in a packed bed scrubber *Resource-Efficient Technologies* **2** S165-S70

[2] Wang Q, Luo J, Zhong Z and Borgna A 2011 CO2 capture by solid adsorbents and their applications: current status and new trends *Energy & Environmental Science* **4** 42-55

[3] Zhuang L, Chen S, Lin R and Xu X 2013 Preparation of a solid amine adsorbent based on polypropylene fiber and its performance for CO2 capture *Journal of Materials Research* **28** 2881-9

[4] Zubair N A, Nasef M M, Mohamad N A, Abouzari-Lotf E, Ting T M and Abdullah E C 2020 Kinetic studies of radiation induced grafting of N-vinylformamide onto polyethylene/polypropylene fibrous sheets and testing its hydrolysed copolymer for CO2 adsorption *Radiation Physics and Chemistry* **171** 108727

[5] Nasef M M 2014 Radiation-Grafted Membranes for Polymer Electrolyte Fuel Cells: Current Trends and Future Directions *Chemical Reviews* **114** 12278-329

[6] Sithambaranathan P, Nasef M M and Ahmad A 2015 Kinetic behaviour of graft copolymerisation of nitrogenous heterocyclic monomer onto EB-irradiated ETFE films *Journal of Radioanalytical and Nuclear Chemistry* **304** 1225-34

[7] Ting T M, Nasef M M and Sithambaranathan P 2017 Kinetic investigations of emulsion- and solvent-mediated radiation induced graft copolymerization of glycidyl methacrylate onto nylon-6 fibres *Journal of Radioanalytical and Nuclear Chemistry* **311** 843-57

[8] Rojek T, Gubler L, Nasef M M and Abouzari-Lotf E 2017 Polyvinylamine-Containing Adsorbent by Radiation-Induced Grafting of N-Vinylformamide onto Ultrahigh Molecular Weight Polyethylene Films and Hydrolysis for CO2 Capture *Industrial & Engineering Chemistry Research* **56** 5925-34

[9] Ueki Y, Saiki S, Hoshina H and Seko N 2018 Biodiesel fuel production from waste cooking oil using radiation-grafted fibrous catalysts *Radiation Physics and Chemistry* **143** 41-6

[10] Yang Y, Li H, Chen S, Zhao Y and Li Q 2010 Preparation and Characterization of a Solid Amine Adsorbent for Capturing CO2 by Grafting Allylamine onto PAN Fiber *Langmuir* **26** 13897-902

[11] Nasef M M, Abbasi A and Ting T M 2014 New CO2 adsorbent containing aminated poly(glycidyl methacrylate) grafted onto irradiated PE-PP nonwoven sheet *Radiation Physics and Chemistry* **103** 72-4

[12] Sherazi T A, Ahmad S, Kashmiri M A and Guiver M D 2008 Radiation-induced grafting of styrene onto ultra-high molecular weight polyethylene powder and subsequent film fabrication for application as polymer electrolyte membranes: I. Influence of grafting conditions *Journal of Membrane Science* **325** 964-72

[13] Huang R Y M and Xu Y F 1989 Pervaporation separation of acetic acid-water mixtures using modified membranes. Part II. Gammaray-induced grafted polyacrylic acid (PAA)-nylon 6 membranes *Journal of Membrane Science* **43** 143-8

[14] Trivedi I M, Mehta P C, Rao K N and Rao M H 1975 Gamma ray-induced graft copolymerization of acrylamide and acrylic acid to nylon 6 fabric *Journal of Applied Polymer Science* **19** 1-14

[15] Hwang M-L, Song J-M, Ko B-S, Sohn J-Y, Nho Y-C and Shin J 2012 Radiation-induced grafting of vinylbenzyl chloride onto a poly(ether ether ketone) film *Nuclear Instruments and Methods in Physics Research Section B: Beam Interactions with Materials and Atoms* **281** 45-50

[16] Ponce-González J, Ouachan I, Varcoe J R and Whelligan D K 2018 Radiation-induced grafting of a butyl-spacer styrenic monomer onto ETFE: the synthesis of the most alkali stable radiation-grafted anion-exchange membrane to date *Journal of Materials Chemistry A* **6** 823-7

[17] Herman H, Slade R C T and Varcoe J R 2003 The radiation-grafting of vinylbenzyl chloride onto poly(hexafluoropropylene-co-tetrafluoroethylene) films with subsequent conversion to alkaline anion-exchange membranes: optimisation of the experimental conditions and characterisation *Journal of Membrane Science* **218** 147-63

[18] Kavaklı P A, Kavaklı C, Seko N, Tamada M and Güven O 2007 Radiation-induced grafting of dimethylaminoethylmethacrylate onto PE/PP nonwoven fabric *Nuclear Instruments and Methods in Physics Research Section B: Beam Interactions with Materials and Atoms* **265** 204-7

[19] Dąbrowski A 2001 Adsorption — from theory to practice *Advances in Colloid and Interface Science* **93** 135-224

IOP Conf. Series: Materials Science and Engineering **808** (2020) 012029 doi:10.1088/1757-899X/808/1/012029

Effect of monoethanolamine concentration on CO_2 capture by poly (chloromethyl styrene) grafted fibrous adsorbent

N A Mohamad[1,2,*], N A Zubair[1,2], E A Lotf[1,3], M M Nasef,[1,4,] A Ahmad[1,3], R R Ali[1,2], T A T Abdullah [1,3], T M Ting[5]**

[1]Advanced Materials Research Group, Center of Hydrogen Energy, Universiti Teknologi Malaysia, Jalan Semarak 54100, Kuala Lumpur, Malaysia.
[2]Malaysia-Japan International Institute of Technology (MJIIT), Universiti Teknologi Malaysia, Jalan Sultan Yahya Petra, 54100 Kuala Lumpur, Malaysia.
[3]Department of Chemical Engineering, Faculty of Chemical and Energy Engineering, Universiti Teknologi Malaysia, Johor Bahru, 81310, Johor, Malaysia.
[4]Chemical Engineering Department, Universiti Teknologi Petronas, 32610 Seri Iskandar, Perak, Malaysia.
[5]Radiation Processing Technology Division, Malaysian Nuclear Agency, 43000 Kajang, Selangor, Malaysia

*nashikinmohamad@gmail.com, **mohamed.nasef@utp.edu.my

ABSTRACT. The CO_2 adsorption performance of poly (chloromethyl styrene) (PCMS) grafted on polyethylene coated polypropylene (PE/PP) fibrous polymer substrate and followed by amination with monoethanolamine (MEA) was investigated using gravimetric sorption system. The chemical, structural and morphological changes in the aminated adsorbents were evaluated using Fourier transform infrared spectroscopy (FTIR) and X-ray diffraction (XRD), respectively. The amination was carried out using three different compositions of MEA (50, 80 and 100%) diluted in water to optimize the yield of loaded amine. The highest percent of amination obtained was about 71% and the amine content increase with the rise in MEA concentration. The CO_2 adsorption capacity increase with MEA concentration with the highest capacity recorded was1.63 mmol/g with pure CO_2 gas at 30 bars and room temperature.

1. Introduction

The development of technologies for the improvement of life style has increased the demand for more power production, which in return increases the usage of fossil fuels and subsequently increase the carbon dioxide (CO_2) emission in the atmosphere [1]. It is reported that the content of CO_2 in atmosphere is predicted to exceed 500 ppm by 2020 if no urgent combating action taken [2]. This situation has raised the concern over the removal of the CO_2 from upstream to downstream of power production processes before releasing them to the atmosphere at the allowable concentration. This can eventually contribute to the reduction of the global warming and climate change in addition to minimization of the health risk such as respiratory illnesses and asthma [1,3,4]. Therefore, several technologies for CO_2 capture have been explored including chemical absorption, physical adsorption, cryogenic methods, membrane separation, and biological fixation [5]. However, chemical absorption process is generally recognized as the most effective and matured technology for CO_2 capture and the absorption is carried by alkanolamines such as monoethanolamine (MEA), diethanolamine (DEA) and methyldiethanolamine

(MDEA), all of which react with CO_2 forming carbamate that can be decomposed to regenerate the alkanolamines [6]. In fact, aqueous MEA solution is the most frequently used alkanolamines absorbent due to its high reactivity with CO_2, low solvent cost, and can be regenerated easily [7,8]. Unfortunately, absorption with amine solutions is associated with equipment corrosion, high power consumption leading to high operational cost and poor overall thermal efficiency [9]. Alternatively, physical adsorption of CO_2 by porous adsorbents offers low energy consumption, ease of regeneration, and superior cycling capability [10]. Recent research efforts revealed new adsorbent materials such as ultrahigh porosity of metal organic framework (MOF) with very high CO_2 adsorption capacity that can reach 54.53 [11] and 58.15 mmol/g[12]. However, physical adsorption still has the disadvantage of lower CO_2 selectivity than chemical absorption in addition to processing issues including the pressure drop, dustiness, clogging, mass loss, challenges in handling and transportation, all of which hampered the industrial application of MOFs [13,14].

Therefore, chemical adsorption with amine-containing adsorbent is preferred as the performance can be sustained due to the strong bonding between the substrate and the CO_2 molecules itself leading to high CO_2 selectivity due to Lewis acid and Lewis base interaction [15]. Such strong bonding between amine groups and the adsorbent is important in minimizing the losses of amine during the adsorption and desorption cycles. Radiation induced grafting (RIG) is a facile method that can impart new chemical characteristics to polymer substrates. This is a promising modification technique because of its ability to permanently modify the substrate or adsorbent without changing their inherent properties [16,17]. Thus, this technique can be used to functionalize the inert polymer substrates with acrylic or vinyl monomers such as glycidyl methacrylate (GMA) or chlormethyl styrene (CMS), which is a conventionally known as vinylbenzyl chloride [18,19]. The objective of this study is to prepare fibrous adsorbent with different amine contents based on polyethylene coated polypropylene (PE/PP) by RIG of CMS monomer and the subsequent amination with different concentration of monoethanolamine (MEA). The CO_2 adsorption performance of the obtained adsorbents was investigated using gravimetric sorption system.

2. Materials and methods

2.1 Chemicals and Reagents

The bi-component fibers of polyethylene coated polypropylene (PE/PP) nonwoven sheet was grafted with poly(chloromethyl styrene) to yield grafted fibers (PE/PP-*g*-PCMS) following the procedure reported in literature [16] and has a degree of grafting of 137%. The irradiation dosage from electron beam was set at 100kGg for 2 hour of reaction with CMS concentration of 5wt% at 40 °C of temperature . MEA (purity ≥99%) was purchased from Merck and diluted to different concentrations (100, 70 and 50%) with deionized water (DI), which obtained from Barnstead Nanopure Diamond Lab Water Purification System (ThermoFisher, Waltham, USA). DI water was also used for samples washing purposes together with ethanol (purity of 99.6%) from Chemiz. Pure CO_2 (99.8%) and N_2 s (99.995%) gases were supplied by Linde Malaysia (Sdn Bhd) for sorption tests.

2.2 Apparatus and measurements

The changes in chemical composition after grafting and amination were analyzed using a Nicolet iS50 FT-IR spectrometer with spectra obtained in a frequency range of 500-4000 cm^{-1} using 32 scans and 4 cm^{-1} resolution. The crystalline structural changes of the samples were investigated by X-ray diffraction (XRD) using a PANalytical Empyrean analyzer at a Bragg's angle in the range of 5-70°. Both analysis was carried out using pristine, grafted PE/PP (PE/PP-*g*-PCMS) and aminated PE/PP (PE/PP-*g*-PCMS-MEA) samples. CO_2 adsorption capacity of the aminated samples were analyzed using a gravimetric sorption analyzers-isoSORP®, Rubotherm-TA instruments operated in a pressure range of vacuum to 30 bar and at 30 °C.

2.3 Amine Functionalization

A 0.1 g of PE/PP-*g*-PCMS was placed inside a 100 ml round bottom flask containing 30 ml of MEA solution. The reaction was carried out under reflux at 100 °C for 2 h of reaction. The MEA-functionalized substrates (PE/PP-*g*-PCMS-MEAx), where x refers to three different contents (loadings) of MEA amine MEA concentrations of 01=100%, 02 = 80% and 03 = 50% were used. The samples were removed and washed repeatedly with DI water and ethanol then dried at 60 °C for 5 hours. The weight changes in PE/PP-*g*-PCMS-MEAx samples were recorded before and after the reaction with an accurate weight balance to calculate the percentage of amination. The following equation was used to calculate the amine percentage:

$$\text{Percent of amination (\%)} = \frac{(\mathrm{Wa} - \mathrm{Wg})/W_{MEA}}{(\mathrm{Wa} - \mathrm{Wo})/W_{CMS}}\ x100\% \qquad (1)$$

Where W_{CMS} and W_{MEA} are the molecular weights of CMS and MEA (g/mol), respectively. W_a and W_g are the weight (g) of the PE/PP-*g*-PCMS substrate before and after amination. The W_o is the weight of PE/PP substrate before grafting.

2.4 CO_2 adsorption tests

0.1 g of PE/PP-*g*-PCMS-MEAx substrates was placed inside the sample container and the adsorption equilibrium of carbon dioxide (CO_2) were performed in a magnetic suspension balance (MSB) of gravimetric sorption analyser-isoSORP® from Rubotherm-TA instruments. Details of the basic principles, components and operational procedure of MSB was described elsewhere [20,21]. Pretreatment was carried out at temperature of 80°C for 2 h under vacuum condition to remove the moisture and impurities from the substrate followed by buoyancy measurements, which were performed at 30°C from vacuum up to 30 bar using purified nitrogen gas with a 500 ml/min flow rate. The last procedure was the adsorption equilibrium measurement, which was measured at 30°C from vacuum to 30 bar with a flowrate of 500 ml/min using pure CO_2. This procedure was repeated to five samples including pristine PE/PP, PE/PP-*g*-PCMS, PE/PP-*g*-PCMS-MEA01, PE/PP-*g*-PCMS-MEA02 and PE/PP-*g*-PCMS-MEA03.

3. Results and discussion

Figure 1 shows a reaction scheme for grafting of CMS onto PE/PP using RIG method and subsequent amination with MEA. The adsorbent precursor was obtained by irradiation of PE/PP substrate with an electron beam to form free radicals. The irradiated samples were subsequently reacted with CMS monomer by replacement of chlorine originated from covalently bonded PCMS grafts [22].

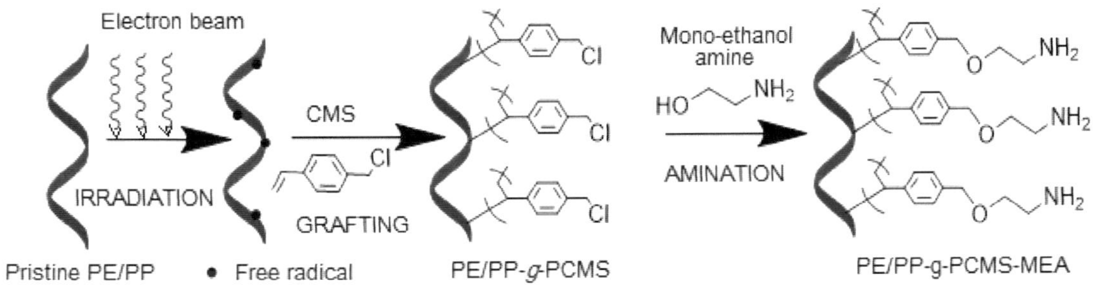

Figure 1. Reaction scheme for grafting of CMS onto PE/PP substrate and subsequent amination

3.1 Structural Characterization

The molecular structure of pristine PE/PP PCMS-grafted and MEA-loaded samples were studied using Fourier transform infrared spectroscopy (FTIR) analysis as presented in Figure 2. The spectrum of pristine PE/PP polymer structure was marked by 2 peaks of antisymmetric and asymmetric stretching vibrations of $-CH_2$ at 2917 and 2847 cm^{-1} [23]. The characteristics peaks at 838, 702 and 676 cm^{-1} from the *p*-substituents on the aromatic ring together with the peaks at 1264 cm^{-1} representing the –C-Cl groups and this clearly prove successful grafting of PCMS chains [22]. Furthermore, the appearance of the peak at 1527 cm^{-1} was assigned for primary amine from MEA, which provide a prove of a successful amine loading [24].

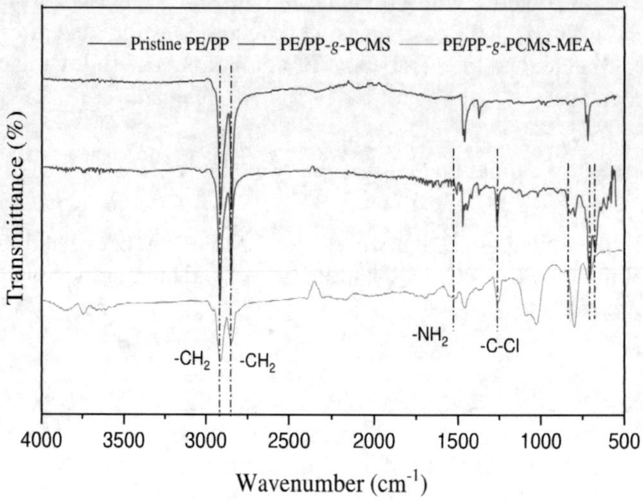

Figure 2. FTIR spectra of pristine PE/PP (black), PE/PP-g-PCMS (blue) and PE/PP-g-PCMS-MEA (red)

Figure 3 shows XRD diffractograms of pristine PE/PP, PE/PP-g-PCMS and PE/PP-g-PCMS-MEA samples. The XRD measurements were carried out to observe the changes in the crystallinity of pristine PE/PP after incorporation of PCMS, which was followed by functionalization with MEA. The position of the peaks occurred at the same angle (2θ), indicating that the structure of pristine PE/PP was well preserved even after grafting and amination procedures [25]. However, all peaks have minor shifts to higher angles suggesting minor changes in the crystal sizes after the grafting and amination. This also proves that the aminated PCMS ligands were introduced to the amorphous regions of PE/PP in a way that led to no significant contribution to the diffraction pattern. [26].

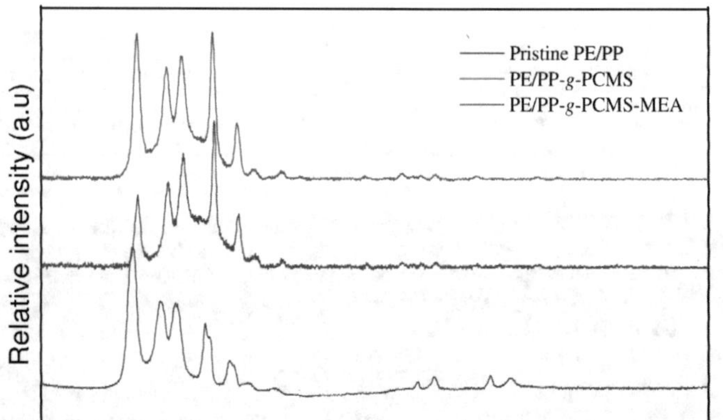

Figure 3. XRD results for pristine PE/PP (black), PE/PP-g-PCMS (blue) and PE/PP-g-PCMS-MEA (red)

IOP Conf. Series: Materials Science and Engineering **808** (2020) 012029 doi:10.1088/1757-899X/808/1/012029

3.2 Effect of variation of monoethanolamine concentration

The effect of MEA concentration on the percent of amination can be seen in Table 1. The MEA concentration that was varied from 50% to 100% aqueous solutions led to an increase in percent of amination from 50 to 71%. This could be attributed to the increase frequency of efficient collisions between amine molecules and methyl chloride from PCMS ligand due to increased concentration of amine molecules in the solution [17]. The presence of a higher content of amine groups was predicted to have a higher CO_2 adsorption capacity due to strong affinity to CO_2 molecules.

Table 1. Percent of amination when treated with different concentration of MEA

Sample	Degree of grafting, DG (%)	Solvent	Concentration of MEA (%)	Percent of amination (%)
PE/PP-*g*-PCMS-MEA01	137	-	100	71
PE/PP-*g*-PCMS-MEA02	137	water	80	58
PE/PP-*g*-PCMS-MEA03	137	water	50	50

3.3 CO_2 adsorption on adsorbents with various MEA contents

The prepared fibrous adsorbent containing MEA was tested for CO_2 adsorption by using gravimetric sorption system at elevated pressure as presented in Figure 4. The results showed an increasing CO_2 adsorption capacity trends with the increase in the pressure for all samples with pristine PE/PP and PE/PP-*g*-PCMS as the reference. Similar trend was reported in literature for another amine-modified adsorbent [27,28]. This is due to the diffusion effect, where the high pressure pushed the CO_2 molecules to have more contact/access to amine sites and enhanced the adsorption capacity [29]. To evaluate the effect of amine content on CO_2 adsorption capacity, three samples of different amine contents: PE/PP-*g*-PCMS-MEA01, PE/PP-*g*-PCMS-MEA02 and PE/PP-*g*-PCMS-MEA03 that were prepared by treated using different concentrations of MEA as presented in Table 1. The CO_2 adsorption capacity showed that amination of PE/PP substrates increase the CO_2 adsorption capacity by 159% for MEA01, 143% for MEA02 and 70% for MEA03. A similar adsorption increasing trend with the increase in MEA concentration in a sequence of MEA01 > MEA02 > MEA03 and capacity of 1.63, 1.53 and 1.07 mmol/g at pressure 30 bar, respectively. These results suggest that the higher amine content gives higher CO_2 adsorption capacity. This is going along with previous observations made on similar fibers with higher content of primary amine confirming that faster adsorption rates and higher adsorption capacities take place with high amine contents [30].

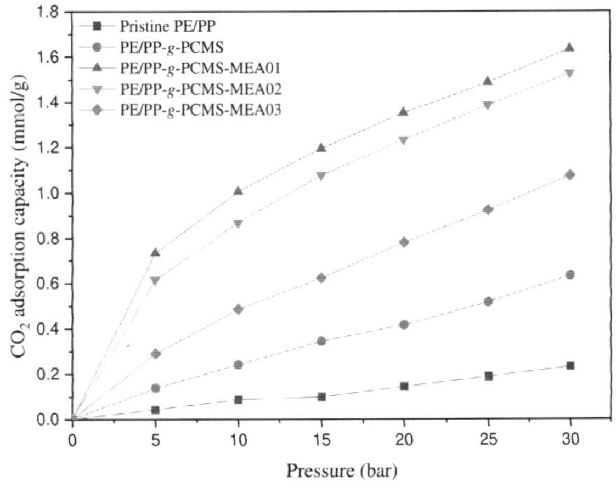

Figure 4. CO_2 adsorption capacity at different pressure for pristine, grafted and aminated PE/PP

4. Conclusion

The effect of MEA amine concentration on the CO_2 adsorption capacity of aminated PCMS containing fibrous adsorbent was successfully carried out. The percent of amination obtained from the experimental work showed that the higher concentration of MEA gives a higher yield of loaded amine with the rising trend in the order of 50% < 80% < 100% with amination percent as follows, 50%, 58% and 71, respectively. The obtained adsorbents demonstrated an increasing CO_2 adsorption capacity values of 1.63 mmol/g for MEA01, 1.53 mmol/g for MEA02 and 1.07 mmol/g for MEA03 at pressure 30 bar. The sample with the adsorption capacity of 1.63 mmol/g has a strong potential for future application in CO_2 removal after conducting the necessary kinetics, thermodynamics and stability investigations.

Acknowledgment

This work was supported by the Universiti Teknologi Malaysia (UTM) and Malaysia Thailand Joint Association (MTJA) under an industrial research grant (Vot:4C172).

References

[1] Ullah R, Patel H, Aparicio S, Yavuz C T and Atilhan M 2019 A combined experimental and theoretical study on gas adsorption performance of amine and amide porous polymers *Microporous Mesoporous Mater.* **279** 61–72

[2] Chu S and Majumdar A 2012 Opportunities and challenges for a sustainable energy future *Nature* **488** 294–303

[3] Liu Z, Liu W, Liu H, Gao T, Zhao H, Li G, Han H, Li Z, Lal R and Ning T 2019 Capture of soil respiration for higher photosynthesis with lower CO_2 emission *J. Clean. Prod.*

[4] Mazari S A, Ali B S, Jan B M, Saeed I M and Nizamuddin S 2015 International Journal of Greenhouse Gas Control An overview of solvent management and emissions of amine-based CO_2 capture technology *Int. J. Greenh. Gas Control* **34** 129–40

[5] Choi W, Seo J, Jang S, Jung J and Oh K 2009 Removal characteristics of CO_2 using aqueous MEA/AMP solutions in the absorption and regeneration process *J. Environ. Sci.* **21** 907–13

[6] Zhang P, Xu R, Li H, Gao H and Liang Z 2019 Mass transfer performance for CO_2 absorption into aqueous blended DMEA/MEA solution with optimized molar ratio in a hollow fiber membrane contactor *Sep. Purif. Technol.* **211** 628–36

[7] Mandal B P, Biswas A K and Bandyopadhyay S S 2003 Absorption of carbon dioxide into aqueous blends of 2-amino-2-methyl-1-propanol and diethanolamine *Chem. Eng. Sci.* **58** 4137–44

[8] Raja Shahrom M S, Nordin A R and Wilfred C D 2019 The improvement of activated carbon as CO_2 adsorbent with supported amine functionalized ionic liquids *J. Environ. Chem. Eng.* **7** 103319

[9] Kamarudin K S N, Zaini N and Khairuddin N E A 2018 Journal of Environmental Chemical Engineering CO_2 removal using amine-functionalized kenaf in pressure swing adsorption system *J. Environ. Chem. Eng.* **6** 549–59

[10] Gao H, Li Q and Ren S 2018 Recent advances on CO_2 capture by porous organic polymers *Curr. Opin. Green Sustain. Chem.* **16** 33–8

[11] Furukawa H, Go Y B, Aratani N, Choi E, O'Keeffe M, Yaghi O M, Ko N, Choi S B, Kim J, Yazaydin A Ö and Snurr R Q 2010 Ultrahigh porosity in metal-organic frameworks *Science (80-.).* **329** 424–8

[12] Xu G, Meng Z, Guo X, Zhu H, Deng K and Xiao C 2019 Molecular simulations on CO_2 adsorption and adsorptive separation in fullerene impregnated MOF-177 , MOF-180 and MOF-200 *Comput. Mater. Sci.* **168** 58–64

[13] Tao L, Xiao P, Qader A and Webley P A 2019 CO_2 capture from high concentration CO_2 natural gas by pressure swing adsorption at the CO2CRC Otway site, Australia *Int. J. Greenh. Gas Control* **83** 1–10

[14] Valizadeh B, Nguyen T N, Smit B and Stylianou K C 2018 Porous Metal–Organic

Framework@Polymer Beads for Iodine Capture and Recovery Using a Gas-Sparged Column *Adv. Funct. Mater.* **28** 1–6

[15] Ünveren E E, Monkul B Ö, Sarıoğlan Ş, Karademir N and Alper E 2017 Solid amine sorbents for CO_2 capture by chemical adsorption: A review *Petroleum* **3** 37–50

[16] Mahmoud Nasef M, Abbasi A and Ting T M 2014 New CO_2 adsorbent containing aminated poly(glycidyl methacrylate) grafted onto irradiated PE-PP nonwoven sheet *Radiat. Phys. Chem.* **103** 72–4

[17] Abbasi A, Mahmoud M, Kheawhom S, Faridi-majidi R, Takeshi M, Abouzari-lotf E and Choong T 2019 Amine functionalized radiation induced grafted polyole fi n nano fi bers for CO_2 adsorption *Radiat. Phys. Chem.* **156** 58–66

[18] Nasef M M and Güven O 2012 Radiation-grafted copolymers for separation and purification purposes: Status, challenges and future directions *Prog. Polym. Sci.* **37** 1597–656

[19] Mohamad N A, Lotf E A, Nasef M M and Ahmad A 2019 A comparison of CO_2 adsorption behaviour of mono- and diamine-functionalised adsorbents *E3S Web Conf.* **90** 8

[20] Chen M, Kang Y, Zhang T, Li X, Wu K and Chen Z 2018 Methane adsorption behavior on shale matrix at in-situ pressure and temperature conditions: Measurement and modeling *Fuel* **228** 39–49

[21] Lee S P, Mellon N, Shariff A M and Leveque J M 2018 High-pressure CO_2-CH_4 selective adsorption on covalent organic polymer *J. Nat. Gas Sci. Eng.* **50** 139–46

[22] Ting T M, Nasef M M and Aw E E L 2019 Kinetics of radiation grafting of glycidyl methacrylate and vinylbenzyl chloride onto polymer fibers *J. Eng. Sci. Technol.* **14** 646–58

[23] Kavaklı C, Barsbay M, Tilki S, Güven O and Kavaklı P A 2016 Activation of Polyethylene/Polypropylene Nonwoven Fabric by Radiation-Induced Grafting for the Removal of Cr(VI) from Aqueous Solutions *Water. Air. Soil Pollut.* **227**

[24] Yu J, Le Y and Cheng B 2012 Fabrication and CO_2 adsorption performance of bimodal porous silica hollow spheres with amine-modified surfaces *RSC Adv.* **2** 6784

[25] Wang Y, Du T, Song Y, Che S, Fang X and Zhou L 2017 Amine-functionalized mesoporous ZSM-5 zeolite adsorbents for carbon dioxide capture *Solid State Sci.* **73** 27–35

[26] Zhang G, Zhao P, Hao L and Xu Y 2018 Amine-modi fi ed SBA-15 (P): A promising adsorbent for CO_2 capture *J. CO_2 Util.* **24** 22–33

[27] Chi Y, Zhang Y, Li G, Zhang Q, Zhao C, Liu S, Yuan L, Liu S and Song Y 2019 CO_2/CH_4 adsorption property on shale from China for ESGR operation *Energy Procedia* **158** 5396–401

[28] Oliveira R J, De Conto J F, Oliveira M R, Egues S M S, Borges G R, Dariva C and Franceschi E 2019 CO_2/CH_4 adsorption at high-pressure using silica-APTES aerogel as adsorbent and near infrared as a monitoring technique *J. CO2 Util.* **32** 232–40

[29] Younas M, Sohail M, Kong L L, Bashir M J K and Sethupathi S 2016 Feasibility of CO_2 adsorption by solid adsorbents: a review on low-temperature systems *Int. J. Environ. Sci. Technol.* **13** 1839–60

[30] Zhuang L, Chen S, Lin R and Xu X 2013 Preparation of a solid amine adsorbent based on polypropylene fiber and its performance for CO_2 capture *J. Mater. Res.* **28** 2881–9

IOP Conf. Series: Materials Science and Engineering **808** (2020) 012030 doi:10.1088/1757-899X/808/1/012030

Silver Nanoparticles on Pullulan derived via Gamma Irradiation Method: A Preliminary Analysis

M.S.N. Salleh[1,2], R.Rasit Ali[1]*, K. Shameli[1], M.Y. Hamzah[3] and J.Z. Chan[1]

[1]Department of Chemical Process Engineering, Malaysia-Japan International Institute of Technology, Universiti Teknologi Malaysia, Jalan Sultan Yahya Petra, 54100 Kuala Lumpur, Malaysia
[2]Faculty of Chemical Engineering, Universiti Teknologi MARA Cawangan Terengganu, Bukit Besi Campus, 23200 Dungun, Terengganu, Malaysia
[3]Radiation Processing Technology Division Malaysia Nuclear Agency, Bangi Kajang Malaysia

*roshafima@utm.my

Abstract. Radiation induced technique was employed during the synthesis of silver nanoparticles (AgNPs) on pullulan. In this study, the Ag-NPs on pullulan by γ irradiated process is performed to reduce Ag+ ions at the ambient temperature without using excessive reducing agents or producing unwanted by-products of the reductant. Moreover, reducing agent can be uniformly distributed in the solution and AgNPs are produced in highly pure and stable form. The results from ultraviolet-visible spectroscopy (UV-vis) and XRD demonstrated that the silver nanoparticles can be synthesize using pullulan. This can be confirmed by absorption band of UV-vis spectrum at 420 nm as well as the XRD pattern at (1 1 1), (2 0 0), (2 2 0), and (3 1 1) planes of silver. In addition, the pullulan also acts as a reducing and stabilizing agent. TEM images showed formed AgNPs are spherical in shape with smooth edges. The TEM also revealed that the increasing radiation dose decreases the particle size and increase the rate of reduction. It was found that the mean diameter of silver nanoparticles was about 3.98 -14.87 nm. In addition, the size of the AgNPs is believed can be tuned by controlling the radiation doses. The γ-rays doses also provide the powerful reduction process in a harmless condition by producing a very stable silver nanoparticles.

1. Introduction

Evidence suggest that silver nanoparticles is among the most important metallic in nanotechnology field. This is due to its unique properties which can be incorporated into antimicrobial applications, biosensor materials, composite fibers, cryogenic super- conducting materials, cosmetic products, and electronic components. In addition, the Ag-NPs can be supported on wide range of solid precursor such as silica, metals/metal oxide, carbo, polymer, chitosan and natural fibers [1].

Silver nanoparticles are conventionally produced by the reduction of silver ions from silver salt precursors, with silver nitrate the most frequently used. Reducing agents can be physical (such as UV irradiation, γ-ray irradiation, microwave irradiation, thermal treatment, photochemical process, and sonochemical process), chemical (such as sodium borohydride (NaBH4), dimethyl formamide (DMF), triethanolamine, hydrazine, etc.), or mixed (hydrothermal)[2]. Moreover, biological materials such as plant extracts, bacteria, fungi, and yeast have been used as mediators for the synthesis of silver nanoparticles on extracellular or intracellular level [3].

In recent years, there has been an increasing interest in the fabrication of metal nanoparticles using various saccharides such as glucose, sucrose, starch, chitosan, and marine polysaccharide [4]. The said approach is regarded as a green method which is safe, biocompatible, nontoxic and environmentally friendly process for the synthesis of metal nanoparticles [3].

The past decade has seen the rapid development of pullulan used in synthesis the silver nanoparticles mainly in conventional chemical reduction techniques. Pullulan is a linear glucosic polysaccharide produced by the polymorphic fungus *Aureobasidium pullulans*, which has long been applied for various applications from food additives to environmental remediation agents. To date, surprisingly synthesizing the Ag-NP on pullulan by γ-irradiation method has not been closely studied. The previous study mainly focusing the usage of pullulan and other starch derivative as a capping agent and stabilizer in producing the Ag-NP. Several studies make a research in utilizing the pullulan to produce metallic nanoparticles and limited their process by applying the conventional chemical reduction process [5–9].

Here, in this study reports a simple γ-irradiation method in producing the AgNPs on pullulan. The radiation method offers many advantages for the preparation of metal nanoparticles. Hydrated electron, resulted from the gamma radiolysis of the aqueous solutions, can reduce metal ions to zero-valent metal particles, avoiding the use of additional reducing agents and the consequent side reactions. This novel gamma irradiation technique was chosen due to its clean and inert process. In addition, this method also provides the faster reaction times, higher yields and improve material properties [9].

2. Materials and method

All material and reagents used in this work were analytical grade and used as received without further purification. Silver Nitrate ($AgNO_3$-99.85%) was used as the silver precursor, which was obtained from Acros Organic (USA). The Pullulan powder (R&M Chemicals, UK) was applied as a solid support for Ag-NPs. All these aqueous solutions were used with double distilled water.

2.1. Synthesis of AgNPs/PL

The initial stage involved the formulation of silver nanoparticles (AgNPs) on Pullulan (PL) by γ-irradiation. The synthesis of AgNPs/PL was start by dispersing 3.0 g of Pullulan in 100 mL double distilled water and consistently stirred for 1 hour at 90°C until clear solution is obtained. The solution of PL is left to cool at the ambient temperature and thereafter 100 mL of aqueous solution of $AgNO_3$ (0.1 mol/L) was added and the mixture was further stirred for 1 hour. The mixture was then divided into six equal parts (20 ml) sample bottle, purged by N_2 for 30 minutes and sealed. Finally, the suspension which contained $AgNO_3$/PL (A0), was irradiated under γ-irradiation with absorbed doses of 5, 10, 15, 20, 25, and 50 kGy (A1-A6). γ-irradiation process was carried out in a ^{60}Co Gammacell irradiator at room temperature (the dose rate is at 35.7 kGy/min) which was provided by Malaysian Nuclear Agency, Bangi.

2.2. Characterization Methods and Instruments

The formation of Ag-NPs was confirmed using UV-Vis Spectroscopy analysis. The produced Ag-NPs were scanned from the 300-1000 nm with UV-Vis Spectrophotometer (UV-2600 Shimadzu, Japan) at medium rate scan. The spectra of the sample were measured using ad designed holder for sample of 2 cm x 2 cm dimension. It is known that UV-vis analysis can be used to determine the peak of the metal atoms. The peak detection is due to the coloured metal ions in response with the absorption of visible light. In addition, the electrons within the metal atoms excited one electronic state to another leading the sharp peak presence at a specific range.

The crystallinity of the nanomaterials was determined by XRD analysis. In this analysis the XRD pattern was started to scan from 10° to 80° at 2Θ angle. The identity of the silver nanoparticles can be confirmed by comparing the XRD pattern with the library. The XRD samples were dropped until certain thickness in the thin film by repeatedly dropping and drying at 60°C. The structure of the produced Ag-NPs was examined using XRD-Empyrean (PANalytical, United Kingdom).

IOP Conf. Series: Materials Science and Engineering **808** (2020) 012030 doi:10.1088/1757-899X/808/1/012030

The TEM sample was prepared by dropping the Ag-NPs on the surface of carbon coater copper grid. The morphology and size of the nanoparticles were determined by this technique. The dried sample was scanned using JEM-2100F transmission electron microscope (JEOL, Japan).

3. Results and discussion

Silver nanoparticles on pullulan (Ag-NPs/PL) were produced by the reduction of $AgNO_3$ into Ag^+ assisted with the fragmentation of large Ag-NPs during γ-irradiation. The prepared sample of silver nanoparticles on pullulan were shown in Figures 1. The colour intensity of the prepared Ag-NPs/PL samples were gradually increased. The A0 AgNP/PL suspensions without γ ray shown the light brown solution. As the γ-rays increased the suspension marks a dark brown solution (A3-A6). Pullulan solution is transparent without colour. The solution of AgNPs/PL are changing from light brown to dark brown solution after gamma irradiation process.

Figure 1. Photograph of Ag-NPs/PL at different γ-irradiation doses; 0, 5, 10, 15, 20, 25, 50 kGy (A0-A6)

3.1 UV-vis analysis

In this study the formation of Ag-NPs with different gamma radiation dose were monitored using UV-Vis spectroscopy. UV-Vis analysis is a top screening analysis in confirming the formation and strength of silver nanoparticles. Figure 2 shows the UV-Vis absorption spectra of the Ag-NPs/PL at different gamma rays dose. This result revealed that the Ag-NPs/PL prepared at 50 kGy, 25 kGy, 20 kGy, 15 kGy, 10 kGy and 5 kGy produced absorption spectra with sharp peaks (surface plasmon resonance, SPR) at 420 nm. On the other hand, the Ag-NPs/PL prepared without gamma radiation, the SPR peaks was not appeared at 420 nm indicating of no Ag-NPs produced. As the irradiation dose increased (from 10 to 50 kGy) the SPR peak position is shifted from 420 nm to 410 nm indicating the formation of smaller particles [10]. The most prominent sharp SPR peak can be obviously seen for Ag-NPs/PL prepared at 50 kGy, 25 kGy and 20 kGy. The peak intensities were higher at higher gamma dose justifying the highest yield of Ag-NPs with a highly uniform particle size [11]. The highest gamma doses was set at 50 kGy to shows the different sample performance between 25 kGy and 50 kGy.

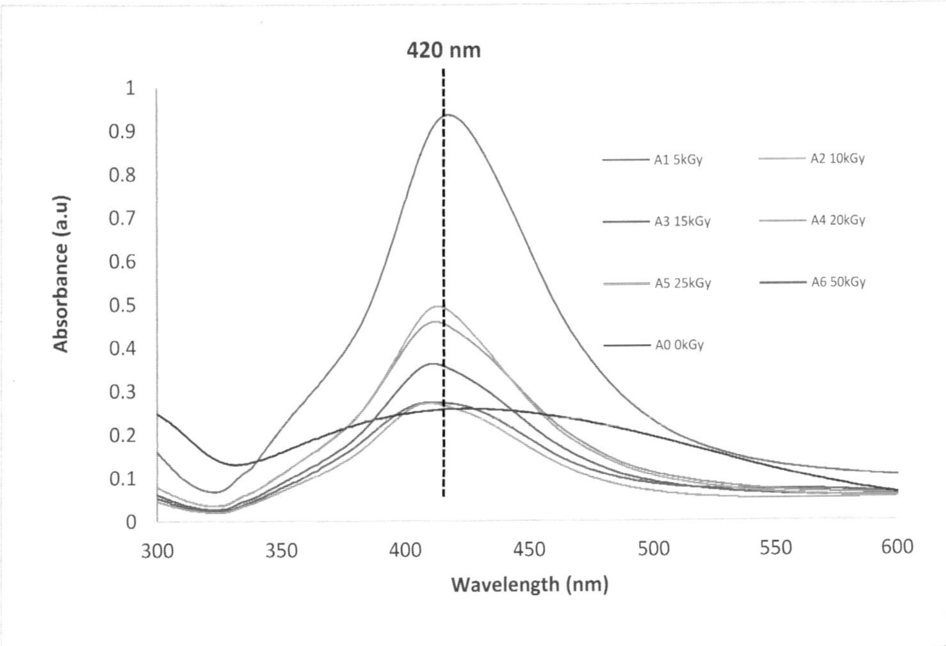

Figure 2. UV-Vis spectrum of the Ag-NPs/PL prepared at 0, 5, 10, 15, 20, 25, and 50 kGy gamma doses.

3.2 XRD analysis

X-ray diffraction (XRD) analysis is one of the most important conformational techniques in the current study. The crystalline region and purity of the AgNPs play an essential role in determining the quality of AgNPs properties. Figure 3 shows the XRD patterns of AgNPs/PL. The purity and crystallinity of synthesized AgNP/PL is confirmed. This can be proved by four (4) diffraction peaks with the 2Θ of 38.20, 44.0, 64.20, and 77.40 corresponding to the (1 1 1), (2 0 0), (2 2 0), and (3 1 1) planes of the face-centered cubic (FCC)[24]. The notable four diffraction peaks are in good agreement with the theoretical standard figures (JCPDS file no. 01-087-0718). The sample of formulation A1-A5, indicated the prominent silver in their crystalline phase without any significant traces of other phases. In addition, the occurrence of nanosized particle of AgNP/PL helps in shaping the wide and sharp XRD patterns peaks [12].

IOP Conf. Series: Materials Science and Engineering **808** (2020) 012030 doi:10.1088/1757-899X/808/1/012030

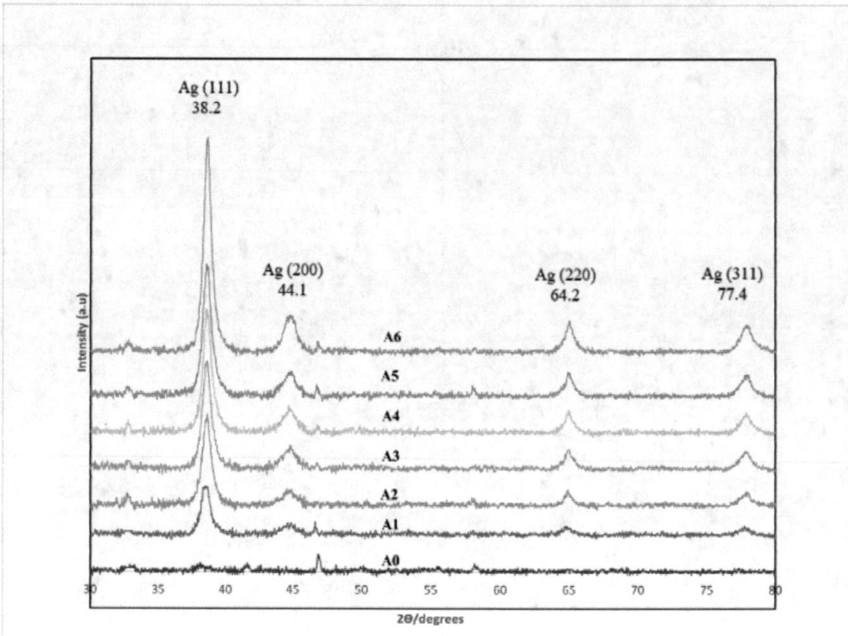

Figure 3. XRD patterns of AgNP/PL at different γ-irradiation doses; 0, 5, 10, 15, 20, 25, and 50 kGy (A0-A6)

3.3 TEM analysis

To evaluate the morphology and particle size distribution of the obtained AgNPs, the Transmission Electron Microscopy (TEM) imaging was conducted. The TEM micrograph clearly revealed that the silver nanoparticles are well dispersed in the pullulan matrix as shown in Figure 4 a), b) [9]. In addition, the AgNPs were also segregates evenly with almost no aggregates are seen. The spherical and oval shape were also observed in the TEM images[2]. It is definitive that the γ-ray reduced the particle size distribution during the irradiation process. The particles size distribution (mean diameter and standard deviation) was noted at 3.98±1.356 nm at 50 kGy dose and 14.87±7.309 at 25 kGy indicating of the evenly AgNPs segregating process [2]. The particles size distribution (mean diameter and standard deviation) was drastically decreased to 3.98±1.356 nm at 50 kGy dose indicating of the evenly AgNPs segregating process [2] The combination of nanostructures at different sizes are leading to the broadness of the size distribution peaks in Figure 4b), proving that the particles at high γ-doses were highly uniform and homogenous. In addition, at 50 kGy γ-ray doses shows the finer size of silver nanoparticles indicative of hydrated radicals (ē aq) are fully produced thus, promoting the reduction of silver ions to silver atoms [13]. It is also found that the pullulan not only function as stabilizer during the synthesis of silver nanoparticles but also as capping agent by providing the template in radiolysis process.

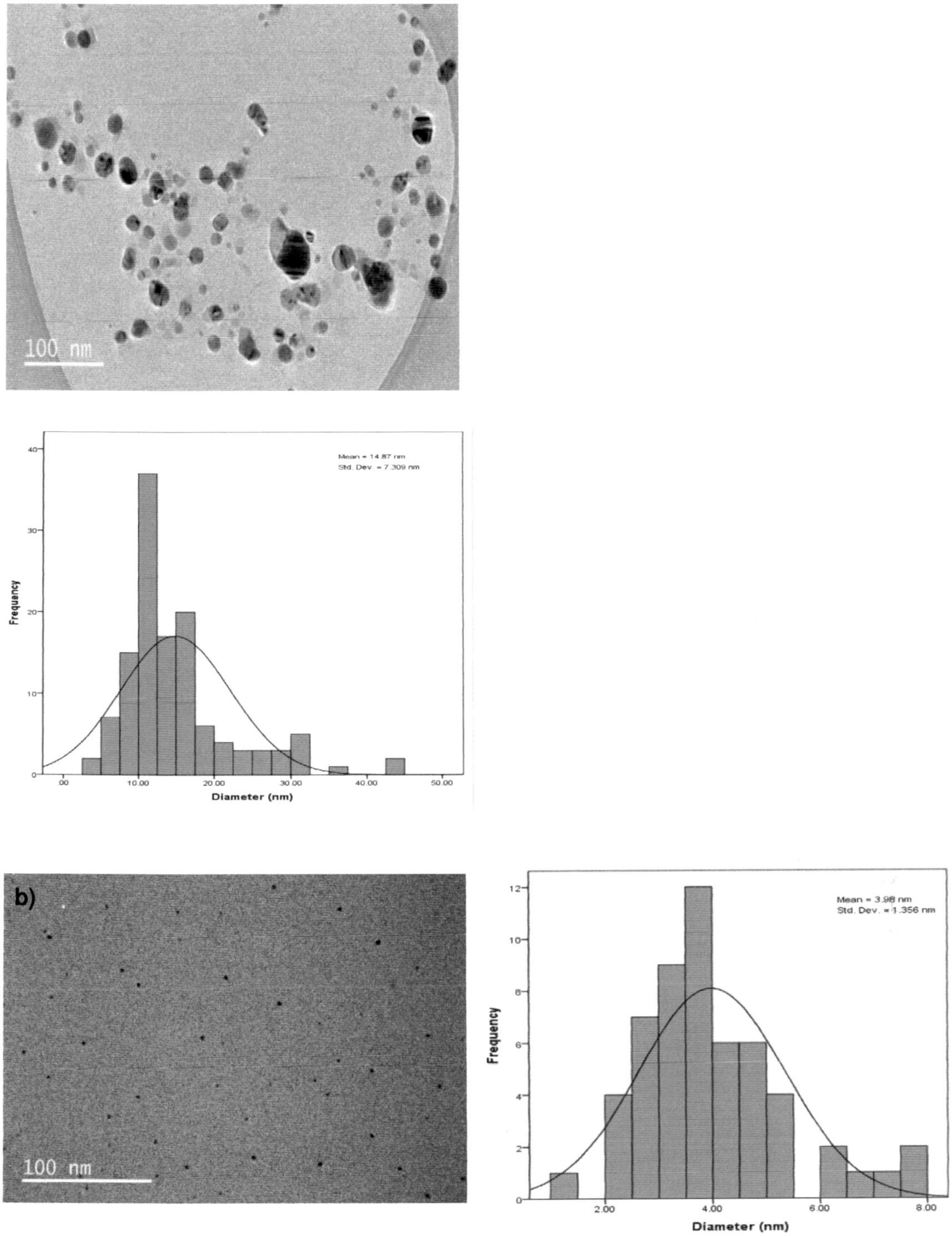

Figure 4. TEM images and their corresponding particle size distribution of AgNP/PL at a) 25 and b) 50 kGy

4. Conclusion

In conclusion, the silver nanoparticles on pullulan was successfully synthesized. The stable AgNPs/PL were prepared without using any chemical reducing agent. This green approach was also believed as a rapid, single step and effective method in producing AgNPs/PL. The UV-visible spectroscopy confirmed the formation of silver nanoparticles by the detection of plasmonic band at 420 nm. The XRD pattern showed that the crystalline structure of the AgNPs for all sample were fcc. TEM imaging ascertained that the AgNP were observed as spherical and oval shape and very well dispersed in pullulan matrix. It is also noted that the γ-irradiation techniques provide Ag-NPs in completely reduced, extremely pure, and very stable states. In addition, there is no undesirable impurities like silver oxide are introduced.

Acknowledgments
The authors are grateful for the financial support from the Malaysia-Japan International Institute of Technology (MJIIT), Universiti Teknologi Malaysia and Ministry of Education (Grant: Q.K130000.2543.19H55 and R.K130000.7743.4J305).

References

[1] Wei D, Sun W, Qian W, Ye Y, Ma X 2019 The synthesis of chitosan-based silver nanoparticles and their antibacterial activity. *Carbohydr Res*. **2009**
[2] Yoksan R, Chirachanchai S 2010 Silver nanoparticle-loaded chitosan-starch based films: Fabrication and evaluation of tensile, barrier and antimicrobial properties. *Mater Sci Eng C*. **2010**
[3] Rhim JW, Wang LF, Hong SI. 2013Preparation and characterization of agar/silver nanoparticles composite films with antimicrobial activity. *Food Hydrocoll*. **2013**
[4] DeGruson ML. Synthesis of bio-based nanocomposites for controlled release of antimicrobial agents in food packaging. The Pennsylvania State University; 2014.
[5] Pinto RJB, Almeida A, Fernandes SCM, Freire CSR, Silvestre AJD, Neto CP, Trindade T 2013 Antifungal activity of transparent nanocomposite thin films of pullulan and silver against Aspergillus niger. *Colloids Surfaces B Biointerfaces*.**103** 143–8.
[6] Fernandes SCM, Sadocco P, Causio J, Silvestre AJD, Mondragon I, Freire CSR 2014 Antimicrobial pullulan derivative prepared by grafting with 3-aminopropyltrimethoxysilane: Characterization and ability toformtransparent films. *Food Hydrocoll*.**35** 247–52
[7] Morsy MK, Khalaf HH, Sharoba AM, El-Tanahi HH, Cutter CN 2014 Incorporation of Essential Oils and Nanoparticles in Pullulan Films to Control Foodborne Pathogens on Meat and Poultry Products. *J Food Sci*.**79** (4).
[8] Coseri S, Spatareanu A, Sacarescu L, Rimbu C, Suteu D, Spirk S, Harabagiu V 2015 Green synthesis of the silver nanoparticles mediated by pullulan and 6-carboxypullulan. *Carbohydr Polym*.**116** 9–17.
[9] Zhang K, Ai S, Xie J, Xu J 2017 Comparison of direct synthesis of silver nanoparticles colloid using pullulan under conventional heating and microwave irradiation. *Inorg Nano-Metal Chem*.**47**(6) 938–45.
[10] Abdelghany AM, Abdelrazek EM, Badr SI, Abdel-Aziz MS, Morsi MA 2017 Effect of Gamma-irradiation on biosynthesized gold nanoparticles using Chenopodium murale leaf extract. *J Saudi Chem Soc*.**21**(5) 528–37.
[11] Hosny AMS, Kashef MT, Rasmy SA, Aboul-Magd DS, El-Bazza ZE 2017 Antimicrobial activity of silver nanoparticles synthesized using honey and gamma radiation against silver-resistant bacteria from wounds and burns. *Adv Nat Sci Nanosci Nanotechnol*. **8**(4).
[12] Kaur A, Preet S, Kumar V, Kumar R, Kumar R 2019 Synergetic effect of vancomycin loaded silver nanoparticles for enhanced antibacterial activity. *Colloids Surfaces B Biointerfaces*. **176** 62–9.
[13] Ghazy O. A., Nabih S., Abdel-Moneam Y. K., and Senna M. M 2017 Synthesis and characterization of silver/zein nanocomposites and their application, *Polym. Compos*.

IOP Conf. Series: Materials Science and Engineering **808** (2020) 012031 doi:10.1088/1757-899X/808/1/012031

Characteristics and carbon dioxide adsorption performance of amine-impregnated KCC-1 with different loading ratio

M S R M Nasir[1], M P Khairunnisa[1], N W C Jusoh[1, 2*] and A A Jalil[3]

[1]Malaysia-Japan International Institute of Technology, Universiti Teknologi Malaysia Kuala Lumpur, Jalan Sultan Yahya Petra, Kuala Lumpur, 54100, Malaysia
[2]Center of Hydrogen Energy, Universiti Teknologi Malaysia Kuala Lumpur, Jalan Sultan Yahya Petra, Kuala Lumpur, 54100, Malaysia
[3]School of Chemical and Energy Engineering, Universiti Teknologi Malaysia, UTM Johor Bahru, Johor, 81310, Malaysia

*nurfatehah@utm.my

Abstract. With our climate condition worsening day by day, the effort to mitigate this effect has been intensified. One of the ways to curb this issue is by reducing the release of carbon dioxide (CO_2) to the atmosphere via CO_2 capture. Among all the material used for CO_2 capture, amine-impregnated porous silica was observed to be a promising adsorbent. Research has shown its potential, but the accessibility issues caused by its narrow pore opening limits its capabilities. KCC-1, the latest member of the porous silica family, possesses unique fibrous morphology that can solve this problem. In this paper, a series of Tetraethylenepentamine (TEPA)-impregnated KCC-1 (KCC-1/TEPA) synthesized at three different loading ratios (KCC-1 to TEPA 2:1, 1:1 and 1:2) is reported. Characteristics study was conducted to determine the effect of loading ratio on its properties, and its CO_2 adsorption capacity was evaluated. FTIR peaks at 3430 cm^{-1}, 3300 cm^{-1}, 2950 cm^{-1}, 2841 cm^{-1}, 1658 cm^{-1}, 1565 cm^{-1} and 1465 cm^{-1} revealed that the impregnation was successful, while XRD diffractogram indicated that KCC-1 structure remained preserved in all impregnated samples. Increased surface area and pore volume reduction was observed with the increase of TEPA-loading ratio. KCC-1/T1:2 reported the highest CO_2 adsorption value among all three impregnated sample with 141 mg/g, while KCC-1/T2:1 recorded the lowest (82 mg/g). Results indicated that increasing TEPA loading ratio increases the number of active sites for CO_2 to attached, hence lead to increase CO_2 capture. Overall, it was shown that amount of CO_2 adsorbed is related to the amount of TEPA loaded to the sorbent.

1. Introduction

The increase of carbon dioxide (CO_2) emission to the atmosphere has been attributed to anthropogenic activities [1]. The main cause is our fossil fuel consumption as sources of energy for our economic and social activities [2]. With the increasing energy demand, more fossil fuel will be consumed, increasing the CO_2 and other greenhouse gases (GHG) emission even further. This in turn creates environmental problem such as global warming, climate change, polar ice cap melting and rising of sea level.

Carbon capture and sequestration (CCS) has been identified as a potential solution for this problem. One of the explored method is the post-combustion capture via dry adsorption. The advantage of this method is it is easy to use, energy efficient and do not cause corrosion to equipment [2, 3]. Among the sorbent that has been studied over the years is amine-impregnated porous silica. Encouraging CO_2

IOP Conf. Series: Materials Science and Engineering **808** (2020) 012031 doi:10.1088/1757-899X/808/1/012031

adsorption values has been reported using porous silicas such as MCM-41 and SBA-15 which impregnated with polyethehylenimine (PEI) and tetraethylenepentamine (TEPA). However, massive loss of sorbents surface area was also reported, which was caused by the impregnation process. This, plus its narrow pore opening, causes accessibility issues that limits its adsorption performance [4].

In 2010, Polshettiwar et al. first reported the successful synthesis of dendritic fibrous nanosilica (DFNS) named as KCC-1 [5]. Unlike other porous silica, KCC-1 has a unique open fibrous morphology that increases the amount of surface area available for adsorption [6]. Singh et al. explored the use of KCC-1 for CO_2 adsorption in 2016 [4]. In that paper, it was reported that KCC-1 with higher nitrogen loading recorded higher surface area reduction and CO_2 adsorption. However, investigation into the effect of higher amine loading to the sorbent's characteristics and adsorption behaviour was not carried out.

The objective of this paper is to study the effect of different TEPA loading ratio to the characteristics and CO_2 adsorption of TEPA impregnated KCC-1. The characteristics of each sample was studied and compared, before their CO_2 adsorption performance was evaluated. Finally, the adsorption behaviour of the adsorption process will also be studied by fitting the adsorption data with several existing isotherm models.

2. Methodology

2.1 Materials

The KCC-1 used in this research was produced by School of Chemical and Energy Engineering, from Universiti Teknologi Malaysia (UTM) Johor Bahru. The materials was synthesized using the microwave-assisted hydrothermal method, as reported in [5]. Typically the synthesis procedure begins with 2.5 g of tetraorthosilicate (TEOS) being dissolved in cyclohexane and pentanol (23 g and 1.144 g) mixture. Afterwards, 1 g of cetylpyridinium bromide (CPB), 0.6 g of urea and 30 mL water will be added into the solution. After 30 minutes of stirring the solution will be treated with microwave radiation for 4 hours. Afterwards, the product will be cooled to room temperature (25 °C), before centrifugation to separate the solid product. The product will be washed with distilled water and acetone before it was air dried at 110 °C. Finally, it will be calcined at 550 °C for 6 hours.

2.2 Sample preparation

The steps for the amine impregnation of KCC-1 was modified from [4]. To synthesize KCC-1 with TEPA at 1:1 loading ratio, 0.5 g of KCC-1 was sonicated in 25 mL of ethanol for 30 minutes to disperse the particles. Separately, 0.5 g of TEPA was dissolved in 30 mL of ethanol. The two mixture then was mixed together and stirred at 25 °C for around three hours. Finally, the mixture was oven-dried at 85 °C until it was fully dried. These steps then were repeated using different weight of KCC-1 and TEPA to achieve 1:2 and 2:1 loading ratio. The samples was denoted as KCC-1/Tx:y, where x:y refers to the ratio of KCC-1 to TEPA for that sample.

2.3 Characteristics Analysis

2.3.1 Fourier Transfer Infra-red (FTIR) analysis. FTIR spectroscopy analysis was carried out to determine the functional groups present in the particles before and after impregnation. The instrument used was the IRT Tracer-100 from Shimadzu of Japan at wavelength scan of 400-4000 cm-1. The sample was prepared by mixing it with potassium bromide (KBr) at ratio 1:100 before being pressed to form a pellet. Prior to sample analysis, a background scan with blank KBr pellet was carried out first to eliminate peaks that might occur from KBr.

2.3.2 X-ray diffraction (XRD) analysis. X-ray diffraction (XRD) analysis was carried out to determine the crystallinity of the particles before and after impregnation. This was done using Advance D8 X-ray

Powder Diffractometer from Bruker of United States. All samples was subjected to the analysis at 2θ angle from 2° to 90° with 0.053 step size value.

2.3.3 Surface area analysis. The surface area analysis was carried out using nitrogen (N_2) adsorption-desorption analysis using the NOVAtouch surface area and pore size analyser by Quantachrome, Anton Paar of United States. About 0.1 g of sample initially degassed for 60 minutes at 100°C for the impregnated samples and 300°C for pristine KCC-1 before subjected to N2 gas adsorption-desorption cycle. Data obtained from the adsorption-desorption isotherm was used to find the Brunauer-Emmett-Teller (BET) surface area and total pore volume.

2.4 CO_2 adsorption test

The sorbent performance for CO_2 adsorption was measured using Sorption Measuring Instrument with Magnetic Suspension Balance system by Rubotherm from Germany. First 0.5 g sample was loaded into the magnetic suspension balance, before pre-treated at 85°C in vacuum conditions until it weight stabilized to remove adsorbed ambient moisture. Buoyancy measurement of the sample was then carried out at 40°C and 30 bar to determine the weight and volume of the sample. For the CO_2 adsorption performances, the test was conducted at 150 mL min^{-1} of 100 % CO_2 gas flow, at 40°C temperature until maximum pressure of 30 bar was reached. The amount of CO_2 adsorbed was measured at every 5 bar interval.

3. Results and Discussion

Figure 1 shows the FTIR spectrometer of KCC-1 and all three TEPA-impregnated KCC-1. In all the samples, peaks corresponding to KCC-1 was spotted, which was at wavelength of 1250 cm^{-1}, 1050 cm^{-1}, 800 cm^{-1}, and 450 cm^{-1}. For all the TEPA-impregnated samples, broad overlapping peaks at around 3400 cm^{-1} and 3300 cm-1, which indicates the presence of both primary and secondary amine, was observed. The same corresponding to presence of amine group were also observed, with peaks at 2935 cm^{-1} and 2831 cm^{-1}, which belongs to the stretching of CH_2, and at 1660 cm^{-1} which can be attributed to N-H bending of amine. Meanwhile peaks at 1566 cm^{-1}, 1477 cm^{-1} and 1300 cm-1 are other characteristics peak of TEPA [7], which indicated the presence of TEPA in the surface of KCC-1 after impregnation, meaning that the impregnation process was successful. Increasing intensity of these peaks was also observed when the loading ratio was increased, indicating the higher amount of TEPA present in the sample.

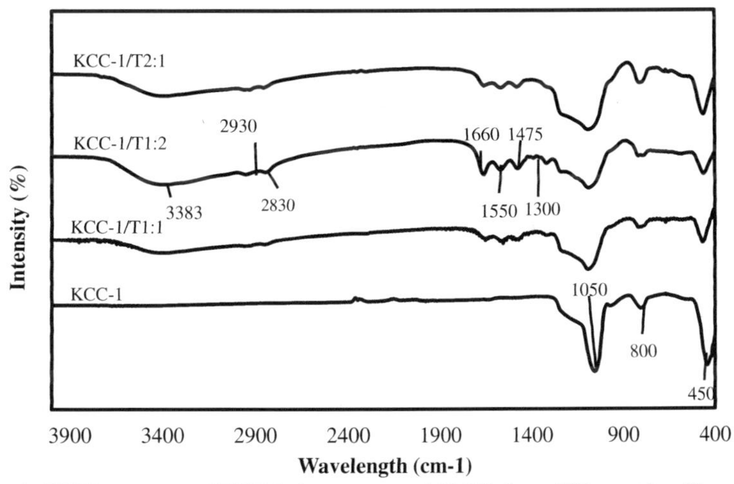

Figure 1. FTIR spectra of TEPA-impregnated KCC-1 at different loading ratio

Figure 2 shows the XRD diffractogram for for KCC-1 with different TEPA loading ratio. All three samples recorded the same broad peak at 2θ value around 23°. The results were very similar with other previously reported XRD analysis of KCC-1 [8, 9]. The broad peak shows that all the samples had an amorphous structure. There were no additional peaks observed with all three TEPA-impregnated samples, therefore it is safe to assume that the structure of KCC-1 is preserved after impregnation. The intensity of the peak was also found to be increasing with the increase of TEPA loading ratio. TEPA has been reported to have a XRD peak at around 19° [10], hence the overlapping peaks of KCC-1 and TEPA contributed to this added peak intensity.

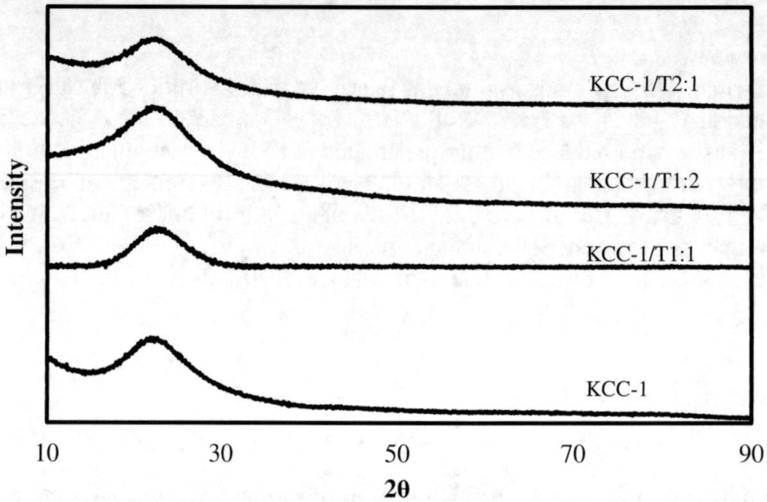

Figure 2. XRD diffractogram of TEPA-impregnated KCC-1 at different weight ratio

Table 1 shows the surface area analysis of KCC-1 impregnated with TEPA at different loading ratio. Massive loss of surface area and pore volume was recorded with all three TEPA-impregnated KCC-1 samples, as compared with KCC-1. The highest loss of surface area and pore volume was recorded with KCC-1/T1:2, with 95% reduction of surface area and 80.7% reduction of pore volume, while

KCC-1/T2:1 has the lowest reduction in both surface area and pore volume with 85.38% and 64% loss respectively. As TEPA-loading ratio increased, more TEPA molecules are attached to the surface and pores of KCC-1, causing the surface area and pore volume to decrease even further.

Table 1: Surface area analysis of KCC-1 and TEPA-impregnated KCC-1 at different loading ratio

Sample	BET surface area (m^2/g)	Total Pore Volume (cm^3/g)
KCC-1	560.295	1.2692
KCC-1/T2:1	81.917	0.4543
KCC-1/T1:1	61.720	0.3344
KCC-1/T1:2	28.082	0.2446

Figure 3 shows the CO_2 adsorption isotherm of KCC1-1 and all three TEPA impregnated samples at different loading ratio. The highest adsorption capacity was recorded with KCC-1/T1:2 sample with 141 mg/g, followed by KCC-1/T1:1 with 125 mg/g, KCC-1 with 102.3 mg/g and KCC-1/T2:1 with 81.8 mg/g. Initially at lower pressure, the adsorption capacity of KCC-1/T2:1 was higher than KCC-1, but as the pressure increase, the adsorption capacity of KCC-1 also increases. At 20 bar it was higher than KCC-1/T2:1, it continued until maximum pressure of 30 bar.

IOP Conf. Series: Materials Science and Engineering **808** (2020) 012031 doi:10.1088/1757-899X/808/1/012031

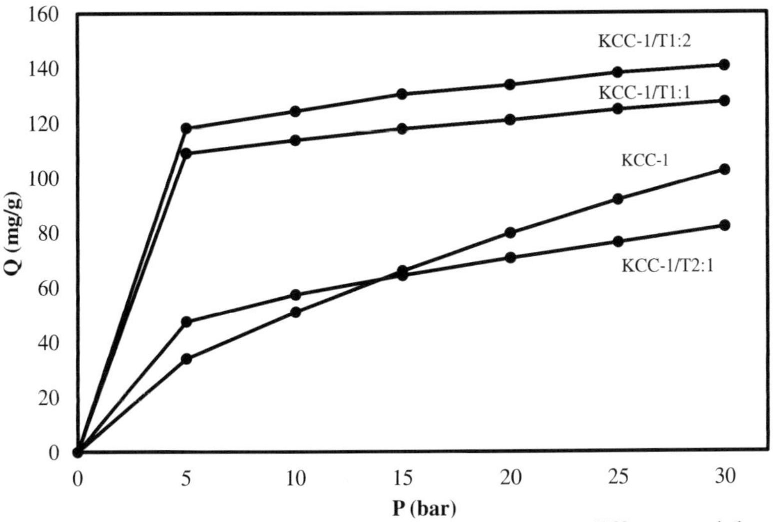

Figure 3. CO_2 adsorption of TEPA-impregnated KCC-1 at different weight ratio

The adsorption of CO_2 with TEPA impregnated KCC-1 involves a specific chemical reaction between primary and secondary amine with CO to formed carbamate ions. It was first proposed by Caplow in 1968 [11] and can be represented by the equation below:

$$CO_2 + 2R_1R_2NH \rightleftarrows R_1R_2NCOO^- + R_1R_2NH_2{}^+ \quad (1)$$

On this basis, by increasing the amount of TEPA loading into KCC-1's surface, we also increased the number of amine groups available to react with CO_2. Hence, this allows for CO_2 adsorption capacity to increase, which is reflected by the KCC-1/T1:2 performance. Similarly, reducing the TEPA loading ratio resulted in smaller number of amine groups present, hence limiting the adsorption performance of KCC-1/T2:1 sample.

Serna-Guerrero et al. [11] argued that the adsorption process with amine-functionalized silica involved contribution from both chemisorption from the previously mentioned mechanism and physisorption process. This was evident in the isotherm shape, where no samples recorded a saturation at higher pressure like common Langmuir isotherm. Instead, here it was observed that the adsorption continues to increase as pressure increases, albeit at slower pace than at lower pressure.

For unimpregnated KCC-1, there are no amine group presence, therefore the adsorption process is purely physisorption. At lower pressure the amount of CO_2 uptake was limited, but when the pressure increased the adsorption capacity also increased [12]. As KCC-1 has a significantly larger surface area compared to TEPA-impregnated KCC-1, this allowed KCC-1 to have higher CO_2 adsorption capacity at 30 bar pressure.

Table 2 below shows the comparison of the CO_2 adsorption performance between this study and other TEPA-impregnated porous silica reported. KCC-1/T1:2 has a much better CO_2 adsorption performance than all other TEPA-impregnated porous silica, including MCM-41, SBA-15 and KIT-6. The fibrous morphology of KCC-1 and high TEPA loading allows for high CO_2 adsorption to be achieved. This indicates that KCC-1/T1:2 has huge potential as an adsorbent for CO_2.

Table 2. Adsorption capacity of KCC-1/T1:2 and previously reported TEPA-impregnated silica

Silica	Adsorption Capacity (mg/g)	Source
TEPA-MCM-41	118.8	[13]
SBA-TEPA(50)	113.5	[14]
50 wt%-TEPA-Si-MCM-41	70.41	[15]
PE-KIT-6(T)-35	127.6	[16]
0.64EB-TEPA/SiO_2	88	[17]
SBA-393-50T	124.52	[18]
KCC-1/T1:2	141	This study

4. Conclusions

In this paper, KCC-1 particles were impregnated with TEPA in three different loading ratio; 1:1, 1:2 and 2:1, where each characteristics and CO_2 adsorption performance was evaluated afterwards. The characteristics indicated that the loading ratio did not radically alter the characteristics of the adsorbent. KCC-1/T1:2 recorded the highest CO_2 adsorption capacity after 30 bar pressure, while KCC-1/T1:1 recorded the lowest. The results of these experiments showed that increasing the TEPA loading ratio help to increase the CO_2 adsorption capacity. This result gives a better understanding on one of the factors influencing the CO_2 adsorption performance of KCC-1/TEPA and can be utilized for further study in this field.

Acknowledgment

The authors would like to acknowledge the contribution from University Teknologi Malaysia Transdisciplinary Research Grant TDRG-Q.K130000.3453.06G61 for providing funds for this research and Centre of Hydrogen Energy (CHE), UTM KL for providing input and assistance.

References

[1] Dutcher B, Fan M and Russell A G 2015 *ACS Appl Mater Interfaces* **7** 2137-48
[2] Younas M, Sohail M, Leong L K, Bashir M J and Sumathi S 2016 *International Journal of Environmental Science and Technology* **13** 1839-60
[3] Lee S-Y and Park S-J 2015 *Journal of Industrial and Engineering Chemistry* **23** 1-11
[4] Singh B and Polshettiwar V 2016 *Journal of Materials Chemistry A* **4** 7005-19
[5] Polshettiwar V, Cha D, Zhang X and Basset J M 2010 *Angew Chem Int Ed Engl* **49** 9652-6
[6] Maity A and Polshettiwar V 2017 *ChemSusChem* **10** 3866-913
[7] Long Q and Wang Y 2015 Energies **8** 12917
[8] Hamid M Y S, Firmansyah M L, Triwahyono S, Jalil A A, Mukti R R, Febriyanti E, Suendo V, Setiabudi H D, Mohamed M and Nabgan W 2017 *Applied Catalysis A: General* **532** 86-94
[9] Wang Y, Hu K, He J and Zhang Y 2019 *RSC Advances* **9** 24783-90
[10] Irani M, Jacobson A T, Gasem K A M and Fan M 2017 *Fuel* **206** 10-8
[11] Caplow M 1968 *Journal of the American Chemical Society* **90** 9
[12] Serna-Guerrero R, Belmabkhout Y and Sayari A 2010 *Chemical Engineering Journal* **161** 173-81
[13] Liu S-H, Hsiao W-C and Sie W-H 2012 *Adsorption* **18** 431-7
[14] Sanz-Pérez E S, Olivares-Marín M, Arencibia A, Sanz R, Calleja G and Maroto-Valer M M 2013 *International Journal of Greenhouse Gas Control* **17** 366-75
[15] Ahmed S, Ramli A, Yusup S and Farooq M 2017 *Chemical Engineering Research and Design* **122** 33-42
[16] Wei J, Mei D, Lin Z, Geng L, Chen S and Liao L 2018 *Nano* **13**
[17] Park S, Choi K, Yu H J, Won Y-J, Kim C, Choi M, Cho S-H, Lee J-H, Lee S Y and Lee J S 2018 *Industrial & Engineering Chemistry Research* **57** 4632-9

[18] Cecilia J A, Vilarrasa-García E, Morales-Ospino R, Bastos-Neto M, Azevedo D C S and Rodríguez-Castellón E 2019 *Adsorption*

IOP Conf. Series: Materials Science and Engineering **808** (2020) 012032 doi:10.1088/1757-899X/808/1/012032

Pullulan mediated zinc oxide microparticles: Effect of synthesis temperature

E D Mohamed Isa[1], N W Che Jusoh[2,3], R Hazan[4] and K Shameli[1,*]

[1]Department of Environmental and Green Technology, Malaysia-Japan International Institute of Technology, Universiti Teknology Malaysia, Jalan Sultan Yahya Petra, 54100 Kuala Lumpur Malaysia
[2]Department of Chemical Process Engineering, Malaysia-Japan International Institute of Technology, Universiti Teknology Malaysia, Jalan Sultan Yahya Petra, 54100 Kuala Lumpur Malaysia
[3]Advanced Materials Research Group, Center of Hydrogen Energy, Universiti Teknologi Malaysia, 54100, Kuala Lumpur, Malaysia.
[4]Malaysian Nuclear Agency, Bangi, 43000 Kajang, Selangor, Malaysia

*kamyarshameli@gmail.com

Abstract. Zinc oxide (ZnO) has been gaining a lot of attention compared to the other semiconductor metal oxide. This is due to their favourable properties such as chemical stability, high electrochemical coupling coefficient, broad range of absorption radiation and high photostability. It also has wide band gap energy (3.37 eV), large exciton-binding energy (60 meV) and high thermal and mechanical stability at room temperature. Besides these desirable properties, the synthesis of ZnO has been widely studied as it is easy to synthesized and its properties such as band gap, shape and size can be controlled through synthesis parameters and methods. In this study, zinc oxide microparticles (ZnO-MCs) is produced with pullulan as mediator via precipitation method. The effect of synthesis temperature on the properties of synthesized ZnO-MCs were also studied. Based on the result obtained, all synthesized ZnO-MCs exhibited hexagonal wurtzite structure. As the synthesis temperature increases, the particles morphology changes from large spherical shape to flower-like morphology. Furthermore, the particle size also decreases with increasing temperature. This result is supported by surface area and pore analysis where the surface area ranging from 6.22 to 22.78 $m^2 g^{-1}$ was obtained as the synthesis temperature increases. All these results indicate that synthesis method and parameters will affect the properties of synthesized ZnO-MCs.

1. Introduction

Zinc oxide (ZnO) is an important II-IV compound semiconductor and it has been widely studied within the context of nanoscience and nanotechnology [1]. The interest on ZnO is due to their unique physical and chemical properties such as chemical stability, high electrochemical coupling coefficient, broad range of absorption radiation and high photostability. Furthermore, it also has wide band gap energy (3.37 eV), large exciton-binding energy (60 meV) and high thermal and mechanical stability at room temperature. All these properties can be tuned according to desire by controlling the synthesis process and parameters. [2].

In these recent years, green synthesis has been the main focus in synthesized ZnO. Green synthesis is defined as synthesis process that utilized natural products such as plant extracts, microorganisms and biopolymers to produce the materials. Biopolymers such as carbohydrate has been used as one of the green materials to produce ZnO. Taghavi et al reported the production of ZnO nanoparticles using arabic gum via sol-gel method [3]. Soraya et al also reported the usage of Tragacanth gum as reducing and stabilizing agent in production of urchin-like ZnO nanorods array. To produce urchin-like ZnO, they utilized combination method of precipitation and ultrasound method [4].

Pullulan is a biopolymer that is produced from starch by growing yeast like fungus *Aureobasidium pullulans*. The advantage of pullulan is it is water soluble. It is made out of maltotriose which consist of three glucose units linked by an α-(1,4) glycosidic bond and the maltotriose units are linked by an α-(1,6) glycosidic bond. Pullulan is non-toxic, non-mutagenic, odourless, biocompatible and biodegradable [5]. It also has wide range of commercial and industrial applications across various fields. Previous study shows that pullulan has been used as stabilizer to produce gold nanoparticles [6]. In a much recent works, pullulan has been used to produce ceria nanoparticles [7]. Currently, there is no works that reported on the usage of pullulan in production of ZnO.

In this study, pullulan is used as stabilizer to produce ZnO microparticles (ZnO-MCs) via precipitation method. The effect of synthesis temperature on the properties of ZnO-MCs were studied.

2. Materials and methodology

2.1 Materials

Pullulan powder (99%) was purchased from China. Chemicals of analytical grade, zinc nitrate hexahydrate ($Zn(NO_3)_2 \cdot 6H_2O$, 98 %) and ammonia solution (30 % NH_3 in water) were purchased from R & M Chemicals, Malaysia.

2.2 Synthesis of ZnO microparticles by precipitation method

4 sets of experiments for the ZnO-MCs were performed by varying the synthesis temperature (room temperature, 40 °C, 60 °C and 80 °C). In a typical run, 5 g of pullulan was dissolved in 150 mL water. 2.97 g of $Zn(NO_3)_2 \cdot 6H_2O$ was added to the solution and the mixture was stirred at 60 °C for 5 minutes. Then 3 mL of ammonia solution was added to the mixture and stirred for another 30 minutes. The resulting precipitate was centrifuged, washed with water and dried overnight in an over at 50 °C. Then, the obtained powders were calcined at 500 °C for 1 hour. The samples were named RTZ, 4TZ, 6TZ and 8TZ according to synthesis temperature of room temperature, 40 °C, 60 °C and 80 °C respectively.

2.3 Characterization

The X-ray diffraction (XRD) analysis of synthesized ZnO-MCs were carried out by using the PANalytical X'pert Pro MPD diffractometer with Cu K_α radiation (λ = 1.5406) in the 2θ range from 20° to 80°. The surface morphology of ZnO-MCs were analysed using field emission electron microscope (FESEM) model Carl Zeiss GeminiSEM 500. The surface area and pore analysis was conducted using Quantachrome Instrument Novatouch. The samples were degas at 150 °C for 4 hours before analysis. Fourier Transform Infrared (FTIR) spectra were recorded over the range of 400-4000 cm^{-1} by using Shimadzu IRTracer-100 with potassium bromide (KBr) pellet method. Diffuse reflectance spectra (DRS) were collected on Shimadzu UV-2600 spectrophotometer with an integrating sphere accessory.

3. Results and discussion

The XRD patterns of synthesized ZnO-MCs at different synthesis temperature are shown in figure 1. The XRD spectra for all synthesizde ZnO-MCs were very similar, all exhibiting characteristic diffraction peaks at 2θ = 31.8°, 34.4°, 36.3°, 47.6°, 56.6° and 62.8° corresponding well to the (100), (002), (101), (102), (110) and (103) crystal planes of hexagonal wurtzite ZnO (JCPDS No. 36-1451) [8, 9]. Besides that, no characteristic peaks for impurities were observed and this indicate that the prepared ZnO-MCs samples have high purity. Furthermore, strong and sharp diffraction peaks indicate that the materials have high crystallinity. It is observed that as the synthesis temperature increases, the peak intensities decrease. This indicates that the particle size decreases with increasing synthesis temperature

IOP Conf. Series: Materials Science and Engineering **808** (2020) 012032 doi:10.1088/1757-899X/808/1/012032

as reported by Olieveira et al [10]. The crystallite size of ZnO-MCs can be calculated from the full width at half maximum (FWHM) of XRD patterns using Scherer formula as shown in equation (1):

$$D = \frac{0.9\lambda}{\beta cos\theta} \quad (1)$$

Where λ is the wavelength of the Cu Kα radiation, β is the FWHM of the peak and θ is the Bragg angle [3, 11]. Using the formula, the average size of ZnO-MCs were 22.76, 24.39, 22.76 and 18.98 nm for RTZ, 4TZ, 6TZ and 8TZ respectively.

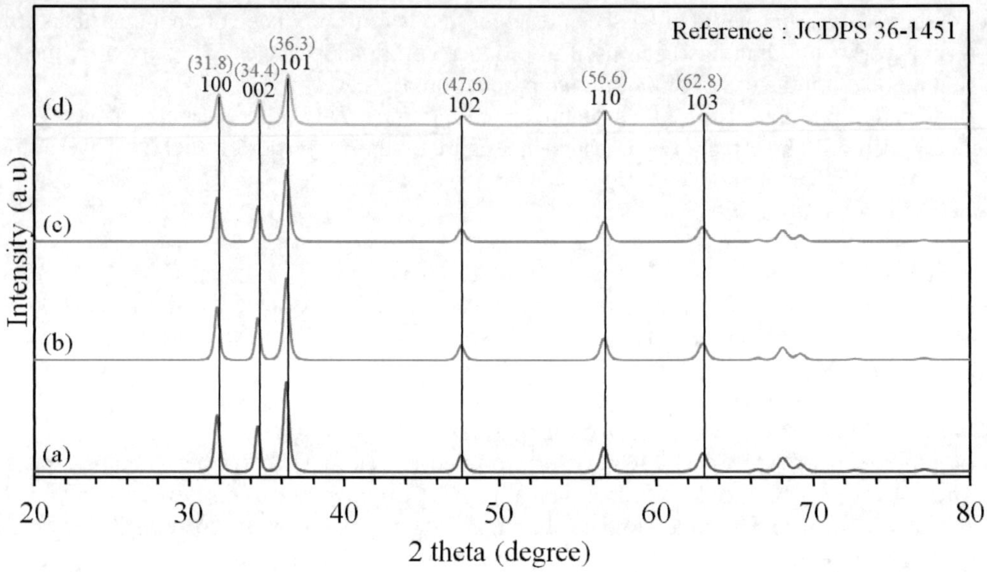

Figure 1. The XRD of synthesized ZnO-MCs. (a) RTZ, (b) 4TZ, (c) 6TZ and (d) 8TZ.

The particles morphology was then further studied using FESEM and Figure 2 shows that all the FESEM images of synthesized ZnO-MCs. RTZ exhibited various morphologies which consist of mixture of spherical and large flower-like structures. As the reaction temperature increases, it can be seen that the particles start to more define flower-like morphology. Besides that, as the temperature increases, the overall size of ZnO-MCs decreases. At higher synthesis temperature, nucleation process is conducive and this leads to lower particles growth rate. This will then lead to smaller particles size. This observation is similar to previous reports [12, 13]. Another unique property of synthesized ZnO-MCs that can be observed through FESEM is the presence of pores within the particles.

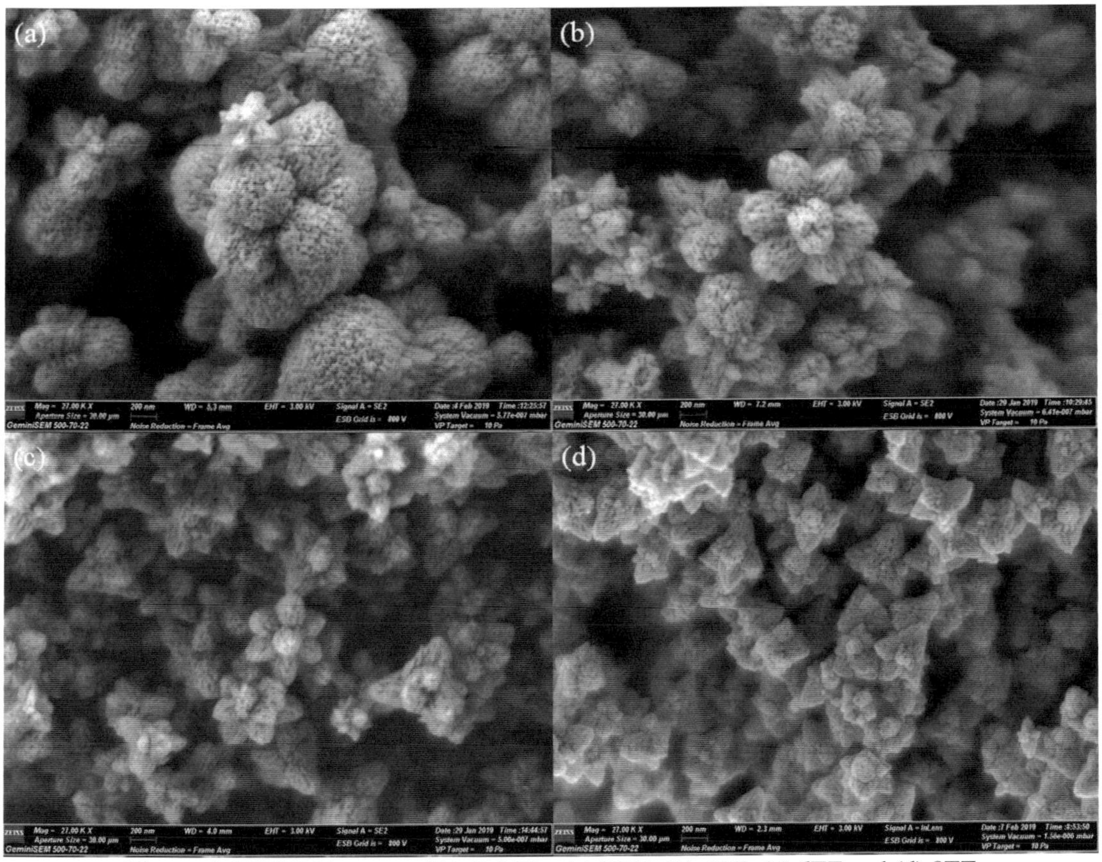

Figure 2. FESEM images of ZnO-MCs. (a) RTZ, (b) 4TZ, (c) 6TZ and (d) 8TZ.

Based on FESEM analysis, it was found that all synthesized ZnO-MCs exhibited some porosity. Therefore, to confirm the presence of pores within the ZnO-MCs, the compounds were analysed through surface area and pore analysis. The nitrogen adsorption-desorption isotherm of all synthesized ZnO-MCs and its respective pore size distribution are shown in Figure 3. It can be seen that all ZnO-MCs exhibit type IV isotherm which indicate the presence of mesopore [11]. This result is in agreement with the FESEM images obtained where there was porosity observed in all synthesize ZnO-MCs. Besides that, from the isotherm, it is also observe that there were hysteresis loops at relatively high pressure and this might be due to presence of large pores of interparticle spaces [11, 14]. From the nitrogen sorption analysis, the BET surface areas were obtained for RTZ, 4TZ, 6TZ and 8TZ where surface area of 20.93, 19.05, 22.78 and 22.60 $m^2 g^{-1}$ were observed respectively. The general trend of increasing BET surface area with increasing synthesis temperature was observed. The is due to the decrease of particles size as smaller particles size have higher surface area value [15].

IOP Conf. Series: Materials Science and Engineering **808** (2020) 012032 doi:10.1088/1757-899X/808/1/012032

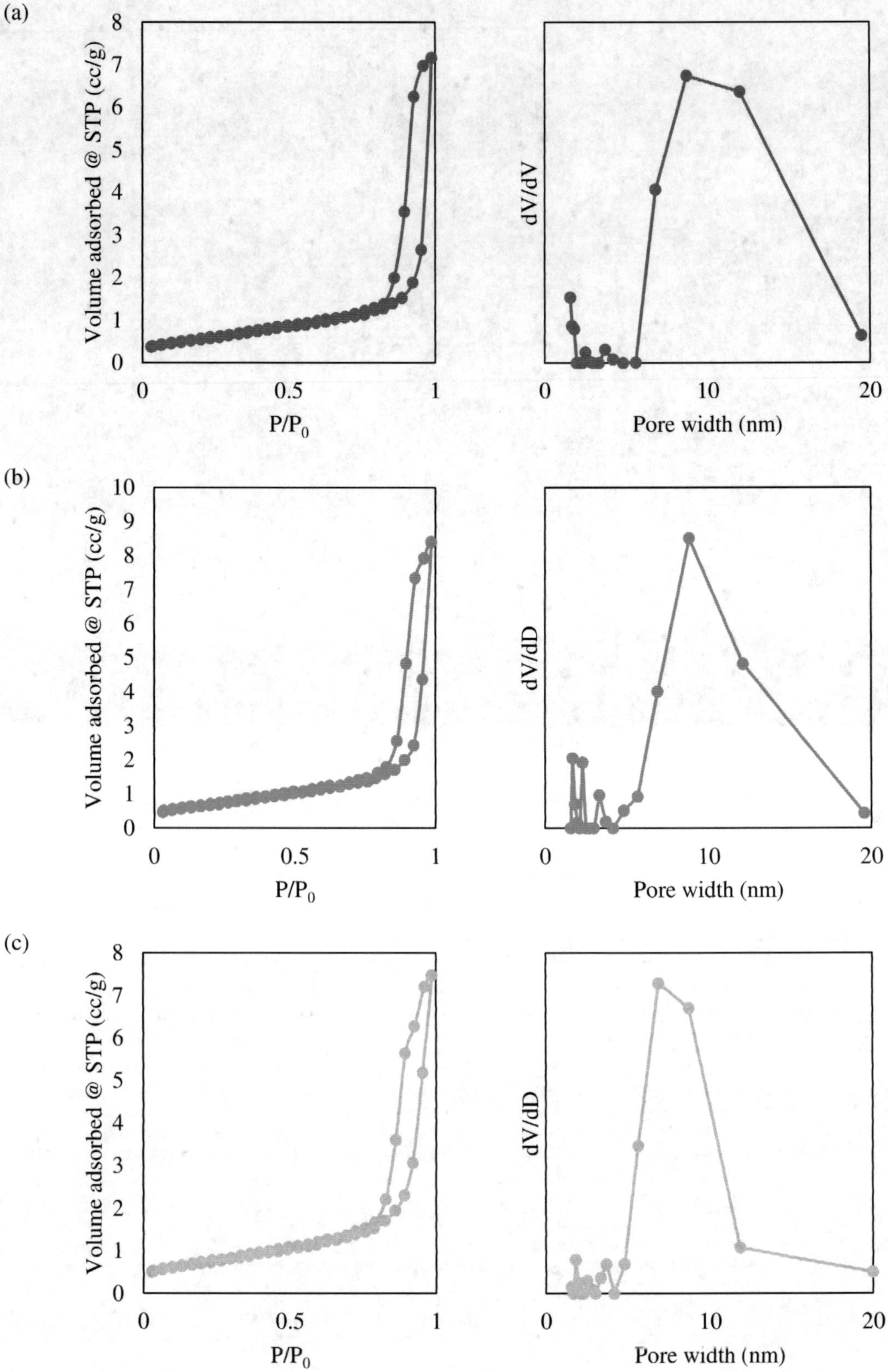

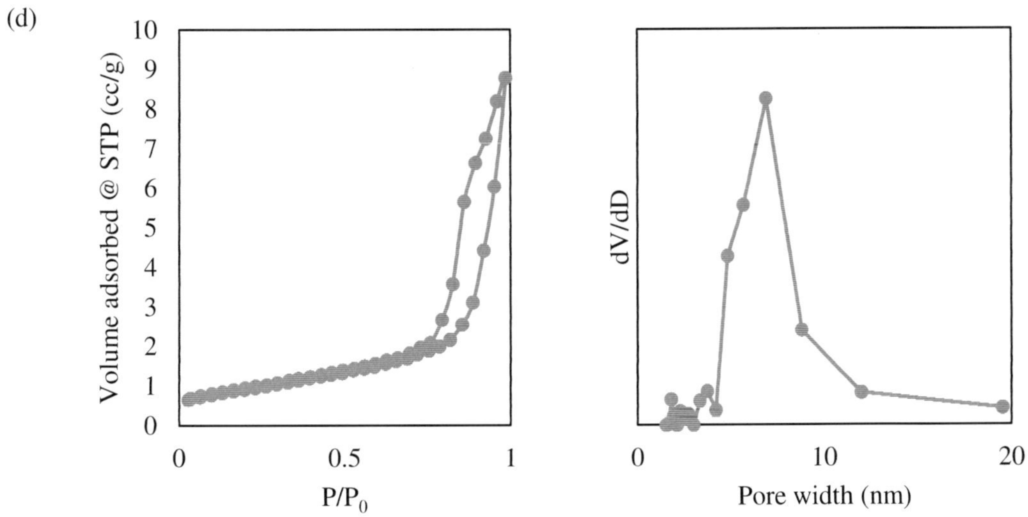

Figure 3. Nitrogen adsorption-desorption isotherm and pore distribution plot of ZnO-MCs. (a) RTZ, (b) 4TZ, (c) 6TZ and (d) 8TZ

The formation of ZnO-MCs was further investigated using FTIR analysis. Figure 4 shows the FTIR spectra of pullulan, synthesized ZnO-MCs before calcination and ZnO-MCs after calcination. In pullulan spectrum, there were two strong absorption bands at 3300 and 2916 cm^{-1} and this is due to OH stretching vibration and sp^3 C-H bond respectively. Single band at 1631 cm^{-1} was assigned to stretching vibration of O-C-O in pullulan. Other features bands of pullulan were also observed around 1358 and 1147 cm^{-1} which correspond to C-O-H bend and C-O-C stretch respectively. Besides that, the typical pullulan band, α-configuration of α-D-glucopyranose units was also observed around 840 cm^{-1}. Furthermore, two bands representing the linkages of pullulan were also found around 925 and 752 cm^{-1}. All the bands observed in the pullulan spectrum are similar to the previously reported works [16, 17].

Through comparison of pullulan's FTIR spectrum with ZnO-MCs before calcination FTIR spectrum, characteristic bands of pullulan could be observed in sample before calcination. It was found that the band at around 840 and 925 cm^{-1} which represent pullulan bands present although at low intensity. This might be due to low presence of pullulan as the sample was washed with water before the analysis. These bands completely disappear in calcined ZnO-MCs spectrum which indicate that the pullulan has been completely removed through calcination. Presence of new peak around 400 cm^{-1} for both before calcination and calcined ZnO-MCs spectra was assigned to Zn-O.

IOP Conf. Series: Materials Science and Engineering **808** (2020) 012032 doi:10.1088/1757-899X/808/1/012032

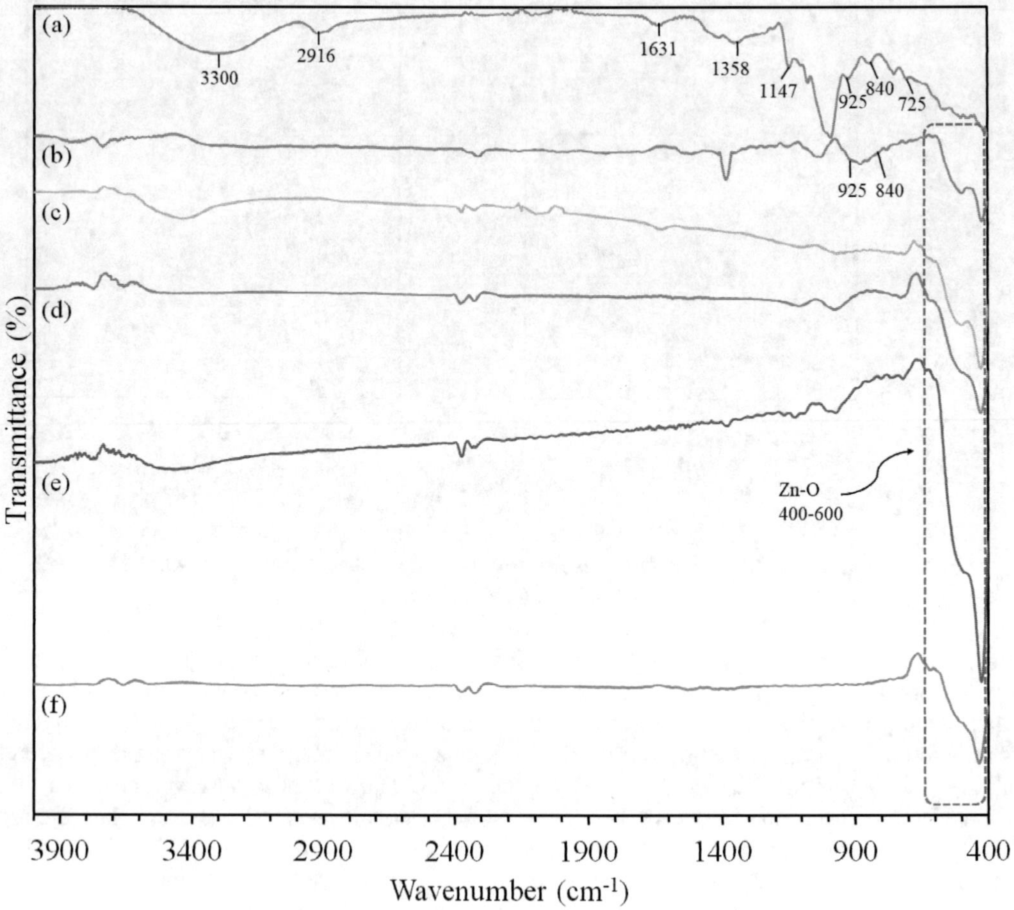

Figure 4. FTIR spectra of (a) pullulan, (b) ZnO-MCs before calcination, (c) RTZ, (d) 4TZ, (e) 6TZ and (f) 8TZ.

The optical properties of synthesized ZnO-MCs were determined through UV-Vis diffuse reflectance. Figure 5 shows the UV-Vis diffuse reflectance spectra of the ZnO-MCs generated at different synthesis temperatures. It can be seen that the reflectance increases at wavelength above 370 nm and this is can be assigned to the direct band gap of ZnO. This is due to electron transitions from the valence band to the conduction band [18]. As ZnO is a direct band gap semiconductor, its band gap energy (E_g) can be determined from equation (2):

$$(\alpha hv)^2 = A(hv - E_g) \tag{2}$$

where α, h, v, E_g and A are the absorption coefficient, Planck's constant light frequency, band hap energy and a constant respectively. The E_g values of the ZnO-MCs samples can be obtained by plotting a graph of $(\alpha hv)^2$ against hv [9]. Through extrapolating the linear portion of the curves to zero, it was found that the E_g values of all the synthesized ZnO were ~3.2 eV.

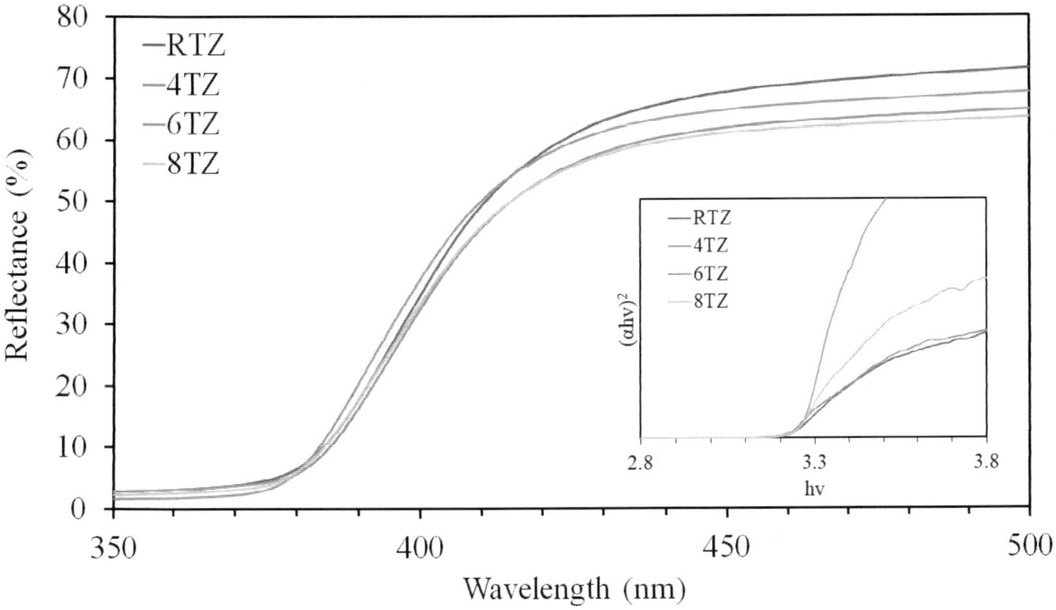

Figure 5. UV-VIS diffuse reflectance spectra of all synthesized ZnO-MCs at different synthesis temperatures. Inset: The corresponding plots of $(\alpha hv)^2$ versus energy (hv) for the band gap energy of ZnO-MCs.

4. Conclusion

In this study, ZnO-MCs were successfully synthesized with pullulan as mediator. The effect of synthesis temperature on the properties of ZnO-MCs were determined. All synthesized ZnO-MCs exhibited hexagonal wurtzite structure. Based on FESEM analysis, it was observed that the particles evolved from large spherical microparticles to flower-like microparticles. Furthermore, as the synthesis temperature increases, the ZnO-MCs particle size decreases which also lead to increase of surface area value. The optical properties of synthesized ZnO-MCs were determined through UV-VIS diffuse reflectance and all the synthesized ZnO-MCs exhibited similar band gap value of ~3.2 eV.

Acknowledgement

The author wish to acknowledged funding by the Malaysian Ministry of Higher Education under the Tier 1 grant (Grant no. #20H33 and #20H55) and express gratitude to the Research Management Centre (RMC) of UTM and Malaysia-Japan International Institute of Technology (MJIIT) for providing an excellent research environment and facilities.

References

[1] Ma Q, Wang Y, Kong J, Jia H and Wang Z 2015 Controllable synthesis of hierarchical flower-like ZnO nanostructures assembled by nanosheets and its optical properties *Superlattices and Microstructures* **84** 1-12

[2] Yıldırım Ö A and Durucan C 2012 Effect of precipitation temperature and organic additives on size and morphology of ZnO nanoparticles *Journal of Materials Research* **27** 1452-61

[3] Taghavi Fardood S, Ramazani A, Moradi S and Azimzadeh Asiabi P 2017 Green synthesis of zinc oxide nanoparticles using arabic gum and photocatalytic degradation of direct blue 129 dye under visible light *Journal of Materials Science: Materials in Electronics* **28** 13596-601

[4] Ghayempour S, Montazer M and Mahmoudi Rad M 2016 Tragacanth gum biopolymer as reducing and stabilizing agent in biosonosynthesis of urchin-like ZnO nanorod arrays: A low

cytotoxic photocatalyst with antibacterial and antifungal properties *Carbohydr Polym* **136** 232-41

[5] Ganeshkumar M, Ponrasu T, Raja M D, Subamekala M K and Suguna L 2014 Green synthesis of pullulan stabilized gold nanoparticles for cancer targeted drug delivery *Spectrochim Acta A Mol Biomol Spectrosc* **130** 64-71

[6] Laksee S, Puthong S, Teerawatananond T, Palaga T and Muangsin N 2017 Highly efficient and facile fabrication of monodispersed Au nanoparticles using pullulan and their application as anticancer drug carriers *Carbohydr Polym* **173** 178-91

[7] Khorrami M B, Sadeghnia H R, Pasdar A, Ghayour-Mobarhan M, Riahi-Zanjani B and Darroudi M 2018 Role of Pullulan in preparation of ceria nanoparticles and investigation of their biological activities *Journal of Molecular Structure* **1157** 127-31

[8] Wang S, Kuang P, Cheng B, Yu J and Jiang C 2018 ZnO hierarchical microsphere for enhanced photocatalytic activity *Journal of Alloys and Compounds* **741** 622-32

[9] He L, Tong Z, Wang Z, Chen M, Huang N and Zhang W 2018 Effects of calcination temperature and heating rate on the photocatalytic properties of ZnO prepared by pyrolysis *J Colloid Interface Sci* **509** 448-56

[10] Oliveira A P A, Hochepied J-F, Grillon F and Berger M-H 2003 Controlled Precipitation of Zinc Oxide Particles at Room Temperature *Chemistry of Materials* **15** 3202-7

[11] Zhang G, Shen X and Yang Y 2011 Facile Synthesis of Monodisperse Porous ZnO Spheres by a Soluble Starch-Assisted Method and Their Photocatalytic Activity *The Journal of Physical Chemistry C* **115** 7145-52

[12] Mohammed Fayaz A, Balaji K, Kalaichelvan P T and Venkatesan R 2009 Fungal based synthesis of silver nanoparticles--an effect of temperature on the size of particles *Colloids Surf B Biointerfaces* **74** 123-6

[13] Khalil M M H, Ismail E H, El-Baghdady K Z and Mohamed D 2014 Green synthesis of silver nanoparticles using olive leaf extract and its antibacterial activity *Arabian Journal of Chemistry* **7** 1131-9

[14] Li X, Shi B, Wang Y, Li M, Liu Y, Gao L and Mao L 2015 Preparation of monodispersed mesoporous silica particles and their applications in adsorption of Au 3+ and Hg 2+ after mercapto-functionalized treatment *Microporous and Mesoporous Materials* **214** 15-22

[15] Yu M, Zhou L, Zhang J, Yuan P, Thorn P, Gu W and Yu C 2012 A simple approach to prepare monodisperse mesoporous silica nanospheres with adjustable sizes *J Colloid Interface Sci* **376** 67-75

[16] Saber-Samandari S, Gulcan H O, Saber-Samandari S and Gazi M 2014 Efficient Removal of Anionic and Cationic Dyes from an Aqueous Solution Using Pullulan-graft-Polyacrylamide Porous Hydrogel *Water Air and Soil Pollution* **225**

[17] Shingel K I 2002 Determination of structural peculiarities of dexran, pullulan and γ-irradiated pullulan by Fourier-transform IR spectroscopy *Carbohydrate Research* **337** 1445-51

[18] Khorsand Zak A, Abd. Majid W H, Mahmoudian M R, Darroudi M and Yousefi R 2013 Starch-stabilized synthesis of ZnO nanopowders at low temperature and optical properties study *Advanced Powder Technology* **24** 618-24

IOP Conf. Series: Materials Science and Engineering **808** (2020) 012033 doi:10.1088/1757-899X/808/1/012033

Effect of size and shape dependent of synthesized copper nanoparticle using natural honey

N A Ismail[1], K Shameli[1] *, N W Che Jusoh[2] and R Rasit Ali[2]

[1]Department of Environment and Green Technology, Malaysia-Japan International Institute of Technology. Universiti Teknologi Malaysia, Jalan Sultan Yahya Petra, 54100 Kuala Lumpur, Malaysia.
[2]Center of Hydrogen, Universiti Teknologi Malaysia, Jalan Sultan Yahya Petra, 54100 Kuala Lumpur, Malaysia.

*kamyarshameli@gmail.com

Abstract
A study on the effect of size and shape of copper nanoparticles (Cu-NPs) by varying the amount of honey has been done using a facile green synthesis method with the presence of ultrasonic assistance. Several amount of different % w/v of honey (0%, 1%, 5%, 10%, 15% and 20% w/v) that contain carbohydrate which are mainly glucose and fructose, and other polyhydroxyl groups act as stabilizing agent and a weak reducing agent supported by ascorbic acid were used to produce the Cu-NPs. The synthesized Cu-NPs were characterized using UV-visible, XRD and HRTEM to prove the size and shape of the nanoparticles. The best amount of honey used to produce Cu-NPs with uniform particle size and shape is at 15 % w/v. The size is 3.81 ± 1.135 nm and it shows a consistence spherical shape using HRTEM analysis image. UV-visible supported the results from the HRTEM. And XRD shows good diffraction pattern for pure Cu-NPs. It proves that honey has the ability to act as stabilizing agent in controlling the size and shape of nanoparticles.

1. Introduction

Nanomaterial has becoming a new promising materials nowadays that exhibits many specialties for a various number of applications such as water treatment, energy storage and biological applications as compared to corresponding bulk materials [1-4]. Due to the small size and high surface area-to-volume ratio, nanoparticles have unique physical properties that attract attentions among the researchers worldwide [5]. Metal nanoparticles such as gold, silver and copper are applicable to use in several industries where each of nanoparticles has different properties depending on the size, shape and other physicochemical properties [6]. Most of the nanoparticles could be used in food packaging [7], wound dressing [8], drug delivery [9] and catalytic application [10].

Copper nanoparticles (Cu-NPs) are known in the field of electronic, sensors and biomedical application aside from silver nanoparticles and gold nanoparticles due to easy availability, low cost, easy of mixing with polymers and their similar properties to those noble metals [11-13]. In synthesizing Cu-NPs, conventional methods is rather costly, slow reaction time and the uses of toxic chemical such as sodium borohydride and hydrazine as reducing agent are involved [12, 14, 15]. However, by using ultrasound irradiation assistance, it can reduce the reaction time, more uniform size distribution and improve phase purity for the nanoparticles [10]. Ascorbic acid, vitamin C which is a non-toxic chemical and reducing agent, also has been used as supported reducing agent in this work to replace the toxic chemicals [16].

A major limitation in preparing Cu-NPs is, it really sensitive to air and easily to oxidize to copper oxides phases which is much more stable state. However, the formation of an oxide layer could decrease the efficiency of Cu-NPs especially for the antibacterial activity properties [5, 17]. Thus, stabilizing agent or capping agent is needed to prevent the agglomeration of nanoparticles to occur as reported in previous literatures [18-20].

Honey is a natural food and a sweet viscous liquid consisting of carbohydrate, enzymes, vitamins, minerals and antioxidant [21, 22]. The monosaccharides, proteins, amino acid and vitamin C in honey which contain polyhydroxyl group could help in reducing and stabilizing the nanoparticles. The use of honey as reducing and stabilizing agent is convenient as it does not require drying, extraction of plant materials and cell culture maintenance [23, 24]. This study is a continuation study based on previous work [24] whereby it is focusing on the effect of size and shape for Cu-NPs by using different amount of honey (% w/v). The synthesized Cu-NPs were then characterized using UV-visible spectroscopy, XRD and HRTEM.

2. Experimental

2.1. Materials

Capilano honey (Australia), double distilled water, copper nitrate trihydrate ($(CuNO_2)_3.3H_2O$), sodium hydroxide (NaOH), and ascorbic acid. All chemical are purchased from R&M chemical, Germany, and used as received without any purification.

2.2. Synthesis of Copper Nanoparticles

Different amount of natural honey from Australia (0, 1, 5, 10, 15 and 20% w/v) was dissolved in appropriate amount of copper nitrate trihydrate ($Cu(NO_3)_2.3H_2O$). The pH of the mixture was adjusted to neutral pH (~ pH 7) with sodium hydroxide and undergo ultrasonic irradiation for 10 minutes with addition of a proper amount of ascorbic acid. The synthesized copper nanoparticle was washed with distilled water and separated by centrifugation at 4000 rpm for 15 minutes. Then, samples are dried in vacuum oven at 60 °C.

2.3. Characterizations of Copper Nanoparticles

All samples were characterized using Ultraviolet-visible (UV-vis) spectroscopy (UV-2600, SHIMADZU). The clean quartz solution cells were used. The sample solution with desired concentration and a blank sample (water) were used for this experiment. The samples were homogenized in the quartz soluti on cells and placed into the UV-vis chamber. The UV-vis spectra were recorded over the range of 220 nm to 800 nm. The structure of Cu-NPs were evaluated using X-ray diffraction (XRD, Philips, X'pert, Cu Ka) in the small-angle range of 20° to 80° (2θ). High Resolution Transmission Electron Microscopy (HRTEM) (model JEM-2100F) was used to examine the size and shape of Cu-NPs.

3. Results and Discussion

3.1. Characterization of Copper Nanoparticles

UV-visible is used to check the formation of Cu-NPs. Figure 1 shows the UV-visible of Cu-NPs for different amount of honey (% w/v). It shows the surface plasmon resonance peaks for all the synthesized Cu-NPs which appeared around 550 to 600 nm. As can be observed from figure 1, only slight difference in peak shifted from red to blue shift which indicate small changes in term of size and shape occur by using different amount of honey. Besides, different shape of nanoparticles also could contribute to the peak of UV-visible where wide peak might be due to different shape of nanoparticles and the peak intensity indicates that same shape of nanoparticles were produced [25]. An increasing of absorbance in the UV-visible shows that an increase for concentration of Cu-NPs and higher concentration of metal nanoparticles were produced during the synthesis process.

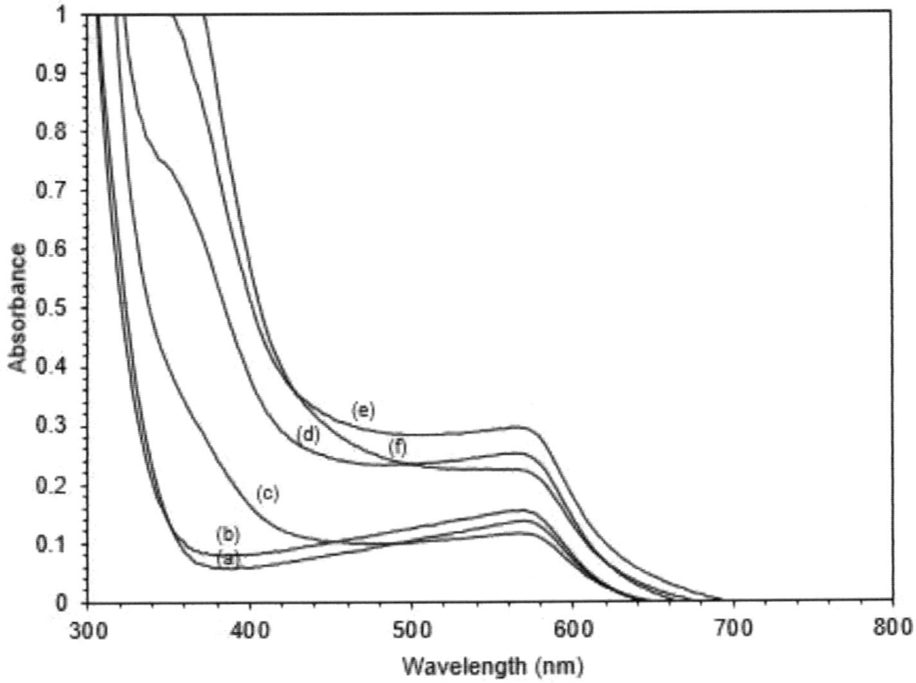

Figure 1. UV-visible of Cu-NPs for different amount of honey 0, 1, 5, 10, 15 and 20% (a-f) respectively.

Meanwhile, the formation of Cu-NPs synthesized by using honey are proved by the XRD analysis where the diffraction pattern matches with the XRD database reference (reference: JCPDS 04-0836) of pure copper nanoparticles as in figure 2. The diffraction peaks for all Cu-NPs honey based synthesized show pure copper without the presence of oxide which are at $2\theta = 43.3°$, $50.5°$ and $74.4°$ which could be assigned to (111), (200) and (220). However, small peak of Cu_2O (111) appeared for Cu-NPs at 0 % that might be because of the oxidation of nanoparticles which start to occur to form the Cu_2O. Scherer equation was used to determine the average crystallite size as in equation (1) where the average size for all the synthesized Cu-NPs were between 20-80 nm. These results are correspond with the previous reported study [1].

$$D = \frac{K\lambda}{FWHM \cos\theta} \tag{1}$$

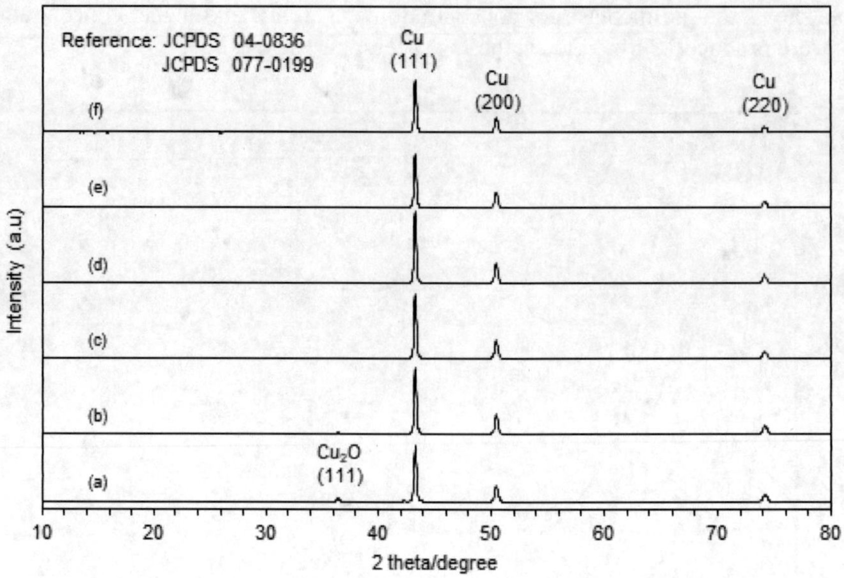

Figure 2. XRD diffraction pattern for Cu-NPs using difference amount of honey 0, 1, 5, 10, 15 and 20% (a-f) respectively.

High resolution transmission electron microscopy was used to monitor the size and morphology of the Cu-NPs. Figure 3 illustrates that the difference of size and shape of each Cu-NPs using different amount of honey. It can be seen that the size of nanoparticles are between 3-100 nm. As can be seen in Figure 3, the size of nanoparticles are decreasing as the amount of honey increase. This is because the presence of the polyhydroxyl groups in honey help in preventing the aggregation of nanoparticles. However, for the Cu-NPs with 5 % and 20 % w/v of honey, the agglomeration of nanoparticles seems to happen and the shape of nanoparticles also inconsistence. These phenomena might be due to the high surface energy of nanoparticles which have high chances for the nanoparticles to attract to each other [26]. This data is correspond to the UV-visible peaks results. The best condition for amount of honey at 0.025 M copper nitrate precursor using honey is at 15 % w/v where the particles have a consistent shape with high amount of Cu-NPs production correspond to the UV-visible results. All results are tally with the UV-visible results obtained.

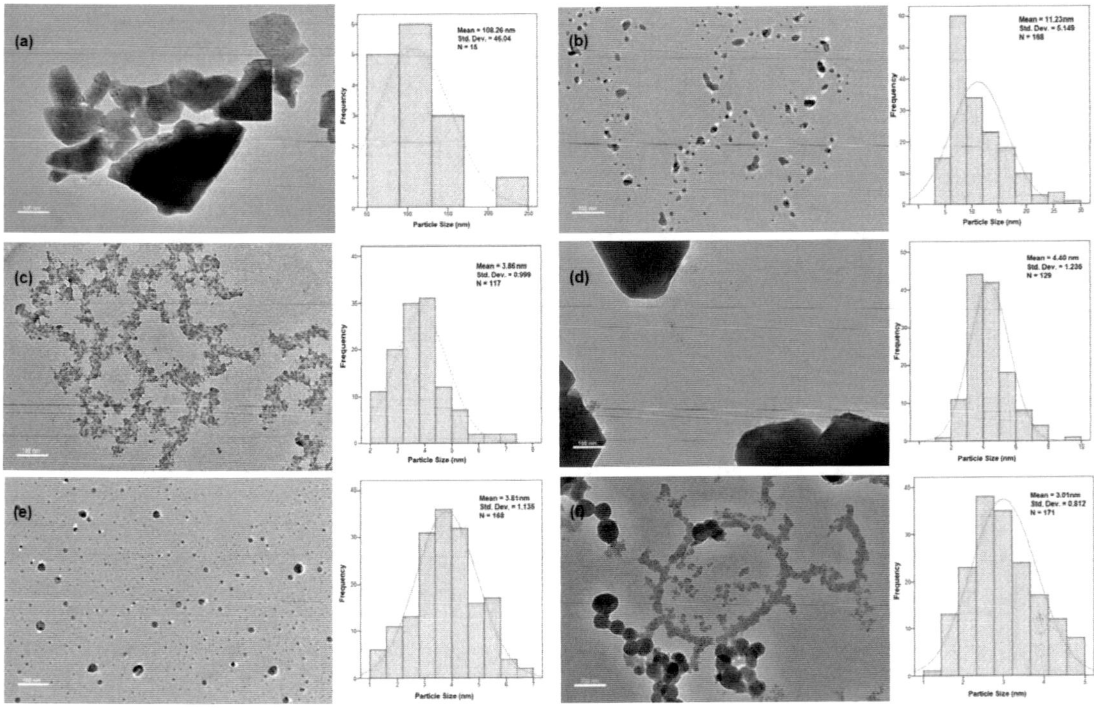

Figure 3. HRTEM of Cu-NPs with different amount of honey 0, 1, 5, 10, 15 and 20% (a-f) respectively.

4. Conclusions

As a conclusion, carbohydrate and polyhydroxyl groups in honey has contributing to control the effect of size and shape of copper nanoparticles. The copper nanoparticles with the range size of 3 to 100 nm are produce depending on the amount of honey used. However, the best amount of honey % w/v for this study is at 15 % w/v with the mean size of 3.81 nm where the shape of nanoparticles are more consistence at this value.

Acknowledgement

The authors wish to acknowledge funding by the Malaysian Ministry of Higher Education under the Tier 1 grants (Grant no. #20H33 and #20H55) and express gratitude to the Research Management Centre (RMC) of UTM and Malaysia-Japan International Institute of Technology (MJIIT) for providing an excellent research environment and facilities.

References

[1] Suárez-Cerda J, Espinoza-Gómez H, Alonso-Núñez G, Rivero I A, Gochi-Ponce Y and Flores-López L Z 2017 A green synthesis of copper nanoparticles using native cyclodextrins as stabilizing agents *Journal of Saudi Chemical Society* **21** 341-8

[2] Zhou G, Xu L, Hu G, Mai L and Cui Y 2019 Nanowires for Electrochemical Energy Storage *Chem Rev*

[3] S. Harikumar P 2016 Antibacterial Activity of Copper Nanoparticles and Copper Nanocomposites against Escherichia Coli Bacteria *International Journal of Sciences* **2** 83-90

[4] Santhosh C, Velmurugan V, Jacob G, Jeong S K, Grace A N and Bhatnagar A 2016 Role of nanomaterials in water treatment applications: A review *Chemical Engineering Journal* **306** 1116-37

[5] Sierra-Ávila R, Pérez-Alvarez M, Cadenas-Pliego G, Comparán Padilla V, Ávila-Orta C, Pérez Camacho O, Jiménez-Regalado E, Hernández-Hernández E and Jiménez-Barrera R M 2015 Synthesis of Copper Nanoparticles Using Mixture of Allylamine and Polyallylamine *Journal of Nanomaterials* **2015** 1-9

[6] Khan I, Saeed K and Khan I 2017 Nanoparticles: Properties, applications and toxicities *Arabian Journal of Chemistry*

[7] Huang Y, Mei L, Chen X and Wang Q 2018 Recent Developments in Food Packaging Based on Nanomaterials *Nanomaterials (Basel)* **8**

[8] Nethi S K, Das S, Patra C R and Mukherjee S 2019 Recent advances in inorganic nanomaterials for wound-healing applications *Biomater Sci* **7** 2652-74

[9] Izadiyan Z, Basri M, Fard Masoumi H R, Abedi Karjiban R, Salim N and Kalantari K 2019 Improvement of physicochemical properties of nanocolloidal carrier loaded with low water solubility drug for parenteral cancer treatment by Response Surface Methodology *Mater Sci Eng C Mater Biol Appl* **94** 841-9

[10] Mosleh S, Rahimi M R, Ghaedi M, Dashtian K and Hajati S 2018 Sonochemical-assisted synthesis of CuO/Cu2O/Cu nanoparticles as efficient photocatalyst for simultaneous degradation of pollutant dyes in rotating packed bed reactor: LED illumination and central composite design optimization *Ultrason Sonochem* **40** 601-10

[11] Din M I and Rehan R 2016 Synthesis, Characterization, and Applications of Copper Nanoparticles *Analytical Letters* **50** 50-62

[12] Din M I, Arshad F, Hussain Z and Mukhtar M 2017 Green Adeptness in the Synthesis and Stabilization of Copper Nanoparticles: Catalytic, Antibacterial, Cytotoxicity, and Antioxidant Activities *Nanoscale Res Lett* **12** 638

[13] Ali Soomro R, Tufail Hussain Sherazi Sirajuddin S, Memon N, Raza Shah M, Hussain Kalwar N, Richard Hallam K and Shah A 2014 Synthesis Of Air Stable Copper Nanoparticles And Their Use In Catalysis *Advanced Materials Letters* **5** 191-8

[14] Gour A and Jain N K 2019 Advances in green synthesis of nanoparticles *Artif Cells Nanomed Biotechnol* **47** 844-51

[15] Abd-Elkareem J I, Bassuony H M, Mohammed S M, Fahmy H M and Abd-Elkader N R 2016 Eco-Friendly Methods of Copper Nanoparticles Synthesis *Journal of Bionanoscience* **10** 15-37

[16] Jain S, Jain A, Kachhawah P and Devra V 2015 Synthesis and size control of copper nanoparticles and their catalytic application *Transactions of Nonferrous Metals Society of China* **25** 3995-4000

[17] DeAlba-Montero I, Guajardo-Pacheco J, Morales-Sánchez E, Araujo-Martínez R, Loredo-Becerra G, Martínez-Castañón G-A, Ruiz F and Compeán Jasso M 2017 Antimicrobial properties of copper nanoparticles and amino acid chelated copper nanoparticles produced by using a soya extract *Bioinorganic chemistry and applications* **2017**

[18] Dinda G, Halder D, Vazquez-Vazquez C, Lopez-Quintela M A and Mitra A 2015 Green Synthesis of Copper Nanoparticles and their Antibacterial Property *Journal of Surface Science Technology* **31** 117-22

[19] Muthukrishnan A M 2015 Green Synthesis of Copper-Chitosan Nanoparticles and Study of its Antibacterial Activity *Journal of Nanomedicine & Nanotechnology* **06**

[20] Khan A, Rashid A, Younas R and Chong R 2015 A chemical reduction approach to the synthesis of copper nanoparticles *International Nano Letters* **6** 21-6

[21] Hoseini S J, Darroudi M, Oskuee R K, Gholami L and Zak A K 2015 Honey-based synthesis of ZnO nanopowders and their cytotoxicity effects *Advanced Powder Technology* **26** 991-6

[22] Oskuee R K, Banikamali A, Bazzaz B S F, Hosseini H A and Darroudi M 2016 Honey-Based and Ultrasonic-Assisted Synthesis of Silver Nanoparticles and Their Antibacterial Activities *Journal of Nanoscience and Nanotechnology* **16** 7989-93

[23] Hosny A M, Kashef M T, Rasmy S A, Aboul-Magd D S and El-Bazza Z E 2017 Antimicrobial activity of silver nanoparticles synthesized using honey and gamma radiation against silver-resistant bacteria from wounds and burns *Advances in Natural Sciences: Nanoscience and Nanotechnology* **8** 045009

[24] Ismail N A, Shameli K, Wong M M, Teow S Y, Chew J and Sukri S 2019 Antibacterial and cytotoxic effect of honey mediated copper nanoparticles synthesized using ultrasonic assistance *Mater Sci Eng C Mater Biol Appl* **104** 109899

[25] Izadiyan Z, Shameli K, Hara H and Taib S H M 2018 Cytotoxicity assay of biosynthesis gold nanoparticles mediated by walnut (Juglans regia) green husk extract *Journal of Molecular Structure* **1151** 97-105

[26] Ashraf M A, Peng W, Zare Y and Rhee K Y 2018 Effects of Size and Aggregation/Agglomeration of Nanoparticles on the Interfacial/Interphase Properties and Tensile Strength of Polymer Nanocomposites *Nanoscale Res Lett* **13** 214

IOP Conf. Series: Materials Science and Engineering **808** (2020) 012034 doi:10.1088/1757-899X/808/1/012034

Photocatalytic Degradation of Malachite Green Dye by Plant-mediated Biosynthesized Zinc Oxide Nanoparticles

S N A M Sukri, E D M Isa, K Shameli*

Department of Environmental and Green Technology, Malaysia-Japan International Institute of Technology, Universiti Teknologi Malaysia, Jalan Sultan Yahya Petra, 54100 Kuala Lumpur, Malaysia.

*kamyarshameli@gmail.com

Abstract. Photocatalytic activity of ZnO-NPs was tested by degradation of Malachite Green dye under UV light irradiation. The ZnO-NPs were biosynthesized using *Punica granatum* (pomegranate) fruit peels extract as the stabilizing agent. Simple sol-gel method and calcination in different temperatures (400, 500, 600 and 700°C) were carried out to obtain pure ZnO-NPs with high photocatalytic properties. In the degradation studies, 20 mg ZnO-NPs were used to degrade Malachite Green dye of 10 ppm initial concentration for a total period of 50 minutes in a 100 ml reaction volume. Results obtained shown that ZnO-NPs calcined in 700°C had the highest removal efficiency at about 99% in 40 minutes. This proves that biosynthesized ZnO-NPs have a high potential to be used as a photocatalyst to degrade textile dyes in a short time for wastewater treatments.

1. Introduction

In the field of catalysis and water treatment, nanomaterials have been gaining much interest because of their unique physico-chemical properties, high surface area to volume ratio, and strong interaction with other materials [1]. Several types of nanomaterials such as copper oxide [2], titanium dioxide [3], magnesium doped zinc oxide [4], silver doped titanium dioxide [5] and iron [6] have been reviewed for their potential in wastewater treatments. Zinc oxide (ZnO) is one of the most extensively studied semiconductors with a large band gap of about 3.3 eV. At room temperature, ZnO is known to have a high excitation binding energy of about 60 meV with stable hexagonal wurtzite structure [7]. Upon irradiation of ZnO-NPs semiconductor with UV light, electrons from its valence band travels to the conduction band and generate hole-electron (h/e) pair which in turn causes the production of hydroxyl radicals and reactive oxygen species (ROS) [8]. These hydroxyl radicals and ROS will oxidize pollutant molecules from dye compounds in the photocatalytic process.

A major pollutant in our water resource comes from the textile industry which discharges large amounts of dye effluents into the water streams daily [9]. These dye effluents have been proven toxic to the environment and human health thus proper treatments to reduce and degrade them has to be developed. Degradation and removal of various types of dyes such as Gentian Violet dye [7], Methyl Orange [10], Basic Violet 3 dye [11] and Methylene Blue [12] by numerous nanomaterials have been studied and reported before. Extensive studies have also been done to investigate the efficiency of ZnO-NPs as catalysts to degrade various textile dyes thanks to their high surface area, non-toxicity, photosensitivity and cost-effectiveness [9]. Malachite Green is one of the most commonly used dyes in the industrial field. Degradation of Malachite Green dye in particular is more difficult due to the presence of complex aromatic ring in their triphenylmethane structure [13]. Malachite Green dye has also been reported to be highly resistant to light and oxidizing agents, and previous studies to remove them using biological treatment and chemical precipitation were not that efficient [13].

This present study reports the successful production of ZnO-NPs mediated using *Punica granatum* (*P. granatum*) fruit (F.) peels extract as reducing and stabilizing agents for the sol-gel and combustion methods. We will investigate the effect of different calcination temperature on the morphology of ZnO-NPs as well as their effectiveness to degrade Malachite Green dye under UV light irradiation.

2. Experimental

2.1 Plant collection and materials
Fruit peels of *P. granatum* were collected from "Green Farm" in Fars province, city of Neyriz, Iran. Zinc nitrate hexahydrate ($Zn(NO_3)_2.6H_2O$, 98 %) of analytical grade was purchased from R&M Chemicals, United Kingdom and used as metal salt precursor. Deionized (DI) water was used to prepare all aqueous solutions. Glassware used in this study were thoroughly cleaned with DI water and dried before use.

2.2 Fruit peel extract preparation
Fresh *P. granatum* F. peels were cleaned using tap water and DI water consequently to remove visible dirt before dried in the oven at 45 °C. The dried peels were ground into fine powder and stored at room temperature for further use. *P. granatum* F. peel extract was produced by extracting the peel powder in DI water at 65 °C for 1 hour in the ratio of 1:10. The extract were then filtered and further centrifuged at 10,000 rpm for 10 min to completely remove any leftover particles. The *P. granatum* F. peel extract was kept at 4 °C for future experiments.

2.3 Biosynthesis of ZnO-NPs
The ZnO-NPs were synthesized using a simple sol-gel and combustion method consequently. In the ratio of 1:10, zinc nitrate hexahydrate ($Zn(NO_3)_2.6H_2O$) was added to *P. granatum* F. peel extract under vigorous stirring. The solution was stirred at 90 °C until the aqueous solvent is completely removed, leaving a gel-like product. The product was then calcined at different temperatures (400, 500, 600, and 700 °C) for 1 hour until it turned into fine white powder form. All ZnO-NPs powder were kept in room temperature and used in the experiments.

2.4 Characterization methods and instrumentation
To evaluate the purity and crystallinity of ZnO-NPs, X-ray diffraction (XRD, Philips, X'pert, Cu Ka) at the small angle range of 2Q (10°- 90°) was used. High Resolution Transmission electron microscopy (HRTEM) (model JEM-2100F) was used to observe and capture images of all biosynthesized ZnO-NPs. The shapes and sizes of ZnO-NPs were then analysed in this study. Information on surface area, pore volume and pore radius were obtained using BET surface analyzer (NovaTouch) by measuring nitrogen adsorption-desorption isotherm at 77 K. All ZnO-NPs were degassed for 2 hours at 150 °C. For photocatalytic experiments, Ultraviolet-visible (UV-vis) spectroscopy (UV-2600, SHIMADZU) was used to measure absorbance values in the wavelength range of 220 to 900 nm.

2.5 Photocatalytic studies
The efficiency of all biosynthesized ZnO-NPs was evaluated by its ability to degrade Malachite Green dye. 20 mg of ZnO-NPs photocatalyst was mixed with Malachite Green dye solution of 10 ppm concentration in a 100 ml reaction volume. The mixture was stirred in dark for 30 minutes to obtain adsorption and desorption equilibrium. After that, the mixture was exposed to UV light irradiation and small amount of samples were taken every 10 minutes for absorbance reading using UV-vis spectrophotometer. The photocatalytic studies were carried out for 50 minutes until complete

degradation of the dye. Results were analyzed to investigate the effectiveness of all ZnO-NPs as a photocatalyst.

3. Results and discussion

To confirm the successful production of ZnO-NPs, XRD analysis was carried out for all samples. Figure 1 shows the Miller indices peaks of all four samples of ZnO-NPs, which can be indexed to hexagonal wurtzite phase structure as supported by JCDPS Cardno. 89-1397 data [14]. As no other foreign peaks can be seen, it is confirmed that the biosynthesized ZnO-NPs are highly pure and crystalline. Calcinating in higher temperatures produce ZnO-NPs with increasing crystallinity as the diffraction peaks become more intense and narrower.

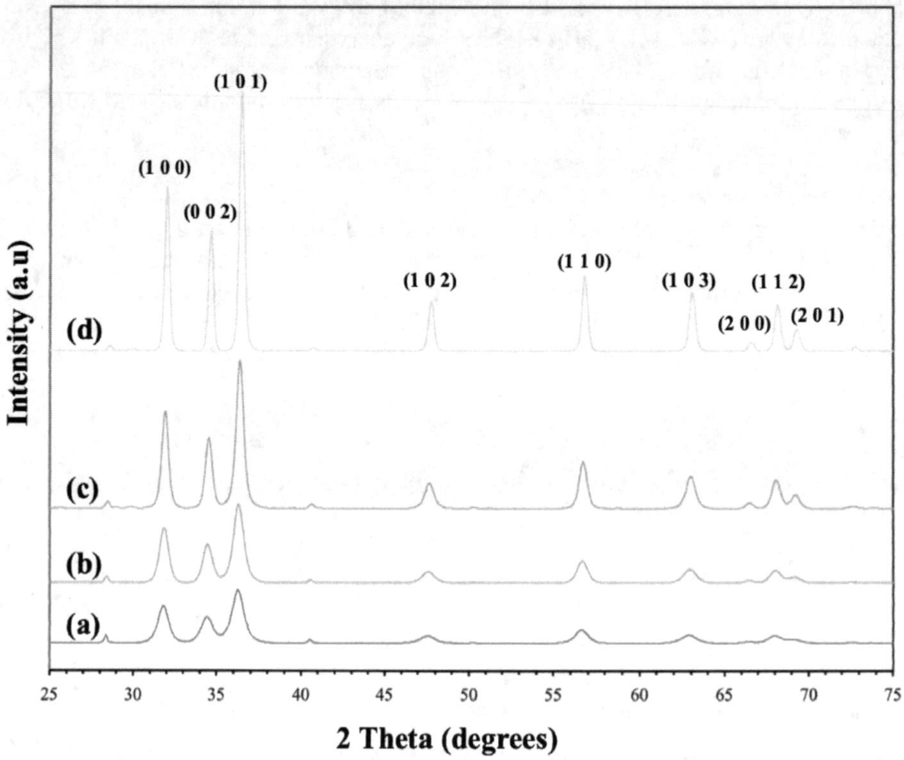

Figure 1. XRD patterns of (a) ZnO-NPs 400, (b) ZnO-NPs 500, (c) ZnO-NPs 600 and (d) ZnO-NPs 700 respectively.

The morphology of all biosynthesized ZnO-NPs was observed using TEM imaging. Figure 2 shows the shapes and sizes of all ZnO-NPs samples. Irregular shapes of ZnO-NPs with slight agglomeration were produced for all calcination temperatures. In lower calcination temperature, mostly spherical particles were synthesized but as calcination temperature increases, more irregular shapes of ZnO-NPs can be observed. This might be due to aggregation of small primary particles to form secondary or even tertiary particles [15]. Therefore, in higher calcination temperature, larger and irregularly shaped ZnO-NPs were produced as can be seen in Figure 2(c,d) as compared to Figure 2(a,b). Particle sizes of ZnO-NPs 400 and ZnO-NPs 500 were measured to less than 50 nm while ZnO-NPs 600 and ZnO-NPs 700 were in the range of 40-70 nm.

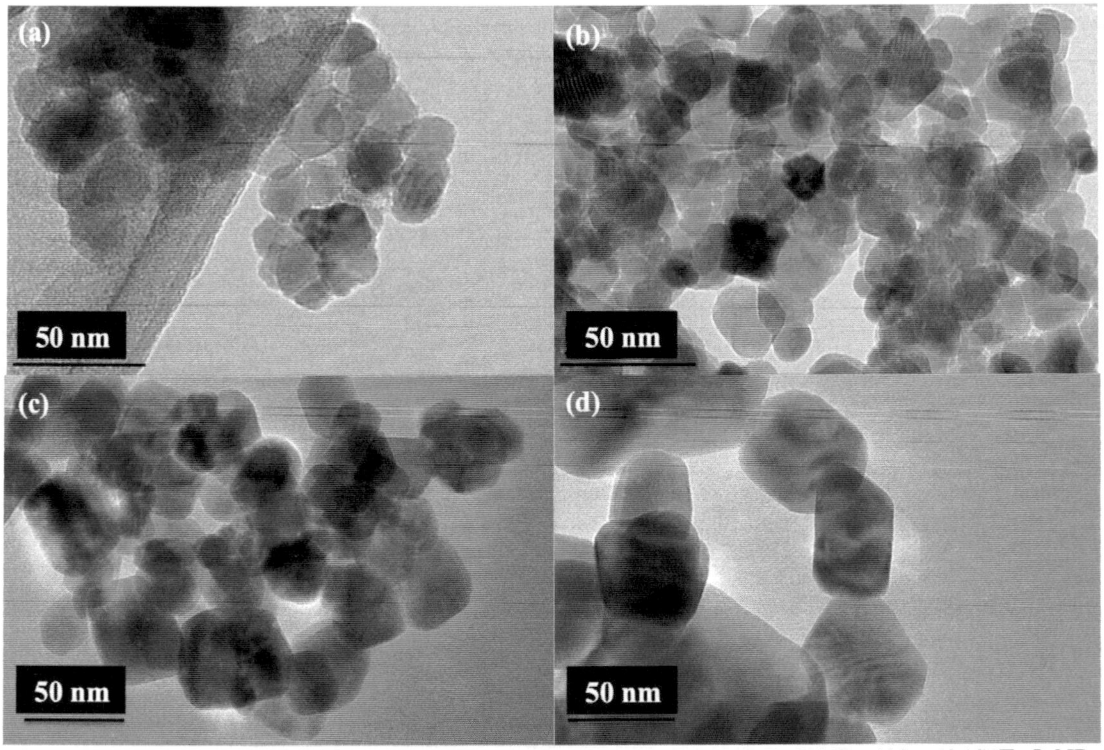

Figure 2. HRTEM images of (a) ZnO-NPs 400, (b) ZnO-NPs 500, (c) ZnO-NPs 600 and (d) ZnO-NPs 700 respectively.

Table 1 lists the surface area, pore volume as well as pore radius of all biosynthesized ZnO-NPs based on BET and BJH analysis. Specific surface area for the biosynthesized ZnO-NPs decreases progressively with increasing calcination temperature. The main factor leading to the reduction of surface area is the increased particle sizes of ZnO-NPs when calcined in higher temperatures, as observed from the HRTEM images in Figure 2. As sizes of ZnO-NPs become larger, pore volume and pore radius also decrease inevitably.

Table 1: Surface area, pore volume and pore radius of all biosynthesized ZnO-NPs.

Sample	Surface area (m^2/g)	Pore volume (cc/g)	Pore radius (nm)
ZnO-NPs 400	52.001	0.158	8.765
ZnO-NPs 500	42.854	0.117	1.245
ZnO-NPs 600	30.444	0.073	1.336
ZnO-NPs 700	17.160	0.021	0.820

The dye degradation activity of all biosynthesized ZnO-NPs samples was carried out against Malachite Green dye under UV light irradiation. The maximum absorbance peak wavelength of malachite green was observed at 617 nm. Figure 3 shows the gradual decrease in maximum absorption peak of Malachite Green with time for all ZnO-NPs samples and a control experiment (without ZnO-NPs photocatalyst). The control experiment as shown in Figure 3(a) shows slow reduction in Malachite Green absorption peak with time. Meanwhile in the presence of ZnO-NPs photocatalyst, upon exposure with UV light, the intensity of absorption decreases until almost zero in about 50 minutes of reaction time signifying a complete degradation of Malachite Green dye for all experiments as plotted in Figure 3(b-e).

IOP Conf. Series: Materials Science and Engineering **808** (2020) 012034 doi:10.1088/1757-899X/808/1/012034

Based on Figure 3(d), ZnO-NPs 700 showed the most rapid reduction of Malachite Green dye as the absorption peak reaches almost zero at only 30 minutes of time. Similarly, ZnO-NPs 600 showed high degradation performance as well as the Malachite Green dye is almost completely degraded at around 40 minutes. Despite having larger particles sizes, ZnO-NPs calcined in higher temperatures of 600 and 700°C showed excellent performance in degrading Malachite Green dye compared to ZnO-NPs with smaller particles sizes.

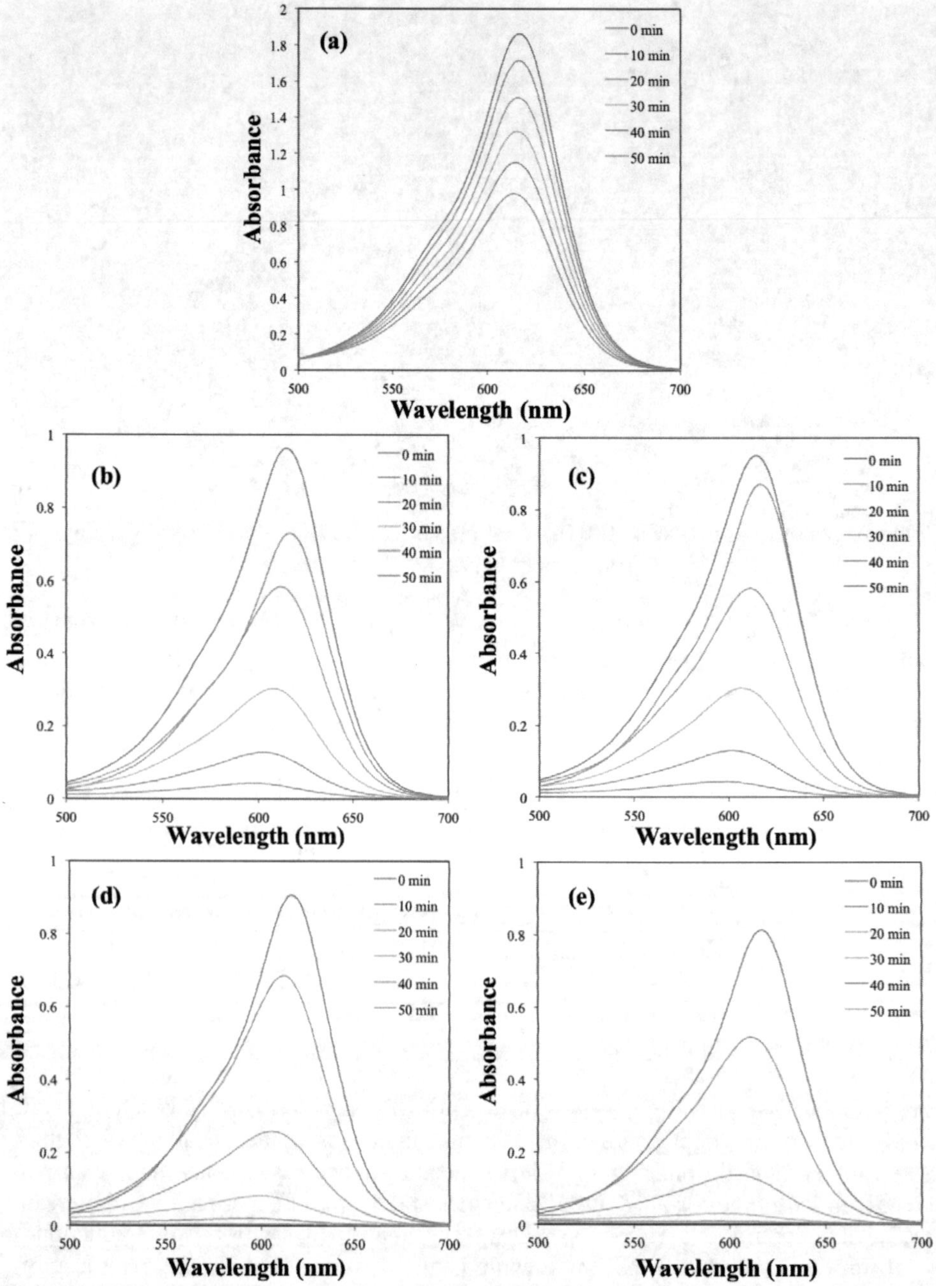

Figure 3. Decrease in absorption peak of Malachite Green dye with time for (a) control (without ZnO-NPs photocatalyst), (b) ZnO-NPs 400, (c) ZnO-NPs 500, (d) ZnO-NPs 600 and (e) ZnO-NPs 700 respectively.

From the absorbance data obtained, a degradation study was carried out and a comparative graph shown in Figure 4 was plotted to analyze the different removal efficiency of all ZnO-NPs photocatalysts. When compared to the control experiment, all four samples of ZnO-NPs photocatalysts showed excellent performance as they successfully degraded Malachite Green dye to almost a 100% rate. The percentage degradations Malachite Green dye by ZnO-NPs calcined at 400°C and 500°C are 96%. For ZnO-NPs calcined at 600°C and 700°C, highest removal efficiency of 99% were achieved in a short time of about 40 minutes. ZnO-NPs 700 performed the best as 10 ppm initial concentration of Malachite Green was degraded in only 30 minutes time. This corresponds to the previous analysis and characterization where ZnO-NPs 700 showed highest removal rate compared to those calcined in lower temperatures despite having larger particle sizes and lower surface area value.

Plausible reasons for this analysis have been discussed in previous studies mentioning the effect of surface defects and charge transfer on dye adsorption abilities of ZnO photocatalysts. When calcination temperature of ZnO-NPs increases, more surface defects such as oxygen vacancies are formed on their surface due to crystal structure disorder. This in turn stimulated greater dye adsorptions which influence their photocatalytic performance [16][17]. Crystallinity enhancement of ZnO-700 compared to other ZnO-NPs also plays an important role in decreasing charge transfer resistance. As charge transfer resistance decreases, electron transfer rate increases thus more hydroxyl radicals and ROS can be generated to oxidize the dye compounds [18].

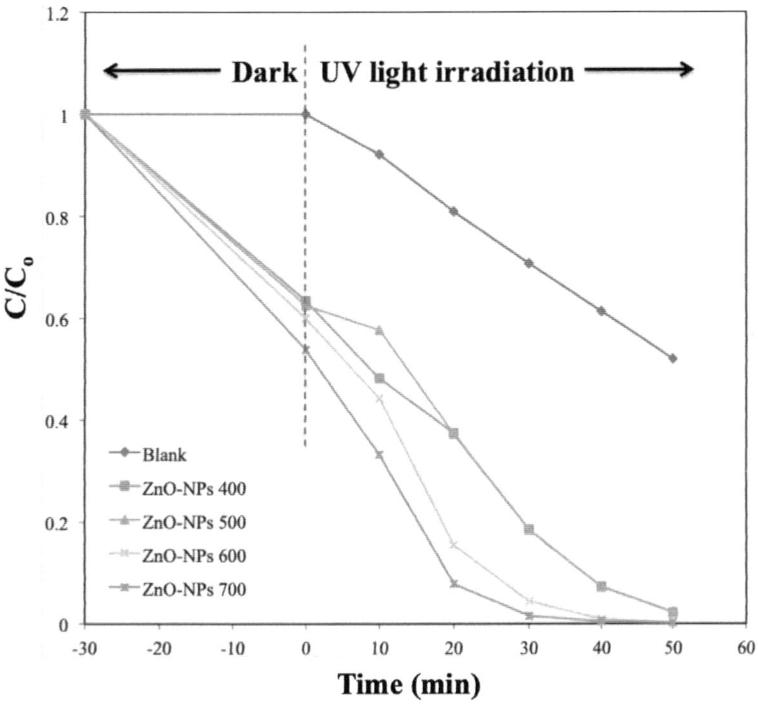

Figure 4. Comparative degradation study of Malachite Green dye with and without the presence of ZnO-NPs photocatalysts.

4. Conclusion

In the present work, plant-mediated biosynthesized ZnO-NPs were successfully produced using a simple sol-gel and combustion method in four different temperatures of 400, 500, 600, and 700°C. The effects of different calcinating temperatures towards the purity, crystallinity, surface area and particle sizes of ZnO-NPs were analysed. Comparative study of degradation efficiency for all biosynthesized ZnO-NPs was also carried out to determine which photocatalyst performed the best. HRTEM imaging showed smaller sized ZnO-NPs when calcined in lower temperatures of 400°C and 500°C with mainly spherical shapes. Higher calcination temperatures at 600°C and 700°C synthesized larger ZnO-NPs with more irregular shapes. As larger particles are produced in higher calcination temperatures, surface area, pore volume and pore radius of ZnO-NPs also decrease progressively. Photocatalytic experiments proved that ZnO-700 has the highest removal efficiency of Malachite Green dye compared to the other ZnO photocatalysts. This might be due to the increase of oxygen vacancies and electron transfer rate in ZnO-NPs calcined in higher temperatures.

Acknowledgements
The authors wish to acknowledge funding by the Malaysian Ministry of Higher Education under the Tier 1 grants (Grant no. #20H33 and #20H55) and express gratitude to the Research Management Centre (RMC) of UTM and Malaysia-Japan International Institute of Technology (MJIIT) for providing an excellent research environment and facilities.

References

[1] K. Saoud, R. Alsoubaihi, N. Bensalah, T. Bora, M. Bertino, and J. Dutta 2015 Synthesis of supported silver nano-spheres on zinc oxide nanorods for visible light photocatalytic applications *Mater. Res. Bull.* **63** 134–140

[2] S. Sathiyavimal, S. Vasantharaj, D. Bharathi, and M. Saravanan 2018 Biogenesis of copper oxide nanoparticles (CuONPs) using Sida acuta and their incorporation over cotton fabrics to prevent the pathogenicity of Gram negative and Gram positive bacteria *J. Photochem. Photobiol. B Biol.* **188** 126–134

[3] L. Yang, L. E. Yu, and M. B. Ray 2008 Degradation of paracetamol in aqueous solutions by TiO_2 photocatalysis *Water Res.* **42** 3480–3488

[4] K. Pradeev *et al.* 2018 Influence of Mg Doping on ZnO Nanoparticles for Enhanced Photocatalytic Evaluation and Antibacterial Analysis *Nanoscale Res. Lett.***13** 229

[5] M. I. Din, R. Khalid, and Z. Hussain 2018 Minireview: Silver-Doped Titanium Dioxide and Silver-Doped Zinc Oxide Photocatalysts *Anal. Lett.* **51** 892–907

[6] G. K. Dinesh, S. Anandan, and T. Sivasankar 2016 Synthesis of Fe / ZnO composite nanocatalyst and its sonophotocatalytic activity on acid yellow 23 dye and real textile effluent," *Clean Technol. Environ. Policy* **18** 1889–1903

[7] K. Kanagamani and P. Muthukrishnan 2019 Photocatalytic degradation of environmental perilous gentian violet dye using leucaena -mediated zinc oxide nanoparticle and its anticancer activity *Rare Met.***38** 277–286

[8] R. Ebrahimi, K. Hossienzadeh, A. Maleki, R. Ghanbari, R. Rezaee, and M. Safari 2019 Effects of doping zinc oxide nanoparticles with transition metals (Ag , Cu , Mn) on photocatalytic degradation of Direct Blue 15 dye under UV and visible light irradiation *J. Environ. Heal. Sci. Eng.* **2019**

[9] N. T. Nandhini, S. Rajeshkumar, and S. Mythili 2019 The possible mechanism of eco-friendly synthesized nanoparticles on hazardous dyes degradation *Biocatal. Agric. Biotechnol.* **19** 2019

[10] A. Yeganeh-Faal, M. Bordbar, N. Negahdar, and M. Nasrollahzadeh 2017 Green synthesis of the Ag/ZnO nanocomposite using Valeriana officinalis L. root extract: application as a reusable catalyst for the reduction of organic dyes in a very short time *IET Nanobiotechnology* **11** 669–676

[11] M. Sorbiun, E. Shayegan, M. Ali, R. Saeid, and T. Fardood 2018 Biosynthesis of Ag , ZnO and bimetallic Ag / ZnO alloy nanoparticles by aqueous extract of oak fruit hull (Jaft) and investigation of photocatalytic activity of ZnO and bimetallic Ag / ZnO for degradation of basic violet 3 dye *J. Mater. Sci. Mater. Electron.* **29** 2806–2814

[12] M. Kheirabadi, M. Samadi, E. Asadian, Y. Zhou, C. Dong, and J. Zhang 2019 Well-designed Ag / ZnO / 3D graphene structure for dye removal : Adsorption , photocatalysis and physical separation capabilities *J. Colloid Interface Sci.*, **537** 66–78

[13] M. Ghaedi, A. Ansari, M. H. Habibi, and A. R. Asghari 2014 Removal of malachite green from aqueous solution by zinc oxide nanoparticle loaded on activated carbon: Kinetics and isotherm study," *J. Ind. Eng. Chem.* **20** 17–28

[14] E. Karaköse, H. Çolak, and F. Duman 2017 Green synthesis and antimicrobial activity of ZnO nanostructures Punica granatum shell extract *Green Process. Synth.* **2017**

[15] K. B. Baharudin, N. Abdullah, and D. Derawi 2018 Effect of calcination temperature on the physicochemical properties of zinc oxide nanoparticles synthesized by coprecipitation *IOP Publ.* **2018**.

[16] I. Sugihartono, A. Retnoningtyas, and C. Rustana 2019 The influence of calcination temperature on optical properties of ZnO nanoparticles The Influence of Calcination Temperature on Optical Properties of ZnO Nanoparticles *AIP Conf. Proc.* **2019**

[17] Q. Zhang and M. Xu 2018 Oxygen Vacancy-Mediated ZnO Nanoparticle Photocatalyst for Degradation of Methylene Blue *Appl. Sci.* **8** 1–12

[18] A. M. Golsheikh, K. Z. Kamali, N. M. Huang, and A. K. Zak 2018 Effect of calcination temperature on performance of ZnO nanoparticles for dye-sensitized solar cells," *Powder Technol.***329** 282–287

IOP Conf. Series: Materials Science and Engineering **808** (2020) 012035 doi:10.1088/1757-899X/808/1/012035

Fabrication and Characterization of PU-g-poly(HEMA) Film for Clotting Time and Platelet Adhesion

Z I Tarmizi[1], R R Ali[1]*, M M Nasef[2], A M Akim[3], Z Eshak[4], S M Noor[5]

[1] Malaysia Japan International Institute of Technology, Universiti Teknologi Malaysia, Jalan Sultan Yahya Petra, Kuala Lumpur, 54100, MALAYSIA

[2] Department of Chemical Engineering, Universiti Teknologi Petronas, Seri Iskandar Perak, 32610, MALAYSIA

[3] Department of Biomedical Sciences, Faculty of Medicine and Health Sciences, Universiti Putra Malaysia, 43400 UPM Serdang, Selangor, MALAYSIA

[4] Imaging Centre, Faculty of Pharmacy, Universiti Teknologi MARA Puncak Alam, Bandar Puncak Alam, 42300, MALAYSIA

[5] Department of Pathology, Faculty of Medicine and Heath Sciences, Universiti Putra Malaysia, 43400 UPM Serdang, Selangor, MALAYSIA

*roshafima@utm.my

Abstract. This paper describes a fabrication of poly (2-hydroxylethyl methacrylate) poly(HEMA) grafted on polyurethane (PU) film prepared by radiation-induced grafting (RIG) copolymerization method using electron beam irradiation for the first time. This method was well known to be fast technique, clean method without involve any chemical initiator, chemically bond the materials, and at the same time is a sterile technique suitable for further potential of biomedical application. This poly(HEMA) grafted on PU film or called as PU-g-poly(HEMA) films was analysed using Fourier-transform infrared (FTIR), scanning electron microscope (SEM), water contact angle analyser (WCA), platelet adhesion and clotting time measurement. As the results, poly(HEMA) was confirmed successful grafted on PU based on the shifting of the functional group, no significant changes in surface morphology, lowering the water contact angle from 78.28° to 70.02°, nearly no platelet adhesion and no excessive disturbance of the clotting time was observed. This means that PU-g-poly(HEMA) was improved its hydrophilicity, thus significantly reduced the platelet adhesion and maintain the normal range of time taken for blood to clot. Therefore, the present PU-g-poly(HEMA) films not only improved hydrophilicity, however, was also compatible with blood. Thus, it may be potential candidates in the biomedical devices or new biomaterial useful for future tissue engineering fields.

1. Introduction

The arrangement of hard and soft segments in the PU polymer backbones make the PU confer good elasticity and biocompatible, resulting it wide application in biomedical application such as catheter [1], pacemaker insulating lead [2], hemoassess bridge fistulas [3] and vascular prosthesis [4]. The vascular graft bigger than 6 mm in diameter was successful to be use as vascular prosthesis however, smaller diameter less than 6 mm produced intimal hyperplasia and thrombosis [5]. Thus, surface modification need to be done to settle the thrombosis problem.

There are two common methods to prepare the materials either using physical blending or chemical method. Physical blending is the most common techniques where two or more components were mixed together to form new blend polymers that supposed to improve some characteristics such as thermal stability [6], ionic conductivity, mechanical strength [7] or porous morphology [8]. Modification of polymer using chemical method was also known as graft copolymerization, can be classified into four main categories by using chemical initiator [9], photo initiator [10], plasma treatment [11] and high energy radiation [12]. Therefore, in this study, surface modification via high energy radiation using electron beam irradiation was conducted due to fast technique, clean, sterile the material with the aim to improve the hydrophilicity of materials [13].

Based on previous study, no one has reported yet the preparation of PU films modified it surface with HEMA using RIG of electron beam. In this present study, the complimentary physical characterization of membranes was performed such as percentage degree of grafting, the morphologic observation using scanning electron microscope (SEM) instrument and the interaction and functional group before and after grafted with HEMA was analysed using ATR-FTIR. Concerning of the properties of film either hydrophilic or hydrophobic, the wettability index measurement was also conducted using water contact angle analysis. Our major objective is to demonstrated the prepared modified surface film can lead to formation of hemocompatible film, so that the PU-g-poly(HEMA) could be suitable and appropriate for blood contact devices or other biomedical applications.

2. Methodology

2.1. Materials

Polyurethane (PU) pellets (SelectophoreTM, Mw: 100 kg/mol) and 2-hydroxylethyl metharcylate (HEMA) monomer (99% purity) were obtained from Sigma Aldrich. Tetrahydrofuran (THF) with 99.5% purity and N, N-dimethyl formamide (DMF) with 99.5% purity was purchased from Merck. All the chemicals were used directly without further purification. The vacutainer sodium citrate tubes were brought from BD Company, USA. Meanwhile, APTT reagent (Dade Actin SL) were purchased from Siemen Healthcare, Erlangen, Germany.

2.2. PU Films Fabrication

PU films (10wt%) was produced by mixing the PU pellets in a mixture of THF and DMF with mixing ratios (1:1) and was stirred using digital magnetic stir plate (WiseStir® SMHS, DAIHAN Scientific, Korea) for overnight at room temperature. The PU solution then was pour into glass petri dish and covered with aluminium foils and kept in fume hood at room temperature to slowly evaporate all the solvents. PU film produced was dried in an oven for 72 hours at 60 °C to remove all residual solvent.

2.3. Surface Modification by Using Radiation-Induced Grafting

PU films produced was cut into 3x3 cm and soaked completely in 10wt% HEMA monomer for overnight in close tightly close bottle (50ml). Then, the PU film was taken out and gently dried with filter paper to remove excess monomer solution. After that, PU thin film was placed inside a thin polypropylene plastics and was sealed under vacuum. The prepared sample was put on the tray that covered with iced and then was radiated using an Electron Beam Processing System (NHV-Nissin High Voltage, EPS 3000, Cockroft Walton type, Japan) at 1 MeV an acceleration voltage, 2 mA beam current and 10 kGy per pass until total radiation dose applied was 50 kGy. After RIG process, the samples were peeled off from the polypropylene plastic and was shaken overnight with distilled water to remove excess monomer and homo-polymers. Finally, PU-g-poly(HEMA) film produced was dried in an oven for 72 hours at a temperature of 60°C.

2.4. Characterization of Materials

2.4.1. Percentage Degree of Grafting Analysis. The calculation for percentage degree of grafting (% DG) for modified surface films (PU-g-poly(HEMA) was described in the equation 1 below:

$$\text{Degree ofgrafting} = W_f - \frac{W_i \times 100}{W_i} \quad (1)$$

where, W_i and W_f refer to dry weight of samples before and after RIG copolymerization process.

2.4.2. Scanning Electron Microscope (SEM) Analysis. The morphology of PU film and PU-g-poly(HEMA) film was observed using scanning electron microscope (SEM) instrument (JSM-6701F, JOEL, USA). All films were coated with platinum prior to analysis for 60 second using auto fine coater instrument (JFC-1600, JOEL, USA).

2.4.3. Fourier Transform Infrared Analysis. The interaction and functional group was analysed using Fourier-transform infrared coupled with attenuated total reflectance (FTIR-ATR) (Frontier FTIR, Perkin Elmer, USA). All samples were scanned in ATR mode with a 16 scan, resolution 16 cm^{-1} and wavenumber range from 650 to 4000 cm^{-1}. The force gauge was applied in the range of 100 to 120 N and all samples were put in desiccant for overnight before analysis to reduce absorbed moisture.

2.4.4. Wettability Index Analysis. Wettability index analysis is the analysis to measure the water contact angle which indicate the characteristic of the polymer either hydrophilic or hydrophobic. Wettability index analysis for all samples were analysed using sessile drop techniques (VCA 3000S TM, AST Products Inc, USA.). The samples were put on slide for easy observation. The 100 μl of syringe size was used and 5μl of distilled water was dropped on the dry sample at room temperature. All the samples were measured ten times and average values were calculated. The contact angle value was presented as means value (n=10) with a standard deviation.

2.4.5. Clotting Time Measurement. Clotting time procedure is to measure how quickly blood clot when giving touch with foreign materials. For preparation of platelet poor plasma (PPP), around 2.7 ml of whole human blood was collected into one blood collection tube with 3.2% sodium citrate solution and was centrifuged at 3000 rpm (10 minutes). After centrifuged, the blood was separated and the plasma produced was transfer into new falcon tube which refer to platelet rich plasma (PRP). Then, the PRP was re-centrifuged at 3000 rpm for 10 minutes and PPP produced was transferred into new falcon tube without disturb the platelet pellet. The sample preparation and analysis using semi-automated blood analyser, each samples were cut into 1cm x 1cm and was put into 24 well plates. 100 μL of PPP was incubated at room temperature for 1 hour. The APTT was determined using a semi-automated blood coagulation analyser (Sysmex CA-50, Sysmex Corporation, Japan).

2.4.6. Platelet Adhesion Analysis. For the preparation of platelet rich plasma (PRP), the total of 2.7 ml of whole human blood was collected into the one blood collection tube with 3.2% sodium citrate solution and centrifuge at 3000 rpm for 10 minutes. The plasma obtained was transferred into new falcon tube which is known as PRP [14]. The unmodified and modified surface films were cut into 1 x 1 cm and were rinsed three times with normal saline and equilibrated for 30 minutes at room temperature. Then, the scaffolds were dried in oven at 60 ^{0} C for 3 hours. The dried scaffolds were incubated with 100 μl of PRP in 24 well plates for one hour at room temperature. After that, the films were taken out and smoothly rinsed with phosphate buffer saline (PBS) to removed un-adhered platelet. Platelets attach on the samples were fixed using formalin (10%) for 30 minutes in room temperature. Then, all films were washed with PBS were dehydrated with graded ethanol from 25%, 50% to 70% for 15 minutes each and were dried overnight in oven for 60 ^{0} C. For the morphology observation, all films were sputtered with platinum for 60 seconds before being observed using SEM instruments.

3. Results and Discussion

3.1. Percentage Degree of Grafting
The successful production of PU-g-poly(HEMA) film using RIG method was detected based on the calculation on the % DG. Therefore, modified surface film produced in this research using 10 wt% of HEMA monomer at 50 kGy absorbed dose was in the range of 16.72 - 16.26 % DG.

3.2. Scanning Electron Microscope (SEM) Analysis
The SEM images of PU films and modified surface PU film (PU-g-poly(HEMA)) were displays on Figure 1. The surface morphology images demonstrated that PU film had smooth surface whereas, PU-g-poly(HEMA) film produced showed rougher surface morphology. The same observation was reported by He et al, (2011), rougher surface morphology on modified surface PU films was produced after prepared using chemical-induced grafting method with benzoyl peroxide as initiator [9].

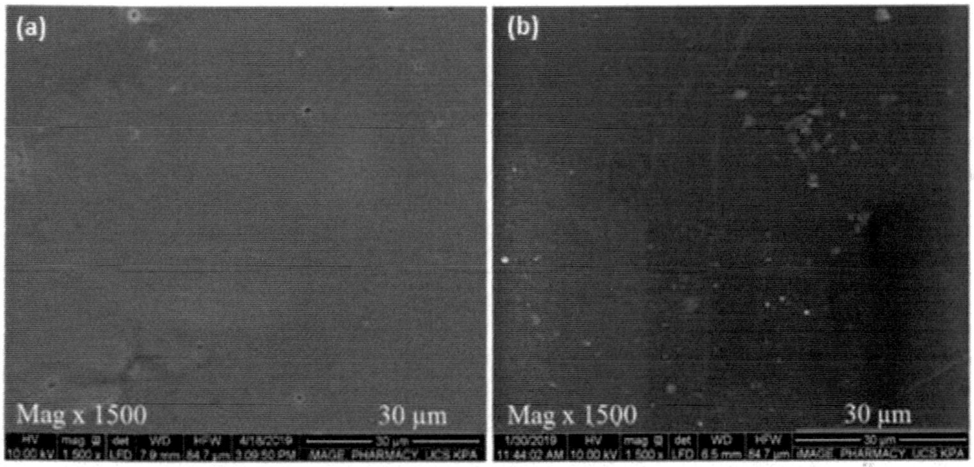

Figure 1. The SEM images shows the surface morphology of (a) PU film and (b) PU-g-poly(HEMA) 16% DG.

3.3. ATR-FTIR analysis
Figure 2 displays the ATR-FTIR images of pristine PU film, pure poly(HEMA) and the modified surface of PU films (PU-g-poly(HEMA) with 16.72% DG). For PU films, the peak at about 3321 cm^{-1} was attributed to the hydrogen bonding interaction of N-H stretching. The vibration band at 2852 cm^{-1} and 2929 cm^{-1} were corresponding to asymmetric and symmetric stretching methylene (-CH_2), followed by the peak at 1691 cm^{-1} correspond to the C=O stretching of the ester carbonyl group of urethane. Meanwhile, the peak at 1527 cm^{-1} and 1099 cm^{-1} were assigned to urethane N-H bending and C-O-C stretching in aliphatic the ether of soft and hard segments of PU [15]. The FTIR spectra of pristine poly(HEMA) exhibited a broad peak at 3320-3600 cm^{-1} corresponded to hydroxyl group, peak at 2848cm^{-1} for aliphatic stretching vibration of CH_2, and the sharp characteristic peak of carbonyl group was located at 1709 cm^{-1} [16]. The other two bands at 1154 cm^{-1} and 1071 cm^{-1} were attributed to carboxylic acid ester and primary alcohol –C-OH, respectively. The modified surface polymer exhibited IR peaks which quite similar to pure PU and poly(HEMA). However, the peak at 1691 cm^{-1} which responsible for urethane C=O stretching was shifted to higher wavenumber 1714cm^{-1}. Besides, the intensity of broad peak at 3320 cm^{-1} to 3600 cm^{-1} which refer to the hydroxyl group was found to become broader and increase due to cooperation of HEMA on the PU polymer backbone. This condition indicates that interaction occurred on the polymeric material surface, confirming the corporation of

poly(HEMA) on the PU film surface. As a conclusion, these results provided evidence for imparting a hydrophilic character to PU film was successfully produced using RIG method (Figure 3).

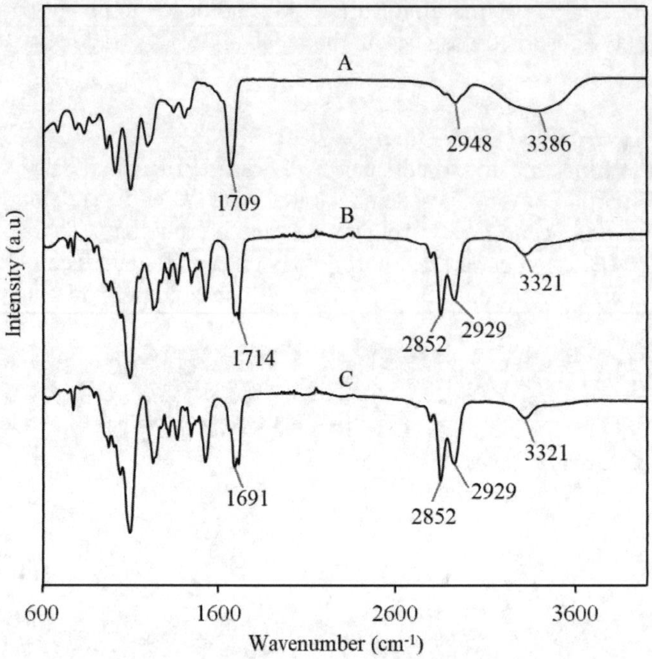

Figure 2. The ATR-FTIR spectra of: (A) PU film, (B) Poly(HEMA) and (C) PU-g-poly(HEMA) film.

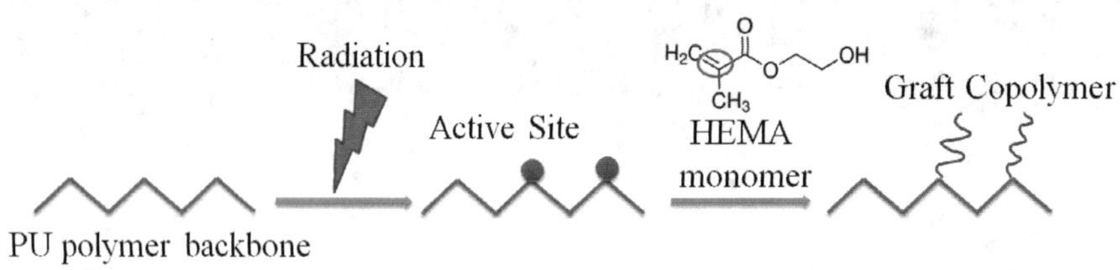

Figure 3. The production method of poly(HEMA) grafted on PU surface by using RIG copolymerization.

3.4. Wettability Index Analysis

Image of water contact angle produced using water contact angle analysis was shown in Figure 4. Smooth surface of PU film was prepared using solvent casting method was found to be 78.28° ± 0.23 and reduced to 70.02° ± 1.50 after grafted with poly(HEMA). This indicate that wettability values were significantly reduced and the hydrophilicity was risen after grafted with poly(HEMA). These findings were presented close value to previous study which is 79°, 89° and 90° for PU film and 42°,57° and 59° for poly(HEMA) grafted on surface films) [17], [9], [18]. Thus, it worth to mentioned that the hydrophilicity of PU scaffold improved after grafting with HEMA. In addition, the even surface of the materials attributed to hydrophilic nature of polymer. The reduced in wettability values was not conducive to rapid wettability of materials likely super-hydrophilic surface. However, with extended time soaked in the solution media, the material was completely wetted by liquid.

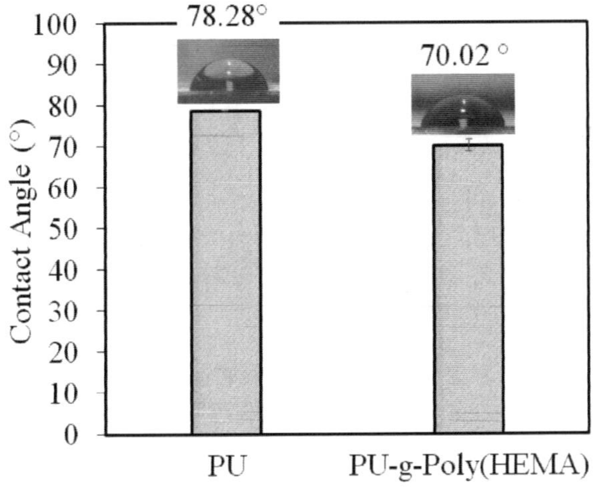

Figure 4. The water contact angle value for PU film before and after RIG surface modification.

3.5. Clotting Time Analysis

APTT mainly conducted to evaluate the effect of factors XII, XI, IX and VIII in the intrinsic pathway. PPP without scaffold was used as control and the graft for APTT results were presented in Figure 5. Based on the graft, the PU films make delay on the clotting time compared to control from 33.8 ± 0.35 to 34.1 ± 1.65 seconds and modified surface PU-g-poly(HEMA) 16.72 % DG was also increases the time taken for blood to clot from 33.8 ± 0.35 to 36.5 ± 1.1 This changes give little effect on coagulation factor where the clotting time for APTT value for normal range was in range of 30 to 40 minutes. The same trend was observed in the previous reported articles, showed that clotting time was risen when treated with bioflocculant (MBS-60) however still in normal range of APTT clotting time [19]. The result from An and co-researchers showed prolonged the clotting time with increasing degree of grafting however 30 min incubation time was conducted [20]. Longer incubation time at 37 °C will cause the loss of factor V and VIII and incubation time more than 10 minutes was not recommended. As a conclusion, the APTT value gathered was still lies in the normal range of APTT, therefore it showed that PU film and PU-g-poly(HEMA) film was compatible with blood.

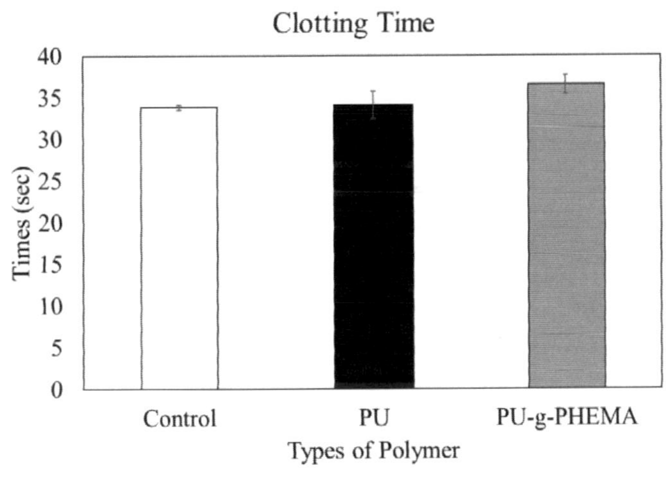

Figure 5. The APTT test for PU film before and after RIG surface modification. Control is the PPP without scaffolds.

IOP Conf. Series: Materials Science and Engineering **808** (2020) 012035 doi:10.1088/1757-899X/808/1/012035

3.6. Platelet Adhesion Analysis

Figure 6 display the SEM image for PU film, and its modified surface (PU-g-poly(HEMA) film. Unmodified surface of PU film showed abundant platelet attachment compared to the modified surface which means that PU-g-poly(HEMA) achieved highly significant reduction in platelet adhesion. These phenomena were probably due to charge repulsion since modified film contained hydrophilic surface and platelet surface are negatively charged. Thus, the platelet adhesion can be restricted by electrostatic repulsion effect between platelet and negatively charge functional groups [21]. The result reported here was in agreement with other previous published paper [22] [23]. Thus, this test demonstrated that the extent of modification was sufficient to significantly reduce platelet adhesion. A decrease platelet adhesion also could reduce not only coagulation processes, but also the possibility of early restenosis.

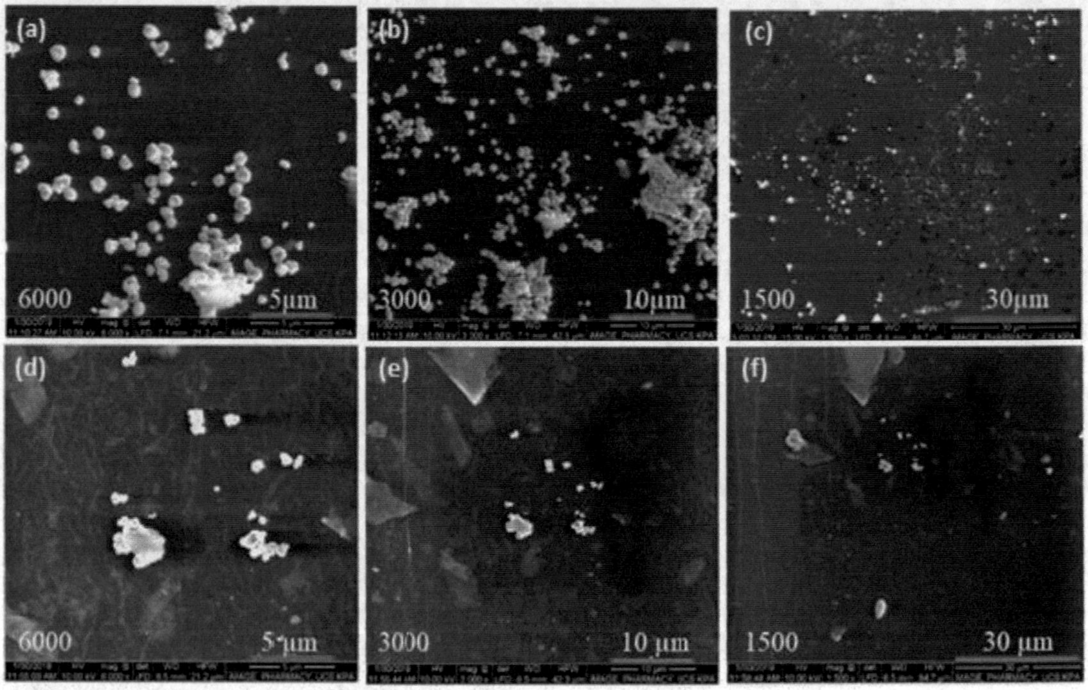

Figure 6. The SEM images of platelet adhered on pure PU film with different magnification (a-c), and on PU-g-poly(HEMA) 16.72% DG film with different magnification (d-f).

4. Conclusion

As a conclusion, the wettability index of PU-g-poly(HEMA) film was reduced from 78.28° to 70.02° revealed that the hydrophilicity of the modified film was successfully improved. Some change on the morphology and ATR-FTIR spectrum supports along the results. The clotting time analysis and platelet adhesion test indicates the film was compatible with blood. Thus, it can be concluded that PU-g-poly(HEMA) film with 16.72 % DG was successfully prepared using RIG method via electron beam radiation and was confirmed improved it hydrophilicity. Therefore, the obtained film was expected have a potential to be used as new material for tissue engineering or off-the-shelf blood contact devices.

Acknowledgement
This research is supported by Universiti Teknologi Malaysia (UTM) research fund under GUP (Vote no: Q.K130000.2543.19H55). Special acknowledgement to Rohani Omar, Amrina Muhammad Amin and Azizan Din for their assistance in this works.

References
[1] Volkow P, Vázquez C, Tellez O, Aguilar C, Barrera L, Rodriguez E, Vilar-Compte D, Zinser J, Calderon E, Perez-Padilla J R, and Mohar A 2003 *Am J Infect Control*, **31** 392
[2] Voogt W G D 1999 Am J Cardiol. **83** 187
[3] Kiyama H, Imazeki T, Kurihara S, and Yoneshima H 2003 *Ann. Vasc. Surg.* **17** 516
[4] Xie X, Eberhart A, Guidoin R, Marois Y, Douville Y and Zhang Z 2010 *J. Biomater. Sci.* **12** 1239
[5] Zhang W J, Liu W, Cui L and Cao Y, 2007 *J. Cell. Mol. Med.* **11** 945
[6] Zakaria Z, Izzah Z, Jawaid M and Hassan A 2012 *BioResources* **7** 5568
[7] Choo K, Ching Y C, Chuah C H, Julai S and Liou N 2016 *Materials* **9** 1
[8] Kim G M, Le K H T, Giannitelli S M, Lee Y J, Rainer A and Trombetta M 2013 *J. Mater. Sci. Mater. Med.* **24** 1425
[9] He C, Wang M, Cai X, Huang X, Zhu H, Shen J and Yuan J 2011 *Appl. Surf. Sci.* **258** 755
[10] Ke Y, Wu G and Wang Y 2014 *Biomed Res. Int.* **2014** 1
[11] Jing X, Mi H Y, Salick M R, Cordie T M, Peng X F and Turng L S 2015 *Mater. Sci. Eng. C.* **49**, 40
[12] Tarmizi Z I, Nasef M M, Nallappan M L, Khairudin N B A, Ali R R and Ahmad A 2018 *AIP Conf. Proc* **1985** 030015-1-9
[13] Nasef M.M 2014 *Chem. Rev.* **114**, 12278
[14] Li R, Wang H, Wang W and Ye Y 2013 *Radiat. Phys. Chem.* **91**132
[15] Zhang Y, Wang J, Fang X, Liao J, Zhou X, Zhou S, Bai F and Peng 2019 *Polymer* **178** 121572
[16] Norhayati A, Zuhaili Y M and Rabiatuladawiah M 2018 *Mater. Today Proc* **5** 22010
[17] Martins M C L, Wang D, Ji J, Feng L and Barbosa M A 2003 *Biomaterials* **24** 2067
[18] Jin Z, Feng W, Zhu S, Sheardown H and Brash J L 2010 *J. Biomed. Mater. Res-Part A*, **95** 1223
[19] Zhao H, Cao G, Chen H, Li H and Zhou J 2017 *Colloids Surfaces B Biointerfaces* **159**, 712
[20] An Z, Li Y, Xu R, Dai F, Zhao Y and Chen L 2018 *Appl. Surf. Sci.* **457** 170
[21] Venault A, Ballad M R B, Liu Y H, Aimar P and Chang Y 2015 *J. Memb. Sci.* **477** 101
[22] Li R, Wu G and Ye Y 2015 *Appl. Surf. Sci.* **356** 1221
[23] Caracciolo P C, Rial-Hermida M I, Montini-Ballarin F, Abraham G A, Concheiro A and Alvarez-Lorenzo C 2017 *Mater. Sci. Eng. C* **75** 1115

Green Synthesis of Gold Nanoparticles using Aqueous Extract of *Clitoria Ternatea* Flower

J Z Chan, R Rasit Ali[1]*, K Shameli[1], M S N Salleh[1,2], K X Lee[1], E D Mohamed Isa[1]

[1] Department of Chemical Process Engineering, Malaysia-Japan International Institute of Technology, Universiti Teknologi Malaysia, Jalan Sultan Yahya Petra, 54100 Kuala Lumpur, Malaysia
[2] Faculty of Chemical Engineering, Universiti Teknologi MARA Cawangan Terengganu, Bukit Besi Campus, 23200 Dungun, Terengganu, Malaysia

*roshafima@utm.my

Abstract. The synthesis of gold nanoparticles (Au-NPs) is being accomplished by the reduction of aqueous gold metal ions, gold(III) chloride trihydrate ($HAuCl_4.3H_2O$) reacted with the aqueous flower extract of *Clitoria Ternatea* (CT). CT flower extract plays an important role in synthesizing Au-NPs. It acts as a reducing (Au^{3+} to Au) and stabilizing agent that can eliminate the usage of chemicals during the production of Au-NPs. Besides that, it also reduces the production of unwanted by-products which would cause hazardous to the surroundings and environment. In this study, an absorption peak of Au-NPs is observed at the range of 540-550 nm from ultraviolet-visible spectroscopy (UV-vis) analysis. Furthermore, the diffraction peaks at $2\theta = 38.44°$, 44.41°, 65.03° and 77.58° respectively which correspond to face-centered cubic structure with (111), (200), (220) and (311) plane confirm the successful synthesis of Au-NPs. According to transmission electron microscopy (TEM), the majority of Au-NPs are spherical in shape and having a mean particle size distribution of 18.16 nm with a standard deviation of 4.67 nm.

1. Introduction

The synthesis of nanoparticles can be done using either chemical or physical method conventionally where chemical synthesis always involves the usage of chemical reagents in order to act as the reducing and stabilizing agent. On the other hand, physical methods of synthesis usually involve the manipulation of surrounding properties or external forces such as temperature and pressure in order to separate the materials from larger size to nanoparticles size. The synthesis of nanoparticles can be achieved by two approaches; one is the top-down approach (bulk to nano size) while the other is bottom-up (atoms to nano size) [1]. With the different of these two approaches, scientists and researchers can produce nanoparticles of different sizes and shapes with different properties. For example, the different dimensions and their physical size of nanoparticles will provide gold nanoparticles that will yield a very different colour. These phenomena are caused by the oscillation of electric fields of a light ray propagating near colloidal nanoparticles which then interact with the free electrons causing a concerted oscillation of electron charge that is in resonance with the frequency of visible light [2].

However, most of the conventional method for synthesized nanoparticles which uses chemicals as a reducing or stabilizing agent and physical method that involve high temperature and pressure are not economically and environmentally friendly due to the toxic by-products being produced during the process that is harmful to the surroundings and environment. For example, using sodium borohydride ($NaBH_4$) as a reducing agent will eventually produce several by-products such as sodium chloride (NaCl), hydrochloric acid (HCl), and borane (BH_3). Thus, the synthesis method that uses natural products that act as reducing agents must be introduced in order to reduce the hazards to the environment and humans [3]. A green alternative of reducing agent can be further divided into two categories, through using microorganisms or using plants. Examples of microorganisms, such as enzyme [4], fungus [5] and algae [6] were reported successfully in the production of Au-NPs. However, synthesized Au-NPs using microorganisms having a much-complicated process especially maintaining their cultures and it is also much more expensive when compared to using plants [3,7]. Thus, plant-mediated synthesis of nanoparticles is developing due to the ease to handle and the ability of mass-production.

Clitoria Ternatea (CT) is commonly known as "*Butterfly pea*" belongs to the family of Fabaceae. The plant is a twining evergreen herb, which will grow up to 3 m high, climbing over any available prop and is distributed in tropical Asia such as Myanmar, Thailand, Indonesia, India, Philippines, and Malaysia. The stems are pubescent and spindly. The compound leaves are made of three to nine oval or elliptical leaflets and the flowers are 2–4 cm long and in various shades of blue [8]. In the olden day, the root is already being used in the treatment of various diseases, like indigestion, constipation, arthritis and eye ailments. It is also employed in cases of ascetics, enlargement of the abdominal viscera, sore throat, skin diseases, and the flower are recommended for the treatment of snake-bite [9]. It is also reported that CT flower contains high amounts of anthocyanin pigments and at the same time many pieces of evidence indicate that CT flower to be rich in antioxidant properties as compared to other flowers and medicine elements. Thus, in ancient human civilization, CT has been used to treat neurological disorders [10]. According to Ghosh 2007, CT flower also showed anti-viral, anti-inflammatory, anti-allergic, and anti-microbial properties, prevent diabetes, protect the cardiovascular system and much more health benefit most probably due to anthocyanin pigment [11]. The natural antioxidants in CT flower include flavonoids, anthocyanin, flavanol glycosides, phenolic acids and procyanidins [11]. Furthermore, the components presented in the water extract from flowers of CT were inositol (38.7%) and pentanal (14.3%) based on the gas chromatogram mass spectrometry (GC-MS) analysis [12].

In this study, Au-NPs were synthesized using CT flower aqueous extract that acts as a reducing and stabilizing agent. This green synthesis technique was chosen due to its clean and hazardous free process. In addition, this method also presents huge potential for mass-production of Au-NPs with fast reaction times, high yield and cheap.

2. Materials and method

All material and reagents used in this work were analytical grade and used as received without further purification. Gold(III) chloride trihydrate ($HAuCl_4.3H_2O$, $\geq$ 99.98%) was used as the gold precursor from Sigma-Aldrich, USA. The CT flowers were purchased from one of the markets in Kuala Lumpur, Malaysia. It is harvested by the owner of the store early in the morning and let it air dry under a ceiling fan indoor to remove excess moisture from the flower for as long as 24 hours. Deionized water is utilized to prepare chemical solutions in new batches. Fresh HNO_3/HCl with the ratio of (3:1, v/v) is used to clean all glassware, followed by deionized water, and dried prior to use.

2.1. Preparation of aqueous CT flower extract

CT flower is collected and will be wash thoroughly with tap water first to remove dust and dirt. The flower is then washed again with deionized water before drying in an oven (EscoIsotherm Forced Convection Laboratory Oven) under 40 °C for 48 hr. This is to ensure that the chemical structure in CT flower will not change due to the high temperature of drying. Besides that, all the flowers are stored at room temperature in an airtight container for further usage. When needed, the flower aqueous is prepared from adding 1 g of CT flower in 50 mL of deionized water, 75°C for 1 hr with continuous stirring at 500 rpm to ensure maximum extraction. The extract will then undergo centrifuge at 3500 rpm for 10 min before being filtered with Whatman Grade 1 filter paper through the suction pump filtration method to further remove solid particles from CT flower extract. Fresh CT flower extract is prepared for all the experiments below.

2.2. Synthesis of Au-NPs
For the synthesis of Au-NPs, approximately 10 mL of CT flower extract was reacted with 10 mM of $HAuCl_4.3H_2O$ gold metal salt with the ratio of (10:1) at room temperature, continuous stirring at 500 rpm. The colour change of the reaction was observed, and the time taken for the changes was noted. The colour of CT flower extract changes colours immediately from dark blue to purple colour which indicates the formation of Au-NPs.

2.3. Characterization of Au-NPs
UV-visible spectroscopy (UV-vis) at regular intervals from the range of 400 to 800 nm (Shimadzu, UV-2600) was used to confirmed the production of Au-NPs. The Au-NPs sample was then transferred to an oven and dried at 70 °C for 2 days in order to remove all the moisture. The dried sample was then collected and analyzed for the structure and composition using powder X-ray diffraction (XRD). The data was recorded using PANalytic EMPYREAN at 45 kV, 40 mA, and over a range of 2θ = 20-80° with a scanning rate of 2θ/min. Transmission electron microscopy (TEM), JEOL JEM-2100F was used to study the morphology, size and distribution of the Au-NPs. NPs size distributions were measured and calculated using program ImageJ Launcher and a histogram was plotted. The samples were prepared by dispersing them in deionized water under sonication for 15 min and then were dropped onto a carbon-coated copper grid and dried at room temperature.

3. Results and discussion
CT flower extract (1.0g, 10 mL) acts as both the reducing and stabilizing agent when reacted with $HAuCl_4.3H_2O$ (10 mM, 1 mL). The most obvious reduction occurred for $HAuCl_4.3H_2O$ was regarding the colour changes as shown in Figure 1 (dark blue to purplish). CT flower is chosen due to the high amounts of anthocyanin pigment which also being used in natural supplements [13]. It also is shown having high reducing, anti-microbial, anti-inflammatory, anti-carcinogenic and antioxidant properties [14–16]. However, CT leaves and roots did not provide enough reducing power to reduce Au^{3+} to Au^0 (Au-NPs).

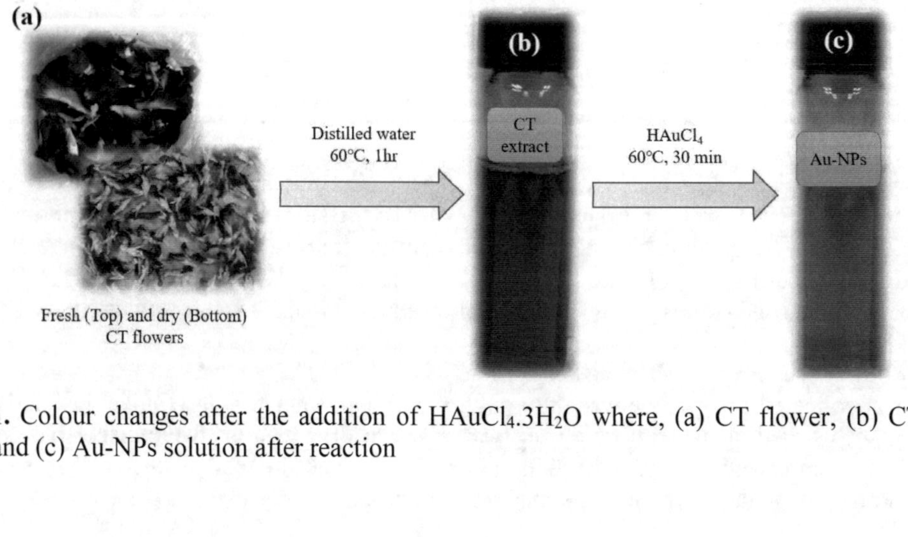

Figure 1. Colour changes after the addition of $HAuCl_4.3H_2O$ where, (a) CT flower, (b) CT flower extract, and (c) Au-NPs solution after reaction

The reaction was rapid when there was enough flower extract to reduce $HAuCl_4.3H_2O$ from Au^{3+} to Au. However, a parameter for reaction time was done in order to ensure the reaction was completed at a specific moment. CT flower extracts possessed a concentration-response relationship in 2,2-diphenyl-1-picrylhydrazyl (DPPH) scavenging activity using ascorbic acid as a positive control. IC_{50} of ascorbic acid (5.34 µg/mL), the IC_{50} of CT flower extract (84.15 µg/mL). Moreover, the Ferric Reducing Antioxidant Power (FRAP) assay calibrated with standard ascorbic acid, the reducing power of the CT flower extract was (0.33 mmol/mg) ascorbic equivalent [17].

3.1. UV-Visible spectroscopy analysis

It is important to know the peaks that can be obtained from pure gold metal salt $HAuCl_4.3H_2O$ alone and CT flower aqueous extract through UV-vis characterization. By doing so, the process of chemical reactions can be monitored closely as only atoms present in the reactants can be ended up in the product. As no new atoms will be created and no atoms are destroyed. Thus, in other words, the reactant peak should be decreased with the formation of new peaks. Initially, pure $HAuCl_4.3H_2O$ with the concentration of 10 mM was measured and a sharp peak is obtained at around 310 nm as shown in Figure 2 (a). Besides that, pure CT flower extract was measured and two peaks at around 575 and 620 nm were obtained as shown in Figure 2 (b). Delphinidin which is an anthocyanidin with the colour of purplish blue was responsible for both of the peaks found in CT flower extract [18].

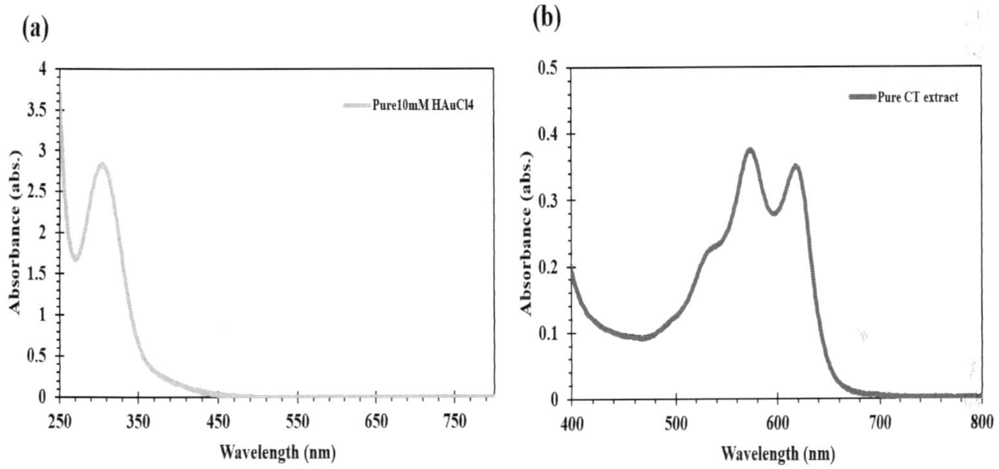

Figure 2. UV-vis absorption spectrum regarding (a) pure $HAuCl_4.3H_2O$ metal salt with a concentration of 10 mM and (b) pure CT extract (1.0g, 1hr) at 60°C.

Generally, Au-NPs has an absorption peak range from 540-550 nm. The UV-vis spectroscopy analysis of CT flower extract reacted with $HAuCl_4.3H_2O$ was shown in Figure 3, with the wavelength range between 400-800 nm. The CT flower extract did not show any absorption peak for all 6 samples at different reaction times (0, 5, 15, 30, 60, 120 min). However, Figure 3 shows the optimum time for the reaction was at 15 min with the highest absorbance at around 0.17 and at the same time the absorption peak also become sharper as compared to other samples. Besides that, the 15 min sample has a better property as the UV-vis absorption peak shifted to the left (blue shift). This proposes a decrease in the size of Au-NPs. The optical properties of Au-NPs are highly dependent on the nanoparticle's diameter. The smaller Au-NPs scatter less light because it has smaller optical cross-sections and due to its albedo [19]. Thus, 15 min sample would be selected to run the XRD and TEM characterization to confirm the crystallinity and purity, and to study the size and shape of AU-NPs respectively.

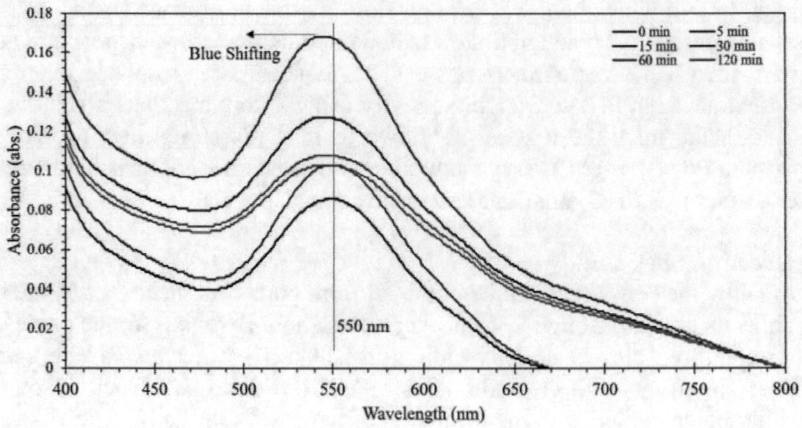

Figure 3. UV-vis absorption spectrum of CT flower extract with $HAuCl_4.3H_2O$ at different time intervals.

3.2. X-ray diffraction analysis

The production of pure crystalline Au-NPs was confirmed again through XRD as shown in Figure 4. After comparing the spectrum with the database, it is found that the entire spectrum was compatible with the JCPDS file no. 00-004-0783 which can be found at X'Pert HighScore. These spectrums give an intense peak at 2θ = 38.44°, 44.41°, 65.03° and 77.58° respectively which correspond to face-centered cubic structure with (111), (200), (220) and (311) plane [3]. Besides that, Figure 4 (b) show the XRD spectrum for pure CT flower extract. It helped to explain the peak at 2θ = 20.88° which was trigged by the organic compound from CT flower extract. The particle size of Au-NPs can be estimated by using the Debye-Scherrer formula as shown in equation 1:

$$d = \frac{k\lambda}{(\beta.cos\theta)} \tag{1}$$

where d is the average crystallite size, k is the Scherrer constant (0.9), λ is the X-ray wavelength (0.154 nm), β is the line broadening in radians, and θ is the Bragg angle [20]. The calculated average Au-NPs size is 17.67 nm at the angle of 38.44°.

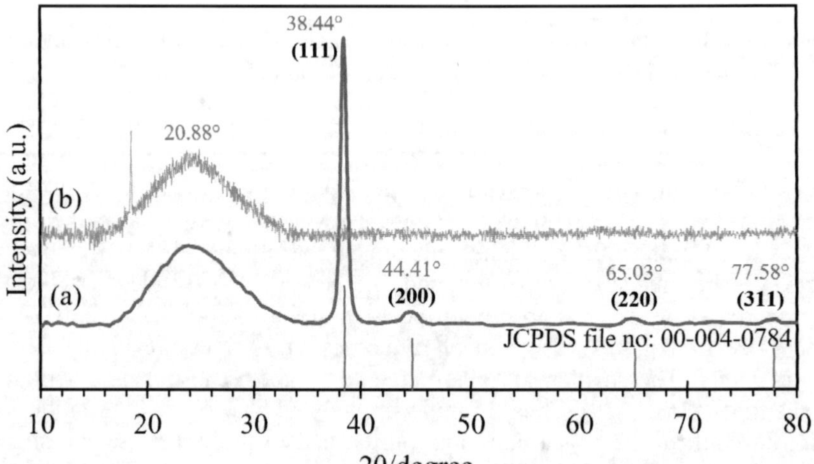

Figure 4. XRD patterns of (a) Au-NPs using CT flower extract, and (b) pure CT flower extract.

3.3. Transmission electron microscopy analysis

The size and the morphology of the Au-NPs synthesized were investigated using the TEM analysis which is represented by Figure 5 (a). The Au-NPs synthesized from CT flower aqueous extract (20 mg/ML, 10mL) reacted with $HAuCl_4.3H_2O$ (10 mM, 1 mL) was well dispersed. It was because CT flower extract did not only have the reducing power (Au^{3+} to Au^0) but also having the stabilizing properties to minimize agglomeration of Au-NPs [21]. Moreover, the TEM image of synthesized Au-NPs also showed that most nanoparticles were in a spherical shape. A histogram of particle size distribution was drawn according to the size of 200 Au-NPs as shown in Figure 5 (b). According to the histogram, the mean particle size was 18.16 nm in diameter with a standard deviation of 4.67 nm. The crystallite size of the synthesized Au-NPs was found to be 17.67 nm in diameter from XRD analysis, which was in an agreement with the result obtained from the TEM that shows a size distribution between 13 – 23 nm in diameter.

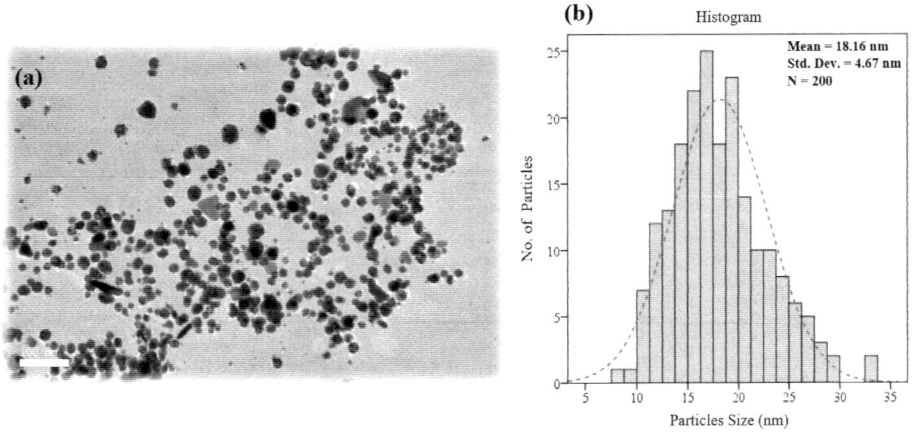

Figure 5. (a) TEM image of Au-NPs forms using CT flower extract at 100 nm, and (b) histogram of the distribution of the particle size of Au-NPs.

4. Conclusion

In conclusion, Au-NPs were synthesized successfully by a green and simple approach using CT flower extract without utilizing any chemical that acts as a reducing or stabilizing agent. Based on the UV-vis analysis studied, the best reaction time for synthesized Au-NPs using CT flower extract was 15 min with the absorbance peak of around 0.17 at 540 nm. Furthermore, according to the XRD analysis, a high purity crystalline of Au-NPs was prepared. The TEM imaging ascertained that Au-NPs were observed as majority spherical in shape with a mean particle size distribution of 18.16 nm in diameter. The synthesized Au-NPs using CT flower extract having the potential to be applied in other applications especially biomedical applications due to its biocompatibility and nontoxicity.

Acknowledgments

The authors are grateful for the financial support from the Malaysia-Japan International Institute of Technology (MJIIT), Universiti Teknologi Malaysia and Ministry of Education (Grant: Q.K130000.2543.19H55 and R.K130000.7743.4J305).

References

[1] Raab C, Simkó M and Fiedeler U 2015 Production of nanoparticles and nanomaterials
[2] S. Aldrich Gold Nanoparticles: Properties and Applications
[3] Kar Xin Lee, Kamyar Shameli, MikioMiyake N K and Nurul Bahiyah Bt Ahmad Khairudin, Shaza Eva BtMohamad and Y P Y 2016 Green Synthesis of Gold Nanoparticles Using

Aqueous Extract of Garcinia mangostana Fruit Peels *Adv. Nat. Sci. Nanosci. Nanotechnol.* **7** 1–7

[4] Narayanan K B and Sakthivel N 2011 Facile green synthesis of gold nanostructures by NADPH-dependent enzyme from the extract of Sclerotium rolfsii *Colloids Surfaces A Physicochem. Eng. Asp.* **380** 156–61

[5] Kitching M, Ramani M and Marsili E 2015 Fungal biosynthesis of gold nanoparticles : mechanism and scale up **8** 904–17

[6] Edhaya Naveena B and Prakash S 2013 Biological synthesis of gold nanoparticles using marine algae Gracilaria corticata and its application as a potent antimicrobial and antioxidant agent *Asian J. Pharm. Clin. Res.* **6** 179–82

[7] Elia P, Zach R and Hazan S 2014 Green synthesis of gold nanoparticles using plant extracts as reducing agents 4007–21

[8] Jain N N, Ohal C C, Shroff S K, Bhutada R H, Somani R S, Kasture V S and Kasture S B 2003 Clitoria ternatea and the CNS *Pharmacol. Biochem. Behav.* **75** 529–36

[9] Division P B, Tissue P, Plant R and Centre R 2004 Effect of cytokinins and auxins on micropropagation of Clitoria ternatea L . *In Vitro* **41** 21–7

[10] Khadatkar S, Manwar J and Bhajipale N 2008 In-vitro anthelmintic activity of root of Clitoria ternatea Linn *Pharmacogn. Mag.* **4** 148

[11] Jamil N, Mohd Zairi M N, Mohd Nasim N A and Pa'ee F 2018 Influences of Environmental Conditions to Phytoconstituents in Clitoria ternatea (Butterfly Pea Flower) – A Review *J. Sci. Technol.* **10** 208–28

[12] Salleh R M, Ong M T and Neda G D 2013 Chemical composition and anti-proliferative properties of flowers of Clitoria Ternatea *Int. Food Res. J.* **20** 1229–34

[13] Pham T N, Nguyen D C, Lam T D, Thinh P Van and Tien X 2019 Extraction of anthocyanins from Butterfly pea (Clitoria ternatea L . Flowers) in Southern Vietnam : Response surface modeling for optimization of the operation conditions Extraction of anthocyanins from Butterfly pea (Clitoria ternatea L . Flowers) in Southern Vietnam : Response surface modeling for optimization of the operation conditions

[14] Ongkowijoyo P, Luna-vital D A, Gonzalez E and Mejia D 2018 Extraction techniques and analysis of anthocyanins from food sources by mass spectrometry : An update *Food Chem.* **250** 113–26

[15] Rupasinghe H P V, Arumuggam N, Amararathna M and Silva A B K H De 2018 The potential health bene fi ts of haskap (Lonicera caerulea L .): Role of cyanidin-3- O -glucoside *J. Funct. Foods* **44** 24–39

[16] Xie L, Su H, Sun C, Zheng X and Chen W 2018 Trends in Food Science & Technology Recent advances in understanding the anti-obesity activity of anthocyanins and their biosynthesis in microorganisms *Trends Food Sci. Technol.* **72** 13–24

[17] Iamsaard S, Burawat J, Kanla P and Arun S 2014 Antioxidant activity and protective effect of Clitoria ternatea flower extract on testicular damage induced by ketoconazole in rats * **15** 548–55

[18] Lee P M and Abdullah R 2011 Thermal Degradation of Blue Anthocyanin Extract of Clitoria ternatea Flower **7** 49–53

[19] Anon 2019 Gold Nanoparticles: Optical Properties *nanoComposix*

[20] Mahdavi M, Namvar F, Ahmad M Bin and Mohamad R 2013 Green Biosynthesis and Characterization of Magnetic Iron Oxide (Fe3O4) Nanoparticles Using Seaweed (Sargassum muticum) Aqueous Extract 5954–64

[21] Sushma N J, Swathi D P G, Deva T M B and Raju P 2016 Facile approach to synthesize magnesium oxide nanoparticles by using Clitoria ternatea — characterization and in vitro antioxidant studies *Appl. Nanosci.* **6** 437–44

IOP Conf. Series: Materials Science and Engineering **808** (2020) 012037 doi:10.1088/1757-899X/808/1/012037

A viable system for carbon dioxide (CO_2) methanation over fibrous silica ZSM-5 for substitute natural gas (SNG)

I Hussain[1], A A Jalil[2,*], N A A Fatah[2], M. Ibrahim[2], M S Azami[1], W Fadlun[2], M A H Aziz[1], HU Hambali[2]

[1] Faculty of Science, Universiti Teknologi Malaysia, 81310 UTM Johor Bahru
[2] School of Chemical and Energy Engineering, Faculty of Engineering, Universiti Teknologi Malaysia, 81310 UTM Johor Bahru, Johor, Malaysia
[3] Centre of Hydrogen Energy, Institute of Future Energy, 81310 UTM Johor Bahru, Johor, Malaysia

*aishahaj@utm.my

Abstract. Catalytic CO_2 methanation offers an attractive and sustainable way for the production of substituted natural gas (SNG), which may be used as a clean alternative energy source than fossil fuels. A metal-free fibrous silica ZSM-5 catalyst (FS@ZSM-5) was synthesized via the microemulsion technique to conduct catalytic CO_2 methanation. The FESEM, BET, and FTIR characterization techniques were used to investigate the surface morphology, pore structure of the catalysts. It was noticed that the commercialized ZSM-5 showed 37% CH_4 selectivity with a rate of methane formation 0.067 mmol $m^{-2}s^{-1}$. While FS@ZSM-5 exhibited high CH_4 selectivity of 66 % with rate of methane formation 0.108 mmol $m^{-2}s^{-1}$ using same conditions (T = 500 °C and GHSV = 36,000 mL h^{-1} g^{-1}). Therefore, the FS@ZSM-5 was proved an efficient and active catalyst for CO_2 methanation activity.

1. Introduction

Climate change is the most contentious and aggravating issue of the 21st century due to greenhouse gases. Carbon dioxide (CO_2), an important greenhouse gas, has contributed much to increase the temperature of our earth planet. Therefore, the recycling of CO_2 has been the key strategy for researchers. The transformation of CO_2 into methane has fascinated substantial interest in recent years due to its high potential for commercialization. The CO/CO_2 methanation reaction was discovered by Sabatier and Senderens in 1902 [1, 2]. CO/CO_2 methanation reactions produce the methane is called substituted natural gas (SNG) and is considered one of the most important energy sources of the 21st century, especially in China, a country that does not have an abundant amount of natural gas [3, 4, 5]. There are many applications of SNG to produce useful products, such as carbon tetrachloride, methanol, and formates. Besides, SNG can also be used in gas turbines, gas engines, and the transportation sector, such as in vehicles. Besides, the existing storage tanks and pipelines infrastructures can be used for transportation SNG to long distances. [5, 6]. The CO_2 methanation is thermodynamically favorable ($\Delta G_{298K} = -130.8$ kJ/mol). For catalytic CO_2 methanation, an efficient, active, and thermally stable catalyst is required to convert CO_2 into CH_4. In the last decades, several studies have been conducted using different types of catalysts [1]. Pandey et al. conducted CO_2 methanation without using support, while, Hamid et al. performed CO_2 methanation without metal [7, 8]. Although several studies have been conducted to design suitable and active catalysts for CO_2 methanation. In spite of this, it is still a great challenge to design a highly efficient and active catalyst for industrial applications due to exothermic nature and coke formation during CO_2 methanation, which are the major issues that may deactivate the catalysts. There are many studies about CO_2 methanation over metal-based catalysts, particularly on transition metals. However, the metal-free catalyst was rarely used in CO_2 methanation. Therefore, CO_2 methanation was conducted using metal-free fibrous silica ZSM-5. In recent times, our research group is active in synthesizing several types of fibrous materials, like KCC-1, discovered by Polystiwar in 2010, such as fibrous silica beta zeolite, fibrous silica mordenite, and fibrous silica TiO_2 and applied in different reactions like isomerization and Photocatalysis [9-13]. Based on the literature, there are very rare studies of CO_2 methanation over metal-free catalysts. Therefore in this present study, a metal-free fibrous silica ZSM5 catalyst was assessed for CO_2 methanation.

2. Experimental

2.1. Catalyst preparation

The microemulsion method was used to synthesize fibrous silica ZSM-5 using a similar to our previously reported protocol [13]. In a typical synthesis procedure, 3.2 g of urea and 9.3 g of cetyltrimethylammonium bromide (CTAB) were mixed together in 190 mL of distilled water and stirred for 20 min to make homogenize solution. 1-Butanol (7mL) and Toluene (130 mL) were added to this mixed solution as oil phase and co-surfactant, respectively. After 30 min stirring, commercial ZSM-5 zeolite seeds (Zeolyst international) were added, and then 10 mL of tetraethylorthosilicate (TEOS) was mixed in a solution with 2 h stirring at room temperature. The solution was shifted into a Teflon bottle from the beaker to keep in oven for 4 h at 120 °C. The centrifugation of the mixture was performed at 4000 rpm to isolate the solid product, which was placed in the oven overnight at 120 °C. Finally, resulting solid material was calcined at 550 °C for 6 h and crushed into a fine powder, entitled "FS@ZSM-5".

2.2. Characterization of catalysts

To examine the catalytic properties of both commercialized ZSM-5 and synthesized FS@ZSM catalysts, different characterizations techniques were used, such as field emission scanning electron microscopy (FESEM) was employed to execute morphology of the catalysts using Zeiss Supra VP35 instrument. The functional groups were studied by the FTIR-KBr method using an Agilent Cary 640 FTIR spectrometer. Besides, the pore size distribution was calculated by non-local density functional theory (NLDFT) methods. Prior to the analysis, the catalysts were degassed 1 h at 300 °C. Then BET surface area was calculated using a Beckman Coulter SA 3100 at 77 K.

2.3. Experimental setup for CO_2 methanation

Catalytic CO_2 methanation was performed in the range of 150-600 °C using a fixed-bed reactor, which was placed in a circular-shaped electric furnace. The catalysts were palletized and sieved first before filling the catalyst into the reactor tube. 200 mg of each catalyst was filled into the reactor tube. Prior to the catalytic performance, air treatment was performed for 1 h to remove the impurities from the surface of the catalyst. Then the reduction was performed for 3 h at H_2 = 20 ml/min to activate the catalysts. The temperature was adjusted and stabilized at 150 °C, and reactant gases were introduced into the column using H_2 /CO_2 = 4 :1, at 1 atmospheric temperature. Products formed during CO_2 methanation activity were monitored by online 7820 N Agilent (gas chromatograph). The CO_2 conversion, methane selectivity were measured by the following Eq. (1-4):

$$CO_2 \text{ conversion } (\%) = \frac{M_{CH_4} + M_{CO_2}}{M_{CO} + M_{CH_4} + M_{CO_2}} \times 100 \quad (1)$$

$$CH_4 \text{ selectivity } (\%) = \frac{M_{CH_4}}{M_{CH_4} + M_{CO_2}} \times 100 \quad (2)$$

$$\text{CH}_4 \text{ yield (\%)} = \frac{X_{CO} \times S_{CH_4}}{100} \tag{3}$$

$$\text{Rate (mmol m}^{-2}\text{s}^{-1}\text{)} = \frac{M_{CH_4}}{S.A \times W_{cat} \times s} \tag{4}$$

3. Results and discussions

Figure 1 indicates the FESEM studies of as-synthesized FS@ZSM-5. It can be seen a round-shaped, spherical, and well-ordered microspheres particles, which look like cockscomb structures of uniform size particles of 100-200 nm. It will be a significant contribution to enhanced mesoporosity in FS@ZSM-5 due to this new morphology. A detailed study of the formation of mesopores will be conducted in the next section. It is suggested that the new morphology may improve the adsorption of H_2 and CO_2 molecules at the surface of FS@ZSM-5 during catalytic activity [14].

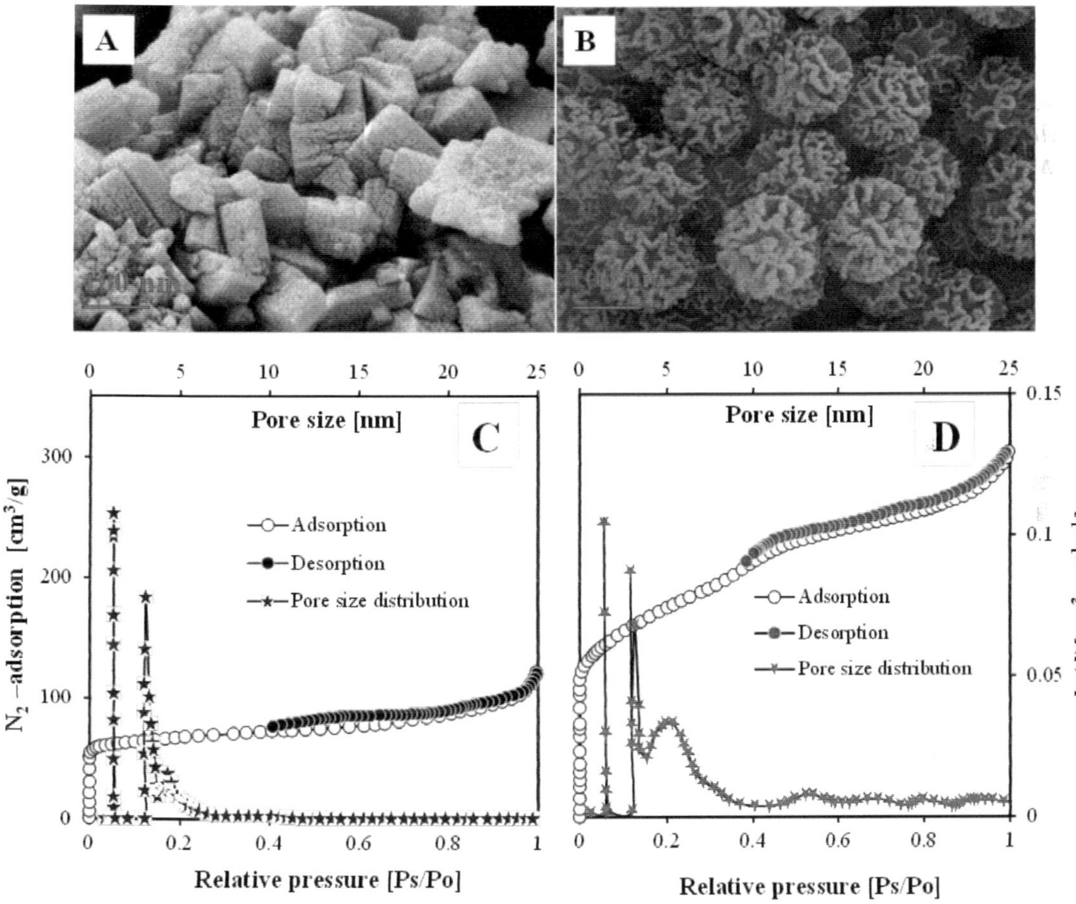

Figure 1. FESEM images of (A) ZSM-5, (B) FS@ZSM-5 and N_2 adsorption-desorption isotherm and NLDFT pore size distribution of (C) ZSM-5 and (D) FS@ZSM-5

Figures C-D presents the textural properties of the catalysts as it can be seen that a reversible type IV isotherm was illustrated by both commercialized ZSM-5 and synthesized FS@ZSM-5 catalysts. The adsorption of nitrogen in both catalysts at lower P/P_0 is due to micropore [8]. It should be noted that at higher relative pressure, the nitrogen adsorption became more significant at P/P_0= 0.4 and 0.9. It was due to intra-and interparticle pores, respectively. Compared to commercialized ZSM-5, the synthesized FS@ZSM-5 (Figure 1D) showed higher intra- and interparticle porosity. Therefore the FS@ZSM-5 demonstrated a tremendous increase in the BET Surface area (615 m^2/g) presented in Table 1. The main pore size distribution appeared in the range of 2-9 nm for ZSM-5 (Figure 1C). While the FS@ZSM-5 illustrated the main pore size distributions in the range of 2-10 nm, it was due to the self-assembling of CTAB molecules during the synthesis of catalysts. The large pores from 10 to 25 nm presented to the inter-fibers distance, which might contributed to increase the BET surface area [8, 14]. These results are consistent with the unique FESEM results.

Table 1. Physicochemical properties of catalysts.

Catalyst	S_{BET} (m^2g^{-1})	Total pore Volume (cm^3g^{-1})	Micropore volume (cm^3g^{-1})	Mesopore volume (cm^3g^{-1})
ZSM-5	257	0.155	0.092	0.063
FS@ZSM-5	615	0.488	0.082	0.406

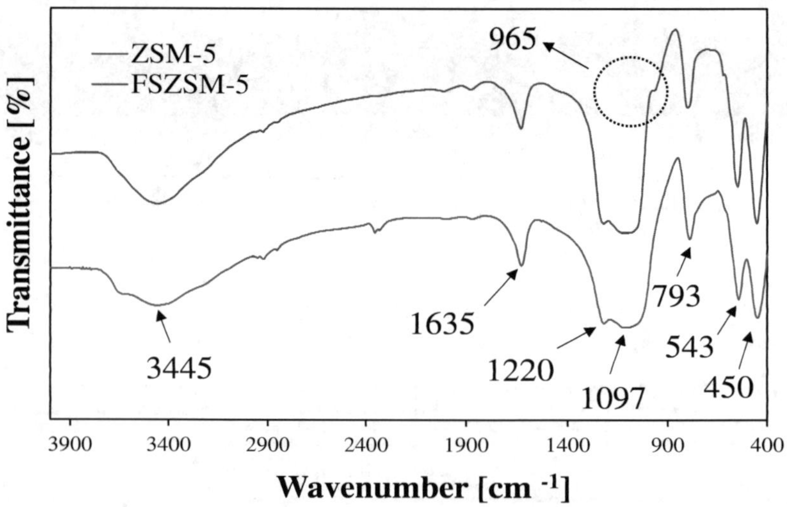

Figure 2 FTIR-KBr spectra of for commercialized ZSM-5 (blue line) and synthesized FS@ZSM-5 (red line)

Figure 2 presents the FTIR-KBr spectra of ZSM-5 and FS@ZSM-5 were performed in the range of 4000 to 400 cm^{-1}. Both catalysts presented an intense band at 3445 cm^{-1}, assigned to hydroxyl groups due to water molecules at the surface of catalysts. The band appeared at 1635 cm^{-1} indicates the deformational vibrations of water molecules. Whereas, the bands at 1220 cm^{-1} and 1097 cm^{-1} show the internal-external and asymmetric stretching of the siloxane groups [14]. The band at 965 cm^{-1} presents the surface hydroxyl groups, which confirmed the fibrous morphology of FS@ZSM-5; similar results were reported by Mariam et al. 2019 [13]. A band appeared at 792 cm^{-1} due to the symmetric stretching of the siloxane groups on the surfaces of the catalyst. The bands at 543 cm^{-1} and 450 cm^{-1} illustrated the internal flexions of the tetrahedrons and five-membered rings, respectively [14].

Figure 3 describes the catalytic CO_2 methanation performance over both commercialized ZSM-5 (Figure 3A) and synthesized FS@ZSM-5 (Figure 3B). As it can be noticed that CH_4 selectivity and rate of methane formation were not significant for both catalysts at low temperatures. It might be due to that the activation energy was not enough to dissociate CO_2 and H_2 energy. However, on increasing the temperature, the CO_2 methanation catalytic activity for both catalysts increased significantly. It should be noted that the FS@ZSM-5 showed higher performance compared to ZSM-5 due to its unique morphology. The commercialized ZSM-5 illustrated 37% CH_4 selectivity with the rate of formation of methane 0.067 mmol $m^{-2}s^{-1}$. While the synthesized FS@ZSM-5 presented 66 % CH_4 selectivity with rate of formation of methane 0.108 mmol $m^{-2}s^{-1}$ at T = 500 °C and GHSV = 36,000 mL h^{-1} g^{-1}.
It should be noted that after 500 °C, the system attained an equilibrium and several side reactions occurred, and the rate of methane formation and CH_4 selectivity lowered significantly, which is consistent with theoretical calculations of CO_2 methanation [14].

In most cases, the catalytic CO_2 methanation is performed over metal-based catalysts. Nevertheless, in the present study, CO_2 methanation activity was conducted on metal-free synthesized FS@ZSM-5, which led to the activation of H_2 and CO_2 molecules to form methane. A similar study was conducted on metal-free KCC-1 to perform CO_2 methanation, where high surface area and intra-inter particles, high basicity increased the CO_2 methanation activity. In another CO_2 methanation study, it was reported that the role of interparticle pores in Au/MCM, which increased the activity in terms of CO conversion [8]. Based on the catalytic properties of synthesized FS@ZSM-5, it is proposed that the high catalytic activity of synthesized FS@ZSM-5 was due to the new morphology after modification of commercialized ZSM-5. The formation of both intra- and inter-particles in synthesized FS@ZSM5, added to high surface area and the high basicity, which assisted the transportation of CO_2 and H_2 and molecules during the catalytic reaction [15-16].

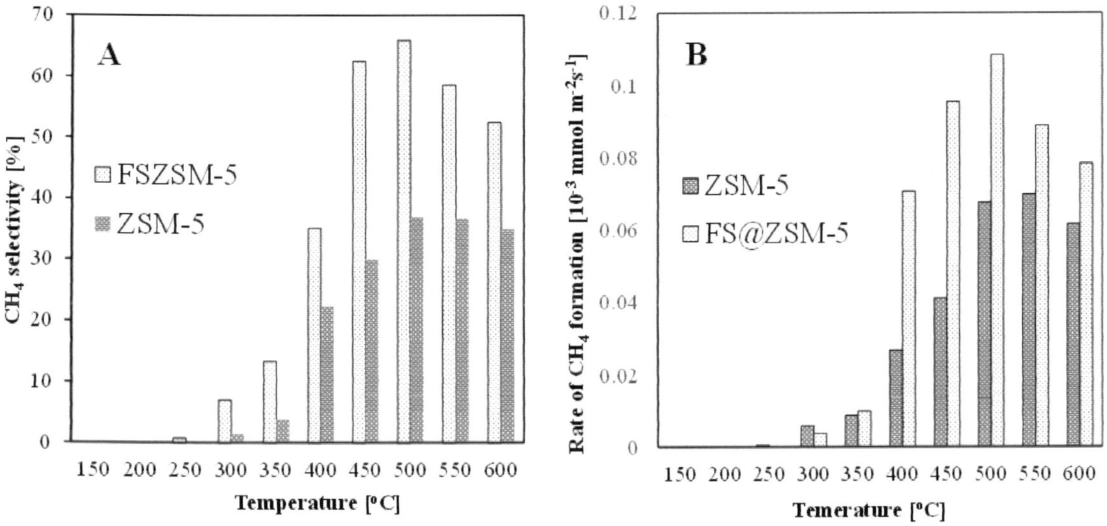

Figure 3. illustrates the catalytic performance for both ZSM-5 and FS@ZSM-5 catalysts (Figure 4A) for the CH_4 selectivity and rate of CH_4 formation (Figure 4B)

4. Conclusion

A metal-free fibrous silica ZSM-5 (FS@ZSM-5) was synthesized successfully via the microemulsion system to conduct catalytic CO_2 methanation. It was found that synthesized FS@ZSM-5 was proved a better catalyst in catalytic performance during CO_2 methanation activity due to its new morphology with surface area and the existence of both intra- and inter-particles distances. This study presents a new perspective of heterogeneous catalysis to clarify CO methanation's applications. The encouraging results of this study may provide a piece of useful information and findings to design a suitable catalyst in CO_2 methanation that can further consider being commercialized in producing SNG.

Acknowledgments
This research work was sponsored and supported by Universiti Teknologi Malaysia through Research University through UTM Transdisciplinary Grant No. 06G53.

References

[1] Bukhari S N, Chong C C, Teh L P, Veit D V, Ainirazali N, Triwahyono S and Jalil A A, Setiabudi H D 2019 *Int. J. Hydrog. Energy.* **44 (37)** pp 20792-20804
[2] Liu S S, Jin Y Y, Yahong H, Jinxian Z and Ren J 2018 *Fuel Process Technol* **177 (274)** pp 266-274
[3] Shuangshuang L, Honggui T, Dandan G, Zhi M and Yuan L 2017 *Catal. Today* **297(298)** pp 298-307
[4] Jun R, Haidong L, Yongyong J, Jiyu Z, Shusen L, Jianying L and Zhong L 2017 *Appl. Catal. B Environ.* **201(561)** pp 561-572
[5] Hussain I, Jalil, A A, Mamat, C R., Siang, T J, Azami, M S and Hambali, H U 2019 *Journal of Energy and Safety Technology (JEST)* **2(1)** pp 15-19
[6] Garbarino G, Wang C, Cavattoni T, Finocchio E, Riani P, Stephanopoulos F M and Busca G 2019 *Appl. Catal. B Environ.* 248 pp 286–297
[7] Frontera P, Macario A, Ferraro M and Antonucci P 2017 *Catalysts* 7(2) pp 59
[8] Hamid MYS, Firmansyah M L, Triwahyono S, Jalil A A, Muktid R R, Febriyanti E, Suendod V, Setiabudi H D, Mohamed M and Nabgan W 2017 *Appl Catal. A: Gen.* **532,** 86 pp 86-94
[9] Polshettiwar V, Cha D, Zhang X and Basset J M 2010 Angew. *Chemie - Int. Ed*, 49. 9652-9656
[10] Ghani N N M, Jalil A A, Triwahyono S, Aziz M A A, A F A Rahman, M Y S Hamid, S M Izan, and Nawawi M G M 2019 *Chem. Eng. Sci.* **193**, pp 217-229
[11] Hambali H U, Jalil A A, Siang T J, Abdulrasheed A A, Fatah N A. A, Hussain I, Azami M S2019 *Journal of Energy and Safety Technology (JEST)* **2(1)** pp 09-13.
[12] Ghani N N M, Jalil, A A Triwahyono S, Aziz M A A, Rahman A F A, Hamid M Y S, Izan S M
[13] M.G.M. Nawawi 2019 *Chem. Eng. Sci.* (193) pp 217-229.
[14] Ibrahim M, Jalil A A, Khusnun N F, Fatah N A A *Int. J. Hydrog. Energy* 2019, https://doi.org/10.1016/j.ijhydene.2019.04.067
[15] Hussaina I, Jalilb A A, Mamat C R, Siang T J, Rahman A F A, Azami M S and Adnan R H 2019 *Energy Convers and management* **199** pp 112056.
[16] Aziz M A A, Jalil A A, Triwahyono S, Muktid R R, Taufiq-Y Y H and Sazegar M R 2014 *Appl. Catal. B: Environ.* **147** pp 359-368
[17] Aziz M A A, Jalil A A, Triwahyono S, Sidik S M 2014 *Appl. Catal. A: Gen.* **486** pp 115-122

IOP Conf. Series: Materials Science and Engineering **808** (2020) 012038 doi:10.1088/1757-899X/808/1/012038

Analysis on Physiochemical Properties of Cellulose Fiber from Rice Straw Waste

Mostafa Yusefi, Roshafima Bte Rasit Ali, Ezzat Chan Abdullah, Kamyar Shameli*

Malaysia-Japan International Institute of Technology, University Teknologi Malaysia, Jalan Sultan Yahya Petra, Kampung Datuk Keramat, 54100 Kuala Lumpur, Malaysia

*kamyarshameli@gmail.com

Abstract. Green biobased polymeric membranes are being increasingly studied for different applications. In this study, freeze dried cellulose fiber with 35% yield was isolated from rice straw. The cellulose fiber was obtained through bleaching and delignification of the rice straw waste using soxhlet instrument and facile method of alkali treatment, respectively. The cellulose fiber was analysed through X-ray powder diffraction (XRD), fourier-transform infrared spectroscopy (FTIR), zeta potential analyzer and scanning electron microscope (SEM). Compared to the straw, all physicochemical properties of treated cellulose fiber increased with zeta potential up to -33.61 mV. FTIR revealed that the treatments on the straw was successful to obtain cellulose fiber with high purity. In addition, the morphological study illustrated cellulose fiber with organized structure.

1. Introduction

Researchers have been concerned using agro-waste for producing eco-friendly materials and applications to reduce environmental issues including toxicity and reduction in landfill space [1, 2]. For example, burning rice straw (RS) as a second highest agro-waste leads to the health risks of air pollution, and effects on organic matter with huge nutrient losses [3]. Similar to other plant-based resources, RS contains high ratio of cellulose fiber [4]. With high demand of using nanotechnology, cellulose fiber ($C_6H_{10}O_5$), one of the unique organic polymers on earth is becoming more popular for different applications. Cellulose fiber contains a high number of glucoses which are linked by β-1,4-glycosidic in linear chain formats. It possesses remarkable structural advantages including high porosity, pore interconnectivity, large surface-to-volume ratio, low density, and hydrophilicity and good physicochemical characteristics [5, 6]. Among various methods of cellulose fiber isolation, alkaline treatment is a facile and economical method to remove lignin, hemicelluloses and pectins from plant-based materials. This treatment caused removing amorphous region while crystal region is protected to obtain cellulose fiber with high crystallinity and purity. In this work, freeze-dried cellulose fiber was extracted from RS by using soxhlet instrument, sodium chlorite, acetic acid and alkali treatment. RS and cellulose fiber were analyzed by XRD, FTIR, zeta potential analyzer and SEM.

2. Experimental

In this work, RS was obtained from Malaysian Agricultural Research and Development Institute (MARDI). All chemicals were analytical grade as purchased from Sigma-Aldrich (St Louis, MO, USA) without further purification. For bleaching process and cellulose fiber isolation from RS, potassium hydroxide (KOH, 85%, EM Science), sodium chlorite ($NaClO_2$, 80%, Fluka) and acetic acid glacial

(CH_3COOH) were used. The aqueous solutions were freshly prepared using double deionized water (DD water) (18.2 MΩ·cm) from a Millipore Direct-Q 5 Ultrapure Water System. All glassware used in this work were cleaned with HNO_3/HCl (3:1, v/v) and DD water followed by the drying process.

2.1. Cellulose fiber Isolation from rice straw

Similar to different biofiber sources, cellulose fiber isolation process from RS was carried out through removing and bleaching the unwanted components and lignin. RS was washed several times with the tap water to remove dirt and dusts, followed seven days air drying in a dark room. Then the dried RS was ground and milled with a 55 micro-mesh screen. To obtain dewaxed rice straw (DRS), a 30 g RS powder was initially dewaxed to dissolve oil, wax and pigment by a 2:1, v/v toluene/ethanol (450 mL) aqueous solution in a soxhlet instrument for 12 h at 70°C using an oil bath. The pasty dark brown color sample was washed with distilled water three times and extra liquid was decanted using a lab suction filtration unit (equipped with a whatman filter paper), followed drying process in a 60°C oven for 24 h. To remove lignin, DRS powder was mixed with a 1000 mL aqueous solution of sodium chlorite (1.4%) as adding drops of acetic acid tuned the pH at 3.0-4.0 and temperature at 70°C with gentle magnetic stirring for 5 h, followed washing and decanting of extra liquid several times which resulted obtaining light yellowish solid sample.

After that, the sample was leached through applying a 600 mL aqueous solution of KOH (5%) at the ambient temperature for 12 hours, followed by pouring 10 fold ice cubes immediately into the solution. The aqueous sample (~300 mL) was centrifuged with speed of 12000 rpm and dried by freez-dryer apparatus (FreeZone 1.0 L Benchtop Freeze Dry System) to finally obtain the white pulpy cellulose fiber sample which was named CF.

2.2. Characterization

X-ray diffraction (XRD) analysis was used to evaluate the crystalline phases of samples at scintag XDS 2000 powder diffractometer. The samples were compressed between two smooth glass films. XRD was performed under ambient condition upon dispersion 2 angles of 5°- 40° with step size of 0.02° and scanning rate of 2 s/step as utilized a Ni-filtered Cu K radiation (= 1.5406 A)°, an operating voltage of 45 kV, and a filament current of 40 mA. Model XPert Pro Panalytical Diffractometer was also used to measure the purity and crystallinity values of the sample [7]. Chemical and super-molecular structural analysis during preparation of CF as a modified natural fiber from RS can be determined by FTIR spectroscopy. FTIR (ThermoNicolet, USA) was employed to measure FTIR Spectra under ambient condition. First the sample was crushed with KBr with the ratio of 1:100, w/w and compressing into a transparent pellets. The spectra was evaluated at the transmittance mode in the range between 4000 to 400 cm^{-1} with a 4 cm^{-1} resolution and an accumulation of 128 scans. Scanning electron microscope (SEM) (XL 30- SFEG, FEI/Philips, USA) was used to measure surface morphology and microstructure of CF sample. A 5-kV accelerating voltage and an 8-mm working distance were applied for observation of the sample. The samples were mounted on the aluminum base with conductive carbon tape and deposited with gold at vacuum (20 mA for 2 min (Bio-Rad SEM coating system).

3. Results and discussion

3.1. X-ray powder diffraction (XRD)

XRD evaluation on RS, DRS and CF are indicated in figure 1. These samples displayed a peak approximately at $2\theta = 16.5°$, 22.5° and 34.6° as are in agreement with normal cellulose-I structure [8]. The cellulose crystals displayed feature allocations of 110, 200, and 004 planes, respectively [9]. Two pecks at 2θ of 16.11° and 22.23° could respectively represent the amorphous and crystal regions of the samples. As was expected the crystallinity and the intensity of peaks increased with applying treatments on the RS. The main elements in the amorphous region of RS are hemicellulose and lignin which were peeled out and be hydrolyzed during the alkaline treatment. Therefore, amorphous region was gradually removed after each treatment while the crystal region was protected[10]. This result shows that the treatment enhanced surface alignment, crystal interfaces and hydrogen bonds of obtained materials.

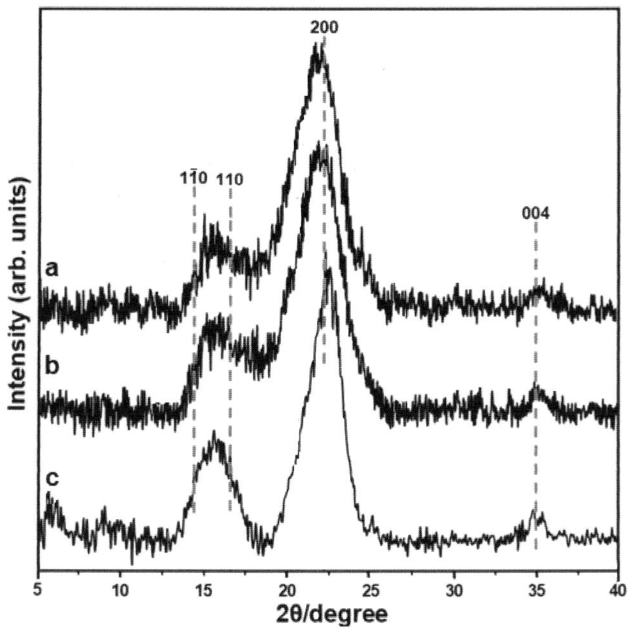

Figure 1. XRD of (a) RS (b) DRS and (c) CF.

3.2. Fourier-transform infrared spectroscopy (FTIR)

FTIR of RS, DRS and CF are indicated in figure 2 a-c, respectively. As can be observed from the figure, CF was successfully extracted from RS after serial treatments including de-waxing, delignification and excretion of silica and hemicellulose. With the bleaching process, the alkali treatment on RS probably created C-H aromatic hydrogen groups. The peak at 1524 cm^{-1} (aromatic skeletal vibrations) was possibly attributed to presence of pyranose ring skeletal C-O-C bonds of cellulose. This peak showed the delignification as its intensity decreased after each treatment that the alkali treatment caused removing the peak. Figure 6c probably showed leaching of the hemicellulose at 1750 cm^{-1} (carbonyl stretching) and removal of silica (Si–O–Si stretching) at 760 and 491 cm^{-1}. As was seen in the Figure 2c, the peaks at 3352, 2891 and 1100 cm^{-1} could demonstrate the stretching vibrations of -OH groups, C-H stretching and cellulose structure, respectively [11]. In the anomeric region (950–700 cm^{-1}), the minor peak at 887 cm^{-1} was pronounced for CF (figure 2c), showing the glycosidic $-C_1$ $-O$ $-C_4$ deformation property of the β-glycosidic bond in cellulose [12]. This probably showed the successful CF isolation process. The CF yield from RS was achieved up to 35% which was almost similar to the result from different studies [13, 14]. Worth nothing to mention that FTIR patterns were similar for cellulose before and after delignification which could show high structural stability with no main damages to cellulose structure [15-18].

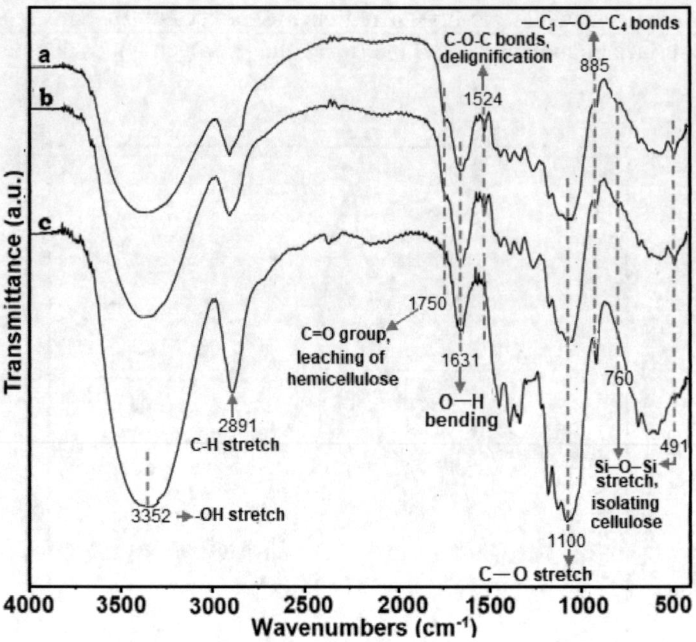

Figure 2. FTIR of (a) RS, (b) DRS, (c) CF.

3.3. Zeta Potential

The suspensions of CF indicated negative zeta potential of -33.61. These results suggested that CF has a good stability in aqueous solution [19]. The colloidal suspension with good stability was resulted from the great electrostatic repulsion that occurred among the fibers.

3.4. The scanning electron microscope (SEM)

SEM images of CF is indicated in figures 3. As ratio of hemicellulose and lignin were reduced, size of fiber gradually decreased by applying dewaxing and delignification treatments. Therefore, CF was smaller with better organized structure compare to DRS. Figure 3b shows that CF is mostly comprised cellulosic fibrils with order and disorder structure[20]. In nature, each fiber contains several to hundreds of microfibers with the manner of adjusting together for providing cellulosic structure [15]. Subsequently, the strong hydrogen bonds between each fiber can be observed [21]. CF contained carbon, oxygen and silicon, while, RS contains high amount of silicon, but it was replaced by oxygen and carbon after the successful dewaxing and alkaline treatments [22]. The treatment on RS caused dissolving of the silica in aqueous ion and subsequently replaced by carbon, introducing improvement of crystallinity and surface area of the extracted CF [22]. From figure 3a, some pores can be seen on the surface area of the CF. CF compared to RS, it has more mesopores, and better uniform pore structure. The porosity on the surface area of CF allowed water and hydrophilic drug penetration into the fiber for its using in drug carrier application.

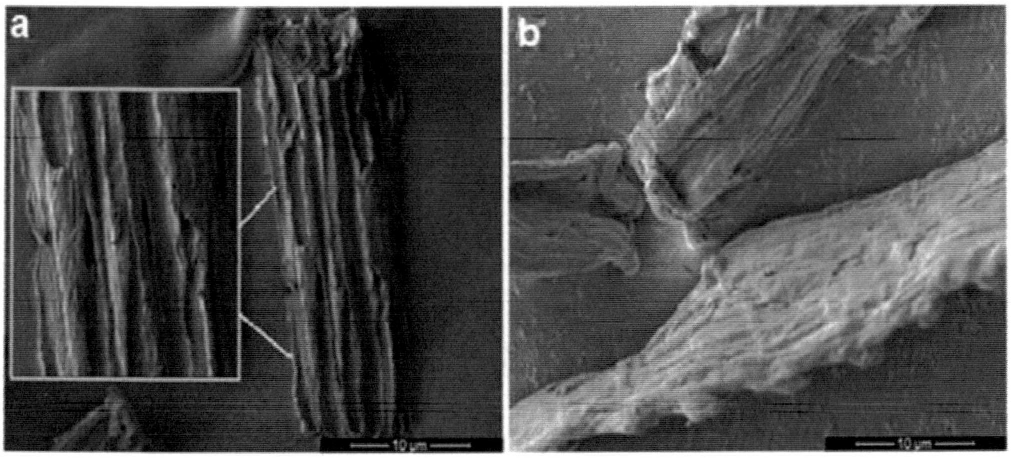

Figure 3. SEM of CF

4. Conclusion

In this study, a facile and eco-friendly method of alkali treatment was used to isolate CF from RS. Physicochemical properties of RS, DRS and CF were systematically evaluated. The XRD analysis of RS after each treatment displayed removal of unwanted substances including lignin to obtain CF with high purity and crystallinity. FTIR results exhibited gradual and parallel shifting and intensity of peaks by the treatments, proving successful bleaching and delignification process. High negative zeta potential of -33 mV proved the colloidal stability of the CF in aqueous solution. SEM images illustrated CF with organized structure. Therefore, this research showed potential of CF isolation from RS. In further study, CF will be used in drug delivery system for cancer therapy.

Acknowledgments

The authors wish to acknowledge funding by the Malaysian Ministry of Higher Education under the Tier 1 grants (Grant No. #20h33 and #20h55) and express gratitude to the research management center (RMC) of Universiti Teknologi Malaysia and Malaysia-Japan International Institute of Technology for providing an excellent research environment and facilities.

References

[1] Yusefi M, Khalid M, Yasin F M, Abdullah L C, Ketabchi M R and Walvekar R 2018 Performance of cow dung reinforced biodegradable poly (Lactic Acid) biocomposites for structural applications *Journal of Polymers and the Environment* **26** 474-86

[2] Yusefi M, Khalid M, Yasin F M, Ketabchi M R, Hajalilou A and Abdullah L C 2017 Physico-mechanical properties of poly (lactic acid) biocomposites reinforced with cow dung *Materials Research Express* **4** 025302

[3] Ram H, Dadhwal V, Vashist K K and Kaur H 2013 Grain yield and water use efficiency of wheat (Triticum aestivum L.) in relation to irrigation levels and rice straw mulching in North West India *Agricultural Water Management* **128** 92-101

[4] Ardanuy M, Claramunt J and Toledo Filho R D 2015 Cellulosic fiber reinforced cement-based composites: A review of recent research *Construction and building materials* **79** 115-28

[5] James R and Laurencin C T 2016 Nanofiber technology: its transformative role in nanomedicine. Future Medicine)

[6] Kim J I, Hwang T I, Aguilar L E, Park C H and Kim C S 2016 A controlled design of aligned and random nanofibers for 3D bi-functionalized nerve conduits fabricated via a novel electrospinning set-up *Scientific reports* **6** 23761

[7] Segal L, Creely J, Martin Jr A and Conrad C 1959 An empirical method for estimating the degree of crystallinity of native cellulose using the X-ray diffractometer *Textile Research Journal* **29** 786-94

[8] Wu R-L, Wang X-L, Wang Y-Z, Bian X-C and Li F 2009 Cellulose/soy protein isolate blend films prepared via room-temperature ionic liquid *Industrial & Engineering Chemistry Research* **48** 7132-6

[9] Chen W, Yu H, Liu Y, Chen P, Zhang M and Hai Y 2011 Individualization of cellulose nanofibers from wood using high-intensity ultrasonication combined with chemical pretreatments *Carbohydrate Polymers* **83** 1804-11

[10] Park S, Baker J O, Himmel M E, Parilla P A and Johnson D K 2010 Cellulose crystallinity index: measurement techniques and their impact on interpreting cellulase performance *Biotechnology for biofuels* **3** 10

[11] Chen X, Yu J, Zhang Z and Lu C 2011 Study on structure and thermal stability properties of cellulose fibers from rice straw *Carbohydrate polymers* **85** 245-50

[12] Jiang F and Hsieh Y-L 2013 Chemically and mechanically isolated nanocellulose and their self-assembled structures *Carbohydrate Polymers* **95** 32-40

[13] Hessien M, Rashad M, Zaky R, Abdel-Aal E and El-Barawy K 2009 Controlling the synthesis conditions for silica nanosphere from semi-burned rice straw *Materials science and engineering: B* **162** 14-21

[14] Lu P and Hsieh Y-L 2012 Preparation and characterization of cellulose nanocrystals from rice straw *Carbohydrate Polymers* **87** 564-73

[15] Tan X Y, Hamid S B A and Lai C W 2015 Preparation of high crystallinity cellulose nanocrystals (CNCs) by ionic liquid solvolysis *Biomass and Bioenergy* **81** 584-91

[16] Yahya M B, Lee H V and Hamid S B A 2015 Preparation of nanocellulose via transition metal salt-catalyzed hydrolysis pathway *BioResources* **10** 7627-39

[17] Hamid S B A, Zain S K, Das R and Centi G 2016 Synergic effect of tungstophosphoric acid and sonication for rapid synthesis of crystalline nanocellulose *Carbohydrate polymers* **138** 349-55

[18] Chen Y W, Tan T H, Lee H V and Abd Hamid S B 2017 Easy fabrication of highly thermal-stable cellulose nanocrystals using Cr (NO3) 3 catalytic hydrolysis system: A feasibility study from macro-to nano-dimensions *Materials* **10** 42

[19] de Oliveira J P, Bruni G P, Lima K O, El Halal S L M, da Rosa G S, Dias A R G and da Rosa Zavareze E 2017 Cellulose fibers extracted from rice and oat husks and their application in hydrogel *Food chemistry* **221** 153-60

[20] Jenkins L and Donald A 1997 Use of the environmental scanning electron microscope for the observation of the swelling behaviour of cellulosic fibres *Scanning* **19** 92-7

[21] Liu D, Zhong T, Chang P R, Li K and Wu Q 2010 Starch composites reinforced by bamboo cellulosic crystals *Bioresource technology* **101** 2529-36

[22] Basta A, Fierro V, El-Saied H and Celzard A 2009 2-Steps KOH activation of rice straw: an efficient method for preparing high-performance activated carbons *Bioresource technology* **100** 3941-7

IOP Conf. Series: Materials Science and Engineering **808** (2020) 012039 doi:10.1088/1757-899X/808/1/012039

Estimation of Particulate Emission Generation in Palm Oil Mill Boiler

N R Jamian[1], M Rashid[1], S M M Muzi [1], N Hussein[2], D Munisamy[3], K M Pa'ad[1], A H M Ali[1] and S Y Yusuf[4]

[1] Air Resources Research Laboratory, Malaysia-Japan International Institute of Technology, 54100 UTM Kuala Lumpur, Malaysia
[2] Department of Environmental Engineering, Faculty of Civil Engineering,81310 UTM Skudai, Malaysia
[3] Centre of American Education, Sunway University, Bandar Sunway, 47500 Selangor, Malaysia.
[4] School of Environmental Engineering, Universiti Malaysia Perlis, Kompleks Pusat Pengajian Jejawi, 302600, Jejawi, Perlis

*ruwaida.kl@utm.my

Abstract. The palm oil industries in Malaysia has been positively growth throughout the year, for it sustainable approach in terms of production and environmental protection. However, this industry also producing considerable amount of particulate emission from the process of burning its biomass waste of palm fibre and shell in the boiler. This study present a method to estimate the particulate emission that are being generated from the boiler. The method were done by analyzing the proximate analysis value of the ash content of palm fibre and shell found in this study. The particulate emission were physically collected from five palm oil mill boilers with boiler capacity ranging from 18-45 tonne/hour. The particulate stack fly ash was collected from the stack gas downstream of a multi-cyclones particulate arrestor according to the USEPA Method 17- Determination of particulate matter emissions from stationary sources. Meanwhile the sample of palm fibre and shell were collected to determine the proximate value. The estimation were then being further analyzed according to the particulate emission that were obtained by the real-stack sampling. In summary, the average moisture content, ash content, volatile matter and fixed carbon for palm shell found in this study are 19.3 ± 5.7%, 2.79 ± 2.2%, 74.3 ± 3.0, and 11.3 ± 4.3, respectively.). The average particulate concentration for these five mills is 2.2±0.9g/Nm^3 and the air pollution control requirement are between 64 and 96% with reference to Clean Air Regulations 2014 of 0.15g/Nm^3 which is very high to be achieve. It also can be estimated that 26% of the ash generated from the combustion of palm fibre and shell in palm oil mill boiler are being released at the stack.

1. Introduction

Biomass fuel from agricultural sector is recognized as having high potential to promote biomass residue to wealth. Five different biomass residues in Malaysia that contributes to this effort, including, forestry, rubber, cocoa, sugarcane and palm [1]. Biomass residue from palm oil production such as empty fruit brunch (EFB), fibre and shell have the highest potential in electricity generation compared to other biomass residue in the country such as coconut, paddy and sugarcane [2]. As interest in developing biomass conversion technology is rapid, detail characteristics on the combustion process of biomass fuel towards emission generation are not well understood. The biomass fuels having wide range of characteristics with poor air pollution control system were identified as some of the factors influencing the particulate emissions during combustion.

A few studies have been done related to the particulate emission from the palm oil mill boiler. the average particle emission concentration obtained at stack was 5.84 ±1.88 g/Nm3 @ corrected to 7% oxygen from a 13,600 kg steam/hour palm oil mill boiler [3]. The study also found that 26% of the particulate matter with diameter less than 10 micron (PM_{10}) concentration is accounted by the finer size fraction of $PM_{2.5}$, which is not easily removed by the existing particulate control device such as the multi-cyclone. In 2009, it is estimated that an average of 51.67 g/s of dust load are being emitted from 417 palm oil mills in Malaysia, making the minimum removal efficiency by the multi-cyclone at 91%. The high particulate emission from palm oil mills are generated from incomplete combustion of the fuels, due to several factors, including high moisture content of the palm fibre and shell, improper mixing of the palm fibre and shell, lack of control in air feeding and air to fuel (A/F) [4]. In addition, ow efficiency of the multi-cyclone in the palm oil mills due to improper design, lack of quality control and also poor maintenance from the millers [5].

In this regard, this study present a method to estimate the particulate emission generated from palm fibre and shell combustion in palm oil mill boiler, based on the ash content and the amount of fuel being burned. This estimation can further predict the stack particulate emission that are being released to the environment.

2. Materials & Methods

2.1. Palm Fibre & Shell Sampling

Figure 1 shows a typical schematic diagram arrangement of the boiler, multi-cyclone and stack of a mill where palm fibre and shell (F&S) are burned in the boiler at a certain ratio. Bottom ash were collected at the bottom of the boiler. Meanwhile, coarse particles collected at the hopper of multi-cyclone, known as retained fly ash and the fine particulates leaving the stack known as stack fly ash were sampled. The bottom ash and retained fly ash were grab sampled on hourly basis for a period of five hours on cumulative basis. Meanwhile, the stack fly ash and gases emission sampling was carried out at the stack sampling port. The sampling method is briefly discussed in the following section.

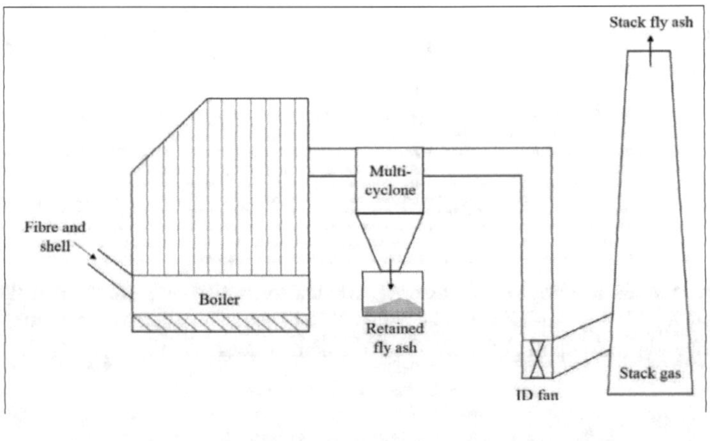

Figure 1. The schematic diagram of the boiler, multi-cyclone, and stack gas in the palm oil mill plan.

2.2. Palm Oil Mill Boiler Characterization

Table 1 summarizes the characteristics of the selected palm oil mills and their boiler operating conditions. Boiler C has the oldest boiler manufactured at 1983 while Boiler D and E operating with boiler manufactured at 2009. In terms of boiler rated steam capacity, Boiler B and D has the largest steam capacity of 45,000kg/hr while Boiler A has the lowest steam capacity of 25,000kg/hr. Meanwhile, the mills running at different steam capacity varies from 17,000kg/hr (Boiler C) to 35,000kg/hr (Boiler D). The boiler is operating on average temperature of 231^{0}C and pressure of 303psi. Boiler B processed the largest amount of FFB per boiler at 54,000kg/hr while Boiler A and C processed the lowest amount of FFB per boiler at 40,000kg/hr. Most of the mills burned the F&S at 80:20 ratio with the F&S feeding varies from 6,000-12,000kg/hr. All boilers were equipped with multi-cyclone as their particulate arrestor.

Table 1. Characteristics of five palm oil mill boilers and their operating conditions

Boiler	**A**	**B**	**C**	**D**	**E**
Boiler manufactured year	2004	2002	1983	2009	2009
Boiler rated steam capacity (t/hr)	25	45	18	45	40
Boiler actual steam capacity (t/hr)	24	26	17	35	30
Boiler steam outlet temperature (^{o}C)	247	247	215	215	215
Boiler pressure (psi)	300	310	300	312	300
FFB processed per boiler (t/hr)	30	54	30	45	40
F&S feeding (t/hr)	8	9	6	12	10
Ratio of F&S	80:20	70:30	80:20	80:20	80:20
Air pollution control system	Multi-cyclone	Multi-cyclone	Multi-cyclone	Multi-cyclone	Multi-cyclone

2.3. Particulate Emission Sampling

The particulate emission sampling was performed at the sampling port in the stack gas downstream of a multiple-cyclones particulate arrestor. Figure 2 presents the particulate sampling train set up that follow the standard USEPA Method 17- Determination of particulate matter emissions from stationary sources were used in this study. Appropriate nozzle sizes were determined in order to get an isokinetic sampling. Set of impingers in ice bath container were used to place the distilled water and silica gel. The probe and the impingers were connected to the isokinetic control console using umbilical cable. The isokinetic control console display the measurement of parameters such as temperature, pressure drop, and air volume. The sampling time was taken on average of 40 minutes for each sample. Glass microfibre thimble canister were used as a collection medium due to its ability to stand for high temperature. The thimble was stored in a desiccator and weighed gravimetrically before and after samplings.

The sampling was carried out isokinetically, where the velocity of the particulate's extraction equal to the actual velocity of the stack gas in the duct within 90-110 % difference. This is to ensure the collected particulate concentration represent the actual condition in the duct. The percent isokinetic sampling rate was calculated using Equation 1.

$$\% \, I = \frac{100 \, T_s \left[K \, V_{wc} + \left(\frac{V_m}{T_m}\right)\left(P_{bar} + \frac{\Delta H}{13.6}\right)\right]}{60 \, P_s \, V_s \, A_n \, t} \tag{1}$$

Where; % I = Isokinetic percent, 90-110%
T_s = absolute stack gas temperature, ^{o}K
K = constant, 0.003454
V_{wc} = volume of water collected, m^3

V_m	= dry gas volume sampled, m^3
T_m	= metered temperature, ^oK
P_{bar}	= barometric pressure, 760 mmHg
P_s	= absolute stack gas pressure, mmHg
ΔH	= pressure drop across orifice meter, mmH_2O
A_n	= area of nozzle used, m^2
V_s	= stack gas velocity, m/s
t	= sampling time, min

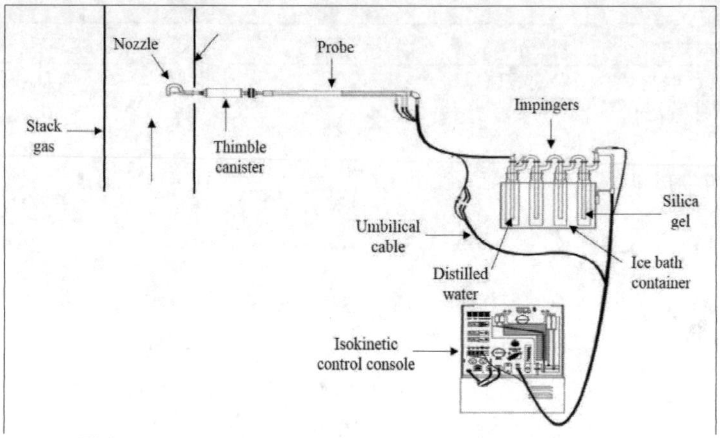

Figure 2. US EPA Method 17 sampling train layout.

2.4. Proximate Analysis
Proximate analysis were done for palm F&S samples from all five mills. The analysis includes, moisture content, volatile matter, ash content and fixed carbon were repeated three times, each. All result were reported in percentage. Moisture content of the palm F&S were was determined according to the methods specified in ASTME871-82(2006), "Standard Method for Moisture Analysis of Particulate Wood Fuels". Volatile matter were determined according to ASTMD872 (2013), "Standard Test Method for Volatile Matter in the Analysis of Particulate Wood Fuels". Ash content were determined according to ASTME1755-01 (2007) "Standard Test Method for Ash in Biomass". Fixed carbon were calculated based on Equation 2.

$$Fixed\ Carbon(\%) = 100 - (Moisture + Volatile\ Matter + Ash) \quad (2)$$

3. Results & Discussion

3.1. Proximate Analysis of Palm Fibre and Shell
Table 2 presents the proximate analysis for palm fibre and shell from five different palm oil mills, obtained in this study compare to the literature, which showed that the average moisture content, ash content, volatile matter and fixed carbon for palm fibre found in this study are 40.3 ± 21.6 %, 6.8 ± 6.8 %, 49.8 ± 24.5, and 3.6 ± 3.9, respectively Three samples of palm fibre and shell from each palm oil mills were analysed for their proximate component as received in order to get the average value. Meanwhile the average moisture content, ash content, volatile matter and fixed carbon for palm shell found in this study are 19.3 ± 5.7%, 2.79 ± 2.2%, 74.3 ± 3.0, and 11.3 ± 4.3, respectively. For a comparison purpose, proximate analysis of palm fibre and shell from five different literature were added in Table 2, which showed the moisture content of palm fibre and shell obtained from this study is slightly higher than reported in the literatures. Meanwhile, the other proximate component of ash content, volatile matter and fixed carbon, showed a closed figure with the average value obtained in the study.

It can be seen that each proximate component diverse in a wide range of value. Generally, the palm fibre has higher moisture content and ash content compared to the palm shell. Meanwhile, the volatile matter for both material is high, compared to other proximate components. In this study, the five palm oil mill boilers are processing the FFB from different plantation, which harvested at different time. Based on the sampling date information in Table 2, the samples of palm F&S for Boiler A and B were taken in June 2014, meanwhile, for Boiler C, D and E, the samples were taken in September 2014. This could influence the wide range of palm F&S characteristics, especially the moisture content.

Table 2. The proximate analysis of palm fibre and shell from five different palm oil mills (as received) and from the literatures (% by mass)

Boilers	**Moisture Content**	**Ash Content**	**Volatile Matter**	**Fixed Carbon**
		Palm Fibre		
A	55.8	5.9	36.8	1.5
B	39.4	5.3	55.0	2.7
C	36.9	7.9	49.9	5.3
D	22.7	7.8	62.5	7.0
E	46.5	6.9	45.1	1.6
Ranges	39.4-55.8	5.3-7.9	36.8-62.5	1.5-7.0
[6]	13.9	3.6	84.8	11.6
[7]	24.0	5.8	78.0	-
[8]	5.2	4.8	75.1	14.9
[9]	6.6	5.0	76.0	12.4
[10]	5.0	11.8	79.0	9.3
Boilers		**Palm Shell**		
A	21.9	7.4	60.0	10.6
B	29.9	1.1	59.5	9.6
C	15.4	0.6	71.7	12.4
D	17.7	2.1	72.4	7.8
E	11.6	2.8	74.3	11.3
Ranges	1.6-29.9	0.6-7.4	59.5-74.3	7.8-12.4
[6]	12.2	1.9	79.2	18.8
[7]	11.0	5.0	74.0	10.0
[8]	8.5	2.6	73.2	15.7
[9]	5.7	2.2	73.7	18.4
[10]	8.4	4.6	75.4	20.0

3.2. Stack Particulate Emission

Table 3 shows the stack emission data obtained in this study where the stack gas temperature were found ranging from 248-286^{0}C and the stack gas volumetric flowrate varied between 22.2 and 38.3m^3/s. Meanwhile, the stack gas velocity varies from 15.4 to 17.2m/s. In addition, the stack gas moisture content varied between 4.3 between 7.0% and the oxygen and carbon dioxide concentration varies between 10.3-17.9% and 2.9-8.5%, respectively. The average particulate concentration for these five mills is 2.2±0.9g/Nm^3 and the air pollution control requirement are between 64 and 96% with reference to Clean Air Regulations 2014 of 0.15g/Nm^3 which is very high to be achieve.

3.3. Prediction of Stack Particulate Emission in Palm Oil Mill Boiler

Figure 3 shows the relationship of predicted particulate concentration released without control from the boiler and the palm fibre and shell mixture ratio. The estimation were done based on palm fibre and shell ash content of 6.8% and 2.8%, respectively, which were obtained from this study. In addition, 15% of clinker formation in the boiler were assumed [11].Based from this graph, it is estimated more particulate are being generated as the amount of palm fibre that are being burned is increases due to the ash content of palm fibre is approximately three times higher than palm shell. This graph can also be used to estimate the percentage of ash released at the stack based on the total ash generated in the boiler. Based from this graph, is it estimated that Boiler A, B, C, D, and E released 26%, 23%, 5%, 45% and 29% stack fly ash from the total ash generated in the boiler. From this result it can be estimated that in average, 26% of ash that are generated in the boiler, were released at the stack of a palm oil mill boiler equipped with an air pollution control system.

Table 3. Stack emission data for five palm oil mill boilers

Parameters	Boilers				
	A	**B**	**C**	**D**	**E**
Stack diameter (m)	1.5	1.5	1.5	1.7	1.7
Stack gas temperature (^{o}C)	268	286	264	248	252
Stack gas volumetric flow rate (m^3/s)	29.2	22.2	29.0	37.5	38.3
Stack gas velocity (m/s)	16.2	17.2	15.6	15.4	15.8
Moisture content (%v/v)	4.3	7.0	4.9	5.4	11.2
Oxygen concentration (%)	17.9	11.2	11.0	10.3	13.5
Carbon dioxide concentration (%)	2.9	8.5	7.3	7.9	6.7
Particulate concentration (g/Nm^3)*	0.5±0.2	1.3±0.0	0.3±0.1	2.9±0.5	1.4±0.1
Particulate concentration (g/Nm^3) @ 7%O_2	2.4±1.0	1.9±0.0	0.4±0.1	3.8±0.7	2.5±0.2
Particulate emission rate (g/s)	7.5±3.0	13.8±0.1	4.2±0.9	53.6±9.8	23.9±2.2

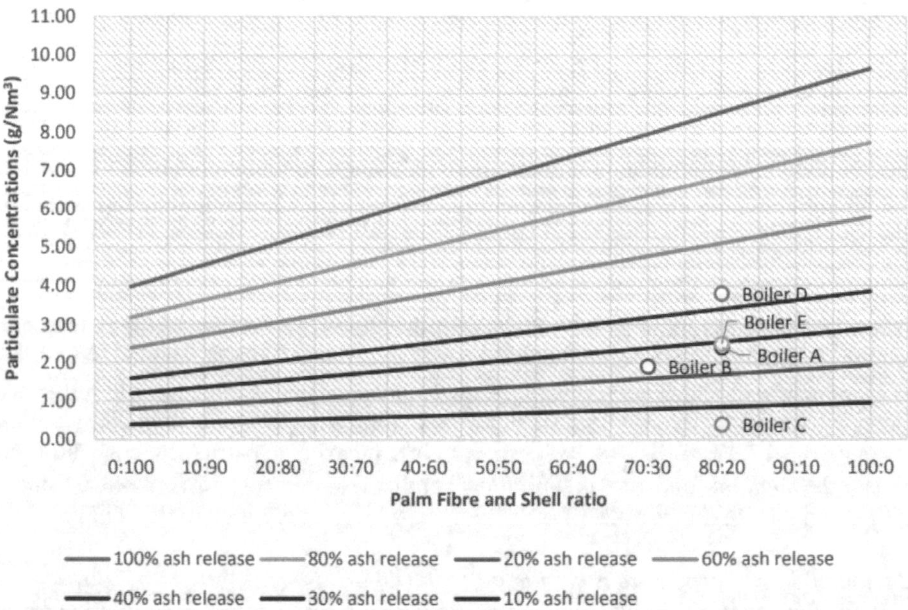

Figure 3 Relationship between estimated particulate concentrations (without control) and palm fibre and shell ratio.

4. Summary

The particulate emission from palm oil mill boilers were estimated based on the ash content of the palm fibre and shell found in this study. The result were further analysed with the particulate emission concentration obtained from the real stack emission sampling. It can be concluded that in average, 26% of ash that are generated in the boiler, were released at the stack of a palm oil mill boiler equipped with an APC system. This study would provide an estimate value of the particulate emission concentration at the stack of a palm oil mill boiler. Further recommendation in order to validate the estimation is by having a particulate emission sampling at the boiler, before the air pollution control system.

Acknowledgments

This work was financially supported by JICA Fund.

References

[1] Chuah T G, Azlina A G K W, Robiah Y, and Omar R 2006 Biomass as the Renewable Energy Sources in Malaysia: An Overview *International Journal of Green Energy*. **33** 323-346.

[2] Shafie S M, Mahlia T M I, Masjuki H H, & Ahmad-Yazid A 2012 A Review on Electricity Generation Based on Biomass Residue in Malaysia *Renewable and Sustainable Energy Reviews*. **16** 5879–5889.

[3] Rashid M, Syahirah M M, NorRuwaida J. and Huda N 2014 Characteristics of Particulate Emission from a Biomass Fired Boiler *Agriculture and Agricultural Science Procedia*. **2** 265 -271.

[4] Kun Y C and Abdullah A M 2013 Simulation of Total Dust Emission from Palm Oil Mills in Malaysia Using Biomass Fuel Composition and Dust Collector Efficiency Models *International Journal of Energy and Environmental Engineering*. 4:19.

[5] DeKock J and Yap D 2000 Performance Criteria of Dust Collectors for Palm Oil Mill Boilers *Proceedingd of the 2000 National Seminar on Palm Oil Milling, Refining Technology, Quality and Environment*. 3-4 July 2000. Genting Highland. 185-194.

[6] Permatasari R, Hui K K, Nazri M, & Ja M 2009 Combustion Characteristics of Palm Wastes in Fluidized Bed Combustion *8th EEEIC International Conference on Environmental & Electrical Engineering*. 10-13 May 2009. Karpacz, Poland.

[7] Garcia-nunez J A, Ramirez-contreras N E, Tatiana D, Silva-lora E, Stuart C, Stockle C, and Garcia-perez M 2016 Resources, Conservation and Recycling Evolution of Palm Oil Mills into Bio-Refineries: Literature Review on Current and Potential Uses of Residual Biomass and Effluents *Resources, Conservation & Recycling*. **110** 99–114.

[8] Faizal H M, Latiff Z A, Wahid M A, and Darus A N 2010 Physical and Combustion Characteristics of Biomass Residue from Palm Oil Mills. *8th IASME/WSEAS International Conference on Heat Transfer, Thermal Engineering and Environment*. 20-22 August 2010. Taiwan, 34-38.

[9] Yang H, Yan R, Chin T, Liang D T, Chen H, and Zheng C 2004 Thermogravimetric Analysis - Fourier Transform Infrared Analysis of Palm Oil Waste Pyrolysis *Energy & Fuels*. **11** 1814–1821.

[10] Wilson L, Yang W, Blasiak W, John G R, and Mhilu C F 2011 Thermal Characterization of Tropical Biomass Feedstocks. *Energy Conversion and Management*. **52(1)** 191–198.

[11] Liew W L, Muda K, and Loh S K 2016 Feasibility Study on Palm Oil Processing Wastes Towards Achieving Zero Discharge. *ARPN Journal of Engineering and Applied Sciences*. **11(4)** 2400–2405.

IOP Conf. Series: Materials Science and Engineering **808** (2020) 012040 doi:10.1088/1757-899X/808/1/012040

Phytotoxicity Study of Native Plants in ex-mining lake water treatment

S F S Draman[1,*], S A Khalid[1], N M Sidek[1,2], S R S Abdullah[2], N Anuar[3]

[1] Faculty of Chemical Engineering, Universiti Teknologi MARA, Terengganu Branch, Bukit Besi Campus, 23200 Dungun, Terengganu, Malaysia
[2] Department of Chemical and Process Engineering, Faculty of Engineering and Built Environment, Universiti Kebangsaan Malaysia, 43600 Bangi, Selangor, Malaysia
[3] Faculty of Chemical Engineering, Universiti Teknologi MARA, 40450 Shah Alam, Selangor, Malaysia

*sfauziah@uitm.edu.my

Abstract. The use of phytoremediation in removing and degrade the contaminants gives great interest for researchers to explore further. This green technology is relatively eco-friendly and cost-effective. Before a full-scale system can be installed to ensure that the remedy is effective by selected plants using phytoremediation process, identification of these plants and its ability to resist toxicity of contaminants is crucial. In this study, the ability of native plants in Malaysia namely *L. articulata*, *E. ochrostacys* and *E. dulcis* to survive when exposed to acid mine drainage water (AMD) are investigated. This was done with free flow systems or free surface (FSF) for 30 days with water from ex mining lake, Tasik Puteri, Bukit Besi, Malaysia. The physical growth of those three plants such as height and withered leaves is monitored every seven day as indication of their survivability. Ferum (Fe) analysis is also carried out on the lower and upper part of the above three plants using Inductively Coupled Plasma – Optical Emission Spectrometry (ICP-OES). Results of the Fe analysis showed that *E.ochrostacys* has the highest accumulation ability compared to *E.dulcis and L. articulata* with a maximum accumulation of 123,584 mg Fe/kg at lower part and 55,151 mg Fe/kg at the upper part. Furthermore, *E.ochrostacys* was also able to tolerate and survive in the water of Tasik Puteri up to 80% survival rate during phytotoxicity study.

1. Introduction

Metal contaminations in ex-mining area are one of the most world environmental issues in which there are many countries all around the world are suffering from. In other words, ex-mining sites are known as one of the most sources of metal pollutant [1, 2]. Due to heavy metal wide sources, toxicity, accumulative behavior and lack of biodegradable properties, heavy metal are considered dangerous for the aquatic environment [3-5]. The type and level of ligands on which the metal could absorb, the water pH, the redox environment of the system and the oxidation state of the mineral components are among the factors that predominately control the solubility of the trace metals in surface waters [6]. When pyritic rocks exposed to air and water, it can produce acid mine drainage (AMD) which is characterized by pH as low as 2 and contain high concentration of metals that is toxic to aquatic organism [7].

Results reported from in situ analysis studies for Tasik Puteri, Bukit Besi shows that the pH is around 3 – 4[8]. According to Department of Environment (DOE) Water Quality Index Classification, the water quality of the lake is classified as class IV that is not suitable for recreational activities but more to irrigation. Therefore, treatment to the lake water is crucial for improving the water quality to achieve permissible Interim National Water Quality Standard (INWQS).

The main factor that possibly contributes to the acidic condition of the lake water is acid mine drainage (AMD). AMD are strong acidic streams rich in dissolved ferrous (Fe^{2+}) and non–ferrous metal sulfates occuring in mine working either operating or abandoned mining sites[9]. AMD is formed from the oxidation of sulphide bearing mineral by the addition of water and oxygen. Then, lead to the formation of an acidic drainage [10]. For Tasik Puteri case, AMD is produced because of the oxidation of pyrite (FeS_2) which is identified as one of the main iron ore in Bukit Besi mining operation.

Few efforts have been made in many conservative technologies (e.g. activated carbon adsorption, reverse-osmosis, nanofiltration, ion exchange, electrodialysis, chemical precipitation and disinfection) as treatment methods[9, 11]. However, those methods are generally expensive, labour and energy-intensive, metal specific and can generate secondary waste or sludge. Thus, an alternative method like phytoremediation should be highlighted since it is low cost, more secure and less destructive [11]. Phytoremediation is defined as the technologies that use selected living plants to clean up, remove, incapacitate or detoxify soil, air, water and sludge, contaminated with hazardous contaminants [12]. Furthermore, phytoremediation is an alternative solution for heavy metal remediation process since it is a cost-effective, long term applicability, environment- and eco-friendly 'green' technology [13]. In this study, three different submerged aquatic plants origin from the riverbank of Tasik Puteri, Bukit Besi are used. The plants are *E. dulcis* ("Purun Tikus"), E. *ochrostacys* ("Purun") and *L. articulata* ("Purun Danau") as shown in Figure 1.

Based on authors' prior knowledge, limited study has been conducted using *E. Dulcis* ("Purun Tikus"), E. *Ochrostacys* ("Purun") and *L. Articulata* ("Purun Danau") to treat contaminated ex mining lake water caused by AMD. Thus, these three native plants will be used in phtyotoxicity study to identify their survivability.

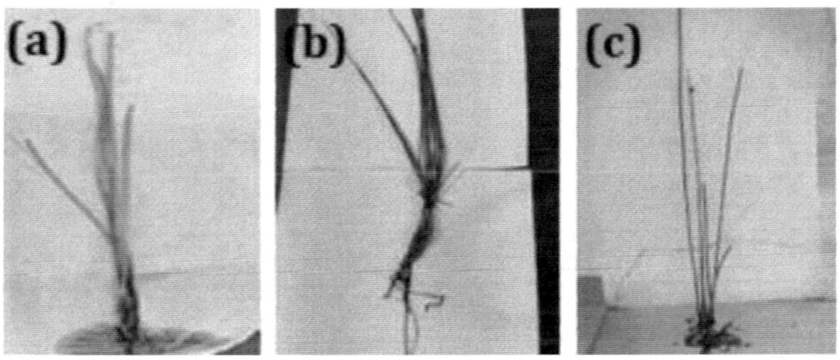

Figure 1. (a) *E.dulcis* ("Purun Tikus") (b) *E.ochrostacys* ("Purun") (c) *L. articulata* ("Purun Danau")

2. Materials and methods

2.1. Plant propagation

E. ochrostacys, *E. dulcis* dan *L. articulata* are submerged aquatic plant. The reproduction of these three aquatic plants is through rhizomes. They have been collected from the riverbank of Tasik Puteri, Bukit Besi. The original habitat of the plants is in the same area around the mining lake. Therefore, it assumed that the growth stages of those plants were same. The plants then replanted in containers measuring 58 cm x 39 cm x 29.5 cm (l x h x w) containing in 3: 2: 1 soil (surface soil: organic material: sand). Since the plants are all aquatic plants, hence throughout the propagation period, the media of the garden soil is watered and the height of water is set on the sub surface so that plants can grow well.

2.2. Phytotoxicity study

Study of phytotoxicity is done for 28 days by using plastic bucket that act as batch reactor with a volume of 3L, a diameter of 27 cm and bucket depth of 27 cm. A total of 1 kg of sand is used and 3 L of water from Tasik Puteri is poured in each bucket and one bucket filled with tap water as control. Throughout the study period, no nutrient addition is done only a litre of water is added to each bucket every seven days to cover the loss of water by evaporation. Observation on the height and physical changes of the plant is recorded for every 7 days. This phytotoxicity study is done on March 2019.

3. Results and discussions

3.1. Plants propagation

Three different plants species are collected at the river bank of Tasik Puteri, Bukit Besi, namely *E. dulcis* ("Purun Tikus"), *E. ochrostacys* ("Purun") and *L. Articulata* ("Purun Danau"). Each aquatic plant used for phytoremediation media taken from locations around the mine/original habitat shown in Figure 2. Meanwhile, Figure 3 shows the example of plant propagation.

Figure 2. Origin plant grown at Tasik Puteri

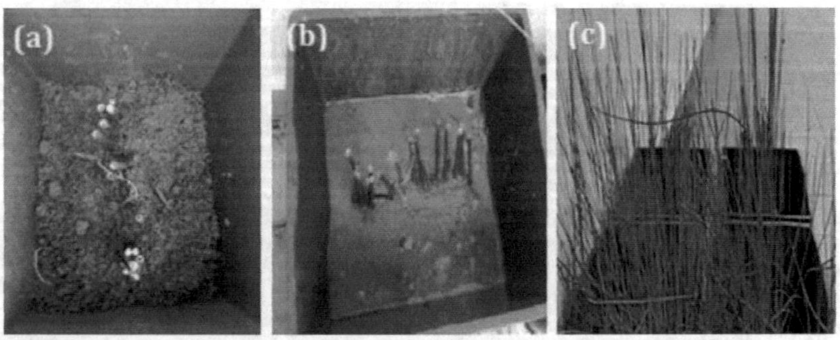

Figure 3. (a) Nursery soil blend (b) Planted parent plant (c) Planted plant aged 1 month

Height of the plant was recorded. Increment in height of the plant can be used to examine the survivability [14]. The physical plant growth is observed every seven days throughout the sampling days to examine the survivability in the new environment. From the graph in Figure 4, all the studied plants are survived in water of Tasik Puteri. However only *E.ochrostacys* show the highest increment in height indicate the highest of survivability.

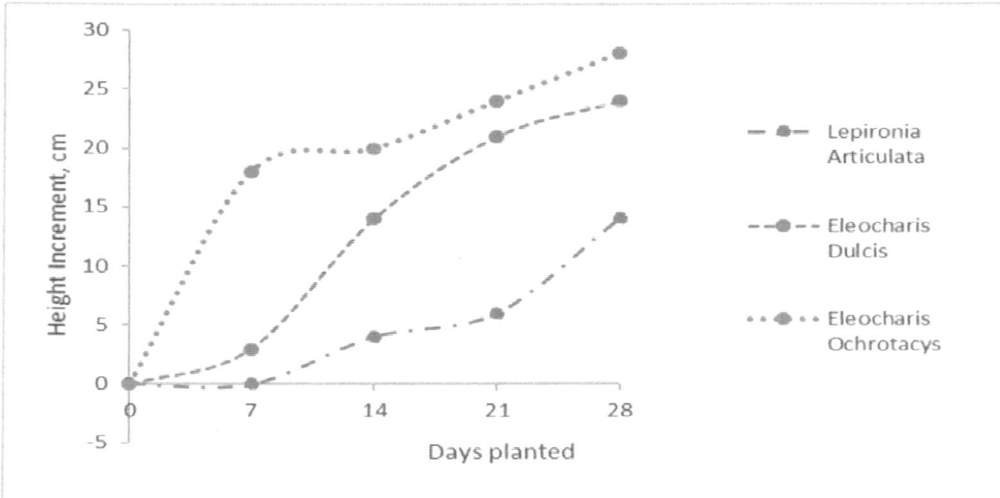

Figure 4. Plant growth

3.2. Phytotoxicity study

Phytotoxicity study was conducted to determine the ability of studied plants to survive in the water of Tasik Puteri. 10 healthy plants of 42 days old were planted in each pail. After 28 days of exposure, the percentage of withered plants in each pail was determined relative to the total number of plants in the pail using the following equation (1) [15].

$$\text{Percentage of withered plant} = \frac{no\ of\ withered\ plant}{total\ plant} \times 100 \quad (1)$$

The survival rate for *E.dulcis*, *E.ochrostacys* and *L. articulata* were 50 %, 80% and 70%, respectively. It was observed that *E. ochrostacys* could survive longer than the other two species plants. Graph in Figure 5 showed the percentage of withered plant for each native plant.

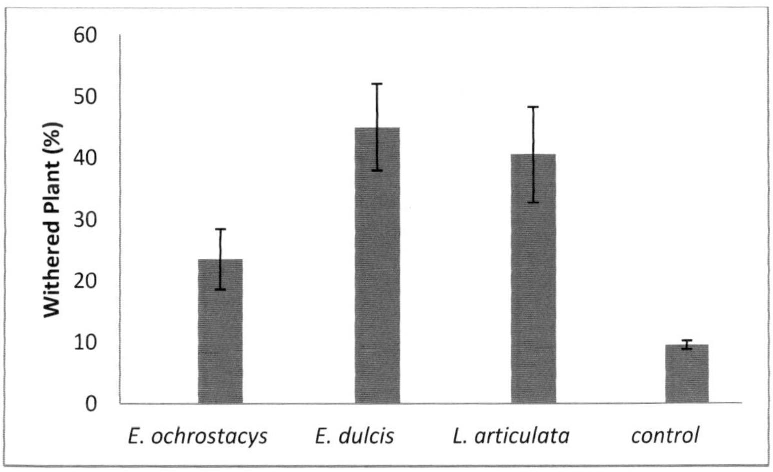

Figure 5. Percentage of withered plant

3.3. Tasik Puteri water treatment

It is proposed that root absorbs the presence of contaminants in water for the studied plants (*E. dulcis, E.ochrostacys* and *L. articulata).* The mechanism of heavy metal removal occurs when an active adsorption process takes place. The existence of carboxyl groups at the root system also encourages a remarkable cation exchange through the cell membrane[16]. Figure 6 show results of heavy metal concentration in the root and stem of the studied plants. The level of iron uptake in lower part is the highest for *E. ochrostacys.* The finding correlated with the results reported by previous researchers [17, 18] that *Eleocharis,* which belongs to Cyperaceae Family, have a high adsorption capacity of toxic metals through the roots. Same pattern for iron uptake result as shown in Figure 6 for the upper part/stem of studied plants indicated that *E. ochrostacys* and *E. dulcis* become the best-studied plants for further phytoremediation study.

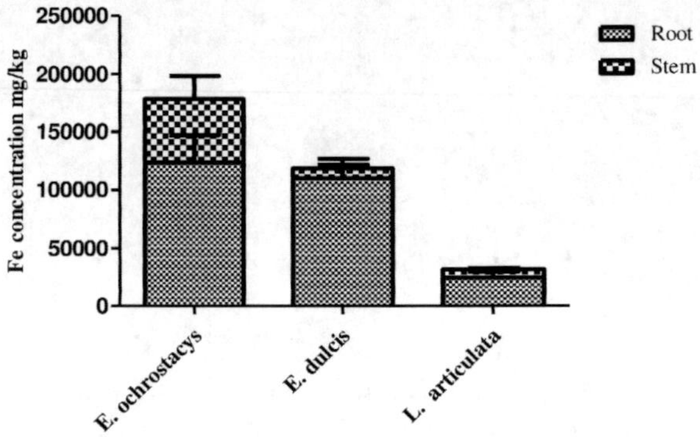

Figure 6. Fe uptake of studied plants

4. Conclusion

Based on heavy metal analysis and phtotoxicity test, it can be concluded that the most suitable plant to be used for phytoremediation is *E.ochrostacys* with the highest accumulation ability of Fe compared to *E.dulcis* and *L.articulata*. *E.ochrostacys* accumulate 123,584 mg Fe/kg at lower part and 55,151 mg Fe/kg at the upper part. Other than that, in phytotoxicity test *E.ochrostacys* was also able to tolerate and survive in the water of Tasik Puteri up to 80% survival rate based on percentage of withered plant and the highest increment based on physical growth. For the future planning, an integrated technique involves phytoremediation and adsorption process, namely as phyto-adsorption is proposed to enhance the phytoremediation process. It can be positively an innovative technique.

Acknowledgments

Authors would like to express our gratitude to Universiti Teknologi MARA for providing the facilities, technical and financial support given throughout the research work by SIG Grant (600-UiTMKD PJI/RMU/ST/DANA SIG 5/2/1 Dst (13/2016) and 600- UiTMKD PJI/RMU/ST/DANA SIG 5/2/2 Dst (01/2019)).

References

[1] Akpor O B 2014 Heavy Metal Pollutants in Wastewater Effluents: Sources, Effects and Remediation. *Advances in Bioscience and Bioengineering* **2** 37
[2] Modoi O C, Roba, C., Török, Z., Ozunu, A. 2014 Environmental risks due to heavy metal pollution of water resulted from mining wastes in NW Romania *Environmental Engineering and Management Journal,* **13** 2325–36
[3] Habeeb M A, Al-Bermani, A. K. K., Salman, J. M. 2016 Environmental study of water quality and some heavy metals in water , sediment and aquatic macrophytas in lotic ecosystem *Iraq* **1** 66–84

[4] Nowrouzi M, Mansouri, B., Nabizadeh, S., Pourkhabbaz, A. 2014 Analysis of heavy metals concentration in water and sediment in the Hara biosphere reserve, southern Iran. *Toxicology and Industrial Health,* **30** 64–72

[5] Simate G S, Ndlovu, S. 2014 Acid mine drainage: Challenges and opportunities *Journal of Environmental Chemical Engineering,* **2** 1785–803

[6] Ndeda L A 2014 Determination of Heavy Metals in Nairobi Dam Water, (Kenya) *IOSR Journal of Environmental Science Ver. IV* **8** 2319–99

[7] Gwenzi W, Mushaike, C. C., Chaukura, N., Bunhu, T. 2017 Removal of Trace Metals from Acid Mine Drainage Using a Sequential Combination of Coal Ash-Based Adsorbents and Phytoremediation by Bunchgrass (Vetiver [Vetiveria zizanioides L]). *Mine Water and the Environment.*

[8] Khalid S A, Draman, S. F. S., Sheikh Abdullah, S. R., Anuar, N., Mohd, M. F. A. R. 2017 In situ analysis of water quality monitoring in ex-mining lake Tasik puteri (Terengganu, Malaysia. *ARPN Journal of Engineering and Applied Sciences* **12** 3251 -4

[9] Lopez J, Reig, M.,Gibert, O., Valderrama, C. ,Cortina, J.L. 2018 Evaluation of NF membranes as treatment technology of acid mine drainage: metals and sulfate removal *Desalination* **440** 122-34

[10] Magagane N, Masindi, V., Munyadziwa, Ramakokovhu, M., Shongwe, M.B., Muedi, K.L. 2019 Facile thermal activation of non-reactive cryptocrystalline magnesite and its application on the treatment of acid mine drainage *Journal of Environmental Management* **236** 499-509

[11] Ismail N I, Abdullah, S.R.S., Idris, M., Hasan, H.A., Al Sbani, N.H. Jehawi, O.H. 2015 Tolerance and survival of Scirpus grosus and Lepironia articulata in synthetic mining wastewater *Journal of Environmental Sciences and Technology* **8** 232-7

[12] Ojoawo S O, Udayakumar, G. and Naik, P. 2015 Phytoremediation of phosphorus and nitrogen with canna x generalis reeds in domestic wastewater through NMAMIT constructed wetland *Aquatic Procedia* **4** 349-56

[13] R S A a A 2018 A review of phytoremediation technology: heavy metals uptake by plants *IOP Conference Series: Earth and Environmental Science* **142**

[14] Macera L G, Pereira, S.R., Lúcia, A. de Souza, T. 2017 Survival and growth of tree seedlings as a function of seed size in a gallery forest under restoration *Acta Botanica Brasilica* **31**

[15] Al-Baldawi I A W, Abdullah, S. R. S., Suja, F., Anuar, N., Idris, M. 2015 The ratio of plant numbers to the total mass of contaminant as one factor in scaling-up phytoremediation process. *Jurnal Teknologi* **74** 111–4

[16] Prihatini N S 2015 A study of iron (Fe) uptake by purun tikus (Eleocharis dulcis) from acid mine drainage in subsurface flow constructed wetland. In: *International Conference: Natural, Mathematical and Environmental Sciences,* pp 1-6

[17] Yamazaki S, Okazaki, K., Kurahashi, T., and Sakakibara, M. 2017 Phytoremediation of arsenic- and molybdenum-contaminated alkaline wastewater by Eleocharis acicularis in winter in Japan. IOP Conference Series: Earth and Environmental Science *IOP Conference Series: Earth and Environmental Science* **71**

[18] Nurfitri A G, Masayuki, S., and Koichiro, S. 2017 Phytoremediation of Heavy Metal-Polluted Mine Drainage by Eleocharis acicularis *Environmental Science: An Indian Journal* **13** 1–11

IOP Conf. Series: Materials Science and Engineering **808** (2020) 012041 doi:10.1088/1757-899X/808/1/012041

Parametric study of CO_2 separation using carbon molecular sieve, zeolite and silica gel

A A Norani, **A Ahmad**, **T A T Abdullah** and **A Ripin**

Centre of Hydrogen Energy, Institute of Future Energy, Universiti Teknologi Malaysia, 81310 Johor Bahru, Johor, Malaysia.

*arshad@utm.my

Abstract. Performances of Pressure swing adsorption (PSA) dependent on the type of adsorbent and efficiency of the regeneration cycle. In this work, a model describing CO_2 mixture flow, mass and heat transfer with multi-component adsorption model are developed in Aspen Adsorption environment to compare the performance of PSA cycles with three different adsorbents (CMS 3K, zeolite 13x and silica gel). Adsorption capacity of CO_2 for three adsorbents was found to be in the order of CMS>zeolite>silica gel. Hence, CMS was chosen as the adsorbents to evaluate PSA performances on different feed flowrate and pressure based on the purity, recovery and productivity of the CO_2 separation. The result shows that, within a certain range, higher adsorption pressure will increase all purity, recovery and productivity of CO2 while higher feeding rate leads to increase recovery but lower purity and productivity.

1. Introduction

Pressure Swing Adsorption (PSA) technology has been widely studied for CO_2 capture at both pilot-scale or laboratory levels in pursuit of more low-energy, environmentally friendly and cost-effective separation techniques [1]. The behaviour of a PSA unit is mainly determined by two important factors that are employing a high capacity adsorbent and effective engineering strategy for regeneration. The studies on PSA process performances aimed at separating CO_2 by experiment and simulation using different adsorbent and working configuration have been reviewed in literature elsewhere [1, 2].

One of the important basis subjects to study the adsorption process performances is by the breakthrough curve [3, 4]. Shafeeyan *et al.* has performed a parametric study of CO_2/H2/N2 gas mixture on commercial activated carbon. The CO_2 partial pressure was observed to be the most influential variable, with high values leading to an increase in both the CO_2 capture capacity and the breakthrough time [5]. However, an increase in the temperature led to a decrease in adsorption capacity and breakthrough time [5,6]. In order to determine the performance of the carbon molecular sieve for separation of binary CO_2/CH4 mixture, Rocha *et al.* have studied the breakthrough curve of CO_2 at 10 and 70 bars. The result shows that at higher pressure, more content of CO_2 has been detected at the outlet which reduces the selectivity and capacity of the adsorbent [7]. More breakthrough experiments were performed to study various parameters and adsorption capacity [8–17].

With the basis of breakthrough curve study, a reliable full-cycle PSA experiment and simulations can be carried out [3]. The performance indicators for PSA cycle depends on the purity, recovery and productivity of the product. Most current reports show that CO_2 PSA process to achieve CO_2 purities of over 90% with recovery ranging from 70% to 90% [18–20]. Cho et al. studied a 2-bed PSA with the 2-stage process at each stage using zeolite 13X to obtain 99.7% CO_2 purity with 78.8% recovery. Delgado

IOP Conf. Series: Materials Science and Engineering **808** (2020) 012041 doi:10.1088/1757-899X/808/1/012041

et al. conducted experimental and simulation study of CO_2 adsorption on activated carbon with 3-bed 12-step. The objective of this work is to study the feasibility of carbon dioxide concentration and recovery using a novel VSA cycle without rinse step, which has the potential of significantly reducing the power consumption of this process. According to the simulated results, it is possible to CO_2 with high purity (>93%) and recovery (>90%) [21]. Webley et al. address the concern on a typical 2-bed cycle that poor due to the lack of room to incorporate more relevant auxiliary intermediate process steps such as pressure equalization, co-current purge, etc. A breakthrough and performance study on 4-bed VPSA that has been developed shows >75% recovery with a purity of > 95% was obtained [18].

The performances of PSA unit is highly dependent on ideal adsorbent with high capacity and selectivity as well effective process configuration and engineering strategy. Therefore, it is very important to identify the best material for adsorbent and to carried out parametric study on different operating parameter in order to evaluate the PSA performances. The current work has focused on building a mathematical model of a single adsorption bed and PSA system in Aspen Adsorption environment. The breakthrough study was conducted to compare the CO_2 adsorption performances of carbon molecular sieve (CMS), zeolite and silica gel in. Based on the comparison, the best adsorbent has been selected for further study on PSA 2-bed 4-step process towards the separation of CO_2 and N_2. Subsequently, parametric study has been done to discover the effect of adsorption pressure and feed flow rate on the CO_2 purity and recovery.

2. Methodology

The PSA system for gas mixture CO_2 and N2 were generated using a built-in property database module from Aspen Adsorption V9 software as shown in Figure 1. The breakthrough study was carried out by a single adsorption unit and 2-bed PSA model simulated to study the performances of PSA. The PSA cycle consists of the 4 steps Skarstorm cycle which are pressurization (PR), adsorption (AD), blowdown (BD) and purge (PG). Table 1 shows the list of properties of column and adsorbents for simulation. The properties of CMS3K, zeolite and silica gel were obtained from the previous study which has been carried out experimentally and validated by simulation.

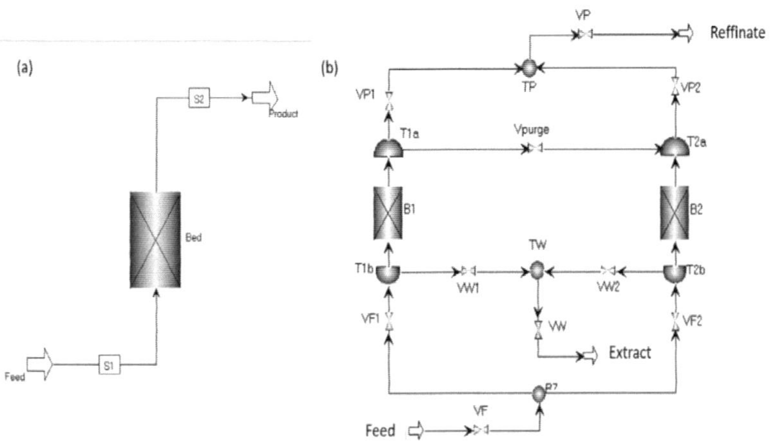

Figure 1. (a) Single adsorption bed for breakthrough study (b) 2-bed 4-steps PSA.

IOP Conf. Series: Materials Science and Engineering **808** (2020) 012041 doi:10.1088/1757-899X/808/1/012041

Table 1. Column and adsorbents properties.

Properties	**Value**		
Column parameter			
Bed length (in)	24		
Bed diameter	16		
Gas composition	CO_2/N2 (0.4/0.6)		
Adsorbent parameter	**ZEOLITE**	**AC**	**CMS**
Inter-particle voidage	0.33	0.35	0.33
Intra-particle voidage	0.54	0.33	0.46
Bulk solid density of adsorbent	756	522	715
Adsorbent particle radius	8.00E-04	0.0012	0.009
CO2 Constant mass transfer coefficients	0.00726	9.199	0.018
N2 Constant mass transfer coefficients	0.00474	1.786	0.112
Isotherm parameter			
IP(1,"CO2")	4.98E-04	0.00554	0.00033
IP(1,"N2")	3.37E-06	0.00375	0.0022
IP(2,"CO2")	0.0105	0.00049	0.0021
IP(2,"N2")	0.871	0.00054	0.00042
IP(3,"CO2")	0.492	0.811	0.43
IP(3,"N2")	0.796	0.937	0.885
IP(4,"CO2")	2160	2300	2850
IP(4,"N2")	1460	1610	1740
IP(5,"CO2")	0.00147	0.0018	0.0031
IP(5,"N2")	0.00106	0.00104	0.0001
IP(6,"CO2")	2040	1720	2450
IP(6,"N2")	1360	0	1970

2.1. Mathematical modeling

A summary of theoretical assumptions employed for the dynamic simulation of CO_2 adsorption on packed bed column is presented below:

- Discretization: Upwind Differencing Scheme 1,
- Mass balance: Convection only
- Momentum balance: Ergun equation
- Kinetic assumption: Linear Lumped resistance
- Isotherm: Extended Langmuir-Freundlich
- Energy balance: Non-isothermal with no conduction
- Reaction: None
- No radial concentration profiles
- No velocity gradients

Mass conservation equation

The overall mass conversation equation can be written as follows:

$$-D_L \frac{\partial^2 y_i}{\partial z^2} + \frac{\partial y_i}{\partial t} + u_z \frac{\partial y_i}{\partial z} + \frac{RT}{p} \frac{1-\varepsilon_b}{\varepsilon_b} \rho_p \left(\frac{\partial q_i}{\partial t} - y_i \sum_{j=i}^{N} \frac{\partial q_j}{\partial t} \right) = 0 \quad (1)$$

where D_L is the axial dispersion coefficient, u_z is the axial physical velocity, y_i and qi are the molar fraction and the adsorbed phase concentration of species i, respectively, ε_b is the bed void fraction, r_p is the pellet density of the adsorbent, R is the universal gas constant and p is pressure in the adsorption bed, t and z are the time and axial position in the bed, respectively.

Energy conservation equation

The energy conservation for the gas and solid phase of an adsorption bed is given by

$$-\varepsilon_b K_L \frac{\partial^2 T}{\partial z^2} + (\varepsilon_t \rho_g C_{pg} + \rho_b C_{ps}) \frac{\partial T}{\partial t} + \rho_g C_{pg} \varepsilon_b u_z \frac{\partial T}{\partial z} - \rho_b \sum_i^N Q_i \frac{\partial q_i}{\partial t} + \frac{2h_i}{R_{Bi}}(T-T_w) = 0 \quad (2)$$

where K_L is the thermal axial dispersion coefficient, C_{pg} is the heat capacity of the gas phase, C_{ps} is the specific heat capacity of the adsorbent, ε_t is the total void fraction of the bed, Q_i is the heat of adsorption of species i, T_w is the wall temperature.

Adsorption isotherms and kinetics

The extended Langmuir-Freundlich model was used to express the adsorption isotherms of a multicomponent gas:

$$q_i^* = \frac{q_{mi} B_i p_i^{n_i}}{1 + \sum_{j=1}^N B_j p_j^{n_j}} \quad (3)$$

Where q_i^* is the equilibrium adsorbed phase concentration, q_{mi} and B_i are the extended Langmuir-Freundlich isotherm parameters, and p_i is the partial pressure of species i.

Momentum conservation equation

The momentum balance in a bed is given by Ergun's equation:

$$-\frac{dp}{dz} = a\mu v_z + b\rho v_z |\vec{v}| \quad (4)$$

with the coefficients:

$$a = \frac{150}{4R_p^2} \frac{(1-\varepsilon_b)^2}{\varepsilon_b^3}, \quad b = 1.75 \frac{(1-\varepsilon_b)}{2R_p \varepsilon_b^3} \quad (5)$$

where m is the dynamic viscosity, v_z is Darcy's velocity, and Rp is the particle radius.

Equation of state

From the ideal gas equation of state, pV = nRT, the molar concentration of the mixture can be written as c = pV = nRT. The molar concentration of component i can be written as

$$y_i = \frac{n_i}{n} = \frac{n_i/V}{n/V} = \frac{c_i}{c} = \frac{c_i RT}{p}, \quad i=1,\cdots,N. \quad (6)$$

The specific heat capacity of the mixture is calculated from the molar fraction weighted average as

$$C_{pg} = \sum_{i=1}^n y_i C_{pi} \quad (7)$$

The density of the mixture can be calculated by

$$\rho = cM = c \sum_{i=1}^N y_i M_i \quad (8)$$

IOP Conf. Series: Materials Science and Engineering **808** (2020) 012041 doi:10.1088/1757-899X/808/1/012041

PSA performance indicator

The performance with measuring some important performance indicators such as product purity, product recovery, product productivity :

$$CO_2 \text{ purity } (\%) = \frac{\int_{t=0}^{t_{BD}} \dot{n}_{CO_2}|_{Extract} dt + \int_{t=0}^{t_{PG}} \dot{n}_{CO_2}|_{Extract} dt}{\sum_{i=0}^{n} \dot{n}_i|_{Extract} dt + \sum_{i=0}^{n} \dot{n}_i|_{Extract} dt} \quad (9)$$

$$CO_2 \text{ recovery } (\%) = \frac{\int_{t=0}^{t_{BD}} \dot{n}_{CO_2}|_{Extract} dt + \int_{t=0}^{t_{PG}} \dot{n}_{CO_2}|_{Extract} dt}{\int_{t=0}^{t_{PR}} \dot{n}_{CO_2}|_{Feed} dt + \int_{t=0}^{t_{AD}} \dot{n}_{CO_2}|_{Feed} dt} \quad (10)$$

$$\text{Productivity } \left(\frac{mol}{kg.s}\right) = \frac{\int_{t=0}^{t_{BD}} \dot{n}_{CO_2}|_{Extract} dt + \int_{t=0}^{t_{PG}} \dot{n}_{CO_2}|_{Extract} dt}{t_{cycle} m_{adsorbent}} \quad (11)$$

3. Result and discussion

The column dynamic breakthrough study was built by using a single column adsorption bed. The gas mixture of CO_2/N_2 has been introduced to the bed with fixed operating condition for all the adsorbents to solely study their adsorption capacity. The operating pressure and temperature were fixed at 5 bar and 298.15K respectively.

Breakthrough occurs when adsorbate reaches the end of the column and leaves with the column effluent. Breakthrough curves are plots of the adsorbate concentration in the column effluent as a function of time. Comparison of CO_2 adsorption on CMS 3K, zeolite 13x and silica gel have been studied by the breakthrough graph. The longest breakthrough time favour adsorption the most because it provides enough residence time for the adsorption process to reach equilibrium. The steepness of the graph also plays important roles as it represents the adsorption capacity. The result of the breakthrough graph in Figure 2. shows that CMS 3K proved to be the best material for separation of CO_2 and N_2, with the longest breakthrough time at 322s and the steepest graph compares to the other two adsorbents which represent high CO_2 adsorption capacity. Zeolite 13x showed the second-largest breakthrough time at 295s and lastly silica gel with the shortest breakthrough time at 124s. Hence, CMS was chosen as the adsorbents for further study of PSA separation of CO_2 and N_2 due to its highest CO_2 adsorption capacity.

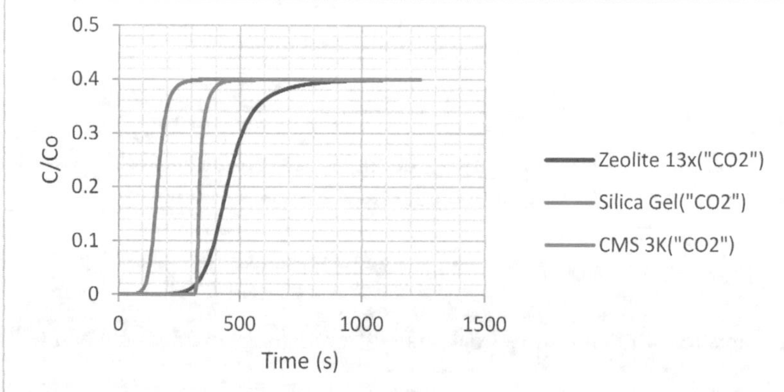

Figure 2. Breakthrough graph for zeolite 13x, silica gel and CMS 3K.

The effect of adsorption pressure and feeding flowrate on the PSA cycle performance are studied while other parameters are being constant. The influence of adsorption pressure was studied by varies the pressure from 1 to 7 bar. Figure 3. shows that as the pressure increases, the purity, recovery and productivity of CO_2 also increase up to 79.3%, 95% and 0.89 mol/s/kg respectively. However, it is noticeable that the purity of CO_2 is low and being stagnant from 4 bars onwards. This is reasonable because of the positive correlation between adsorption pressure and adsorption capacity. When the pressure is high, both adsorbates (CO_2/N_2) tend to adsorb more deeply into the adsorbent. The purity also was influenced by the raffinate gas that been supply from another column at the blowdown stage.

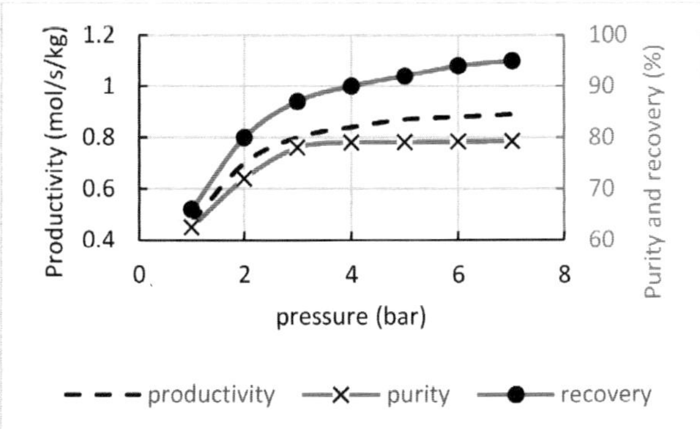

Figure 3. Effect of different pressure on CO2 purity, recovery and productivity.

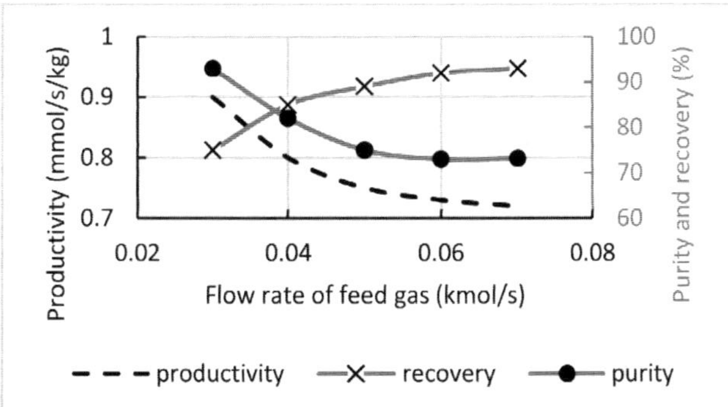

Figure 4. Effect of different feed flow rates of CO2 purity, recovery and productivity.

The feed flow rate of the gas mixture into the column during feed and adsorption stages also an important factor affecting the process performances. An excessive flow rate could deteriorate process performances. With other parameters fixed, the flow rates have been varied by 0.03 to 0.07 kmol/s. Figure 4. shows as the flow rate increases, the recovery increases by up to 97%, inversely the purity and productivity decrease to 63% and 0.88 kmol/s/kg respectively. The faster flow rate indicates more amount of gas is being introduced into the column and more CO_2 can be adsorbed. However, it also means that the adsorbent could saturate faster, results in more CO_2 escaped from the column, decreasing recovery of CO_2.

4. Conclusion

A model describing CO_2/N_2 mixture flow, mass and heat transfer with multi-component adsorption has been developed for studying breakthrough curves and 2-bed 4-step PSA cycle using Aspen Adsorption. The adsorption capacity of CO_2 on CMS 3K, zeolite 13x and silica gel have been studied using a single bed adsorption unit. The comparison of the three adsorbents result shows that CMS 3K had the highest adsorption capacity based on the longest breakthrough time and steepest graph. A parametric study on the PSA cycle is then being conducted using CMS 3K to investigate the effect of adsorption pressure and feed flowrate on purity, recovery and productivity of CO_2. The result shows that, within a certain range, higher adsorption pressure will increase all purity, recovery and productivity of CO_2 while higher feeding rate leads to increase recovery but lower purity and productivity.

References

[1] L. Riboldi and O. Bolland 2017 Overview on Pressure Swing Adsorption (PSA) as CO2Capture Technology: State-of-the-Art, Limits and Potentials *Energy Procedia* **114** 2390–2400,.

[2] A. D. Wiheeb, Z. Helwani, J. Kim, and M. R. Othman 2016 Pressure Swing Adsorption Technologies for Carbon Dioxide Capture *Sep. Purif. Rev.* **45** 108–121.

[3] J. Xiao, L. Fang, P. Bénard, and R. Chahine 2018 Parametric study of pressure swing adsorption cycle for hydrogen purification using Cu-BTC *Int. J. Hydrogen Energy* **43** 13962–13974,

[4] J. C. Knox, A. D. Ebner, M. D. Levan, R. F. Coker, and J. A. Ritter 2018 On the Limitations of Breakthrough Curve Analysis in Fixed-Bed Adsorption

[5] M. S. Shafeeyan, W. M. A. W. Daud, A. Shamiri, and N. Aghamohammadi 2015 Modeling of Carbon Dioxide Adsorption onto Ammonia-Modified Activated Carbon: Kinetic Analysis and Breakthrough Behavior *Energy and Fuels* **29** 6565–6577

[6] P. Lestinsky, M. Vecer, P. Navratil, and P. Stehlik 2015 The removal of CO_2 from biogas using a laboratory PSA unit: Design using breakthrough curves *Clean Technol. Environ. Policy* **17** 1281–1289

[7] L. A. M. Rocha, K. A. Andreassen, and C. A. Grande 2017 Separation of CO_2/CH_4 using carbon molecular sieve (CMS) at low and high pressure *Chem. Eng. Sci.* **164** 148–157

[8] H. Ahn, C. Chun, M. Park, I. S. Ahn, and C. H. Lee 2001 Thermal effects on the breakthrough curve of a hydrogen ternary system at a fixed bed *Sep. Sci. Technol.* **36** 2121–2145

[9] J. H. Park, J. N. Kim, S. H. Cho, J. D. Kim, and R. T. Yang 1998 Adsorber dynamics and optimal design of layered beds for multicomponent gas adsorption *Chem. Eng. Sci.* **53** 3951–3963

[10] S. Garcia, M. . Gil, C. . Martin, J. . Pis, F. Rubiera, and C. Pevida, "Breakthrough adsorption study of a commercial activated carbon for pre- combustion CO_2 capture S. García, M.V. Gil, C.F. Martín, J.J. Pis, F. Rubiera, C. Pevida 1 Instituto Nacional del Carbón, CSIC, Apartado 73, 33080 Oviedo, Spain," pp. 1–25.

[11] L. F. Gomez, R. Zacharia, P. Bénard, and R. Chahine 2015 Simulation of Binary CO_2/CH_4 Mixture Breakthrough Profiles in MIL-53 (Al) *J. Nanomater.* **2015** 1–15

[12] A. Aquino *et al.* 2016 Carbon Dioxide Removal with Tuff: Experimental Measurement of Adsorption Properties and Breakthrough Modeling Using CFD Approach *Energy Procedia* **101** 392–399

[13] R. Ben-Mansour, M. Basha, and N. A. A. Qasem 2017 Multicomponent and multi-dimensional modeling and simulation of adsorption-based carbon dioxide separation *Comput. Chem. Eng.* **99** 255–270

[14] P. Brea, J. A. Delgado, V. I. Águeda, and M. A. 2017 Uguina Modeling of breakthrough curves of N_2, CH_4, CO, CO_2 and a SMR type off-gas mixture on a fixed bed of BPL activated carbon *Sep. Purif. Technol.***179** 61–71

[15] T. L. P. Dantas *et al.* 2011 Carbon dioxide-nitrogen separation through adsorption on activated carbon in a fixed bed *Chem. Eng. J.* **169** 11–19

[16] M. Xu, H. C. Wu, Y. S. Lin, and S. Deng 2018 Simulation and optimization of pressure swing adsorption process for high-temperature air separation by perovskite sorbents *Chem. Eng. J.* **354** 62–74

[17] M. H. Chahbani and D. Tondeur 2001 Pressure drop in fixed-bed adsorbers *Chem. Eng. J.* **81** 1–3 23–34

[18] P. A. Webley, A. Qader, A. Ntiamoah, J. Ling, P. Xiao, and Y. Zhai 2017 A New Multi-bed Vacuum Swing Adsorption Cycle for CO_2 Capture from Flue Gas Streams *Energy Procedia* **114** 2467–2480

[19] Q. Fu, H. Yan, Y. Shen, Y. Qin, D. Zhang, and Y. Zhou 2018 Optimal design and control of pressure swing adsorption process for N_2/CH_4 separation *J. Clean. Prod.* **170** 704–714

[20] R. Zhao *et al.* 2017A comparative study on CO_2 capture performance of vacuum-pressure swing adsorption and pressure-temperature swing adsorption based on carbon pump cycle *Energy* **137** 495–509

[21] J. A. Delgado, M. A. Uguina, J. L. Sotelo, V. I. Águeda, A. Sanz, and P. Gómez 2011 Numerical analysis of CO_2 concentration and recovery from flue gas by a novel vacuum swing adsorption cycle *Comput. Chem. Eng.* **35** 1010–1019

Experimental Study of the Characteristics of Acoustic Cavitation Bubbles Under the Influence of Ultrasonic Wave

Hadi N A H and Ahmad A*

Center of Hydrogen Energy, Faculty of Chemical and Energy Engineering, Universiti Teknologi Malaysia, 81310 UTM Johor Bahru, Johor, Malaysia.

*arshad@utm.my

Abstract. Sonochemistry application in various fields, especially separation industries, are constantly increasing and extensively studied in recent years. In the applications, a detailed understanding of the characteristics and behaviour of acoustic cavitation bubble is crucial. However, acoustic bubbles behave differently in each liquid. Owing to numerous amounts of acoustic bubbles showing dramatic transition during sonication and have complex interaction in a liquid mixture, it is more difficult to observe acoustic bubble consistently and thoroughly. In response to this problem, our study will investigate separately acoustic cavitation bubble form in sonicated water, ethanol and mixture of ethanol-water. Each liquid was sonicated at 20 kHz ultrasound with a transducer attached to the bottom of an acrylic vessel. In order to visualise acoustic bubbles clearly, a high-speed video camera is used to capture the image of acoustic bubbles. Image processing and analysis are by the application of Image Processing Toolbox in MATLAB R2018b. Experimental result reveals that acoustic cavitation bubbles in a mixture of ethanol-water have the largest radius, volume and higher buoyancy force compared to pure ethanol and water. Results obtained coincided well with theory, which indicated that sonicated mixture containing alcohol provide resistance to gas transfer across liquid/bubble interface, leading to an increased bubble radius and hence the buoyancy force.

1. Introduction

Cavitation is a phenomenon of the generation of bubbles due to the decrease in liquid pressure with a consecutive violent collapse of each bubble [1]. These involve bubbles formation, growth, pulsation and collapse of bubbles in liquid during propagation of the sound wave. Chemical effects of ultrasound such as production of free radical and acceleration of a reaction originate in the extreme local condition of collapse bubble has been used in various industries [2]. For many years, the use of ultrasound energy in chemical process industries has developed a growing interest in replacing the conventional method [3] that used a massive amount of energy consumption. Recently, a new approach to the separation process of ethanol-water called "ultrasonic distillation" has been investigated. Many researchers investigated the separation of ethanol from the aqueous solution by ultrasound. It was reported that the application of ultrasonic enhance the total area of vapor-liquid interface and can be served as an input source of thermal energy required for vaporisation [4]. Research done by Yasuda [5] reported that the utilisation of ultrasonic atomization shows an increasing number of ethanol generation rate droplets with increasing ethanol solution. Their works were in alignment with Wakisaka [6] proving that there is a close relationship between evaporation properties in bubbles and cluster-level structure in alcohol-water binary mixtures.

Based from all the results and the data interpretations from researchers, it is plausible such "ultrasonic distillation" separation technique are feasible. However, separation mechanism in terms of acoustic bubble characteristic is not fully elucidated due to difficulties in characterizing the bubbles in a mixture. In this work, the characteristics of acoustic cavitation bubbles in terms of radius, volume, movement and buoyancy force under the influence of ultrasonic sound wave at 20kHz is examined. To investigate the characteristics of acoustic cavitation bubble, a detailed experimental work of three different liquids with distinctly different properties – water, ethanol and ethanol-water mixture were conducted. Acoustic cavitation bubbles were observed using a high-speed video camera. The results obtained were analysed using MATLAB R2018b, and the results from each liquid were compared thoroughly. Because separation of azeotropic mixtures occurs inside the acoustic bubbles, results of this study may be useful in the separation of azeotropic mixture by ultrasonic distillation, which is the next focus of process intensification in chemical separation industry [7].

2. Experimental Setup

Figure 1 shows the experimental setup for the generation and visualisation of the acoustic cavitation bubble. For bubble generation, ultrasound with a frequency of 20kHz sine wave was supplied by a function generator. The signal was then supplied to a bolt-clamped Langevin-type transducer. The transducer was bonded to a 3 mm thick 316 stainless-steel plate that was located at the bottom of rectangular sono-reactor. The reactor was made from acrylic, and the inner dimensions were 70 mm x 70 mm with 200 mm height. The liquid in the reactor was in direct contact with a stainless-steel plate. Non-degassed water at 20 °C was filled into the vessel up to 80mm height. A sound pressure field was developed and grew in the reactor by the ultrasound irradiated from the bottom of the reactor. In order to visualise the acoustic cavitation bubbles, we used a high-speed video camera and a flat LED light, mounted facing each other. The setting of the camera was as follows: image size 1024 x 1024 pixels, frame rate 500 fps, exposure time 5s, and spatial resolution 49.3 um/pixel. The recorded images were processed with MATLAB R2018b. To determine the characteristic of an acoustic cavitation bubble, we investigated bubbles formation in three different liquids which are, non-degassed water, pure ethanol and a mixture of ethanol-water 50:50 by v/v. Table 1 shows the properties of each liquid used in the experiment.

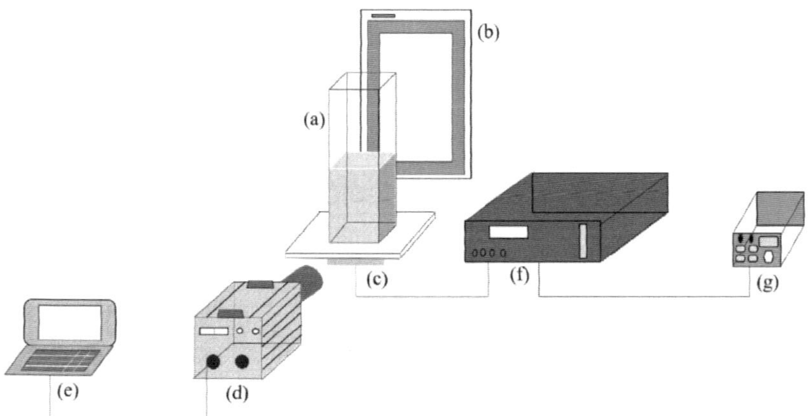

Figure 1. Experimental setup for the visualization of acoustic cavitation bubble. (a) Acrylic reactor (b) LED light (c) Transducer (d) High-speed video camera (e) Computer (f) Amplifier (g) Function generator

Table 1. Liquid properties for water and ethanol

	Water	Ethanol
Temperature (°C)	20	20
Density, ρ (kg/m^3)	1000	785
Dynamic viscosity, μ (Pa s)	1.0 x 10^{-3}	1.1 x 10^{-3}
Speed of sound, c (m/s)	1482	1100
Surface tension, σ (Nm^{-1})	0.079	0.022
Vapor pressure, p_v (Pa)	2.2 x 10^3	5.3 x 10^3
Thermal conductivity, Ķ (Wm^{-1}K^{-1})	0.58	0.014

3. Results and Discussion

3.1. Visualizations of acoustic cavitation bubbles

Growth of bubble in a system involves two processes, namely rectified diffusion which involves gas diffusion into a bubble and bubble coalescence where bubbles coalesce each other if the number concentration of bubbles is large enough [8]. Bubble coalescence can have a significant influence on the population and size of acoustic bubble [9,10]. Figure 2 shows the differences between acoustic cavitation bubbles formed in each liquid. When non-degassed water is sonicated, a few visible bubbles are readily observed as in Figure 2(a). Acoustic bubble population has shown to increase and move vigorously towards the liquid surface as the liquid change from non-degassed water to ethanol (Figure 2(b)) and mixture of ethanol-water (Figure 2(b)). This can be attributed to mainly the coalescence of bubbles to form more bubbles as the rate of growth is faster in comparison to non-degassed water. During sonication, ultrasound generate forces that are capable to drive bubble into contact and increase the frequency of coalescence. This explains the readily observed bubbles when liquid is sonicated. In liquid containing alcohol, bubble coalescence is much more complicated to evaluate. To the author best knowledge, only a few experimental studies concerning acoustic bubble in alcohol have been investigated [11,12,13]. Results obtained are only limited to the generation of cavitation bubble development and research found that alcohol when sonicated provide resistance towards gas transfer process at acoustic bubble/liquid interface. We will discuss these observations more in detail in the following section.

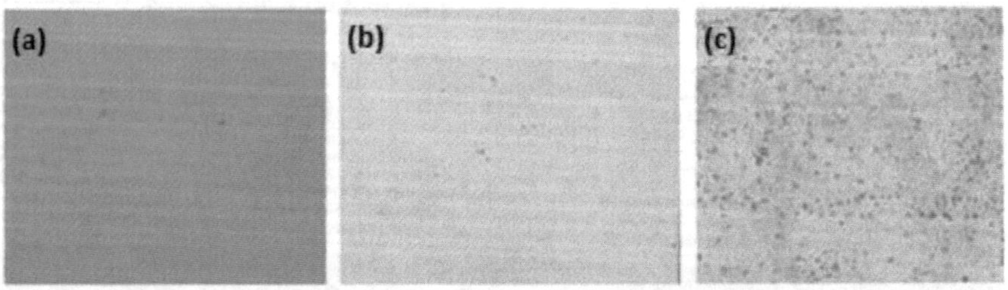

Figure 2. Image of acoustic cavitation bubbles under 20kHz ultrasound, (a) Non-degassed water, (b) Pure ethanol, (c) Ethanol-water mixture

3.1.1. Non-degassed water. Figure 3 shows the movement of the acoustic cavitation bubble in non-degassed water oscillating under acoustic frequency of 20 kHz. The bubbles in figure 3 are more likely to be driven towards the center of sono-reactor. Bubble located at the bottom of frame 1 move slightly upward at frame 2 and remains at almost constant position in frame 3 before moving downward again in frame 4. Movement of bubbles influence each other in a complex way and are related to the growth of the bubble. Under ultrasonic wave, there are two acoustic force that are responsible to drive bubbles towards each other, which are primary Bjerknes forces and secondary Bjerknes forces. Bubbles traveling to the center of sono-reactor are smaller than its resonant radius and directed toward a pressure antinode in a standing wave as shown in Figure 3, where it is a localized region in the liquid volume. Bubbles will continue to expand during compression phase of acoustic cycle and repelled from the antinode. Many experimental works have explained the mechanisms of acoustic cavitation bubble in non-degassed water; thus, we will not explain further on this, but rather focus on bubbles generates in an organic mixture in the following section.

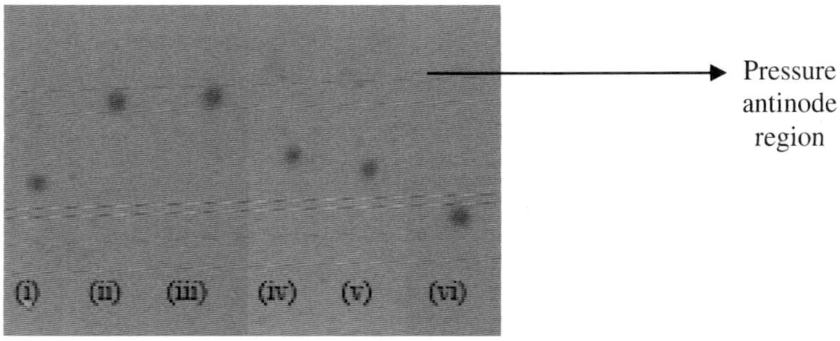

Figure 3. Movement of acoustic cavitation bubble in non-degassed water

3.1.2. Ethanol. Ethanol exhibits a distinct, fast development of cavitation under the influence of ultrasonic wave, making it interesting to discuss. In liquid ethanol, we observed that population of acoustic bubble is more than non-degassed water (figure 2) and generates larger acoustic bubble radius (figure 4). When a bubble is generating from alcohol solution, gas diffusion is strongly retarded. Consequently, lifetime of a bubble becomes much longer and volume of gas increase which is why larger bubble was generated. Sonication towards a high vapor pressure liquid results to a decrease in cavitation collapse thus produce more acoustic cavitation bubbles in sono-reactor. Kenji [14] explained that gas solubilities in organic solvents are much higher compared to water, suggesting that there are different cavitation dynamics occur in the case of organic solvent. A higher vapor content inside the cavitation bubbles leads to more vapor content inside the acoustic bubbles, hence, increases buoyancy force (figure 5).

Figure 6 shows the movement of the bubbles to the liquid surface and the action of buoyancy force when reaching ethanol-air interface. We observed that there are coalescence of small bubbles to form larger bubbles occur during cavitation. Since the gas bubbles can grow large enough, acoustic bubble can leave the system entirely due to buoyancy. Mossaz [15], also in agreement with this, wrote since there are no acoustic bubble streaming, shielding and scattering effect, acoustic bubble movement did not interrupt by ultrasonic sound wave to propagate in liquid. Iwai and Li [16] also reported a lower surface tension in the liquid would decrease the erosion rate of the bubble due to no major collapsing event in it.

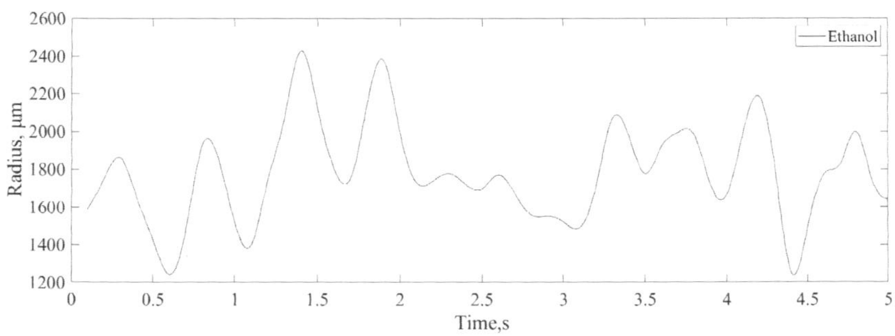

Figure 4. Fluctuation trend of the acoustic bubble radius in ethanol

IOP Conf. Series: Materials Science and Engineering **808** (2020) 012042 doi:10.1088/1757-899X/808/1/012042

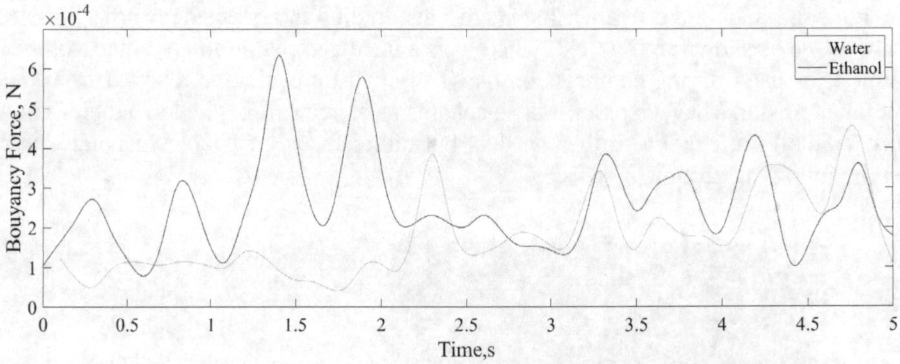

Figure 5. Comparison of acoustic bubble buoyancy force in non-degassed water and ethanol

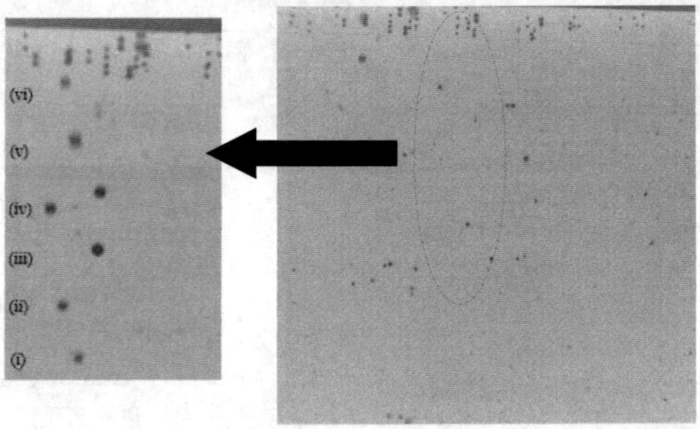

Figure 6. Movement of acoustic bubble in ethanol to liquid-air interface

3.1.3. Ethanol-water solution. Due to the strong demand of its purification from azeotropic mixture, ethanol separation from its aqueous solution is a major subject of discussion in chemical engineering. This section discusses characteristic of acoustic cavitation bubbles in sonicated ethanol-water at 50% concentration. Figure 7 shows time series movement of acoustic bubble in ethanol-water during sonication. We observed that the population of bubbles are much more compare to ethanol and non-degassed water and almost all of the acoustic bubble produced are moving towards the liquid surface as shown in figure 7. The presence of ethanol mixed with water can inhibit the ability of acoustic bubble to coalesce. The inhibition can be explained by the intermolecular interaction with water molecules in ethanol-water mixture. Addition of ethanol in water will break the hydrogen bond network and causes a decrease in surface tension. If alcohol concentration increase, there will be more alcohol rich cluster formed due to gradually ruptured water network in a mixture. Water molecules will losses its hydrogen bond network structure and behave as single-molecule bonded with ethanol molecules [17]. As a result, mixture of ethanol-water has larger volume compare to non-degassed water and ethanol (figure 8).

Alcohol molecules like ethanol, have a hydrophobic nature. Therefore, they adsorb at the liquid/bubble interface. During the expansion of cavitation bubble, adsorption of solute molecule at liquid/bubble interface will evaporate into bubble interior. Molecules will then decompose and accumulate as product thus, reducing bubble temperature when collapse. However, these adsorptions, too, will affect rectified diffusion of gas in the acoustic bubble. Diffusion of gas out of the bubble during contraction will be retarded, causing bubble to be large in volume (figure 8) and lead to higher gas content left in acoustic bubble. Bubbles that become larger may leave the system by buoyancy. Due to the formation of larger bubbles, fluid flow velocity around large microbubbles varies along the reactor (figure 9). Velocity of the fluid inside the reactor is the highest at the center of the sono-reactor. This is due to the presence of localized, high fluid flow created by large cavitation bubble and high -speed acceleration from the propagation of ultrasound through the medium.

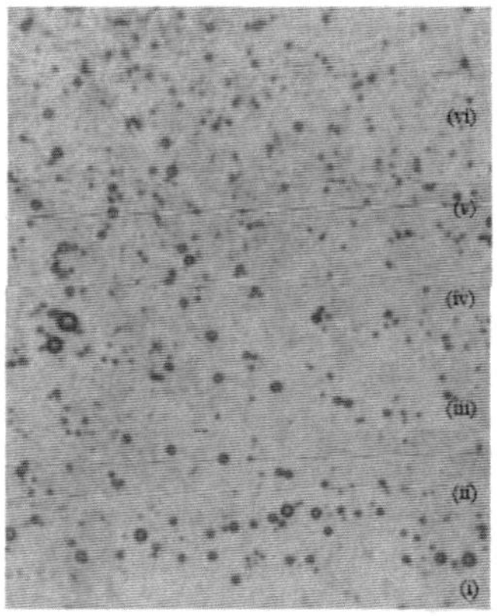

Figure 7. Time series of acoustic bubble movement in ethanol-water mixture

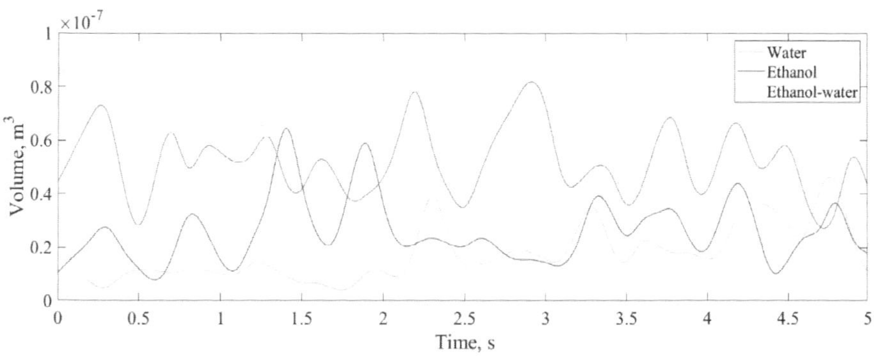

Figure 8. Comparison of acoustic bubble volume in (a) non-degassed water, (b) pure ethanol, (c) mixture of ethanol-water

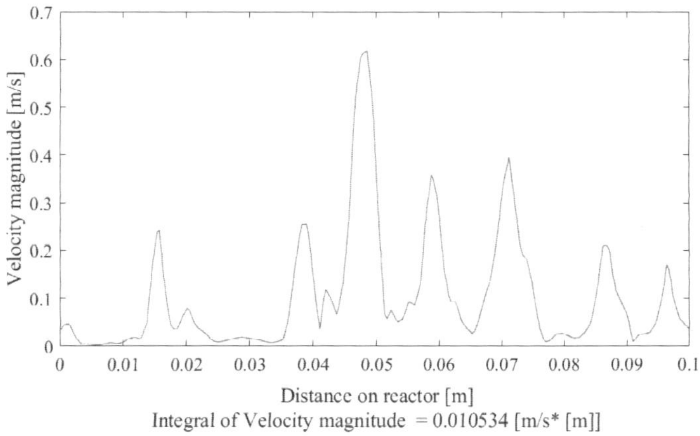

Figure 9. Velocity of fluid along the reactor

4. Conclusion

The formation of acoustic cavitation bubbles and its movement are mainly due to the liquid properties and intensity of ultrasonic sound waves used. In this paper, we studied acoustic cavitation bubbles characteristic in three different liquids which are non-degassed, ethanol and mixture of ethanol-water under a 20kHz ultrasonic sound wave. Results showed that their characteristics are very different due to variations in liquid properties, i.e., surface tension, viscosity, vapour pressure, which play a significant role in the establishment of cavitation bubble. Key findings of the study are:

- The formation of acoustic cavitation in non-degassed water is much less visible compared to the other liquid. There is more acoustic cavitation bubble present in pure ethanol and ethanol-water mixture, and in both of the liquid, acoustic bubbles oscillate vigorously moving towards the liquid-air interface.
- Movement of acoustic bubbles in non-degassed water is primarily due to primary and secondary Bjerknes force. Acoustic bubble experiences less primary Bjerknes Force (attractive force) is repelled from pressure antinode.
- In ethanol, acoustic cavitation bubble can sustain a longer lifespan due to lower surface tension. Surface tension of bubble-gas interface plays a major role in internal bubble pressure.
- Distinctive results in the case of ethanol-water mixture are observed. The addition of ethanol in water provides resistance to gas transfer across liquid/bubble interface, leading to an increased in bubble radius and hence, the buoyancy force.

Results presented by a large area of research, including this paper indicates there might be a possibility of breaking azeotrope using this technique. Due to the complexity in the phenomenon of acoustic cavitation bubble as well as the solution property of ethanol and water mixture, separation mechanism is still under discussion investigation. The results to be uncovered would ascertain whether or not it is feasible to design a system that would consistently breaks azeotropes in organic mixture.

Acknowledgments
The authors are grateful to Universiti Teknologi Malaysia for financial assistance for this research using Transdisciplinary Research Grant VOT No:05G99.

References

[1] Thanh Nguyen T, Asakura Y, Koda S and Yasuda K 2017 Dependence of cavitation, chemical effect, and mechanical effect thresholds on ultrasonic frequency *Ultrason Sonochem* **39** 301-6

[2] Merouani S, Hamdaoui O, Rezgui Y and Guemini M 2013 Effects of ultrasound frequency and acoustic amplitude on the size of sonochemically active bubbles – Theoretical study *Ultrasonics Sonochemistry* **20** 815-9

[3] Reay D, Ramshaw C and Harvey A 2013 *Process Intensification (Second Edition),* ed D Reay, *et al.* (Oxford: Butterworth-Heinemann) pp 205-49

[4] Spotar S, Rahman A, Gee O C, Jun K K and Manickam S 2015 A revisit to the separation of a binary mixture of ethanol–water using ultrasonic distillation as a separation process *Chemical Engineering and Processing: Process Intensification* **87** 45-50

[5] Yasuda K, Mochida K, Asakura Y and Koda S 2014 Separation characteristics of alcohol from aqueous solution by ultrasonic atomization *Ultrason Sonochem* **21** 2026-31

[6] Wakisaka A, Matsuura K, Uranaga M, Sekimoto T and Takahashi M 2011 Azeotropy of alcohol–water mixtures from the viewpoint of cluster-level structures *Journal of Molecular Liquids* **160** 103-8

[7] Gogate P R 2017 Intensification of chemical processing applications using ultrasonic and microwave irradiations *Current Opinion in Chemical Engineering* **17** 9-14

[8] Ashokkumar M 2011 The characterization of acoustic cavitation bubbles – An overview *Ultrasonics Sonochemistry* **18** 864-72

[9] Muramatsu H and Saito T 2017 The relationship between bubble motion and particle flocculation pattern under 20-kHz-ultrasound radiation in water *Chemical Engineering Science*

[10] Mizushima Y, Nagami Y, Nakamura Y and Saito T 2013 Interaction between acoustic cavitation bubbles and dispersed particles in a kHz-order-ultrasound-irradiated water *Chemical Engineering Science* **93** 395-400
[11] Suzuki K, Kirpalani D M and Nii S 2011 Influence of Cavitation on Ethanol Enrichment in an Ultrasonic Atomization System *J. Chem. Eng. Jpn.* **44** 616-22
[12] Suzuki K, Arashi K and Nii S 2012 Determination of Droplet and Vapor Ratio of Ultrasonically-Atomized Aqueous Ethanol Solution *J. Chem. Eng. Jpn.* **45** 337-42
[13] Tzanakis I, Lebon G S, Eskin D G and Pericleous K A 2017 Characterizing the cavitation development and acoustic spectrum in various liquids *Ultrason Sonochem* **34** 651-62
[14] Kenji Okitsu T S, Norimichi Takenaka, Hiroshi Bandow, Rokuro Nishimura, Yasuaki Maeda 2006 Acoustic Multibubble Cavitation in Water.pdf *The Journal of Physical Chemistry Letters* **110** 20081-4
[15] Mossaz S, Colombet D and Ayela F 2017 Hydrodynamic cavitation of binary liquid mixtures in laminar and turbulent flow regimes *Experimental Thermal and Fluid Science* **80** 337-47
[16] Iwai Y and Li S 2003 Cavitation erosion in waters having different surface tensions *Wear* **254** 1-9
[17] Mizukoshi Y, Nakamura H, Bandow H, Maeda Y and Nagata Y 1999 Sonolysis of organic liquid: effect of vapour pressure and evaporation rate *Ultrasonics Sonochemistry* **6** 203-9

AUTHOR INDEX